PERIODIC TABLE
OF THE ELEMENTS

Atomic weight
Atomic number
Melting point (°C)
Element
Electron Configuration

GROUP IA	IIA	IIIB	IVB	VB	VIB	VIIB	VIII			IB	IIB	IIIA	IVA	VA	VIA	VIIA	VIIIA
1 1.0079 **H** 1s¹																	2 4.00260 **He** -272 1s¹
3 6.941 **Li** 180.5 (He) 2s	4 9.01218 **Be** 1278 (He) 2s²											5 10.81 **B** 2300 (He) 2s²2p	6 12.011 **C** -3500 (He) 2s²2p²	7 14.0067 **N** -209.9 (He) 2s²2p³	8 15.9994 **O** -218.4 (He) 2s²2p⁴	9 18.998403 **F** -219.6 (He) 2s²2p⁵	10 20.179 **Ne** -248.7 (He) 2s²2p⁶
11 22.98977 **Na** 97.8 (He) 3s	12 24.305 **Mg** 650 (He) 3s²											13 25.98154 **Al** 660 (Ne) 3s²3p	14 28.0855 **Si** 1415 (Ne) 3s²3p²	15 30.9738 **P** 44.1 (Ne) 3s²3p³	16 32.05 **S** 112.8 (Ne) 3s²3p⁴	17 35.453 **Cl** -100.9 (Ne) 3s²3p⁵	18 39.948 **Ar** -189.2 (Ne) 3s²3p⁶
19 39.0983 **K** 63.6 (Ar) 4s	20 40.08 **Ca** 848 (Ar) 4s²	21 44.9559 **Sc** 1541 (Ar) 3d4s²	22 47.90 **Ti** 1660 (Ar) 3d²4s²	23 50.9415 **V** 1890 (Ar) 3d³4s²	24 51.996 **Cr** 1857 (Ar) 3d⁵4s	25 54.9380 **Mn** 1244 (Ar) 3d⁵4s²	26 55.847 **Fe** 1535 (Ar) 3d⁶4s²	27 58.9332 **Co** 1495 (Ar) 3d⁷4s²	28 58.70 **Ni** 1453 (Ar) 3d⁸4s²	29 63.546 **Cu** 1083 (Ar) 3d¹⁰4s	30 65.38 **Zn** 419.6 (Ar) 3d¹⁰4s²	31 69.72 **Ga** 29.8 (Ar) 3d¹⁰4s²4p	32 72.59 **Ge** 937 (Ar) 3d¹⁰4s²4p²	33 74.9216 **As** 817 (Ar) 3d¹⁰4s²4p³	34 78.95 **Se** 217.4 (Ar) 3d¹⁰4s²4p⁴	35 79.904 **Br** -7.2 (Ar) 3d¹⁰4s²4p⁵	36 83.80 **Kr** -156.6 (Ar) 3d¹⁰4s²4p⁶
37 85.4678 **Rb** 38.9 (Kr) 5s	38 87.62 **Sr** 769 (Kr) 5s²	39 88.9059 **Y** 1522 (Kr) 4d5s²	40 91.22 **Zr** 1852 (Kr) 4d²5s²	41 92.9064 **Nb** 2468 (Kr) 4d⁴5s	42 95.94 **Mo** 2617 (Kr) 4d⁵5s	43 (98) **Tc** (Kr) 4d⁵5s²	44 101.07 **Ru** 2310 (Kr) 4d⁷5s	45 102.9055 **Rh** 1966 (Kr) 4d⁸5s	46 106.4 **Pd** 1552 (Kr) 4d¹⁰	47 107.868 **Ag** 962 (Kr) 4d¹⁰5s	48 112.41 **Cd** 321 (Kr) 4d¹⁰5s²	49 114.82 **In** 156.6 (Kr) 4d¹⁰5s²5p	50 118.69 **Sn** 232 (Kr) 4d¹⁰5s²5p²	51 121.75 **Sb** 630 (Kr) 4d¹⁰5s²5p³	52 127.60 **Te** 449 (Kr) 4d¹⁰5s²5p⁴	53 126.9045 **I** 113.5 (Kr) 4d¹⁰5s²5p⁵	54 131.30 **Xe** -111.9 (Kr) 4d¹⁰5s²5p⁶
55 132.9054 **Cs** 28.4 (Xe) 6s	56 137.33 **Ba** 725 (Xe) 6s²	57 138.9055 **La** 921 (Xe) 5d6s²	72 178.49 **Hf** 2227 (Xe) 4f¹⁴5d²6s²	73 180.9479 **Ta** 2996 (Xe) 4f¹⁴5d³6s²	74 183.85 **W** 3440 (Xe) 4f¹⁴5d⁴6s²	75 186.207 **Re** 3180 (Xe) 4f¹⁴5d⁵6s²	76 190.2 **Os** 3045 (Xe) 4f¹⁴5d⁶6s²	77 192.22 **Ir** 2410 (Xe) 4f¹⁴5d⁷	78 195.09 **Pt** 1772 (Xe) 4f¹⁴5d⁹6s	79 196.9665 **Au** 1064 (Xe) 4f¹⁴5d¹⁰6s	80 200.59 **Hg** -38.9 (Xe) 4f¹⁴5d¹⁰6s²	81 204.37 **Tl** 303.5 (Xe) 4f¹⁴5d¹⁰6s²6p	82 207.2 **Pb** 327.5 (Xe) 4f¹⁴5d¹⁰6s²6p²	83 208.9804 **Bi** 271 (Xe) 4f¹⁴5d¹⁰6s²6p³	84 (209) **Po** 254 (Xe) 4f¹⁴5d¹⁰6s²6p⁴	85 (210) **At** -71 (Xe) 4f¹⁴5d¹⁰6s²6p⁵	86 (222) **Rn** -71 (Xe) 4f¹⁴5d¹⁰6s²6p⁶
87 (223) **Fr** 700 (Rn) 7s	88 226.0254 **Ra** 700 (Rn) 7s²	89 227.0278 **Ac** 1050 (Rn) 6d7s²															

58 140.12 **Ce** 799 (Xe) 4f²6s²	59 140.9077 **Pr** 931 (Xe) 4f³6s²	60 144.24 **Nd** 840 (Xe) 4f⁴6s²	61 (145) **Pm** 1168 (Xe) 4f⁵6s²	62 150.4 **Sm** 1077 (Xe) 4f⁶6s²	63 151.96 **Eu** 822 (Xe) 4f⁷6s²	64 157.25 **Gd** 1313 (Xe) 4f⁷5d6s²	65 158.9254 **Tb** 1356 (Xe) 4f⁹6s²	66 162.50 **Dy** 1412 (Xe) 4f¹⁰6s²	67 164.9304 **Ho** 1474 (Xe) 4f¹¹6s²	68 167.26 **Er** 1529 (Xe) 4f¹²6s²	69 168.9342 **Tm** 1545 (Xe) 4f¹³6s²	70 173.04 **Yb** 819 (Xe) 4f¹⁴6s²	71 174.967 **Lu** 1663 (Xe) 4f¹⁴5d6s²
90 232.0381 **Th** 1730 (Rn) 6d²7s²	91 231.0359 **Pa** (Rn) 5f²6d7s²	92 238.029 **U** 1132 (Rn) 5f³6d7s²	93 237.0482 **Np** 640 (Rn) 5f⁴6d7s²	94 (244) **Pu** 641 (Rn) 5f⁶7s²	95 (243) **Am** (Rn) 5f⁷7s²	96 (247) **Cm** (Rn) 5f⁷6d7s²	97 (247) **Bk** (Rn) 5f⁸6d7s²	98 (251) **Cf** (Rn) 5f¹⁰7s²	99 (252) **Es** (Rn) 5f¹¹7s²	100 (257) **Fm** (Rn) 5f¹²7s²	101 (256) **Md** (Rn) 5f¹³7s²	102 (259) **No** (Rn) 5f¹⁴7s²	103 (260) **Lr** (Rn) 5f¹⁴6d7s²

Materials Science for Engineering Students

About the Cover

The front cover shows some of the 15,000 spun aluminum disks that constitute the façade of the Selfridges building in Birmingham UK. This is one example of the utilization of metals for their environmental stability and aesthetics. Other examples are the titanium-clad Guggenheim Museum in Bilbao, Spain, and the traditional copper claddings that acquire an attractive green patina through weathering. All three metals rely on the fact that chemical "attack" by the environment forms a dense film on the surface that protects it from further degradation. The use of copper has a very long history, but the use of aluminum required 20th century metallurgical technology, and titanium became widely available only recently.

Materials Science for Engineering Students

Traugott Fischer

Stevens Institute of Technology

AMSTERDAM • BOSTON • HEIDELBERG • LONDON • NEW YORK • OXFORD
PARIS • SAN DIEGO • SAN FRANCISCO • SINGAPORE • SYDNEY • TOKYO
Academic Press is an imprint of Elsevier

Academic Press is an imprint of Elsevier
30 Corporate Drive, Suite 400, Burlington, MA 01803, USA
525 B Street, Suite 1900, San Diego, California 92101-4495, USA
84 Theobald's Road, London WC1X 8RR, UK

Library of Congress Cataloging-in-Publication Data
Application submitted.

British Library Cataloguing-in-Publication Data
A catalogue record for this book is available from the British Library.

ISBN: 978-0-12-373587-4

For information on all Academic Press publications,
visit our Web site at: http://www.books.elsevier.com

Printed in Canada.

08 09 10 6 5 4 3 2 1

To my students.

Contents

Preface

This book is based on the experience acquired by teaching a course on materials to the engineering students at Stevens Institute of Technology. It appeared to me that it is possible to improve on existing textbooks despite my great respect for quite a few of them. So I have attempted to innovate in subject matter and in the presentation.

Much innovation has occurred in engineering materials and in the way they are used. Light sources are evolving in the direction of reduced energy use and large-area mobile image displays; electric batteries are ubiquitous and will acquire even more importance with the spreading of hybrid vehicles; the quest for renewable and nonpolluting energy sources will undoubtedly stimulate progress in plastic solar cells; medical advances spur the increasing use of biomaterials; and nanomaterials stimulate thousands of start-up companies intent on capitalizing on their novel properties. These developments cannot be ignored by engineers and receive a succinct treatment in the book. Obviously the introduction of these new topics cannot be done by increasing the amount that must be learned. Therefore, the traditional structural materials, especially metallurgy, are treated in less detail than in most other books.

I endeavored to present the material the way we learn naturally: we first observe a fact or a phenomenon, and then we try to understand it. Having found the explanation, we generalize it by formulating an abstract theory and explore whether it has wider applications. This is generally known as "inductive presentation." Applying it in my lectures, I observed that it stimulates the students' curiosity and raises their interest. I have also introduced new concepts through concrete examples; once that is done, abstracting a general law becomes a pleasure of the mind.

Students often express difficulty in learning a subject because they do not see the use of it. I have read somewhere that we only learn efficiently when we feel the need to know. I have tried to follow this principle and present the "real engineering" topics first and the necessary science "just in time," as it is needed. Why, for example, would an engineer bother with crystal structure? Rather, where would the engineer need to know it? The answer is that only crystalline metals can be shaped by plastic deformation and this deformation is easier in certain structures than in others. So plastic deformation is presented first, and the crystal structure is introduced just as its need is apparent. This principle is followed throughout the book.

The breadth of the field is a real challenge for a course on Materials. Most Engineering curricula allocate one three-credit semester course, consisting of 42 hours of instruction to this subject. With this limited volume of information one can either cover a small area in depth or provide a "shallow" coverage of the entire field and teach a "survey course." I would not underestimate the merit of the latter approach; depth can also be acquired through the relationships between lightly touched separate concepts. The worst way of teaching this material would be to follow the book from its beginning, hope to be able to teach it all, but end where time runs out. I strongly recommend taking some time to plan what one wants to cover in the available time and to eliminate what one does not consider essential. I recommend a complete treatment of Chapter 1, which is the foundation on which the rest is built. For the remainder of the material, I would emphasize the fundamental concepts and drop the details wherever possible. Also, principles normally taught in other courses are included here but could be skipped; this is the case of Chapter 2, the section on crystallography of Chapter 3, the science of diffusion in Chapters 6 and 7, the introductory physics in Chapters 11, 13, and 14, and possibly the electrochemistry of Chapter 15. Many instructors may also choose to ignore the last three chapters. A more detailed list of suggestions can be found in the Web site for instructors at www.textbooks.elsevier.com. The Web site also contains the solutions of all the problems and other resources, such as Powerpoint slides of the figures and hyperlinks to some useful web sites. Solutions to most of the numerical problems are printed at the end of the book.

This book owes much to my students who taught me how to teach; by their interest, their enthusiasm, by their difficulties, and by their occasional boredom, they taught me what motivates us to learn, what works and what does not work in teaching. I am indebted to my colleague Professor Milton Ohring at the Stevens Institute of Technology. He was kind enough to give me the Word document of his entire book *Engineering Materials Science* to use as I wish. Although the style of this book is quite different from his, I have taken many of the figures from his book and reproduced some of its sections word for word. I also owe thanks to Professor Bernard Gallois for inspiring the "just in time science" concept and for much advice. I am indebted to Professor Hong Liang of Texas A&M University and her students who have read the different chapters of the book and given me constructive criticism.

I wish to express my gratitude to Dr. Marwan Al-Haik, Dept. of Mechanical Engineering, University of New Mexico, Dr. James Falender, Dept. of Chemistry, Central Michigan University, Dr. Gregg M. Janowski, Dept. of Materials Science and Engineering, University of Alabama at Birmingham, and Dr. Robert G. Kelly, Professor of Materials Science & Engineering, University of Virginia. Professor Glenn McNutt, College of Engineering, Embry-Riddle Aeronautical University, Dr. Lew Reynolds, Dept. of Materials Science and Engineering, North Carolina State University, Professor Mark L. Weaver, Dept. of Metallurgical Materials Engineering, University of Alabama, and Professor Henry Du of Stevens Institute of Technology, who have carefully read the manuscript, have saved me from errors and omissions and have made valuable suggestions. My thanks also to the editors at Elsevier, to Joel Stein who gave me much early encouragement and advice, to Steve Merken, Acquisitions Editor, who helped and guided me during the preparation of the manuscript, and to Jeff Freeland, Senior Project Manager, who produced this book. And, of course, my gratitude goes to my wife whose patience and support were essential for this endeavor.

Traugott Fischer

The Classes of Materials

The book consists of five parts. In the first part, we take a general look at materials and examine the fundamental origins of their properties. Part 2 describes structural materials, which are used for their strength in the construction of buildings and machines. Part 3 treats functional materials, which include electric conductors and insulators; semiconductors and the devices they engender; lenses; light detectors and lasers; magnets; and batteries. In Part 4 we examine the chemical interaction of materials with the environment in corrosion and, as biomaterials, with the human body. Part 5 describes nanomaterials and the techniques we use to measure the structure and fundamental properties of materials. We recommend that the study begin with Part 1. The other parts of the book can be studied in any desired order.

In Part 1, we find that materials can be divided into three broad classes: metals, ceramics, and polymers (composites are combinations of these classes and semiconductors are specialized ceramics). The mechanical, electrical, and optical properties of materials inside one class are quite similar, but they are very different from those of the other classes. It turns out that these properties can be explained by the behavior of the electrons in the material as the atoms assemble to form the solid. This first part sets the foundation for the rest of the book.

Types of Materials, Electron Energy Bands, and Chemical Bonds

Look around you; you will easily distinguish three classes of materials: metals, ceramics, and polymers. Metals can be bent to shape but are strong; they conduct electricity and are never transparent. Ceramics, like stones, china, and glass, cannot be bent but fracture; they are electric insulators and some are transparent. Polymers (plastics) are lighter than the other materials and relatively soft; they do not keep the new shape when bent; they are used as electric insulators and can be transparent (most eyeglasses are plastic). Semiconductors are selected ceramics that are highly purified and processed to achieve unique electronic properties. Fiberglass and other composites are artificial mixtures of two or three of the material classes mentioned above.

Can we understand where these different properties come from? By studying the structure of atoms and the behavior of electrons when atoms are assembled to make materials, the differences between these classes become clear. In fact, these differences have much to do with Pauli's exclusion principle; at the end of the book, you will be able to answer the following challenge: "Draw a stone bridge and a steel bridge and explain their different shapes in terms of Pauli's exclusion principle." We can also classify the materials according to their utilization. One distinguishes structural materials, which are used for their strength in the construction of buildings and machines; functional materials, which include the conductors, insulators, semiconductors, optical materials, and magnets; and biomaterials, which are selected and processed for their compatibility with the human body and are used in the fabrication of prostheses and medical devices. In all these classes of utilization, one finds metals, ceramics, polymers, and composites.

LEARNING OBJECTIVES

After studying this chapter, the student will be able to:

1. Identify the main classes of materials and describe their distinguishing properties.

2. Describe the electronic structure of an atom in terms of nucleus and electrons.

3. Describe a chemical bond in terms of the lowering of the energy of shared valence electrons.

4. Define electron energy levels and energy bands.

5. Distinguish between an energy band and a chemical bond.

6. Relate the electrical, optical, and mechanical properties of metals and ceramics to the behavior of the valence electrons in terms of Pauli's exclusion principle.

7. Describe metallic, covalent, and ionic bonds.

8. Describe secondary bonds and distinguish between van der Waals and permanent dipole bonds.

9. Relate the mechanical, optical, and electrical properties of polymers to the nature of the secondary bonds.

10. Distinguish between thermoplastic and thermoset polymers in terms of the bonds between molecular chains.

11. Relate elasticity and thermal expansion to the force potential of the chemical bond.

1.1 THE CLASSES OF MATERIALS

Take a paper clip; it is made of **metal**. It is solid, but you can bend it and it will keep its new shape. Most of our tools are made of metals because they are strong, hard, and do not break readily under high stress. Let us observe how most metallic objects are fabricated: many consist of sheets or bars of metal that have been bent, stamped, or forged into their final shapes; they have also been cut, sawed, or drilled and can be fabricated to high precision.

Now look at a porcelain dish, which is a **ceramic**. You would not think of trying to bend it: it is very strong but brittle. A knife does not scratch a porcelain plate: porcelain is harder than steel but it will break when dropped to the floor. Ceramics remain strong at very high temperatures: ceramic bricks line the inside of furnaces and form the crucibles used in melting metals. These materials can only be machined by abrasion with diamond.

Consider a plastic soda bottle. First, notice how light the plastic materials, scientifically known as **polymers**, are; this is because they are hydrocarbons, consisting mostly of carbon and hydrogen atoms, which are very light. The plastic is much softer than a metal or a ceramic; you can bend it, but it will not keep the new shape the way a metal does. Most plastics cannot be used at high temperatures; they soften or liquefy at temperatures not much higher than that of boiling water.

Let us examine an electric cord. Its core is copper, a metal, which is an excellent conductor of electricity. In fact, all metals are good electric conductors. This wire is surrounded by a polymer insulator that can withstand high voltages without passage of any measurable current. Now look at a spark plug. The sparks are generated by a high voltage applied between two metallic electrodes. These are surrounded

Table 1.1 Distinctive Properties of Metals, Ceramics and Polymers.

	Metals	Ceramics	Polymers
Mechanical Properties	Strong Deformable Impact resistant	Hard Not deformable Brittle	Soft Cannot be shaped like metals
Electrical Properties	Electric conductors	Insulators	Insulators
Optical Properties	Always opaque	Can be transparent	Can be transparent
Composition	Elements of columns 1 to 3 of the periodic table and all transition metals and rare earths.	Diamond, Si, Ge and compounds: oxides, carbides, nitrides, sulfides, selenides of metals.	Hydrocarbons: molecules of C with H and some other elements.

by a white ceramic insulator that can withstand a high voltage and tolerates the high temperatures of the engine. Ceramic insulators are also used to insulate high-voltage power lines from the metallic towers. **Metals are good electric conductors; ceramics and polymers are electric insulators**.

Windows, glasses, diamond, sapphire (Al_2O_3), quartz (SiO_2), and many other ceramics are transparent. Other ceramics are translucent, which means that light passes through them but is scattered by internal defects; some ceramics are colored; they are transparent to some wavelengths of light but not to others.

Plastics can be transparent, translucent, or colored as well. The lenses of eyeglasses are now most often made of a polymer because of its low weight. Plastic soda bottles and many other polymers are transparent or translucent. No metal is transparent. Light of all wavelengths is absorbed and reflected by metals. Metals are used as mirrors.

Our observations are recapitulated in Table 1.1.

What about semiconductors and composites such as fiberglass or graphite tennis rackets? These materials do not constitute different classes by nature. Semiconductors are ceramics (mostly silicon and compounds such as GaAs, GaP, etc.) or polymers; they are highly refined and processed to obtain their remarkable electronic properties. Composites are artificial mixtures of different materials that combine the properties of their constituents (fiberglass composites combine the strength of glass fibers with the light weight of polymers; metal-ceramic composites possess the hardness of ceramics and the fracture resistance of metals). These will be discussed in later chapters.

Can we understand what makes a material a metal, a ceramic, or a polymer?

The major properties of the classes of materials are a direct consequence of the behavior of their valence electrons. In order to show that, we start by reviewing the structure of atoms and the periodic table, then we examine the behavior of the electrons when atoms join to form a molecule, and finally we examine the behavior of the electrons in a solid consisting of a very large number of atoms. With this knowledge, we analyze how the behavior of these electrons generates the distinctive properties of metals, ceramics, and polymers.

1.2 **THE STRUCTURE OF ATOMS**

All matter is made of atoms. We know that every atom is formed of a small, positively charged nucleus that is surrounded by electrons. The nucleus is composed of protons and neutrons. Each proton carries the electric charge $q = +1.6 \times 10^{-19}$C; the neutron carries no charge. Each electron carries the negative charge $q = -1.6 \times 10^{-19}$C. A neutral atom contains the same number of protons and electrons. If this number is not equal, the atom carries a positive or negative charge and is called an ion.

The number of protons in its nucleus characterizes every element; this is the **atomic number**. The proton and the neutron are about 1,840 times heavier than the electron; **the number of protons and neutrons in the nucleus of an atom determines the atomic mass of an element**. Not all atoms of a given element have the same number of neutrons. Atoms of a particular element that have different numbers of neutrons are referred to as **isotopes**. Because most elements are mixtures of different isotopes, the atomic mass characterizing a given element is not necessarily an integer. Iron, for instance, has atomic number 26 and atomic mass 55.85. Iron atoms contain 26 protons and 26 electrons. About 85% of iron atoms contain 30 neutrons and 15% of them have 29 neutrons. The **molecular weight** of a compound is the **sum of the atomic masses** of the elements it contains. The molecular weight of water is 18.016 = 16.00 (O) + 2 × 1.008 (H); the molecular weight of alumina (Al_2O_3) is = 2 × 26.98 (Al) + 3 × 16 (O) = 101.98.

A **mole** is the amount of material that weighs the atomic or molecular weight in grams; it is composed of 6.023×10^{23} atoms (or molecules if it is a compound); this is **Avogadro's number**. A mole of iron weighs 55.85 g and contains 6.023×10^{23} atoms. A mole of alumina weighs 101.98 g and contains 6.023×10^{23} molecules.

1.2.1 **The Electrons in an Atom**

The **electrons** are bound to the positively charged nucleus by the Coulomb attraction. Electrons are waves as well as particles. They are placed in concentric shells designated by the principal quantum number n as sketched in Figure 1.1. The binding energy of the electron is the work that must be done to remove it from the atom. Electron binding energies for a few elements are listed in Table 1.2. The shapes and binding energies of the electron orbitals are those of the standing waves that exist in the electric potential of the nucleus and the other electrons. The symmetry of the atomic orbitals is shown in Figure 1.2; it is determined by the quantum numbers l and m.

The **K shell**, with principal quantum number $n = 1$, contains only one orbital of s symmetry ($l = 0$). This orbital is denoted $1s$; it is depicted in Figure 1.2.

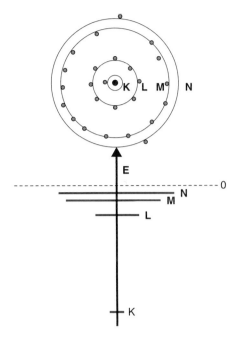

■ **FIGURE 1.1** Schematic arrangement of the electrons in iron. Top: Schematic of the energy shells. Note that the electrons are more correctly drawn as charge clouds as in Figure 1.2 because of their wave nature. Bottom: Binding energies of the electrons. Zero energy is chosen as the energy of an electron that can barely escape the atom. A positive energy is the kinetic energy of the electron in vacuum; the bound electrons have negative energies.

The **L shell**, corresponding to the principal quantum number $n = 2$, has one s orbital ($2s$) and three p orbitals (denoted $2p$) with quantum numbers $l = 1$ and $m = -1$, 0, and $+1$. The binding energies of electrons in the L shell are about 10 times smaller than in the K shell. $2p$ orbitals are shown in Figure 1.2.

The **M shell**, with principal quantum number 3, has one $3s$ orbital, three $3p$ orbitals, and five $3d$ orbitals. The shapes of $3d$ orbitals are shown in Figure 1.2.

The **N shell** has one $4s$, three $4p$, five $4d$, and seven $4f$ orbitals.

1.2.2 **The Pauli Exclusion Principle and the Number of Electrons in Each Orbital**

The observation that no more than two electrons, with different spins, occupy any electron orbital led Pauli to establish his famous **exclusion principle**, which states that no two electrons can exist in the same place with the same set of quantum numbers (i.e., with the same orbital and the same spin). Thus, the K shell contains two electrons (with different spins) in the $1s$ orbital; the L shell has space for eight electrons, two in $2s$ orbitals and six in $2p$ orbitals; the M shell can contain a maximum of 18 electrons, two in $3s$, six in $3p$, and 10 in $3d$ orbitals, and the N shell can accommodate 32 electrons (including 14 $4f$ electrons).

Table 1.2 Binding Energies of Electrons in Selected Elements.

Atomic Number	Carbon 6	Aluminum 13	Silicon 14	Iron 26	Gold 79	
1s	283.8	1,559.6	1,838.9	7,112.0	80,724.9	K shell
2s	6.4	117.7	148.7	846.1	14,352.8	L shell
2p	6.4	73.3	99.5	721.1	13,733.6	"
		72.9	98.9	708.1	11,918.7	"
3s		2.2	3.0	92.9	3,424.9	M shell
3p		2.2	3.0	54.0	3,147.8	"
					2,743.0	"
3d				3.6	2,291.1	"
					2,205.7	"
4s				3.6	758.8	N shell
4p					643.7	"
					545.4	"
4d					352.0	"
					333.9	"
4f					86.6	"
					82.8	"
5s					107.8	O shell
5p					71.7	"
					53.7	"
5d					2.5	
6s					2.5	P shell

From American Institute of Physics Handbook, pp. 7–158 to 165.

Iron, for example, with atomic number 26, possesses two $1s$ electrons with a binding energy of 7112 eV, two $2s$ electrons with binding energy 846.1 eV, six $2p$ electrons with binding energies 721.1 and 708.1 eV, two $3s$ electrons with binding energy 100.7 eV, six $3p$ electrons with binding energy 54 eV, six $3d$ electrons with binding energy 3.6 eV, and two $4s$ electrons with binding energy 3.6 eV. Its electronic structure is denoted as

$$1s^2 2s^2 2p^6 2s^2 3p^6 3d^6 4s^2$$

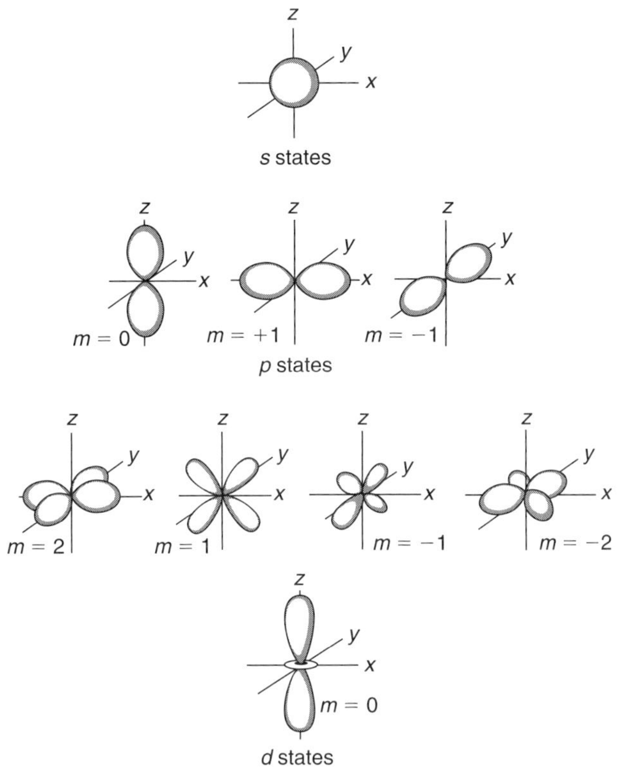

■ **FIGURE 1.2**   Symmetry of the charge distribution in electron orbitals of an atom. The *s* orbitals have spherical symmetry. The *p* orbitals have two charge lobes or regions of high electron density extending along the axes of a rectangular coordinate system. The *d* orbitals have four charge lobes.

We note that, in iron, the $4s$ orbital is fully occupied with two electrons but the $3d$ orbitals, which could accommodate 10 electrons, are only partially occupied.

1.2.3 Valence and Core Electrons

The electrons with the lowest binding energies ($E_B < 10\,\text{eV}$) have the largest orbitals and interact with other atoms to form chemical bonds; these are the **valence electrons**. The atoms in the inner shells (smaller principal quantum number n) are tightly bound; they are the **core electrons**. The core electrons of every element have a precise set of binding energies unique to that element (see Table 1.2). The measurement of these binding energies, by absorption or emission of X-rays or by photoelectron emission, permits the identification of the elements contained in a solid. These methods will be described in Chapter 19.

Elements with similar occupation of their outermost shell have similar chemical properties. For instance, all elements that have one *s* electron in their highest shell, namely, hydrogen, lithium, sodium, potassium, rubidium, and cesium, show similar chemical behavior. This led Mendeleyev to establish the **periodic table** shown in Figure 1.3. Elements with similar occupation of their outermost shell occupy a column in the table. Thus the first column contains the Alkali metals which possess one *s* electron in the valence shell. The second column contains the Alkaline Earths with two *s* electrons. The column IIIB contains boron, aluminum, gallium, indium, and thallium which have two *s* and one *p* electrons; column VIIB contains the halogens that have one electron shy of a full shell; and, finally, the VIII column contains the noble elements with full valence shells, which render them chemically inert. The columns IIIA to VIIIA as well as IB and IIB are the transition metals; they are characterized by a progressive filling of the *d* electron orbitals. In a given column, the occupation of the valence orbitals is the same for all elements; the latter differ in the number of core electrons.

Now what happens to the shapes and energies of the orbitals when two atoms approach each other? This is most easily examined first with the hydrogen molecule, H_2.

1.3 ATOMIC AND MOLECULAR ORBITALS OF ELECTRONS

The hydrogen atom possesses one electron in the $1s$ orbital. When two hydrogen atoms approach each other, each electron feels the attraction of both nuclei. The atomic orbitals are replaced by **molecular orbitals** that extend over the whole H2 molecule, as shown in Figure 1.4. Two molecular orbitals are formed when the atoms approach to form a molecule. One orbital, shown as A in Figure 1.4, concentrates electron charge between the nuclei and has a lower energy than the original atomic orbitals; this is the **bonding orbital** (C) because the total energy is lowered when the hydrogen atoms approach each other to form an H_2 molecule. In the other orbital (B), the electronic charge density between the atoms is decreased. This orbital has a high energy that decreases when the atoms are separated. This is the **antibonding orbital**. Since an orbital can accommodate two electrons (with different spins), the electrons from both atoms occupy the lower lying bonding orbital and the energy of the H_2 molecule is lower than that of the two atoms separated. This binds the atoms together; separating the atoms would require the work of lifting the electrons from the bonding orbital back to the higher energy of the atomic orbitals. This work necessary for separating the atoms is the **chemical bond energy.**

The helium atom has two electrons. Formation of a He_2 molecule would place two electrons into the bonding orbital and two into the antibonding orbital. The energy cost of the antibonding orbital is larger than the energy gain of the bonding orbitals. The energy of the four electrons would be higher than it is in the isolated atoms; therefore the He_2 molecule does not exist.

Lithium possesses three electrons. Two electrons are in the K shell and have a binding energy of 54.75 eV; their orbitals are very small and do not overlap with those of neighboring atoms, they are core electrons. The third electron is in the L shell, with smaller binding energy and a larger orbital. When two lithium atoms approach, only the L electrons of neighboring atoms overlap and form bonding and antibonding orbitals, similar to those in hydrogen.

Key

29	63.54
1083	**Cu**
(Ar) 3d^{10}4s	

Atomic number → 29
Atomic weight → 63.54
Melting point → 1083
Symbol → **Cu**
Electronic configuration → (Ar) 3d^{10}4s

Metal

Nonmetal

Intermediate

■ **FIGURE 1.3** Periodic table of elements.

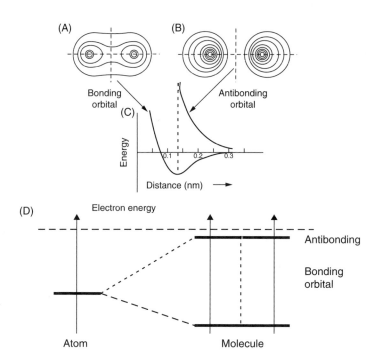

■ **FIGURE 1.4**   The molecular orbitals of the hydrogen molecule. (A) Bonding orbital. (B) Antibonding orbital. (C) Energy of the bonding and antibonding orbitals as a function of the distance between the nuclei. (D) Energy of the atomic orbital in the single atom and the two molecular orbital in the molecule. (The dotted line represents the difference between the energies of the two orbitals shown in C and in Figure 1.5 in the case of two atoms.)

1.4 THE ELECTRONIC STRUCTURE OF THE SOLID: ENERGY BANDS AND CHEMICAL BONDS

What happens when three atoms get together? The valence electron orbitals of the three atoms combine to form three molecular orbitals, of distinct shapes and energies; these extend over the entire molecule (Figure 1.5). Each orbital can accommodate two electrons of different spins. When 10 atoms form a particle, their orbitals overlap and form 10 combinations of different shapes and energies that extend over the whole particle; they can accommodate 20 electrons.

So far we have designated the shapes or trajectories of the atoms as orbitals because they describe orbits around the nucleus. In a macroscopic solid the wave function that determines the shape and trajectory of an electron extends over the whole solid; we need a new language. We will designate the location, trajectory, momentum, and energy of an electron as the **state** it is in. We can use the same word to describe the orbitals in atoms. In the atom, therefore, we can say that the electrons are in the $1s$, $2s$, $2p$, and so on, state. We keep the term orbital to describe electrons in the individual atoms since their trajectory is an orbit around the nucleus.

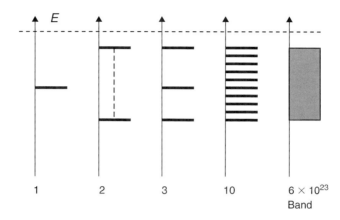

■ **FIGURE 1.5**   Energy levels for valence electrons for (1) a single atom, (2) a diatomic molecule, (3) a system of three atoms, (10) a particle with 10 atoms, (6×10^{23}) a solid of 1 mole. The latter energy band contains exactly 6×10^{26} distinct levels. Each energy level can be occupied by two electrons (with different spins).

In a solid with a very large number of atoms, a given valence orbital (such as $3p$ for aluminum) of each atom combines with those of its neighbors in as many different ways as there are atoms in the solid and form the $3p$ states in the solid. The electron states are distinct, but the energy difference from one to the next is immeasurably small: these states form a quasi-continuous **energy band**, a few electron volts in width that **contains exactly one state for every atom in the solid**. (The orbitals of the core electrons are too small to overlap; their band energies are not spread into a band.) Let us take the example of sodium. Figure 1.5 illustrates how the $3s$ orbital of a single sodium atom, for example, (far left) becomes two states in a molecule with two atoms, three distinct states in a particle with three atoms, 10 states when 10 atoms form a particle, and the $3s$ band in a mole of sodium.

In accordance with the **Pauli exclusion principle, not more than two electrons, with different spins, can occupy a given state**; therefore, **an energy band has space for exactly two electrons per atom** in the solid.

As in the case of hydrogen, **the average energy of the electron states in the valence band of a solid is lower than the energy of the atomic orbital of separated atoms**. This lowering of the electron energies is responsible for the **cohesion**, or **bond energy** that attaches the atoms to each other.

Chemical bonds formed by the **sharing of electrons among the atoms** are called **primary bonds**. The shapes of the valence states and the occupation of the bands by valence electrons give rise to three types of primary bond: the **metallic bond**, the **covalent bond**, and the **ionic bond**.

In a solid, the valence electrons are shared among all the atoms. Their energy levels are spread into an energy band that can fit exactly two electrons per atom. The average energy of the valence electrons is lower than in the isolated atom. This difference forms the chemical bond.

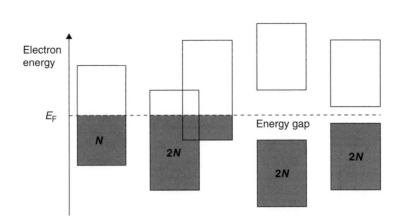

■ **FIGURE 1.6** Energy bands in metals and ceramics. Gray fields: occupied energy levels. White fields: unoccupied energy levels. (A) Metal with less than two valence electrons per atom; the valence band is partially filled with electrons. (B) Metal with two valence electrons per atom and overlapping bands; both are partially filled. (C) Insulating ceramic with full valence band separated by large energy from upper, empty band. (D) Semiconductor with full energy band separated by small energy from upper, empty, conduction band.

We will now show that a solid with a partially empty valence band presents the characteristic properties of a metal. When all levels of the valence band are occupied by electrons, the solid has the properties of a ceramic or a polymer; it is an insulator when the energy gap is large or a semiconductor when the gap is relatively small (i.e., smaller than ~3.5 eV). See Figure 1.6.

1.5 **METALS**

Let us take the example of aluminum. Its electronic structure is $1s^2 2s^2 2p^6 3s^2 3p^1$. Its core electron levels are fully occupied, but it possesses only one electron per atom in the $3p$ band. The valence electrons of the entire solid occupy the states, with the lowest energies in accordance with Pauli's exclusion principle: two electrons in the lowest state, then two in the next higher, and so on, until all electrons are accommodated. Since there is only one $3p$ electron per atom, the $3p$ band is not full. This is illustrated in Figure 1.7. We will now show that this partial filling of the valence band is responsible for the distinctive electrical, optical, and mechanical properties of metals.

The highest occupied energy level at absolute zero temperature is called the Fermi level, E_F. Above this level, a quasi continuum of empty states is available into which the electrons can be excited. Electrons can be excited from an occupied (color) level into an empty (yellow) level with little expense in energy. The electrons in such solids can easily modify their shapes, their velocities, and their energies: they form a **"gas of free electrons."** In particular, the motion (kinetic energy) of the electrons can be modified by an applied electric field (color arrow in Figure 1.8) to create an electric current: a metal is an electric conductor. When light of any photon energy strikes the metal, it can be absorbed by electrons

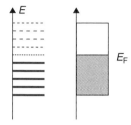

■ **FIGURE 1.7**   Energy band for a metal having one valence electron per atom. The particle with 10 atoms has one valence electron per atom: the five lowest (color) levels contain two valence electrons each; the upper (dotted) levels are empty. The energy band of the solid with many atoms is half full: the lower levels (color) contain two electrons each; the upper levels (white) are empty.

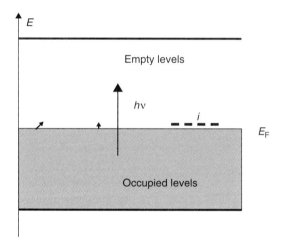

■ **FIGURE 1.8**   Properties of a metal. In the partly filled band, an electron can acquire any amount of energy. color arrow on left: it can be accelerated by an electric field and acquire a small increment in kinetic energy to create an electric current: metals are electric conductors. Arrows: photons can be absorbed and give their energy to an electron: metals absorb all light. New energy level I (dotted line): new bonding orbitals can be created with little expense in energy: atoms can slide easily into new positions through intermediate positions marked *i* in Figure 1.9. Metals can be deformed plastically.

that are thereby excited to higher energies: the material absorbs all light (vertical arrows in Figure 1.8); it is never transparent. As in the hydrogen molecule (Figure 1.4), the average energy of the electrons in the partially filled band is lower than in the isolated atom; therefore the free electrons in this band are responsible for the metallic bond. Since the electrons are free, the metallic bond maintains the atoms bound to each other but does not prescribe their position.

In permanent (plastic) deformation of solids, illustrated in Figure 1.9, atoms slide past each other into new, permanent positions. In our example, the plane of atoms on the right of the dotted line slides upward along the black arrow. Atom 1, for instance, slides into position 2 (black arrow). While sliding,

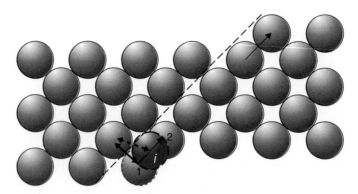

■ **FIGURE 1.9**   Permanent deformation of a solid: the atoms glide past each other into new crystalline positions. The movement of one atom is demonstrated; it moves from position 1 to position 2 (black arrow) through a number of intermediary positions marked *i*. The double arrow represents a bonding electron orbital in the original position. The dotted arrow is a bonding orbital when the atom is in the intermediate position; this orbital does not belong to the equilibrium energy band marked in color in Figure 1.7, but is a novel orbital with a different energy marked as *i* in Figure 1.8.

it must occupy intermediate positions (marked *i*). The valence electrons bonding permanent atoms to each other (double arrow) occupy the lower levels in the band (color in Figure 1.8). While the atom moves into an intermediate position, the bonding orbitals must assume a new shape (dotted double arrow in Figure 1.9), but this can occur because the electrons are free: atoms can slide past each other and metals can be plastically deformed with a small expense of energy.

> The distinguishing features of metals: electrical conduction, absorption of light of all wavelengths, and their ability to be deformed plastically, are all consequences of the unfilled valence band. The metallic bond is formed by the free electrons in a partially filled valence band.

We understand why aluminum has an unfilled band: it has one 3*p* electron. But magnesium has **two** 3*s* electrons, yet it is a metal. How can it have an incompletely filled energy band? It can have an incomplete band because in most elements the width of the highest lying bands is larger than their separation in energy. For instance, Table 1.2 gives the same binding energy of 2 eV for the 3*s* and 3*p* electrons for aluminum. The bands overlap in energy, as shown in Figure 1.6B. Electrons "spill over" from the 3*s* to the 3*p* band so that neither band is full. Because of this overlap of the energy bands, most elements in the periodic table are metallic; this is indeed shown in Figure 1.3.

In order of tonnage, iron and steels are the most widely used metals. Next, but far behind, are aluminum and copper, together with their numerous alloys. The list of commercially employed metals is large and includes nickel, titanium, zinc, tin, lead, manganese, chromium, tungsten, and the precious metals gold, silver, and platinum. Metals, particularly steels and superalloys (containing Ni, Cr, Co, Ti), fill many critical technological functions for which there are no practical substitutes. They are the substances of which machine tools, dies, and energy generation equipment are made. In bridges, buildings, and other large structures, metals, alone or in combination with concrete, provide the safety inherent in their resistance to fracture. The reactors and processing equipment used to produce polymers, ceramics, and semiconductors are almost exclusively constructed of metals. One of the great metallurgical achievements in the twentieth century was the modern jet engine. Its progress has gone hand in hand with the development of superalloys capable of withstanding severe stresses and resisting oxidation at high temperatures. By giving patients a new lease on life, metallic orthopedic implants that replace damaged bones represent a great advance in modern medical practice. Many are made of alloys similar to those of jet engines. The functional applications of metals are mostly in the area of electric conductors, mirrors, and electric batteries. Copper is used for the transmission of electric power and in electric motors and transformers. Silver and gold find applications in jewelry, but also in electric contacts, especially in computers. The magnetic metals, iron, cobalt, and nickel, and many rare earths, find extensive functional applications in transformers, electric motors, and actuators, and in electronic data storage.

1.6 **CERAMICS**

We will now show that solids with a valence band that is fully occupied by electrons present the distinguishing properties of ceramics. By virtue of Pauli's exclusion principle, none of the electrons in a full band can change its shape or energy because two electrons already occupy every possible state. Electrons can be excited to the empty higher-lying band, but this requires a large amount of energy. In the language of physics, the electrons are tightly bound. The consequences are illustrated in Figure 1.10.

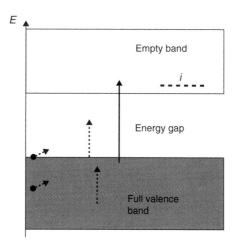

■ **FIGURE 1.10**   The properties of a solid with a full valence band. Impossible processes are marked in dotted arrows. Short arrow: the kinetic energy of valence electrons cannot be altered, because the final state does not exist (energy gap) or is already occupied. Dotted vertical arrows: a photon of visible light cannot be absorbed because the absorbing electron would have to be excited into the energy gap or into an occupied state. Full vertical arrow: ultraviolet light of sufficient photon energy can be absorbed. Dotted energy level marked *i*: The formation of orbital for a transition position of atoms (Figure 1.9) would require excitation of a valence electron into a high-energy empty band: plastic deformation would require excessive stresses so the material breaks.

Because all states in the band are occupied, electrons cannot change their velocity or energy in the presence of an electric field: no current can flow and the material is an electric insulator. The ceramic cannot be deformed plastically: to change the position of any atom would require novel orbitals (dotted double arrow in Figure 1.9) which do not exist in the valence band; the formation of a new bonding orbital i would require the excitation of electrons into the higher, empty, band. This, in diamond for instance, with a gap of 8.5 eV requires a large amount of energy and extremely large stresses: the material breaks before it changes shape. The energy gaps of other ceramics, such as silica (SiO_2), alumina (Al_2O_3), and silicon nitride (Si_3N_4), are somewhat smaller, but still large enough to prevent electric conduction or plastic deformation.

Now to the optical properties: no electron excitations can occur inside the valence band where there is no empty level. Light can only be absorbed if the photon energy is larger than the band gap. Visible light has a photon energy of between 2 and 3 eV. Ceramics, with a band gap larger 3 eV, are typically transparent to visible light. (Color in ceramics will be discussed in Chapter 13.)

> In ceramics, the energy band of valence electrons is completely filled. It is separated by a large energy gap from the next higher band of energy levels, which are empty. As a consequence, ceramics are electrical insulators, can be transparent to visible light, and cannot be deformed plastically but are very hard.

Silicon and germanium are ceramics with a relatively small energy gap between the filled and the empty energy bands as illustrated in Figure 1.6D. This gap is 1.15 eV in silicon and 0.76 eV in germanium. The III-V compounds, such as GaAs, GaP, and so on (see the periodic table), have band gaps that do not exceed 3.5 eV. These materials are **semiconductors** and will be examined in Chapter 11. Semiconductors are transparent to infrared light but absorb visible light.

The only solid elements with full valence bands are diamond, silicon, germanium, sulfur, and phosphorus. All other ceramics are compounds of metallic and nonmetallic elements. Stones are natural ceramics, and so are clay and porcelain, which are compounds of alumina (Al_2O_3), silica (SiO_2), and water, and contain various mixtures of other oxides. Clay and porcelain have been used for many centuries for the fabrication of dishes, vases, sanitary equipment, construction bricks, and refractory materials. Glasses are ceramics consisting mainly of silicon oxide with the addition of sodium, potassium, or boron. These additions are responsible for the amorphous structure of glasses and their softening at relatively low temperatures. They will be discussed in Chapter 8.

High performance ceramics are synthesized; they include alumina (Al_2O_3), silica (SiO_2), other oxides, such as TiO_2, ZrO_2, Na_2O, and Li_2O; carbides WC, TiC, SiC, and BC; nitrides Si_3N_4, TiN, and BN; and borides TiB_2. Since they are compounds of metals with the lighter elements N, C, and O, their density is usually lower than that of metals. The synthetic ceramics have recently assumed critical importance in applications where extreme hardness, wear resistance, or the maintenance of strength and chemical stability at high temperatures are required. Ceramics, as a rule, have a high melting temperature. They are used as refractory linings in furnaces and in crucibles for the melting of metals.

The low weight of silicon nitride is also exploited in turbocharger turbine wheels for their low inertia (Figure 1.11) and in high-speed ball bearings, for instance in dentist drills.

■ **FIGURE 1.11**   Silicon nitride turbine for the turbocharger of an automobile.

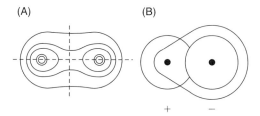

■ **FIGURE 1.12**   (A) Covalent bond. (B) Ionic bond (e.g., NaCl). The atom on the right has a higher electronegativity than the one on the left. The circles are the atomic orbitals before bonding: the thicker line indicates the molecular orbital; the positive ion decreases in size, the negative ion increases.

Ceramics and semiconductors find many applications as functional materials. They are used as insulators, piezoelectric actuators and force sensors, optical materials, light detectors and emitters, lasers, electric rectifiers, and transistors.

1.6.1 Covalent, Ionic, and Mixed Bonds

As in the case of molecules and metals, the chemical bond of ceramics results from the fact that the average energy of electrons in the valence band is lower than in the isolated atoms. Separating the atoms requires work that raises the energy of the valence electrons.

In solid diamond, silicon, and germanium, where all atoms are the same, the electrons are **shared equally** between the atoms as shown in Figure 1.12A. All atoms remain neutral. This constitutes the covalent bond.

In NaCl, there is a shift of electronic charge from Na to Cl. Sodium has full K and L shells and a single $3s$ electron that is weakly bound to its core; chlorine misses one electron to complete its M shell and attracts the electrons more strongly than sodium. The power to attract electrons in a chemical bond, the **electronegativity**, has been measured and is shown in Figure 1.13. It is smallest at the left of the periodic table and increases as we go to the right of the table; it is largest for the halogens, which attract electrons strongly in order to complete their shell. The shape of the molecular orbital in a compound is sketched in Figure 1.12B. The valence electrons are still shared between the atoms but shift toward the atom with higher electronegativity. As a result, chlorine carries a negative charge and is enlarged; the sodium atom carries a positive charge and diminishes in size. The result is an ionic bond. The electron charge transferred from the positive to the negative ion never amounts to a full electron; as sketched in Figure 1.12B, some electron charge remains on the positive ion and only a fraction of an electronic charge is transferred to the other atom. The fraction of electronic charge that is transferred is the **ionicity** of the bond; it is approximated by the equation

$$\% \text{ ionicity} = [1 - \exp - 0.25(X_A - X_B)^2] \tag{1.1}$$

where X_A and X_B are the electronegativities of the elements shown in Figure 1.13. These were calculated first by Linus Pauling and later by J.C. Phillips.

Thus, a pure ionic bond, in which the electron is totally transferred from one atom to the other, does not exist. The most ionic bond is that of cesium fluoride, with an ionicity of 95%. In this compound, the cesium atom retains only 5% of its original electronic charge. Sodium chloride has an ionicity of 68%. At the other extreme, SiC possesses 12% ionicity and GaAs 9.5%; the chemical bond in GaAs is 90.5% covalent and 9.5% ionic. The covalent bond in diamond, silicon, and germanium, of course, has zero ionicity.

The degree of ionicity of the chemical bonds has practical implications for the properties of the solids. In a covalent solid, the shapes of the molecular orbitals govern the positions of the atoms. This is especially important in diamond, silicon, SiC, and the III-V semiconductors because the orbitals of carbon and silicon are formed from sp^3 hybrids.

The sp^3 hybrid is a new atomic orbital that is formed by the combination of the s and the three p orbitals. Four such linear combinations can be formed; they form orbitals that extend from the atom in the four directions shown in Figure 1.14. These hybrids are responsible for the crystal structure of diamond, silicon, and germanium, shown in Figure 1.15A. They also determine the positions of the oxygen atoms in SiO_2, illustrated in Figure 1.15B. This **tetrahedral** placement of the neighbors of silicon and carbon atoms occurs in the structure of glass, of SiC and Si_3N_4, and all silicates and carbides.

In a highly ionic solid, where the valence electrons are no longer shared but transferred almost totally to the more electronegative ion, the ions are attracted by coulombic forces between the electrical charges. The shape of the electron orbitals does not control the bond, but the electric neutrality of the material and the relative sizes of the ions determine the positions of the atoms in the ionic solid.

In practice, one considers as ionic the solids whose structure and chemical properties are determined by their ionic character. Any solid with ionicity larger than 50% is considered ionic. Compounds with small ionicity, such as SiC, the compound semiconductors such as GaAs, GaP, and Si_3N_4, are considered

620.11

IA	IIA	IIIB	IVB	VB	VIB	VIIB	VIII	VIII	VIII	IB	IIB	IIIA	IVA	VA	VIA	VIIA	0
1 H 2.1																	2 He –
3 Li 1.0	4 Be 1.5											5 B 2.0	6 C 2.5	7 N 3.0	8 O 3.5	9 F 4.0	10 Ne –
11 Na 0.9	12 Mg 1.2											13 Al 1.5	14 Si 1.8	15 P 2.1	16 S 2.5	17 Cl 3.0	18 Ar –
19 K 0.8	20 Ca 1.0	21 Sc 1.3	22 Ti 1.5	23 V 1.6	24 Cr 1.6	25 Mn 1.5	26 Fe 1.8	27 Co 1.8	28 Ni 1.8	29 Cu 1.9	30 Zn 1.6	31 Ga 1.6	32 Ge 1.8	33 As 2.0	34 Se 2.4	35 Br 2.8	36 Kr –
37 Rb 0.8	38 Sr 1.0	39 Y 1.2	40 Zr 1.4	41 Nb 1.6	42 Mo 1.8	43 Tc 1.9	44 Ru 2.2	45 Rh 2.2	46 Pd 2.2	47 Ag 1.9	48 Cd 1.7	49 In 1.7	50 Sn 1.8	51 Sb 1.9	52 Te 2.1	53 I 2.5	54 Xe –
55 Cs 0.7	56 Ba 0.9	57–71 La–Lu 1.1–1.2	72 Hf 1.3	73 Ta 1.5	74 W 1.7	75 Re 1.9	76 Os 2.2	77 Ir 2.2	78 Pt 2.2	79 Au 2.4	80 Hg 1.9	81 Tl 1.8	82 Pb 1.8	83 Bi 1.9	84 Po 2.0	85 At 2.2	86 Rn –
87 Fr 0.7	88 Ra 0.9	89–102 Ac–No 1.1–1.7															

FIGURE 1.13 Electronegativities of the elements.

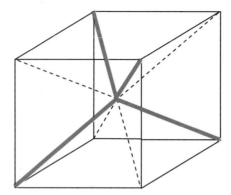

■ **FIGURE 1.14** The four directions of the sp^3 hybrids. These hybrid orbitals are responsible for the structure of diamond and silicon, as shown in Figure 1.15.

(A)   (B)

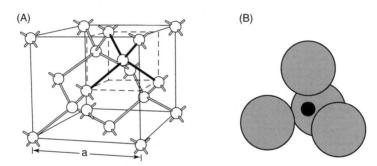

■ **FIGURE 1.15** Left: structure of diamond and silicon. Right: structure of SiO_2; small black atom is silicon; the larger gray atoms are oxygen.

Table 1.3 Ionicity of Some Ceramics.	
Material	**Percent Ionicity**
CaF_2	89
MgO	73
Al_2O_3	63
SiO_2	51
Si_3N_4	30
ZnS	18
SiC	12

covalent; their structure and mechanical properties are governed by the shape of the sp^3 orbitals. Intermediate ionicity gives rise to **mixed** or polar covalent bonds. Covalent materials are usually harder and more brittle than ionic materials. Table 1.3 shows the ionicity of some important ceramics.

Table 1.4 Chemical Bond Energy of Some Materials.

Material	Bond Type	Chemical Bond Energy		Melting Temperature °C	Thermal expansion $\alpha(10^{-6}\,°C^{-1})$
		KJ/mol	eV/atom		
Metals					
Al	Metallic	334	3.4	660	22
Cu		338	3.5	1083	17
Fe		406	4.2	1538	12
W		849	8.8	3410	4.4
Ceramics					
Diamond	Covalent	713	7.4	4350	1
Si		450	4.7	1410	3
WC	Polar Covalent			2776	4
MgO		1000		2800	13.5
SiO_2		879		1710	0.5
LiF		612		842	
Al_2O_3		1674		2980	8.8
Si_3N_4		749		1900	2.5
Polymers				Maximum utilization temperature	
Polyethylene	van der Waals	~10	~0.1	110	150
PVC	Permanent Dipole	~50	~0.5	200	
Epoxy	Covalent Crosslinking			250	

The metallic, covalent, and ionic bonds, in which electrons are shared between the atoms, are the **primary bonds**. These bonds are strong and account for the strength of the metals and ceramics and their high melting temperatures. The bond strengths of some materials are reproduced in Table 1.4. Generally, the ionic and covalent bonds are stronger than metallic bonds; therefore ceramics are stable at high temperatures.

1.7 POLYMERS AND SECONDARY BONDS

It is immediately obvious that polymers are different from metals and ceramics. They are much lighter, softer, and less rigid than either metals or ceramics; they melt at relatively low temperatures. Polymers are usually transparent or translucent (or they have strong colors because dyes have been incorporated in them); they are excellent electric insulators.

Organic polymers consist of long chains of carbon atoms to which hydrogen or other atoms are attached. The simplest polymer is polyethylene, which consists of a chain of ethylene molecules attached to each other by opening the C=C double bond.

$$
\begin{array}{c}
\mathrm{H\ \ H\ \ \ H\ \ H\qquad\quad H\ \ H\ \ H\ \ H} \\
\mathrm{|\ \ \ |\ \ \ \ |\ \ \ |\qquad\quad |\ \ \ |\ \ \ |\ \ \ |} \\
\mathrm{C{=}C\ +\ C{=}C\ \rightarrow\ -C-C-C-C-} \\
\mathrm{|\ \ \ |\ \ \ \ |\ \ \ |\qquad\quad |\ \ \ |\ \ \ |\ \ \ |} \\
\mathrm{H\ \ H\ \ \ H\ \ H\qquad\quad H\ \ H\ \ H\ \ H}
\end{array}
$$

When this process is repeated a large number of times, a very long molecule is obtained that can be described as

The repeat unit
$$
\begin{array}{c}
\mathrm{H\ \ H} \\
\mathrm{|\ \ \ |} \\
\mathrm{-C-C-} \\
\mathrm{|\ \ \ |} \\
\mathrm{H\ \ H}
\end{array}
$$
is called a **mer**.

$$
\begin{array}{c}
\mathrm{H\ \ H} \\
\mathrm{|\ \ \ |} \\
\mathrm{-[C-C]_N-} \\
\mathrm{|\ \ \ |} \\
\mathrm{H\ \ H}
\end{array}
$$

The number N of mers is the **degree of polymerization**.

The chemical bonds that bind the carbon atoms to other carbon and to hydrogen are covalent and formed by the sp^3 hybrids of carbon. The position of the H atoms is governed by the geometry of the sp^3 hybrids. Figure 1.16 shows a fragment of a polyethylene molecule. Figure 1.16 reproduces the angle made by the sp^3 hybrid bonds. While this angle is rigid, the bond rotates easily around its own axis (dotted line in Figure 1.16). Therefore, the polymer molecules are not straight, but flexible and entangled as shown in Figure 1.17. The energy band containing these valence electrons is completely filled; therefore polymers are electric insulators and often transparent.

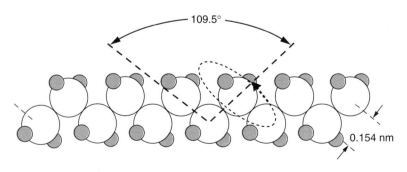

■ FIGURE 1.16   Portion of a polyethylene molecule. The white atoms are carbon; the dark atoms are hydrogen.

When two polyethylene molecules approach each other, they attract each other by weak secondary bonds. There is no sharing of valence electrons; the valence electrons in one of the molecules repel the electrons of the other so that their orbitals are slightly deformed. Small electric dipoles are induced in the two approaching molecules and attract each other. This attraction constitutes the **van der Waals bond**; its strength is about 10 kJ/mol or 0.1 eV/atom (see Table 1.4), much smaller than that of the primary bonds of metals and ceramics.

Polyvinyl chloride (PVC) is harder than polyethylene, indicating a stronger bond between the chains. Figure 1.18 shows the bonding between two PVC molecules. The structure of PVC is similar to that of polyethylene, except that one of the four hydrogen atoms is replaced by chlorine. Chlorine has a high electronegativity and attracts electron charge, creating a mixed covalent-ionic bond inside the molecule. The result is the formation of a **permanent electric dipole** in each mer that is much stronger than the induced dipole in polyethylene. The bonds formed by the attraction of permanent dipoles are stronger than the van der Waals bond but still much weaker than the primary bonds of metals and ceramics. When one of the charged atoms in a permanent dipole bond is hydrogen, one speaks of a **hydrogen bond**. (Such bonds, for instance, are responsible for the cohesion of water and ice.) The strengths of various bonds are shown in Table 1.4. Van der Waals, permanent dipole, and hydrogen bonds, which do not involve the sharing of electrons, are **secondary bonds**.

■ **FIGURE 1.17**   Entangled polymer molecules.

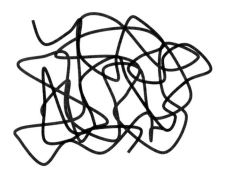

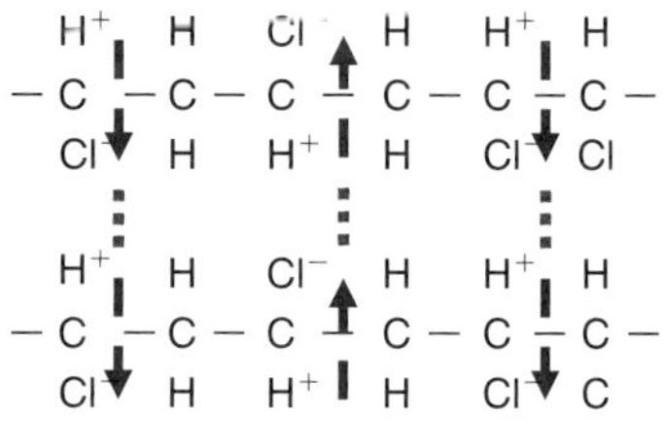

■ **FIGURE 1.18**   Bonding between two PVC molecules. The ionic bond of chlorine forms a permanent dipole in the molecule. Attraction between the permanent dipoles forms a relatively strong secondary bond. Since a hydrogen atom is involved, this is also called a hydrogen bond.

The mechanical strength of polymers is derived from the weak secondary bonds between the molecules and from the **entanglement of the long chains**. (Think of the cohesion of a platter of spaghetti.) Because of the weakness of the secondary bonds, polymers remain solid only at moderate temperatures, not much higher than that of boiling water.

The polymers with secondary bonds between the chains are **thermoplastics**: when they are heated, their thermal energy overcomes the weak bonds between chains; the materials become progressively softer until they liquefy. For this reason, they are easily recycled.

In a second class of polymers, called **thermosets**, primary bonds are formed between the chains. These primary bonds are called **crosslinks**. Thermoset polymers are stronger and stable to higher temperatures than thermoplastics. They do not soften at high temperatures but lose their hydrogen and transform into char when the temperature is excessive; thermosets cannot be recycled. Their advantage is that they solidify at room temperature by a chemical reaction between two liquids; epoxy glue and the resin in fiberglass are two examples. Polymers will be discussed in more detail in Chapter 9.

1.8 **BOND ENERGY AND THE DISTANCE BETWEEN ATOMS**

For all types of bonds, the potential of forces binding the atoms together can be represented by Figure 1.19, which plots the potential energy U and the attractive force as a function of the distance d between the nuclei of neighboring atoms. (Zero potential energy U is chosen as that of two atoms at a very large distance from each other.)

The force F binding the two atoms is equivalent to $-dU/dr$ and is shown colored in Figure 1.19. The equilibrium distance r_o between the atoms corresponds to the minimum potential energy where the force is zero. This equilibrium distance (0.255 nm in Figure 1.19) serves to define the radii R of the atoms as $r_o = R_1 + R_2$. The minimum potential represents the bond energy E_B (0.583 eV in the figure).

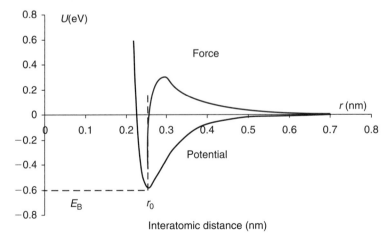

■ **FIGURE 1.19**   Potential energy (black) and force (colored) between two atoms. It represents the Lennard-Jones potential of copper.

Accurate calculations of the potential $U(d)$ are exceedingly complex. The potential can be modeled by the empirical Lennard-Jones formula

$$U(r) = 4\varepsilon_{LJ}[(\sigma/r)^{12} - (\sigma/r)^{6}] \tag{1.2}$$

The adjustable parameters ε_{LJ} and σ are chosen so that the formula reproduces the known equilibrium distance and bond energy of the metal. Figure 1.19 was obtained from the Lennard-Jones formula for copper, with $\varepsilon_{LJ} = 0.583\,\text{eV}$ and $\sigma = 0.277\,\text{nm}$.

1.8.1 Elasticity

When the separation is larger than r_o, an attractive force pulls the atoms closer together; when the distance is smaller, $r < r_o$, a repulsive force tends to separate the atoms. The bond between the atoms thus acts as a spring with spring constant $k_s = -(dF/dr)$ at $r = r_o$. This is the tangent to the force curve $F(r)$ (Figure 1.20). The elasticity of the bond is responsible for the elasticity of the solid: when a tensile stress is applied to the body, the interatomic spacing r between atoms increases in the direction of the stress and so does the length of the body. Compressive stresses cause a decrease in atomic distance and length of the body. Young's modulus of a solid is proportional to the spring constant k_s.

1.8.2 Thermal Expansion

Materials expand and contract as their temperature rises and falls. This phenomenon has its origin in the shape of the interatomic force potential of Figure 1.19. At any temperature, the atoms vibrate around their equilibrium position (Figure 1.21). An increase in temperature corresponds to a higher thermal energy kT and larger vibration amplitude. The interatomic force potential $U(r)$ is asymmetric, and the average distance between the atoms increases as the temperature and vibration energy increase: the material expands.

Thermal expansion is used in mercury thermometers and in bimetallic thermostats. It also obliges the constructors of steel bridges to anchor the bridge on one side and place it on rollers on the other.

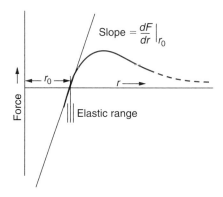

■ **FIGURE 1.20** The spring constant of the interatomic force. It is proportional to Young's modulus E.

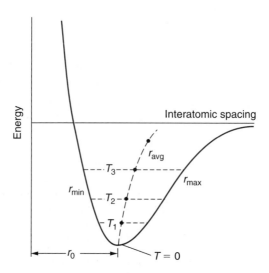

■ **FIGURE 1.21**   Thermal expansion of a solid. At temperatures $T_3 > T_2 > T_1$, the atoms vibrate with increasing amplitude. The asymmetry of the interatomic force potential causes an increase of the average atomic distance with increasing temperature.

The bond energy and force curves (Figures 1.19 and 1.20) are similar for all types of bonds. Solids with large bond energy have a smaller distance r_0 and a deeper potential E_B. Therefore Young's modulus is larger and the thermal expansion coefficient is smaller for solids with larger bond energy. Table 1.4 shows the bond energies, Young's modulus, and coefficients of thermal expansion of selected solids.

1.9 STRUCTURAL MATERIALS, FUNCTIONAL MATERIALS, AND BIOMATERIALS

The computer is a good example with which we can observe the role of different materials. The materials forming the box that contains the computer, the hinges, the keys, and the disk drive are selected for their strength and their resistance to fracture. The same applies to the table on which it sits and the building in which it is located. These are structural applications, and the materials selected and processed for optimal performance are **structural materials**. Structural materials depend on the strength of the chemical bond between the atoms.

The power cord is made of two materials: a plastic electric insulator that prevents electric shock surrounds a copper wire that conducts the current powering the computer. The soul of the computer is a semiconductor chip modified to contain millions of transistors that process the electrical signals. The hard disk is a magnetic material that stores the information bits in the form of small magnets. The screen is made of materials that emit light and form the text and images we see. Note that in these applications, the material performs a specific function. (The material itself is the machine.) These are **functional materials**. The various functions of the materials are carried out by their electrons; the performance of functional materials is governed by their electron energy-band structure.

Biomaterials are selected for their compatibility with the human body, namely because they are not toxic, do not corrode, do not cause excessive blood clotting, and are not rejected by the body.

All three classes, metals, ceramics, polymers, and their composites, are represented among the structural, functional, and biomedical materials, but the requirements on the material are different and so is the processing that aims at optimal performance.

■ SUMMARY

1. Materials are made of atoms that consist of a nucleus surrounded by electrons. The nucleus contains protons and neutrons. The protons carry a positive electric charge of 1.6×10^{-19} Coulomb (C). The electrons have low mass and carry a charge of -1.6×10^{-19} C. The atomic number of an element is the number of protons it contains. The atomic mass is the mass of the protons and neutrons.

2. The electrons are waves as well as particles. Their shapes and energies correspond to standing waves in the atom and assume discrete, precise values characterized by four quantum numbers. The shape (symmetry) of the electron waves is denoted by the second quantum number. Only one electron exists in an atom with a given set of quantum numbers (Pauli's exclusion principle).

3. When atoms are in close proximity, their outermost electrons, called valence electrons, are no longer confined to each atom but extend over the entire molecule or solid and form molecular orbitals also called states. In a solid, there are as many such states as there are atoms; each has a distinct energy. The precise energy level of an electron in an atom (e.g., the $3p$ level of Al) widens into an energy band in the solid (e.g., the $3p$ band of solid Al). A band has space for exactly two electrons (of different spins) per atom in the solid.

4. The average energy of the valence electrons in a solid is lower than in isolated atoms; this difference forms the chemical bond energy.

5. In an elemental solid where all atoms are the same, the valence electrons are shared equally and form covalent bonds: all atoms are electrically neutral.

6. In a compound, electron charge is transferred from atoms with lower electronegativity (usually to the left in the periodic table) to atoms with higher electronegativity (usually to the right in the periodic table). This transfer forms positive and negative ions in the solid. When the transfer is 0.5 electron charge or more, the bond is called ionic. When the transfer is less than 0.5 electron charge, the bond is mixed covalent-ionic.

7. In group IV elements (diamond, Si, Ge) and in compounds, the band of valence electrons is full. According to Pauli's exclusion principle, all possible orbitals are occupied: the electrons cannot change their energy, shape, or velocity: the solid is a ceramic. Ceramics are electrical insulators, often transparent (do not absorb light) and rigid.

8. In most solid elements, the band of valence electrons is not full. These electrons can acquire higher energy, velocity, and change shapes; they constitute a gas of free electrons. Solids with a partially filled valence band conduct electricity, absorb all radiation, and can deform plastically: they are metals.

9. Polymers are long hydrocarbon molecules. Their valence band is full: they form covalent C–C and C–H bonds and ionic bonds with other elements. No electron transfer exists between such molecules; they are bound to each other by weak secondary bonds. The weakest bond is the van der Waals bond: it is caused by slight deformation of the electron orbitals (polarization) because of electrostatic interaction between electrons. The polar secondary bond is caused by attraction between permanent dipoles in the molecules. Permanent dipoles are the result of ionic bonds inside the molecules. Secondary bonds are much weaker than primary bonds.

10. Polymers with only secondary bonds between the chains are thermoplastics. They soften at moderately high temperatures ($>100°C$). Polymers that have some primary bonds (called crosslinks) between the chains are thermosets. They do not soften when heated, but lose their hydrogen and become char.

■ KEY TERMS

A
absorption
   photons, 9
aluminum, 14
atom, 6
atomic mass, 6, 29
atomic number, 6
Avogadro's number, 6

B
band overlap, 16
Biomaterials, 29
bond energy, 26
bond energy, 17

C
ceramic, 4, 17
chemical bond, 10, 12
composite, 5
covalent bond, 13, 19, 20, 33
crosslinks, 26

D
degree of polymerization, 24

E
Elasticity, 27
electron, 6
   binding energy, 6
   core, 9

orbital, 7
quantum number, 6
shell, 6
valence, 9
electronegativity, 20, 25, 29
energy band, 8

F
filled energy band, 12
free electron gas, 12
functional materials, 28

H
hydrogen, 10
hydrogen bond, 25

I
interatomic distance, 26
ionic bond, 13, 20, 25, 33
ionicity, 20

M
Mendeleyev, 10
metal, 4, 14
metallic bond, 13, 15, 33
mixed bond, 22
mole, 6
molecular orbitals, 10, 12, 19,
   20, 29
   antibonding, 10

bonding, 10
molecular weight, 6

P
Pauli exclusion principle, 7
periodic table, 5, 10, 16, 18,
   20, 29
permanent electric dipole, 25
polar covalent, 22
polyethylene, 24
polymer, 4, 23
primary bond, 13

S
secondary bonds, 25
semiconductor, 5
sp^3 hybrid, 20, 24
structural materials, 28

T
thermal expansion, 27
thermoplastics, 26

V
van der Waals bond, 25, 30

X
X-ray
   absorption, 9

■ REFERENCES FOR FURTHER READING

[1] Borg RJ, Dienes GJ. *The Physical Chemistry of Solids*. Academic Press, Boston, 1992.

[2] Smart L, Moore E. *Solid State Chemistry: An Introduction*. Chapman and Hall, London, 1992.

[3] Pauling L. *The Nature of the Chemical Bond and the Structure of Molecules and Crystals; An Introduction to Modern Structural Chemistry*, 3rd ed. Cornell University Press, Ithaca, NY, 1960.

■ PROBLEMS AND QUESTIONS

1.1.   Select three objects in your immediate surrounding. And for each:

    a. Name the materials its components are made of. (If you cannot name the specific material, name the class to which it belongs.)

    b. Describe as much as you can how each of its components was manufactured.

    c. State whether this component could have been made of a material from a different class (metal, ceramic, polymer).

    d. Describe how this different choice would have required a different manufacturing method.

1.2.   Hold a glass bottle, a soda can, and a plastic soda bottle.

    a. Compare their weights and the amount of material used.

    b. Describe why so much or so little material was necessary.

    c. Describe how they were manufactured. (If you cannot describe all operations, describe the ones that are clearly identifiable.)

1.3.   Sports equipment continues to undergo great changes in the choice of materials utilized. Focus on the equipment used in your favorite sport and describe:

    a. What materials change or substitution has occurred.

    b. Reasons for the change.

1.4.   Sketch a stone bridge, a steel bridge, a wood bridge, and a reinforced concrete bridge. Describe how their shape is related to the properties of the material. (These relationships will become clearer in the later chapters.)

1.5.   What materials are used in an electric motor?

1.6.   In your automobile, what material is the engine block made of? How was it manufactured? Is there any machining involved? Could it have been made of a material from **another class**? Why or why not?

1.7.   What material is the body of your automobile made of? How was it manufactured? Could it have been made of another material? How would this other material have changed the manufacturing process?

1.8.   a. How many atoms are there in $1\,cm^3$ of pure silicon?

    b. How many atoms are there in a pure silicon wafer that is 25 cm in diameter and 0.5 mm thick?

    c. If the wafer is alloyed with 10^{16} phosphorous atoms per cm^3, what is the atomic fraction of phosphorus in silicon?

1.9.   Aluminum has an atomic density of 6.02×10^{22} atoms per cm^3. What is its mass density?

1.10.   How many grams of Ni and Al are required to make 1 kg of the compound Ni_3Al?

1.11.   Create an energy level diagram for silicon similar to the bottom of Figure 1.1. Do it to scale.

1.12. A common form of the potential energy of interaction between atoms is given by the Lennard-Jones potential $U(r) = 4\varepsilon_{LJ}[(\sigma/r)^{12} - (\sigma/r)^6]$.

    a. Derive an expression for the equilibrium distance of separation in terms of ε_{LJ} and σ.

    b. Derive an expression for the energy at the equilibrium separation distance.

1.13. Provide a reasonable physical argument for each of the following statements.

    a. The higher the melting temperature of the solid, the greater the depth of the potential energy well.

    b. Materials with high melting points tend to have low coefficients of thermal expansion.

    c. Materials with high melting points tend to have a large modulus of elasticity.

1.14. Draw a plot of Young's moduli vs. melting temperature of all metallic elements. Can you provide a reason why the relation is not perfect?

1.15. In what ways are electrons in an isolated copper atom different from electrons in a copper penny?

1.16. What is so special about the electronic structure of carbon that enables it to form over a million organic compounds with hydrogen, oxygen, nitrogen, and sulfur?

1.17. State whether ionic, covalent, metallic, or van der Waals bonding is evident in the following solids. (Where applicable distinguish between intramolecular and intermolecular bonding.)

    a. Mercury.

    b. KNO_3.

    c. Solder.

    d. Solid nitrogen.

    e. SiC.

    f. Solid CH_4.

    g. Aspirin.

    h. Rubber.

    i. Na_3AlF_6.

    j. PbTe.

    k. Snow.

1.18. The George Washington Bridge is 4,760 ft long between anchorages. The roadbed of the bridge is made of steel. Estimate the difference in length of the roadbed between a very hot summer day (100°F) and a cold winter day (0°F). Use the thermal expansion coefficient of iron. (1 ft is 0.3 m; conversion of degrees: $T(°C) = [T(°F) - 32] \times 5/9$).

1.19. Identify the main classes of materials and describe their distinguishing properties.

1.20. Describe the structure of an atom. What determines its atomic number? What determines its atomic weight? Which of its components are involved in chemical reactions and why? Which of its components can be used in nondestructive chemical analysis and why?

1.21. Describe a chemical bond in terms of the behavior of electrons.

1.22. Define electron energy levels and energy bands.

1.23. Distinguish between an energy band and a chemical bond.

1.24. Describe the difference between an electric conductor and an insulator in terms of its electronic band structure.

1.25. Are there metals that are transparent when 1 cm thick? Why or why not?

1.26. Describe the properties and behavior of electrons forming a metallic bond.

1.27. Describe the properties and behavior of electrons forming a covalent bond.

1.28. Describe the properties and behavior of electrons forming an ionic bond.

1.29. Describe the properties and behavior of electrons forming ionic bonds in most real materials.

1.30. Describe the properties and behavior of electrons forming a van der Waals bond.

1.31. Describe the properties and behavior of electrons forming a hydrogen bond.

1.32. Distinguish between thermoplastic and thermoset polymers in terms of the bonds between molecular chains.

1.33. What is the origin of elasticity in materials?

1.34. What causes the thermal expansion of materials?

1.35. Draw the potential of interatomic forces. Identify the bond strength and the size of the atoms involved.

1.36. Name a structural material.

1.37. Name a functional material. To what basic class of materials does it belong?

Part 2

Structural Materials

Structural materials are the ones we use in the construction of buildings and machines where we rely on their response to applied forces. In buildings and machines, we rely on the ability of the material to resist deformation under loads; in springs, we utilize the elastic deformation of the materials; in manufacturing, we exploit the possibility of shaping metals by rolling, forging, bending, and cutting. We take advantage of the hardness of ceramics and their ability to conserve their strength at high temperatures, and we take advantage of the low cost and low weight of polymers when the mechanical requirements are not severe.

In what follows, we will first examine how various materials respond to applied stresses, their elastic and plastic deformation, how they fracture, and how they fail by fatigue under repetitive stressing and by creep at high temperatures. Then, in separate chapters, we examine the mechanical properties of metals, ceramics, polymers, and composites and how these materials are processed to obtain a specific desired performance. By learning how materials deform and break, we will learn not only how to design to avoid failure but also how to select and process them for optimum performance in various applications.

The Strength of Materials

A large variety of materials are offered on the market. The engineer must be able to understand the specifications of available materials and select them judiciously in design. Materials must also be tested upon reception and often further processed to obtain the desired performance. In this chapter we review the mechanical properties of materials and how they are measured. We examine elastic stiffness, strength, ductility, and resistance to fracture, fatigue resistance under cyclic loading, and creep at high temperatures.

LEARNING OBJECTIVES

After studying this chapter, the student will be able to:

1. Distinguish between forces applied to a solid and the resulting stresses. Define tensile, compressive, and shear stresses.

2. From the data provided in a table, draw a stress-strain curve and identify elastic and plastic deformation, yield stress, ultimate tensile stress, ductility, and work hardening.

3. Describe residual stresses, their effect on the strength of structural materials, and use them to advantage.

4. Select and use the appropriate hardness test.

5. Distinguish between ductile and brittle fracture.

6. Describe a stress concentrator (also called stress raiser) and avoid the danger it presents.

7. Distinguish between the resistance to deformation and to fracture, and name their units.

8. Explain why hard materials are more brittle than soft metals.

9. Explain why ceramics are unreliable in tensile stresses.

10. Describe fatigue, its mechanism, and its consequences. Obtain adequate fatigue resistance by appropriate design of the pieces and processing of the material.

11. Measure and describe creep and name the important engineering situations where it plays a major role.

2.1 **STRESSES AND STRAINS**

When we pull on a bar of material with a force F, its length increases by an amount Δl. This is shown in Figure 2.1. The force required for a given deformation is proportional to the cross section A of the bar; under a given force, the elongation Δl of the bar is proportional to its original length l. We wish to express the properties of a material in a manner that is independent of its shape and size, so we define the **stress** σ as the force per unit area

$$\sigma = F/A \tag{2.1}$$

and the **strain** as the relative elongation

$$\varepsilon = \Delta l/l \tag{2.2}$$

The international unit for stress is the Pascal (Pa) or Newton per square meter (N/m^2) and the American unit is the pound per square inch (psi). Practical stresses are too large for these units to be convenient, and one expresses the stress in megapascals

$$1\ \text{MPa} = 10^6\ \text{N/m}^2 = 1\ \text{N/mm}^2$$

or in thousands of pounds per square inch (ksi).

$$1\ \text{ksi} = 6.9\ \text{MPa}$$

It is easier to visualize the megapascal as one Newton per square millimeter than a million Newtons per square meter.

The strain is dimensionless; it is often expressed in percent (%).

Instead of pulling on the bar, we can compress it. Compressive stresses have the same units. By convention, one uses positive stresses and strains for tensile deformation and negative values for compression.

Figure 2.2 illustrates the application of a **shear stress** and the resulting **shear deformation**. Such stresses occur, for instance, in any rotating shaft driven by a motor. A shear stress is caused by a couple of forces F_1 and F_2 acting in opposite directions. A single force couple (F_1, F_2) would disrupt the equilibrium (it would cause the piece to rotate instead of deforming); therefore a second couple (F_3, F_4) is applied to achieve a balance of moments. To obtain the resultant shear stresses, we divide the forces by the areas to which they are applied. The relevant area for the force F_1 is shaded in Figure 2.2.

The shear stress is defined as

$$\tau_{yz} = F_1/A \tag{2.3}$$

The second couple of forces generates the shear stress τ_{zy}, which has the same magnitude

$$\tau_{yz} = \tau_{zy} \tag{2.4}$$

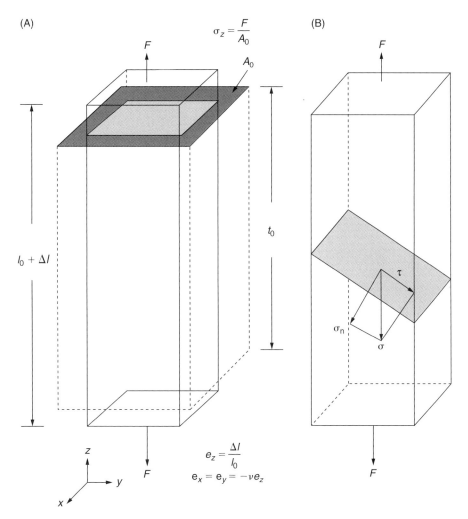

(A)

$$\sigma_z = \frac{F}{A_0}$$

(B)

FIGURE 2.1 The stresses and elastic strains caused by a tensile force F on a solid. Figure B shows that on any surface in the solid, real or imaginary, the applied stress can be decomposed into a normal stress perpendicular to the surface and a shear stress parallel to it. The shear stress τ is largest when the surface makes an angle of 45° to the applied force. (Note: the elastic deformations on this figure are exaggerated; in practice, elastic strains in solids, except rubbers, do not exceed 1%.)

The resultant deformation is defined as the shear strain

$$\gamma = \Delta l/l = \tan \alpha \tag{2.5}$$

where α is the angle by which the solid is deformed.

The stresses are defined by the surfaces to which they apply. It is possible to consider other surfaces, oriented in different directions, on which the stresses are different. The right side of Figure 2.1 shows

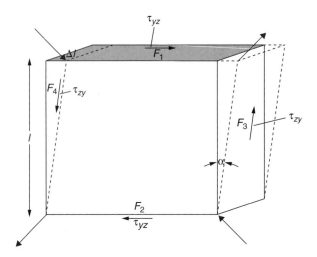

■ **FIGURE 2.2**   Shear stresses and shear strains. (The colored arrows show these shear stresses decomposed into tensile and compressive stresses in other directions.) Note that in shear deformation Δl is perpendicular to l.

an inclined surface on which the original tensile stress is decomposed into a tensile stress σ_n normal to the new surface and a shear stress τ that is parallel to it. The shear stress in Figure 2.2 can also be decomposed into a tensile stress along one diagonal of the cube (from bottom left to top right) and a compressive stress along the other. These are shown colored in the figure. The analysis of the stresses in solids is an important aspect of mechanical design and is treated in specialized texts on the strength of materials.

2.2 **ELASTIC DEFORMATION**

When the applied stresses are not too large, elastic deformation takes place. The stress is proportional to the strain

$$\sigma = \varepsilon E \tag{2.6}$$

This is **Hooke's law**. E is the **elastic modulus** also called **Young's modulus**. Elastic deformation is reversible: when the stress is removed, the strain disappears and the solid regains its original shape. Hooke's law applies to compression as well as tension.

A solid that is elongated with strain ε_z becomes smaller in the two other directions (the bar in Figure 2.1 becomes thinner as it elongates).

$$\varepsilon_x = \varepsilon_y = -\nu\,\varepsilon_z \tag{2.7}$$

where ν is called **Poisson's ratio.** Note the minus sign: the strains due to Poisson's ratio are of opposite sign to the strain that is parallel to the stress. Negative strains denote shrinkage.

Elastic shear stresses τ are also proportional to shear stains γ

$$\tau = \gamma G \tag{2.8}$$

where G is called the **shear modulus.** It is related to Young's modulus as

$$G = \frac{E}{2(1 + \nu)} \tag{2.9}$$

There is no Poisson's ratio for shear deformation: application of a shear stress τ causes the shear deformation γ shown in Figure 2.2 and no other.

Table 2.1 gives the elastic modulus, the shear modulus, and Poisson's ratio for a number of important structural materials. Note that the elastic and shear moduli are expressed in gigapascals (GPa). $1 \, GPa = 10^9 \, Pa = 1{,}000 \, MPa$.

Elastic deformation is caused by the elasticity of the chemical bond. We have seen in Chapter 1, Section 1.8, that the chemical bond establishes an equilibrium distance between the atoms. If a tensile force is applied to the atoms, their distance increases; if a compressive force is applied, the distance between the atoms decreases. The bonds between the atoms act like springs. **When we apply a tensile stress to a solid, its bonds are stretched, and the distance between its atoms increases. If we apply a compressive stress, the distance between the atoms decreases. When the stresses are removed, the bonds resume their equilibrium lengths. This is true in elastic deformation only.**

The most obvious application of elasticity is in springs. Note that the shapes of the springs (coils and leaf springs) are designed in such a way that relatively large displacements of the ends are obtained with local strains remaining below 1%. Elastic shear deformations are utilized in torsion bar suspension of cars. Other applications are the measurement of forces; actually, one measures the elastic strains and deduces the stress from Equation (2.6) and the force from Equation (2.1). Elastic deformation is also important in the functioning of much sporting equipment. It is also an important factor in fastening by screws (together with friction). Stiffness (a high elastic modulus) is sought in precision machining where elastic deformation of the machine causes errors: this is also why the final stage of precision machining is accomplished with small depths of cut (i.e., small tool forces).

2.3 PLASTIC DEFORMATION OF METALS

When the stress exceeds a certain limit, the response of the body is no longer elastic and depends on the nature of the material. Above a stress called the yield stress, the deformation of metals is large and permanent. This is plastic deformation.

The ability to be deformed plastically is a valuable quality of metals. Most metallic objects are manufactured by plastic deformation: bars and sheets are made by rolling, car bodies by stamping, bottles

Table 2.1 Elastic Properties for Selected Engineering Materials at Room Temperature.

Material	Elastic modulus, E (10^6 psi, GPa)		Shear modulus, G (10^6 psi, GPa)		Poisson's ratio ν	Density Mg/m³ g/cm³
Metals						
Aluminum alloys	10.5	72.4	4.0	27.6	0.31	2.7
Copper alloys	17	117	6.4	44	0.33	8.9
Nickel	30	207	11.3	77.7	0.30	8.9
Steels (low alloy)	30.0	207	11.3	77.7	0.33	7.8
Stainless steel (8–18)	28.0	193	9.5	65.6	0.28	7.9
Titanium	16.0	110	6.5	44.8	0.31	4.5
Tungsten	56.0	386	22.8	157.3	0.27	19.3
Ceramics						
Diamond	145	1,000				3.51
Alumina (Al2O3)	53	390	—	—	—	3.9
Zirconia (ZrO2)	29	200				5.8
Silicon carbide	65	450	—	—	—	2.9
Titanium carbide	55	379	—	—	—	7.2
Tungsten carbide	80	550	31.8	—	0.22	15.5
Quartz (SiO2)	13.6	94	4.5	—	0.17	2.6
Pyrex glass	10	69.0	—	—	—	
Fireclay brick	14	96.6	—	—	—	
Plastics						
Polyethylene	0.058–0.19	0.4–1.3	—	—	0.4	0.91–0.97
PMMA	0.35–0.49	2.4–3.4	—	—	—	1.2
Polystyrene	0.39–0.61	2.7–4.2	—	—	0.4	1.1
Nylon	0.17	1.2	—	—	0.4	1.2
Other materials						
Concrete-cement	6.9	45–50	—	—	—	2.5
Common bricks	1.5–2.5	10.4–17.2	—	—	—	
Rubbers		0.01–0.1	—	—	0.49	
Wood (parallel to grain)		9–16				0.4–0.8
Wood (perpendicular to grain)		0.6–1.0				0.4–0.8

Property values for ceramics and polymers vary widely depending on structure and processing. Ranges and average values are given. (Data taken from many sources.)

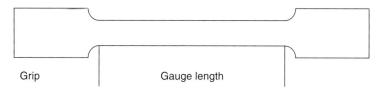

■ **FIGURE 2.3**   Tensile test bar. Deformation is measured in the thinner section.

and pans by deep drawing, and wires are made by drawing. Even cutting and turning make use of the plastic deformation of the metal by a cutting tool. Most of these shaping processes are performed at room temperature. Plastic deformation is also an important factor in safety: we shall see that plastic deformation provides metals with their resistance to fracture. Ceramics do not deform plastically but fracture when subjected to high tensile stress. Polymers do not deform plastically, they deform by viscous flow like liquids: Their response to stresses depends on temperature and deformation rate in a complex manner and will be described in Chapter 9.

2.3.1 **The Tensile Test**

The tensile test is the most widely used measure of a material's response to applied stresses. In the tensile test, a bar or a plate of the material is prepared in accordance with American Society for Testing and Materials (ASTM) standards. The specimen is more massive at the ends and has a reduced section known as the gauge length (see Figure 2.3). It is then gripped in a tensile testing machine (Figure 2.4) and loaded in uniaxial tension with increasing force until it fails. A load cell measures the force F applied to the sample as the grips separate and an extensometer attached to the test bar measures the increase Δl of the gauge length.

The tensile testing machine is used to measure the plastic deformation and fracture of materials. The elastic deformations of metals and ceramics are very small, ($\varepsilon < 1\%$) and are not obtained accurately with this machine. In reality, Young's modulus is determined by measuring the velocity of sound propagation in the material.

2.3.2 **The Stress-Strain Curve**

Figure 2.5 shows schematic stress-stain curves for ceramics, metals, and polymers, and Figure 2.6 presents some examples of measured curves.

The steep straight line at left represents elastic deformation. Its slope corresponds to Young's modulus according to Equation (2.6). This deformation is reversible: the sample recovers its initial shape when the stress is removed.

When the stress increases beyond the **yield stress** σ_y, metals undergo a much larger **plastic deformation**. Plastic deformation is irreversible. When the stress is removed, only the elastic deformation is reversed and plastic deformation remains. This is shown as ε_P in Figure 2.5.

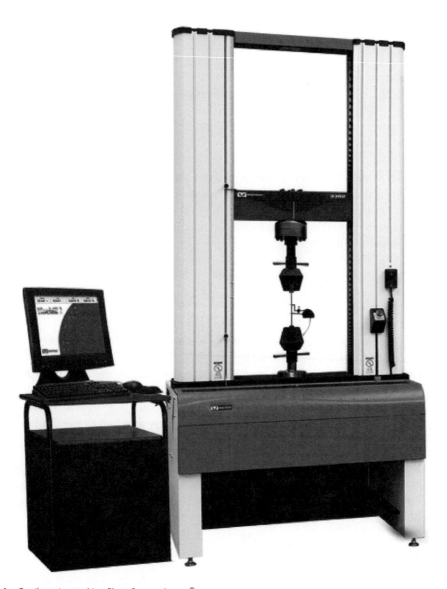

■ **FIGURE 2.4**   Tensile testing machine. Photo Courtesy Instron®.

In certain metals, the transition to plastic deformation is not abrupt and the exact elastic-plastic transition is difficult to pinpoint; see, for example, the gradual transition in AISI 4142 and AISI 1095 steels in Figure 2.6. For this reason the **offset yield stress** is defined as the stress required for a 0.2% plastic strain (or $\varepsilon = 0.002$). It is obtained, as shown in Figure 2.7, by drawing a line parallel to the elastic line, starting at 0.2% strain, and noting the stress level at which it intersects the stress-strain curve.

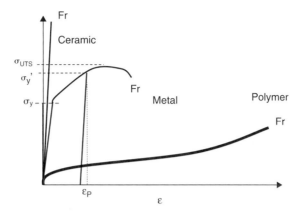

■ **FIGURE 2.5** Schematic stress–strain curve of a ceramic, a metal, and a polymer. In the elastic range, the slope is Young's modulus and the strain is reversible: removal of the stress removes the strain. In the plastic region of a metal, deformation is permanent. Ceramics do not deform plastically, but break above a certain stress. Elastomers (rubbers) are capable of large elastic and plastic deformation. F indicates fracture of the material.

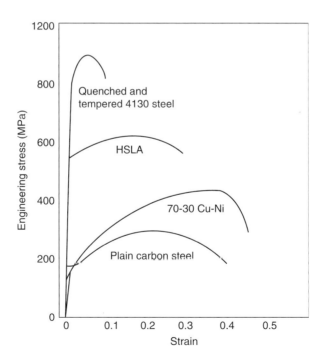

■ **FIGURE 2.6** Measured stress–strain curves of some steels and a 70-30 Cu-Ni brass.

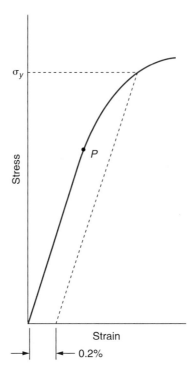

■ **FIGURE 2.7**   Construction to determine the offset yield stress. It is obtained by drawing a line parallel to the elastic stress-strain curve, offset by $\varepsilon = 0.002$; the intersection with the stress-strain curve corresponds to σ_y.

In most metals, the stress-strain curve is similar to that of steel AISI 1095 in Figure 2.6: increasing the plastic strain requires an increase in stress. This is also illustrated for the case of the metal in Figure 2.5. If we remove the stress after an amount of deformation, the permanent deformation ε_P remains. When we apply the stress again to the deformed sample, the elastic deformation range is increased to the new yield strain $\sigma_y{}'$. The metal has become stronger because of the deformation; this is **strain hardening.** The mechanism of strain hardening and its uses will be discussed in the next chapter. The stress eventually reaches a maximum, called the **ultimate tensile stress** σ_{UTS}, or **tensile strength** σ_T. From this point, the force required for further deformation decreases. If we observe the sample, we find that the decrease in force is not due to a weakening of the material, but to **necking**: a localized deformation decreases the cross section at some point. Even if the stress σ necessary for deformation keeps increasing, the force $F = \sigma A$ decreases. All further plastic deformation takes place in this neck.

Ultimately, the sample breaks. The strain at which the sample breaks is called **elongation at fracture** ε_F or simply **elongation**. The mechanical properties of a metal are thus defined by the elastic modulus E, the yield strength σ_y, the tensile strength σ_{UTS}, and the elongation ε_F.

2.3.3 **Ductility**

In the manufacture of metallic objects, one needs to ask the question: "How much can I deform or shape a metal before it breaks?" The relevant property is the **ductility** of the material. The measure of ductility given by the tensile test is the elongation at fracture

$$\varepsilon_f = \frac{l_f - l_o}{l_o} \tag{2.10}$$

Since most practical deformation does not involve tensile deformation, but often rolling or forging, a more common definition of ductility is the **percent reduction in area**

$$\%RA = \frac{A_o - A_f}{A_o} \cdot 100 \tag{2.11}$$

The two measures are related because, in plastic deformation, the volume of the material remains constant:

$$V = A_f l_f = A_o l_o$$

A material with ductility less than 5% is considered **brittle**.

2.3.4 **Resilience and Toughness**

In many instances, for instance in impact loading, the relevant question is not what stresses can be applied, but what energy the material can absorb. **Resilience** (Figure 2.8A) is the ability of a material to absorb and release elastic strain energy. The energy absorbed in deformation is calculated as $E = \int F dl$. If we replace the force F by $\sigma = F/A$ and the deformation Δl by the strain $\varepsilon = \Delta l/l$, we obtain the energy per unit volume, namely the **energy density**

$$E_R = \int_0^{\sigma_y} \sigma d\varepsilon = \frac{1}{E} \int_0^{\sigma_y} \sigma d\sigma = \tfrac{1}{2} \sigma_0^2 E \tag{2.12}$$

that the material can absorb by elastic deformation only. It is represented by the area of the triangle below the elastic part of the stress-strain curve.

Toughness (Figure 2.8B) is the **total energy per unit volume the material can absorb before it fractures**. It is measured as the total area under the stress-strain curve. It is an important property of material as it is associated with safety. Note that ductility is a total deformation; toughness is an amount of energy absorbed.

In the construction of bridges, buildings, and airplanes, for instance, one utilizes ductile and tough metals. On the contrary, in cutting tools, where "keeping the edge," that is, avoiding plastic deformation, is more important than avoiding fracture, one utilizes the hardest possible materials, which tend to be brittle.

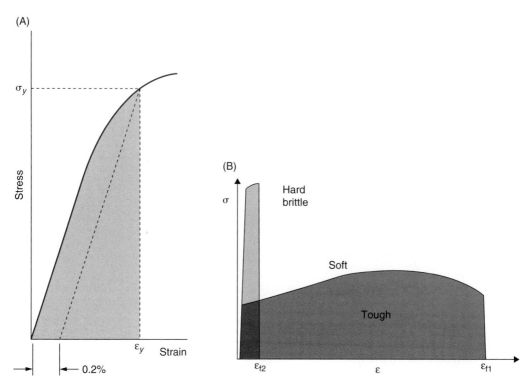

■ FIGURE 2.8   (A) The resilience of a material. (B) The toughness of a brittle and a soft metal. While the brittle material is stronger (it resists plastic deformation), it absorbs less energy than the weaker metal.

2.3.5 True Stress-True Strain and Engineering Stress and Strain

In Figures 2.5 to 2.7, the stress $\sigma = F/A$ is defined as the applied force divided by the **initial** cross section of the sample. It is obvious that these do not represent the true stresses and strains when a neck is forming. In the neck, the tensile force F is constant but the cross section A varies; therefore the local **true stress** varies and is largest at the narrow point of the neck as shown in Figure 2.9.

The **true strain** ε_T in the neck is also larger than the values $\varepsilon = \Delta l/l$ indicated in Figures 2.5 and 2.6. True stress and true strain in the neck cannot be measured and are approximated by calculations.

True stress and strain are important in scientific investigations because they are an accurate representation of the material's behavior. They are also important in manufacturing, where large deformations, for instance by rolling or forging, do not cause necking. In these cases, reduction in area, Equation (2.10), is a more useful measure of ductility than elongation at fracture.

While the stress-strain curves of Figures 2.5 to 2.7 do not represent the real stresses and strains in the sample, they are easily measured and are generally used to characterize materials. These are the **engineering stress-strain** curves. Figure 2.10 compares an engineering stress-strain curve and a true stress-strain curve of steel.

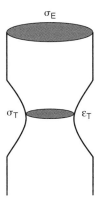

■ **FIGURE 2.9**   True stress and true strain in the neck region. Reduced cross section leads to higher local stress. Almost all deformation takes place in the neck.

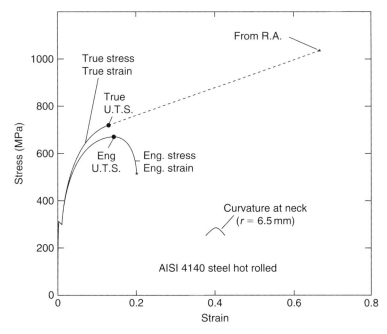

■ **FIGURE 2.10**   Engineering and true stress-strain curves for AISI 4140 hot rolled steel. The two dotted lines represent different calculations. From M.A. Meyers and K. K. Chawla, Mechanical Metallurgy: Principles and Applications, p. 565, Prentice Hall, Englewood Cliffs, NJ (1984).

2.4 **RESIDUAL STRESSES**

Residual stresses are not caused by the application of an external force but remain in the material when the external stress is removed. Uneven rapid cooling or extensive plastic deformation leaves internal stresses in which part of the material is compressed and a neighboring region is in tension. These stresses add themselves vectorially to the stresses caused by external forces. If a piece of material is subjected to a tensile external stress, for instance, a tensile residual stress adds itself to the applied stress and causes failure of the piece at an unexpected low load. A piece of glass with high residual stresses can break without an applied load. Compressive residual stresses reduce the total stress in a piece subjected to tensile forces and increase the resistance of the material to failure. We shall see in later chapters that the judicious introduction of compressive stresses increases fatigue resistance, produces safe glass for car windows (tempered glass), and increases the strength of concrete (prestressed concrete).

Residual stresses are measured by X-ray diffraction which determines the distances between atoms (see Chapter 19). Changes in these distances due to elastic strains are a measure of the stresses according to Hooke's law, Equation (2.6).

2.5 HARDNESS

The hardness of a material is its resistance to scratching or to indentation by a hard object. It is measured by the penetration of a hard sphere, a cone, or a pyramid that is pressed into the surface with a known force. This is a simple, rapid, nondestructive measurement and is by far the most widely performed of the mechanical tests.

Several hardness test methods have been devised, some for the convenience of the measurement and others for the quality of the information they provide. These are illustrated in Figure 2.11.

The **Rockwell hardness tests** are the most rapid and most popular. They measure the depth of penetration of the hard indenter under a selected load. Several Rockwell hardness tests exist, with different indenter geometries and applied loads so that accurate hardness measurements can be obtained for a wide range of materials, from the softest aluminum to the hardest steel. The indenters are a diamond cone or a steel sphere of 1/16, 1/8, ¼ or ½ in diameter. Applied loads are 60, 100, and 150 kg (1 kg = 9.81 N; the kilogram force is used because the Rockwell test predates the introduction of the International Units). When measuring Rockwell hardness, one first presses the indenter into the surface with a load of 10 kg; then one increases the load to 60, 100, or 150 kg and measures the increment in depth of penetration. The hardness is reported by an R followed by the letter corresponding to the test performed. For instance, a ball bearing has hardness RC 60.

Modern hardness instruments apply the load automatically for the correct length of time and compute the hardness value. Each measure requires only a few seconds. Rockwell hardness values have been established arbitrarily and are not simply related to the properties measured in the tensile test. They are specified in ASTM Standard Test E 18, "Standard Test Methods for Rockwell Hardness and Rockwell Superficial Hardness of Metallic Materials."

When performing the **Brinell hardness test**, one presses a sphere into the surface with the force P and measures the diameter of the indent left after the load has been removed. If D is the diameter of the sphere and d the diameter of the indentation, the Brinell hardness is computed as

$$B_{\mathrm{H}} = \frac{2P}{\pi D[D - \sqrt{D^2 - d^2}]} \tag{2.13}$$

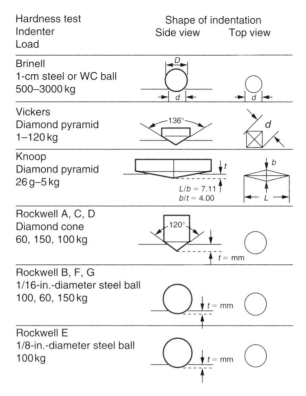

■ **FIGURE 2.11**   The hardness tests.

The **Vickers Hardness test** is the preferred method for the scientific characterization of materials. In this test one presses a diamond pyramid into the surface with a load that is chosen for the material and the type of information sought. Loads from 1g to 1,000g are generally used, although loads as high as 20 kg are applied for special measurements. After removal of the load, one measures the diagonal d of the square indentation left by the diamond. The hardness is the ratio of the load to the surface of the indentation

$$HV = \frac{1.854P}{d^2} \tag{2.14}$$

where P is the load applied and d the diagonal of the indentation.

Vickers hardness is usually reported in kg/mm², although it is often given without indication of the units. In the scientific literature, Vickers hardness is sometimes reported in the international units of MPa or GPa. Note that 1 kg/mm² = 9.81 MPa. It is usual to specify the load P when reporting Vickers hardness, such as $HV_{1\,kg}$ 500. (This is 500 kg/mm² measured at a load of 1 kg.)

A modern Vickers hardness tester applies the selected load automatically. It is equipped with an optical microscope for the measurement of the indentation diagonal and computes the hardness.

The **Knoop hardness test** is a variant of the Vickers test, designed to measure the hardness of narrow samples (such as cross sections of thin sheets or coatings). For this purpose the pyramid is replaced by a diamond knife as shown in Figure 2.11. The measurement is performed on the same instrument, with the same method, as the Vickers test. Knoop hardness is computed as

$$HK = \frac{1.42P}{L^2} \tag{2.15}$$

The most obvious application of hardness is in a material's resistance to scratching, to wear, and to machining. Since hardness is measured by a permanent indentation, it is also a measure of a material's resistance to plastic deformation. As a rule of thumb, the Vickers hardness value is about three times the tensile strength. This correspondence, however, is not exact because the deformation in hardness measurements has a complex geometry and strain hardening takes place.

2.6 FRACTURE

Metals fracture in ductile and brittle modes and sometimes in a combined ductile-brittle manner. Ductile fracture is accompanied by large strains and extensive plastic deformation in the region of the crack tip. Brittle fracture propagates rapidly and leaves surfaces that can often be fitted together; it is the most feared mode of failure because it occurs usually without prior warning of impending catastrophe.

2.6.1 Ductile Fracture of Metals

Ductile metals **neck** during plastic deformation. When the latter is large enough, the neck breaks and the fracture surface displays the **cup-cone** fracture morphology shown in Figure 2.12A. Microscopic examination of the fracture surface shows dimples on the flat crater bottom loaded in tension (Figure 2.13) and a cone lip at the periphery. The process of ductile fracture may be broadly viewed in terms of the sequential nucleation and growth of microvoids schematically indicated in Figure 2.14. The earliest stage of fracture creates isolated microscopic cavities. These nucleate at inclusions or second phase particles that are harder than the metal and do not deform plastically. The microvoids grow with further increase of the stress and finally coalesce and form a crack (Figure 2.14A).

The crack spreads outward toward the periphery of the neck (Figure 2.14B). Finally, an overloaded outer ring of material is all that is left to connect the specimen halves, and it fails by shear at 45° with respect to the tensile force, which is the direction of maximum shear stress τ.

In high purity FCC and BCC metals free of inclusions, necking to ~100% area reduction (i.e., to a point) is possible. In this case, the material deforms plastically and separates by shear.

2.6.2 Brittle Fracture of Metals and Ceramics

Figure 2.12B shows a metal bar that failed by brittle fracture. Higher magnification reveals that **transgranular cracking** often occurs during brittle fracture, especially at low temperatures. In such cases the crack propagates **across** the grain interiors (Figure 2.15A). When cracks propagate **along** grain boundaries, we speak of **intergranular fracture**. Such failures (Figure 2.15B) reveal clearly outlined grains that stand out in relief like the stones of a wall.

(A)   (B)

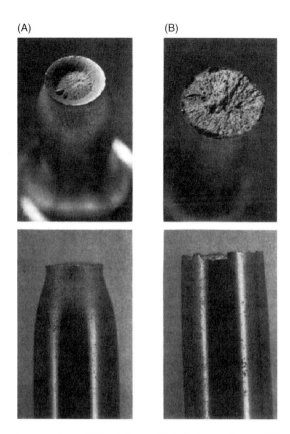

■ FIGURE 2.12   Fracture of metallic tensile test specimens. (A) Ductile fracture. (B) Brittle fracture. Courtesy G.F. Vander Voort, Buehler LTD.

2.6.3 **Cracks and Stress Concentrations**

We are not capable of predicting the initiation of a crack, but once a crack exists, fracture mechanics describes its propagation. Cracks at the surface or the interior of objects under load generate very large local increases in stress. This fact is illustrated in Figure 2.16 where a body (plate), containing elliptical-shaped surface indentations and interior holes, is pulled by uniaxial forces, F.

Far away from the holes and indentations, the force F generates a constant stress

$$\sigma_a = \frac{F}{A}$$

where A is the cross section of the plate. At the tip of the crack of half-length c and radius of curvature ρ, the local tensile stress is

$$\sigma_{\text{tip}} = \sigma_a 2 \left(\frac{c}{\rho} \right)^{1/2} \tag{2.16}$$

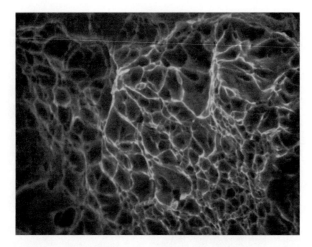

■ **FIGURE 2.13**   Dimples in the ductile fracture surface of a metal. Courtesy G.F. Vander Voort, Buehler LTD.

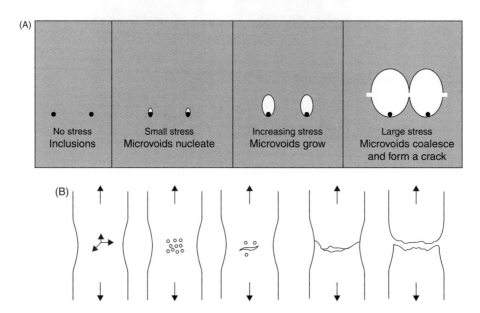

■ **FIGURE 2.14**   (A) Formation and coalescence of microvoids in ductile fracture. (B) Model of ductile fracture. From left to right: application of stress, nucleation of microvoids, coalescence of microvoids, formation of elliptical crack, shearing of collar and separation.

σ_{tip} can be very large at the tip of sharp cracks where the radius of curvature ρ is small. A **stress concentration factor,** k_σ, can be defined as the ratio of amplified local stress at the crack tip to the background stress, that is, $k_\sigma = \sigma_{tip}/\sigma_a$. For flat elliptical surface cracks k_σ is

$$k_\sigma = 2\,(c\,/\rho)^{1/2} \tag{2.17}$$

(A)

(B)

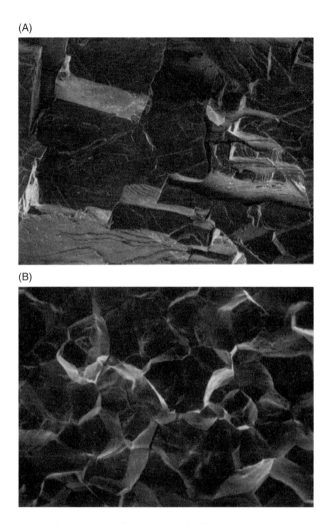

■ **FIGURE 2.15** (A) Scanning electron microscope image of transgranular brittle fracture in Fe-2.5 wt% Si tested at −195°C. (B) Scanning electron microscope image of intergranular fracture in a nickel-based superalloy. Courtesy of G.F. Vander Voort, Buehler LTD.

Here c is half the length of the major axis and ρ is the radius of curvature at the crack tip. **Longer and sharper cracks raise the stress concentration k_σ.**

Now we can understand why hard materials are more brittle than soft ones. This is illustrated in Figure 2.17. The left figure shows a sharp crack at the surface of a hard material. The stresses at the crack tip are large, following Equations (2.16) and (2.17). The stress is large enough to propagate the crack. In a soft metal, these stresses cause plastic deformation that increases the radius ρ and decreases the stresses below what is necessary for crack propagation.

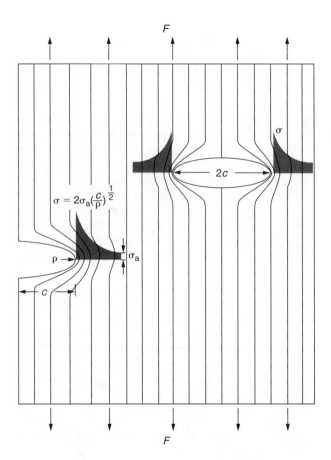

■ **FIGURE 2.16**   The concentration of tensile stresses around internal and surface cracks in a uniformly stretched solid. Parallel lines denote uniform stress.

2.6.4 **Fracture Toughness**

Figure 2.16 shows that we can know the average stress and the crack length and that the crack radius ρ is variable and depends on the properties of the material. Therefore one defines the fracture resistance as the maximum combination of stress and crack length that does not lead to crack propagation. It is known as the **critical stress intensity factor** or **fracture toughness** K_{1C}.

$$K_{1C} = Y\sigma\sqrt{\pi c} \tag{2.18}$$

When $K < K_{1C}$, the crack is stable. But when $K > K_{1C}$, the crack is unstable and will propagate. Y is a numerical factor that depends on the geometry of the crack. $Y = 1$ for the through crack in the body

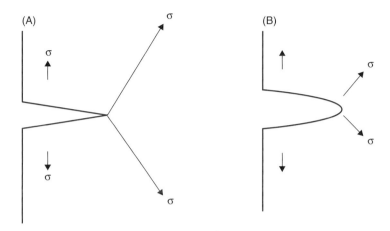

■ **FIGURE 2.17**   Influence of hardness on fracture resistance. (The arrows represent tensile stresses.) (A) sharp crack in a hard material: the stresses at the tip are large and the crack propagates. (B) crack in a soft metal: the large stresses at the crack tip cause plastic deformation and increase the radius ρ at the crack tip, reducing the stress below the threshold for crack propagation.

of the specimen (Figure 2.16) and $Y = 1.1$ for a crack at the surface of a plane specimen. The fracture toughness has dimensions MPa$\sqrt{}$m or ksi$\sqrt{}$in. 1 ksi$\sqrt{}$in ≈ 1.1 MPa$\sqrt{}$m.

Values of K_{1C} for a number of materials are entered in Table 2.2. The high values of K_{1C} for metals are due to the plastic zone and the increase of the radius ρ at the crack tip. Among the important trends gleaned from this table are the following:

1. Glasses and simple metal oxides have the lowest values of K_{1C}. They are typically 10–100 times smaller than the fracture toughness of ductile metals.

2. K_{1C} for some of the toughened ceramics are much higher than for glasses but are still well below (by a factor of ~10) the level for metals.

EXAMPLE 2.1   *What is the ratio of the critical crack sizes in aluminum relative to aluminum oxide when stressed to their maximum respective elastic stresses?*
ANSWER   At the maximum elastic stress, $\sigma = \sigma_y$ (the yield stress), and

$$c = 1/\pi \, (K_{1C}/\sigma_y)^2$$

From Table 2.2, for Al: $K_{1C} = 44$ MPa$\sqrt{}$m and $\sigma_y = 345$ MPa, and for Al$_2$O$_3$: $K_{1C} = 3.7$ MPa$\sqrt{}$m and $\sigma_y = 270$ MPa. (This is not a yield stress. Alumina does not deform plastically; it is the stress at which alumina fractures.)

For aluminum, the crack length is 5.3×10^{-3} m $= 5.3$ mm.

In aluminum oxide, the crack length is 6×10^{-5} m $= 60 \, \mu$m, which is 85 times shorter.

Table 2.2 Room Temperature Yield Strength and Plane Strain Fracture Toughness Data for Selected Engineering Materials.

Material	Yield strength		K_{1c}	
	MPa	ksi	MPa$\sqrt{m}$	ksi$\sqrt{in}$
Metals				
Aluminum alloy (7075-T651)	495	72	24	22
Aluminum alloy (2024-T3)	345	50	44	40
Titanium alloy (Ti-6Al-4V)	910	132	55	50
Alloy steel (4340 Tempered 260°C)	1,640	238	50	45.8
Alloy steel	1,420	206	87.4	80
Ceramics				
Concrete	–	–	0.2–1.4	0.18–1.27
Soda-lime glass	–	–	0.7–0.8	0.65–0.75
Aluminum oxide	–	–	2.7–5.0	2.5–4.5
Polymers				
Polystyrene	–	–	0.7–1.1	0.6–1.0
Polymethyl methacrylate	54–73	8–10.5	0.7–1.6	0.6–1.5
Polycarbonate	62	9.0	2.2	2.0
Tungsten carbide	–	–	13	12

The different responses to flaw size of metals relative to ceramics can be qualitatively understood from this example. Large cracks in metals can often be detected during inspection and possibly repaired by sealing them shut. But the flaws of critical size in ceramics are small and easy to overlook, a combination that makes ceramics fracture-prone in service.

1. **A crack or a notch is a stress concentrator.** The local stress is many times larger than the average stress; if the crack is sharp, that is, if the radius of curvature ρ is small, the local stress σ_{tip} is large enough to rupture bonds. A crack propagates even at moderate average stress σ_a. The avoidance of stress concentrators is an important consideration in mechanical design.

2. The propagation of a crack can be arrested by increasing the radius ρ. In a plate, for instance, a crack can be arrested by drilling a hole of sufficient diameter.

■ **FIGURE 2.18** Specimens for the measurement of fracture toughness. Dark: metal. Light: ceramic. Courtesy of P.S. Han, MTS Systems Corporation.

3. In soft metals (with a low yield stress) plastic deformation caused by the local stress σ_{tip} at the crack tip increases the radius ρ (i.e., blunts the crack) and reduces the stress concentration. In hard metals, the high yield strength prevents crack blunting and maintains a high stress concentration at the crack tip. As a rule, **hard materials are brittle and soft materials are ductile**.

4. Ceramic materials are very hard and do not deform plastically under tensile stress; they break abruptly after elastic deformation. **Ceramic materials and glasses are brittle; their fracture toughness is low**.

5. Ceramic materials contain small cracks as a consequence of their processing. The length of these cracks cannot be controlled and is not easily measured. Therefore the average (design) stress σ_a that leads to fracture in a ceramic cannot be determined with certainty. **Ceramics are unreliable in tension**; they should be designed in such a way that they are subjected to compressive stresses only.

2.7 THE MEASUREMENT OF FRACTURE RESISTANCE

2.7.1 Fracture Toughness

To determine K_{1C} values of materials, specimens are manufactured in the form of plates containing machined cracks of known length as shown in Figure 2.18. As tensile loads are applied, the specimen halves open and the resulting crack extension is continuously monitored (Figure 2.19). Noting the critical crack size and stress necessary to induce fracture, one obtains the value for K_{1C} with Equation (2.18). Ductile metals require large specimens for reliable fracture toughness measurements; brittle materials can be measured with smaller specimens.

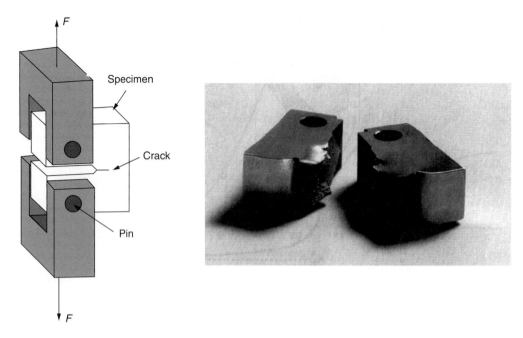

FIGURE 2.19 **FIGURE 2.19** Measurement of fracture toughness and broken sample. Courtesy of P.S. Han, MTS Systems Corporation.

2.7.2 Indentation Toughness of Ceramics

For routine measurements of toughness, the fabrication of ceramic samples as shown in Figure 2.18 is expensive. A less reliable but more convenient measurement is afforded by high-load Vickers indentations normally used for hardness measurements. When a material is brittle and the Vickers indentation is made at sufficiently high load (minimum 20 N, but generally 100 N or 10 kg), radial cracks form at the tip of the indent as shown in Figure 2.19. Empirical equations provide the fracture toughness from a measurement of the crack length and the size of the hardness indent. One such equation (Antsis et al. 1981) is

$$K_{1C} = 0.016 \, (E/H)^{1/2} P/c^{3/2} \tag{2.19}$$

Here K_{1C} is the fracture toughness, E the elastic modulus, H the hardness determined from the indentation by Equation (2.14), P is the load applied, and c the crack length measured form the center of the indent. K_{1C} is expressed in MPa$\sqrt{}$m or ksi$\sqrt{}$in.

Such measurements allow one to measure local variations of hardness and toughness in a piece of material. The method cannot be used with metals that are too ductile to form radial cracks. See Figure 2.20.

2.7.3 Charpy and Izod Measurements of Notch Toughness

The widely used Charpy impact test is a standard way to assess toughness. In this test a standard bar specimen, with a square cross section and a V-shaped notch cut into it, is strained very rapidly to fracture by means of a swinging pendulum-like hammer (Figure 2.21).

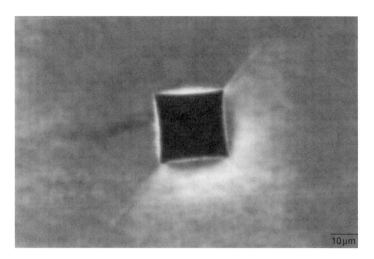

■ **FIGURE 2.20**   Vickers indentation in a ceramic. The size of the indentation determines the hardness. The length of the radial cracks is used to measure the toughness.

The difference between the initial and final potential energies of the hammer (measured by the initial and final hammer heights) is the impact energy absorbed by the specimen. Charpy and Izod test methods differ only in the position of the sample as shown in Figure 2.21.

Charpy and Izod impact toughness values measure the energy needed to fracture the specimen and are expressed in foot-pounds (ft-lb). They provide a measure of the toughness discussed in Section 2.5. These tests are more convenient but less reliable than the fracture toughness tests of Figures 2.18 and 2.19.

2.7.4 **The Rupture Strength of Ceramics**

Ceramics are brittle; they do not deform plastically but fracture under tensile stress. The strength of ceramics is often measured in a bend test shown in Figure 2.22.

A thin plate of ceramic is placed between four cylinders to which forces are applied as shown in Figure 2.22. These forces tend to bend the plate elastically and generate a tensile stress on the lower surface. The stress at which the plate fractures has several names: flexural strength, modulus of rupture, fracture strength, or bend strength. Flexural strengths of several ceramics are shown in Table 2.3. Note that the flexural strengths, given in MPa or ksi, are indicated with a wide uncertainty. This is easy to understand with a look at Equation (2.16). The toughness of ceramics is a reasonably well-defined quantity. It involves a stress (which is measured in the bend test) and a crack length. In a ceramic, these cracks are small, have random lengths, and cannot be measured; the variation in crack lengths accounts for the dispersion in fracture stresses.

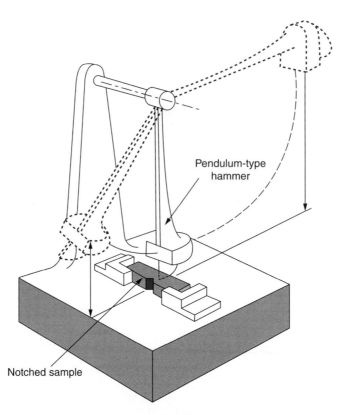

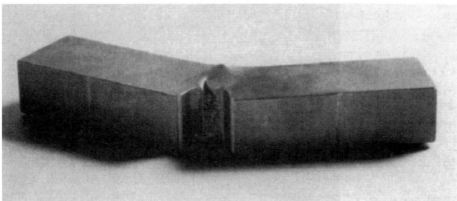

■ **FIGURE 2.21**   Impact toughness test. (Top): Schematic of a Charpy impact testing machine. The hammer of weight W released from height h breaks the sample as it swings and stops at height h′. The difference in potential energies $W(h - h')$ is the impact toughness. From C.R. Barrett, W.D. Nix, and A.S. Tetelman: The Principles of Engineering Materials, Prentice-Hall, Englewood Cliffs, NJ (1973).(Bottom): Ductile Charpy specimen after impact. Courtesy of P.S. Han, MTS Systems Corporation.

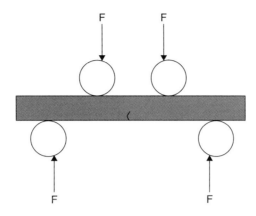

■ **FIGURE 2.22**   Four-point bend test.

Table 2.3 Flexural Strength and Modulus of Elasticity for Selected Materials.				
	Flexural strength		**Modulus of Elasticity**	
Material	**MPa**	**ksi**	**GPa**	**10^6 psi**
Silicon nitride (Si_3N_4)	250–1,000	35–145	304	44
Zirconia (ZrO_2)	800–1,500	115–215	205	30
Silicon carbide (SiC)	100–820	15–120	345	50
Aluminum oxide (Al_2O_3)	275–700	40–100	393	57
Glas-ceramic (Pyroceram)	247	36	120	17
Mullite ($3Al_2O_3$-$2SiO_2$)	185	27	145	21
Spinel ($MgAl^2O_4$)	110–245	16–36	260	38
Magnesium oxide (MgO)	105[b]	15[b]	225	33
Fused silica (SiO_2)	110	16	73	11
Soda-lime glass	70	10	69	10

[a] *Partially stabilized with 3 mol% Y_2O_3.*
[b] *Containing approximately 5% porosity.*

2.8 **FATIGUE**

Machine elements such as wheels, shafts, and ball bearings that are subjected to cyclic or repeated loads can function as designed for quite a long time and break suddenly, without any change in load. This is fatigue failure. Its prevention is obviously important in the construction of aircraft wings and helicopter rotors, but it also plays a role in bridge construction.

Fatigue is the failure (by fracture) of structures that are subjected to repeated or cyclic loading. By cyclic loading we mean almost any repeated stress variation, for example, (a) axial tension-compression, (b) reversed bending, and (c) reversed torsion or twisting. Internal flaws such as inclusions of softer or harder material, machined grooves, threads, or deep fillets serve as stress concentrators that nucleate cracks. The resultant cracks propagate a little at each cycle and grow with time to macroscopic dimensions until the component breaks when a crack of critical size is reached. This sequence of events has been repeated in components of rotating equipment such as motor and helicopter shafts, train wheels and tracks, pump impellers, ship screws and propellers, and gas turbine discs and blades. Even surgical prostheses implanted into the human body have suffered fatigue failure in service. **Under cyclic loading, failure can occur at stresses that are significantly below the tensile or yield stress σ_y of the material**.

Figure 2.23 shows the fatigue failure of a shaft. A tiny circular crack has been nucleated at point C. The propagation of this crack with the repeated applications of stress generated circular or semicircular

■ **FIGURE 2.23**   Fracture surface of a 4 1/2 in diameter, 4320 steel drive shaft that has undergone fatigue failure. Courtesy G.F. Vander Voort. Buehler LTD.

ridges away from the crack nucleus. "Beach marks" denote the instantaneous positions of the advancing crack and give the fracture surface a clamshell appearance. Finally, only a small area of the cross section at the shaft keyway remained to support the load, and it ruptured by a ductile overload fracture at B.

Figure 2.24 shows striations in a nickel-based superalloy; they represent the positions of the crack at each stress cycle.

In the fatigue testing machine (Figure 2.25), a bar specimen that narrows in diameter toward the middle, is mounted horizontally and rotated at high speed with a motor. A hanging load tilts the grips, stressing its surface alternately in tension and compression during rotation. This test provides a sinusoidal loading about a zero mean stress level (Figure 2.25B). For the case of arbitrary sinusoidal loading, maximum (σ_{max}), minimum (σ_{min}), mean (σ_{mean}) stress, and stress range (σ_r) values are defined in Figure 2.25C. The number of rotations to failure is counted at a chosen stress amplitude. Then a new specimen is mounted, a different load is chosen, yielding a new stress amplitude, and the number of cycles to failure is determined once again, and so on. In this tedious manner sufficient data points are accumulated to generate an *S-N* (or stress-number of cycles) curve.

The *S-N* response for assorted materials is displayed in Figure 2.26. Here *S* is the stress and *N* the number of cycles to failure. The horizontal lines show that most steels can rotate indefinitely below a stress level known as the **endurance** or **fatigue limit** (the curve horizontal). Some high strength steels and many nonferrous metals (e.g., Al and Cu), on the other hand, do not exhibit an endurance limit.

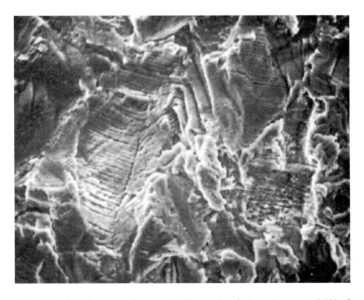

■ **FIGURE 2.24**   Fatigue striations in a nickel-base superalloy viewed in the scanning electron microscope at 2,000x. Courtesy G.F. Vander Voort. Buehler LTD.

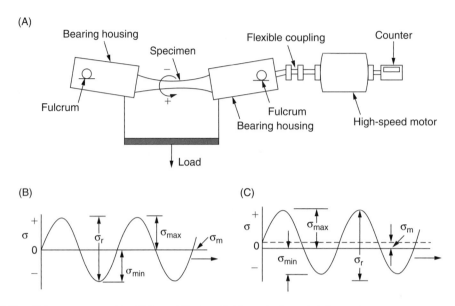

■ FIGURE 2.25 (A) Fatigue testing machine. (B) Sinusoidal loading. (C) Arbitrary loading with definition of mean, maximum, minimum stress, and stress amplitude.

Ceramics and polymers display similar fatigue characteristics. To guard against fatigue, stress levels should not exceed $1/3$ σ_{UTS}.

2.9 CREEP

At temperatures of about half the melting point $(0.5T_M)$ and above, materials undergo time-dependent plastic straining when loaded. This phenomenon is known as creep; it can occur at stress levels less than the yield strength. The extension of a component may eventually produce a troublesome loss of dimensional tolerance and ultimately lead to catastrophic rupture. The need to avoid turbine blade creep due to the high temperatures of a jet engine is an often cited example. High pressure boilers and steam lines, nuclear reactor fuel cladding, and ceramic refractory brick in furnaces are components and systems that are also susceptible to creep. Since room temperature is more than half the absolute melting temperature of solder, creep in solder is a problem to be contended with in electronic circuits.

Creep tests are performed on round tensile-like specimens that are stressed by fixed suspended loads while being heated by furnaces that coaxially surround them. The typical response obtained in a creep test is shown in Figure 2.27, where the specimen elongation or strain is recorded as a function of time. There is an initial **elastic** extension or strain the instant load is applied. Then a plastic straining ensues in which the creep strain rate $d\varepsilon/dt$ decreases with time. This **primary creep** period then merges with

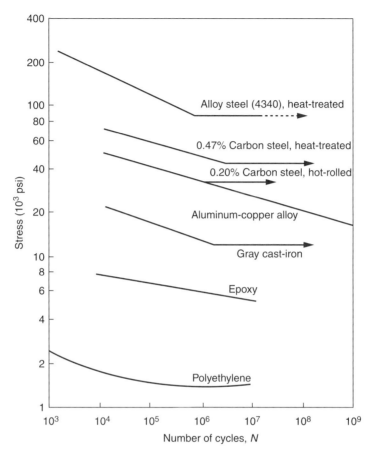

■ **FIGURE 2.26**   S-N data for various metals and polymers.

the **secondary creep** stage where the strain rate is fairly constant. Alternately known as **steady-state creep**, this region of **minimum creep** normally occupies most of the test lifetime. Finally the strain rate increases rapidly in the **tertiary creep** stage and leads to rupture of the specimen.

Engineers perform two types of tests. The first aims to determine the steady-state creep rate over a range of stresses and temperatures. These tests are performed at the same stress but at different temperatures or at the same temperature but different stresses, as shown in Figure 2.28. Specimens are usually not brought to failure in such tests; accurately predicting their extension is of interest here. The second type of test, known as the creep rupture test, is conducted at higher stress and temperature levels in order to accelerate failure. Such test information can either be used to estimate short-term life (e.g., turbine blades in military aircraft), or be extrapolated to lower service temperatures and stresses (e.g., turbine blades in utility power plants) to predict long-term life.

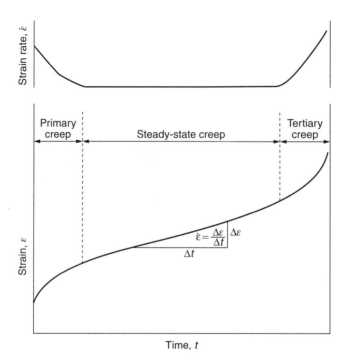

■ **FIGURE 2.27**   Schematic of creep. Bottom: elongation as a function of time. Top: strain rate as a function of time.

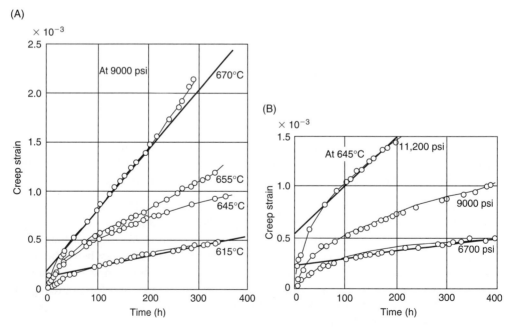

■ **FIGURE 2.28**   Creep strain versus time in a 0.5 wt% Mo, 0.23 wt% V steel. (A) Constant stress, variable temperature. (B) Constant temperature, variable stress. From A.J. Kennedy, Processes of Creep and Fatigue in Metals, Wiley, NY (1963).

2.9.1 **Steady-State Creep**

Much engineering design is conducted on the basis of steady-state creep data. Since the creep strain rate is accelerated by both temperature and stress, a useful phenomenological equation that succinctly summarizes the data of Figure 2.28 and reflects the role of these variables is

$$d\varepsilon/dt = A\,\sigma^m \exp - (E_c/kT) \qquad (2.20)$$

where A and m are constants; m typically ranges in value from 1 to 5. The exponential term is the Boltzmann factor that reflects the thermally activated nature of creep. E_c is the creep activation energy; it is close to the activation energy for bulk diffusion.

EXAMPLE 2.2   *Determine E_c and m for the data shown in Figure 2.28.*
ANSWER   The steady-state strain rate is the slope of the strain-time response. Limiting the calculation to only the maximum and minimum slopes for the superimposed lines, the following results are tabulated:

	Test Temperature (°C)		Stress (psi)	Strain Rate (h^{-1})
1.	670	(= 943 K)	9,000	$(2.5 - 0.28) \times 10^{-3}/(370 - 0) = 6 \times 10^{-6}$
2.	615	(= 888 K)	9,000	$(0.56 - 0.2) \times 10^{-3}/(400 - 0) = 9 \times 10^{-7}$
3.	645	(= 918 K)	11,200	$(1.5 - 0.6) \times 10^{-3}/(200 - 0) = 4.5 \times 10^{-6}$
4.	645	(= 918 K)	6,700	$(0.5 - 0.25) \times 10^{-3}/(400 - 0) = 6.25 \times 10^{-7}$

To determine E_c, data from tests 1 and 2 are utilized. From Equation (2.17),

$\dot{\varepsilon}(1)/\dot{\varepsilon}(2) = (\exp - E_c/kT_1)/(\exp - E_c/kT_2)$

or

$E_c = kT_1 T_2/(T_1 - T_2)\ln[\dot{\varepsilon}(1)/\dot{\varepsilon}(2)]$

Substituting, $E_c = 8.31$ J/mol-K $(943 \times 888/55) \times \ln[6 \times 10^{-6}/9 \times 10^{-7}] = 240$ kJ/mol.

Similarly, using test data 3 and 4,

$\dot{\varepsilon}(3)/\dot{\varepsilon}(4) = (\sigma_3/\sigma_4)^m$

and

$m = \ln(\dot{\varepsilon}(3)/\dot{\varepsilon}(4))/\ln(\sigma_3/\sigma_4)$

After substitution

$m = \ln(4.5 \times 10^{-6}/6.25 \times 10^{-7})/\ln(11,200/6,700) = 1.01$

Knowing E_c and m enables $\dot{\varepsilon}$ to be evaluated at any other temperature and stress level.

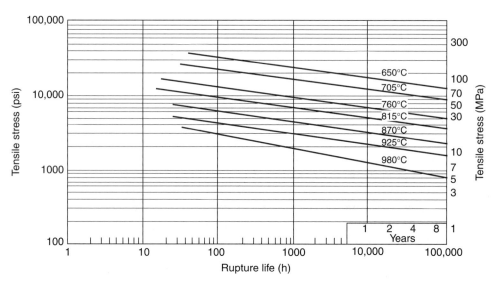

■ **FIGURE 2.29**   Creep-rupture data of Incoloy 800, an iron-based alloy containing 30 wt% Ni and 19 wt% Cr.
Source: Huntington Alloys Inc.

2.9.2 **Creep Rupture**

Creep or stress rupture test data are normally plotted as the stress versus rupture time (t_R) on a log-log plot as shown in Figure 2.29. Each data point represents one test at a specific temperature and stress level. Given sufficient testing at different temperatures, a complete profile of the material response is obtained.

■ SUMMARY

1. A force applied to a solid induces stresses in its interior. In a simple geometry, such as a bar, the stress is the force divided by the cross section of the piece. Tensile stresses increase the length of the material; compressive stresses decrease its length. Shear stresses result from a couple of forces in opposite directions. Stresses are expressed in megapascals, where 10^6 N/m^2 = 1 MPa, or in thousands of pounds per square inch (ksi). The resultant deformations are dimensionless, often expressed in percent (%).

2. Elastic deformation is instantaneous and reversible. It is expressed by Hooke's law: $\sigma = \varepsilon E$ for tensile stresses and $\tau = \gamma G$ for shear stresses. $E \approx 210$ GPa for iron and steel and $E = 70$ GPa for aluminum. The maximum elastic deformation in metals and ceramics is lower than $\varepsilon = 1\%$. It is much larger for polymers and $\varepsilon \approx 1,000\%$ for elastomers (rubbers).

3. In the case of metals, the yield stress is the stress at which plastic deformation starts; the ultimate tensile stress is the largest stress that a material can sustain before fracture; the ductility is the largest deformation of the material before fracture; and work hardening is the increase in strength due to plastic deformation.

4. Residual stresses remain in the material in the absence of external forces. They are caused by uneven cooling or plastic deformation.

5. In hardness tests, one measures the penetration of a hard object pressed into the surface with a known force. Different techniques measure the depth of penetration (Rockwell hardness) or the lateral size of the indent (Brinell, Vickers, Knoop hardness).

6. Ductile fracture occurs after extensive plastic deformation. It occurs by the formation and coalescence of voids and shearing.

7. Brittle fracture occurs by the propagation of cracks that usually nucleate at the surface.

8. Resilience is the maximum energy a solid can absorb elastically. Toughness is the maximum energy a metal can absorb by plastic deformation.

9. Fracture toughness is the resistance to fracture. The units of fracture toughness are $MPa\sqrt{m}$ or $ksi\sqrt{in}$; $1\,ksi\sqrt{in} = 1.1\,MPa\sqrt{m}$.

10. Cracks, indentations, and sharp interior corners constitute stress concentrators that can lead to crack propagation.

11. The fracture toughness of ceramics is well defined; the fracture stress is not a precise measure because fracture is caused by small cracks of unknown length.

12. Fatigue is the failure of materials due to repeated, cyclic application of a stress. Fatigue failure can occur at stresses **below the yield stress**.

13. Steels and some other materials exhibit a fatigue limit. This is a stress below which no fatigue failure occurs.

14. Creep is the slow plastic deformation due to diffusion of atoms at high temperatures ($T > \frac{1}{2}$ melting temperature). Creep occurs at stresses **below the yield stress**. Note the difference: plastic deformation is not dependent on time, and the stress determines the **amount** of deformation; creep is a viscous-like deformation, and the stress governs the **rate** of deformation.

■ KEY TERMS

B
Brinell hardness, 50
brittle fracture, 52
Brittle fracture
   intergranular, 52
   transgranular, 52
brittle materials, 47

C
Charpy, 60

creep, 66
creep rupture, 70
critical stress intensity, 56

D
ductile fracture, 52
Ductility, 47

E
elastic deformation, 40

elastic modulus, 40
elongation at fracture, 46
endurance, 65
engineering stress–strain, 48

F
fatigue, 64
fatigue limit, 65
fracture, 52
fracture toughness, 56, 59

■ REFERENCES FOR FURTHER READING

[1]   Antsis, G.R. Chantikul, P. Lawn B.R. and Marshall, D.B. A Critical Evaluation of Indentation Techniques for Measuring Fracture Toughness; I: Direct Crack Measurements, *J. Amer. Ceram. Soc.* 64(9), 533–538 (1981).

[2]   Ashby, M.F. *Materials Selection in Mechanical Design*, 3rd ed., Butterworth-Heinemann, Woburn, MA (2005).

[3]   Ashby, M.F. Jones, D.R.H. *Engineering Materials, An Introduction to Their Properties and Applications*, 3rd ed., Butterworth-Heinemann, Woburn, MA (2005).

[4]   Handbook, A.S.M. *Mechanical Testing and Evaluation*, ASM International, Materials Park, OH (2000).

[5]   Boyer H.E. (ed.), *Atlas of Stress Strain Curves*, Publisher, Materials Park, OH (1986).

[6]   Courtney, T.H. *Mechanical Behavior of Materials*, 2nd ed., Waveland Press, Long Grove, IL (2005).

[7]   Dowling, N.E. *Mechanical Behavior of Materials*, 3rd ed., Prentice Hall, Englewood Cliffs, NJ (2005).

■ PROBLEMS AND QUESTIONS

We use International and American units in these problems. Use the following conversions: Stresses, 1 ksi ≈ 7 MPa; Temperatures, $T_C = 5/9 (T_F - 32)$.

2.1.   a. What is the final length of a 2 m long bar of copper 10 mm in diameter, stressed by a 5,000 N force?

   b. If a steel bar of the same diameter has the same force applied to it, how long must it be to extend the same amount as the copper bar in part a?

2.2. Wire used by orthodontists to straighten teeth should ideally have a low modulus of elasticity and a high yield stress. Why?

2.3. During the growth of silicon single crystals, a small seed measuring 0.003 m in diameter is immersed in the melt and solid Si deposits on it. Small seeds minimize the likelihood of defects propagating into the growing crystal. What is the length of a 0.25 m diameter crystal that can be supported by such a seed if the yield stress of silicon is 1.5 GPa? The density of silicon is 2.33 g/cm³ (2.33 Mg/m³).

2.4. Steel pipe employed for oil drilling operations is made up of many segments joined sequentially and suspended into the well. Suppose the pipe employed has an outer diameter of 7.5 cm and an inner diameter of 5 cm, a density of 7.90 Mg/m³ (7.9 g/cm³), and a yield strength of 900 MPa. If a safety factor of two is assumed,

   a. What is the maximum depth that can be safely reached during drilling?

   b. What is the maximum elastic strain in the pipe and at what depth does it occur?

   c. At what depth is the elastic strain minimum?

2.5.   a. Write an expression for the maximum elastic strain energy per unit weight of a material.

   b. How much elastic strain energy (in units of J/kg) can be stored in steel if the yield stress is 300,000 psi or 0.01 E?

   c. A 12 V automobile battery weighs 15 kg and can deliver 150 A. hr of charge. How much specific energy (J/kg) is stored by the battery?

2.6. A 1,500 kg load is hung from an aluminum bar of dimensions 1 m × 0.01 m × 0.01 m and stretches it elastically along the long axis. What is the change in volume of the bar?

2.7. A tensile bar has an initial 2.000 in gauge length and a circular cross-sectional area of 0.2 in². There is no necking. The loads and corresponding gauge lengths recorded in a tensile test are:

Load (lb)	Gauge length (in)
0	2.0000
6,000	2.0017
12,000	2.0034
15,000	2.0042
18,000	2.0088
21,000	2.043
24,000	2.30
25,000	2.55
22,000	3.00

Plot the data and determine the following:

   a. Young's modulus.
   b. 0.2% offset yield stress.
   c. Ultimate tensile stress.
   d. Percent elongation.
   e. Percent reduction of area.
   f. Provide these data in international units.

2.8. What is the difference between:

   a. Stress intensity and critical stress intensity?
   b. Strength and toughness?
   c. Cracks in metals and cracks in ceramics?
   d. Transverse-rupture strength and tensile strength?

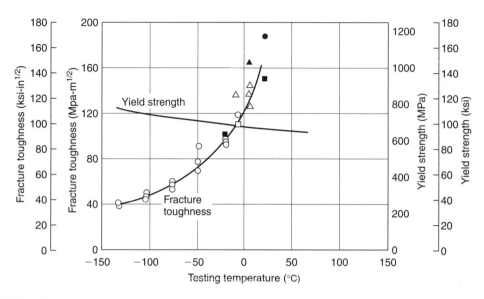

■ **FIGURE 2.30** Fracture toughness and yield strength as a function of testing temperature.
From Metals Handbook, American Society of Metals, Metals Park, OH (2000).

2.9. Draw a stone bridge and a steel (railroad) bridge and explain the difference of their shapes in terms of their material properties.

2.10. A plate of steel has a yield stress of 1,000 MPa. The plate fractured when the tensile stress reached 800 MPa, and it was therefore hypothesized that a surface crack was present. If the fracture toughness for this steel is 60 MPa$\sqrt{}$m, what crack size is suggested?

2.11. Show that 1 ksi$\sqrt{}$in $\approx$ 1 MPa$\sqrt{}$m.

2.12. Big Bird Helicopter Company requires a stainless steel that is tough enough for use in the Tropics (120°F) as well as the Arctic (−60°F). They peruse this book and find the data of Figure 2.30. At the designed use stress, the component can tolerate a surface flaw of 0.080 in (2 mm) at 120°F. Manufacturing inspection equipment can only detect flaws that are larger than 0.050 in (1.25 mm). Will a fracture unsafe situation arise at −60°F for the same loading? ($T_C = 5/9$ [$T_F - 32$]).

2.13. A steel cylinder containing CO_2 gas pressurized to 10 MPa is being transported by truck when someone fires a bullet that pierces it, making a 1 cm hole.

   a. Explain the conditions that might cause the steel to fracture violently with fragments flying off at high velocity.

   b. Similarly, explain the conditions that would cause the cylinder to release the gas relatively harmlessly.

2.14. Steady-state creep testing of electrical solder wire yielded the indicated strain rates under the test conditions given.

Test	Temperature (°C)	Stress (MPa)	Strain rate (sec^{-1})
1	22.5	6.99	2.50×10^{-5}
2	22.5	9.07	6.92×10^{-5}
3	46	6.99	2.74×10^{-4}

From these limited data, determine the creep activation energy (E_c) and stress exponent, *m*.

2.15. Incoloy 800 tubes are selected for use in a pressurized chemical reactor. If they are designed to withstand wall stresses of 14 MPa at a temperature of 900°C, predict how long they will survive.

2.16. Suppose the fatigue behavior of a steel is characterized by a two-line response when plotted on an *S*-log*N* plot, namely (1) *S*(MPa) = 1,000 – 100 log*N*; between *N* = 0 to *N* = 10^6 stress cycles; and (2) *S*(MPa) = 400; for *N* > 10^6 stress cycles.

    a. Sketch the *S*-log*N* plot.
    b. What is the value of the endurance limit?
    c. How many stress cycles will the steel probably sustain prior to failure at a stress of 460 MPa?
    d. Elimination of surface scratches and grooves changed the endurance limit to 480 MPa. (Assume plot 1 is not altered.) How many stress cycles will the steel probably sustain prior to failure at a stress of 470 MPa? How many stress cycles will the steel probably sustain prior to failure at a stress slightly above 480 MPa?

2.17. Assume the same stress develops in WC and Al_2O_3 tool bits during identical machining processes. Since crack formation is a cause of tool bit failure, what is the ratio of the critical flaw dimensions in Al_2O_3 that can be tolerated relative to WC?

2.18. The fracture toughness of a given steel is 60 MPa$\sqrt{}$m, and its yield strength is given by σ_y(MPa) = 1,400 – 4 *T* where *T* is the temperature in degrees Kelvin. Surface cracks measuring 0.001 m are detected. Under these conditions, determine the temperature at which there may be a ductile-brittle transition in this steel.

2.19. The quenched and aged Ti-6Al-4V alloy used in the Atlas missile has yield points of 1,600 MPa at −195°C, 1,150 MPa at −57°C, and 840 MPa at 22°C. Suppose small cracks 0.5 mm long were discovered on cryogenic storage containers made from this alloy whose fracture toughness is 50 MPa$\sqrt{}$m. Estimate the ductile-to-brittle fracture transition temperature of this alloy. Would it be safe to expose these containers to liquid nitrogen temperatures (77 K)? Assume the fracture toughness is independent of temperature.

2.20. An aluminum aircraft alloy was tested under cyclic loading with $\Delta\sigma$ = 250 MPa and fatigue failure occurred at 2 × 10^5 cycles. If failure occurred in 10^7 cycles when $\Delta\sigma$ = 190 MPa, estimate how many stress cycles can be sustained when $\Delta\sigma$ = 155 MPa.

2.21. Draw the stress-strain curve of a steel with the following properties: Elastic modulus = 210 GPa, yield stress 140 MPa, UTS 220 MPa, elongation at fracture 35%. Draw the curve so it resembles Figure 2.6.

2.22. Define a residual stress, some of its possible origins, and the positive and negative effects it can have on the mechanical behavior of a material in service. Provide an example.

2.23. Describe the Rockwell and the Vickers hardness tests.

    a. What is done and what is measured?
    b. Which one is more rapid? Which one gives more scientific information?

2.24. Name and describe the two modes of fracture. Describe which one is more feared and why.

2.25. What must be done in design to increase the resistance to fracture?

2.26. Define the strength of a material and the units used to measure it.

2.27. Define fracture toughness of a material and the units used to measure it.

2.28. Distinguish between toughness and fracture toughness.

2.29. Engineers are careful to avoid stress raisers. What are they and why must they be avoided?

2.30. A hard steel is strong but breaks easily. Explain why.

2.31. The strength of a steel is given as 225 MPa. That of a ceramic is indicated as 200–800 MPa. Explain the difference in notation.

2.32. You wish to replace a metallic machine element with a ceramic one. Can you always use the same shape? If not, what must you avoid?

2.33. Define fatigue. At what stresses can fatigue failure occur?

2.34. Define the fatigue limit.

2.35. How is fatigue resistance measured?

2.36. How does one design to increase fatigue resistance?

2.37. What is creep?

2.38. At what stresses can creep failure take place?

Deformation of Metals and Crystal Structure

In permanent deformation of metals, planes of atoms in crystals glide over each other and perfect metallic crystals deform easily. So what does one do to obtain hard tool steels? Which metals can be deformed extensively in manufacture without fracturing or tearing? The answer to these questions lies in the structure of crystals and in the nature of crystalline defects. We need a language that describes the various crystal structures, the position of atoms, the properties and orientation of the gliding planes. Some materials exist in different crystal structures; this polymorphism can be exploited to advantage. In fact, the polymorphism of iron is responsible for the existence of steel. Defects in crystal structures are responsible for many of the most useful properties of materials, and the management of crystal defects is an important part of materials processing. In this chapter, we examine the plastic deformation of metals, we learn to describe crystals and their defects, and, finally, we describe the easy deformation of metals by the movement of dislocations. We will use this knowledge to achieve an understanding of the desired strength and the ductility of metals in the next chapter.

LEARNING OBJECTIVES

After studying this chapter, the student will be able to:

1. Define the elastic and plastic deformations and their origin.

2. Distinguish between a crystal structure and an amorphous structure.

3. Describe the concept of a unit cell.

4. Describe FCC, BCC, and HCP structures. Explain why FCC and HCP are both the densest packing of atoms.

5. Describe polymorphism and give an example.

6. Draw and recognize atomic coordinates in a cubic lattice.

7. Draw and recognize directions in a cubic lattice.

8. Draw and recognize atomic planes in a cubic lattice.

9. Draw the glide planes and glide directions in FCC and BCC structures.

10. Name and describe the point defects in a solid.

11. Describe and draw edge and screw dislocations.

12. Describe grain boundaries and their structure.

13. Distinguish crystal structure and microstructure.

14. Describe plastic deformation in terms of dislocation motion.

3.1 THE PLASTIC DEFORMATION OF METALS

In the previous chapter, we saw that all materials deform elastically under moderate stresses. This elastic deformation has its origin in the elasticity of the chemical bond, as described in Section 1.8.1: the distance between the atoms increases under a tensile stress and decreases under compression. This deformation is reversible: the atoms resume their equilibrium positions when the stress is removed. When the stress exceeds a certain value, metals undergo a much larger deformation and keep the new shape when the stress is removed. This is plastic deformation.

We already know from Chapter 1 that plastic deformation is possible in metals because the chemical bond of metals involves the free electrons of an unfilled valence band; it binds the atoms close to each other but does not prescribe their relative positions. The following observations provide some additional clues to the mechanisms of plastic deformation.

1. Only crystalline metals can be deformed plastically. Amorphous metals (metglasses) are as hard and brittle as glass.

2. The plastic deformation of metals requires surprisingly small shear stresses. Unless they are processed specifically for high strength, pure metals are quite soft.

3. When one deforms a single crystal, one observes that plastic deformation occurs by slip of selected crystal planes. This is shown in Figure 3.1.

4. Plastic deformation of metals can occur at room temperature. The required stresses depend only slightly on temperature, in stark contrast to the viscous flow of liquid and glasses.

5. Plastic deformation can be very rapid as evidenced by modern stamping methods. The stresses for rapid deformation are only slightly larger than for slow deformation.

Plastic deformation is intimately related to the crystal structure of materials. This was already suggested by Figure. 1.9 in Chapter 1. Before we can understand plastic deformation and learn the various methods by which we can control the mechanical properties of metals, we must briefly review the crystal structure of materials.

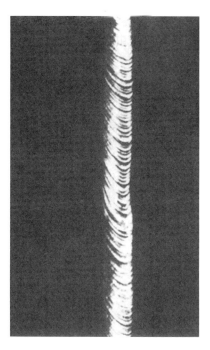

■ **FIGURE 3.1**   Plastic deformation of a zinc crystal. The rod was initially smooth and cylindrical. The new shape was caused by the sliding of planes over each other. These planes are all parallel. From C.F. Elam, *The Distortion of Metal Crystals*, Oxford University Press, London (1953).

3.2 **THE CRYSTAL STRUCTURE OF METALS**

Despite external appearance, materials like metals are crystalline: their atoms are arranged in regular, periodic, three-dimensional structures called **crystals**. The crystalline nature of metals is demonstrated by high-resolution electron microscopy as in Figure 3.2 and by X-ray and electron diffraction (Chapter 19).

We have seen in Chapter 1 that the chemical bond of metals is **nondirectional**: the free electrons of the valence band bind the atoms to each other without prescribing their relative positions. So why are metals crystalline? The answer is that a regular stacking of atoms in a crystal structure permits the densest possible packing of material just as you can fit more wood blocks into a box when you pile them neatly (into a "crystal structure") than when you throw them in (an "amorphous structure"). The densest packing of atoms forms the Hexagonal Close Packed (HCP) or the Face Centered Cubic (FCC) crystals. Because of weak directional forces, some metals crystallize in the body centered cubic structure, which is not quite as dense. A disordered, amorphous, arrangement of the atoms would require more space.

If you pack spheres on a plane in the densest possible way, you will form the hexagonal structure shown in Figure 3.3. Let us now place another close-packed plane on top of the first one so that the planes will be as close as possible. The atoms of the second layer will fit into the cusps between three atoms of the first layer. As Figure 3.3 shows, there are two ways we can stack such close-packed layers on top of each other.

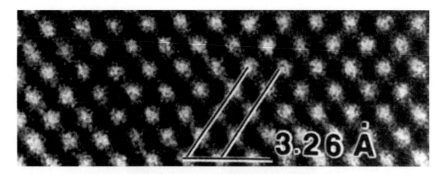

■ **FIGURE 3.2** Transmission microscope picture showing the positions of atoms in germanium. From A. Bourret and J. Desseaux, *Journal de Physique* C 6, 7 (1979).

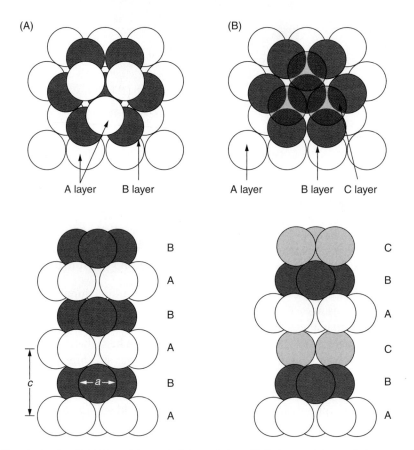

■ **FIGURE 3.3** The two ways of stacking hexagonal close-packed planes of spheres. (A) The hexagonal close-packed structure where the third layer is directly over the first (ABA structure). (B) The face centered cubic structure in which three layers occupy different positions; the fourth layer is placed directly over the first (ABCA structure).

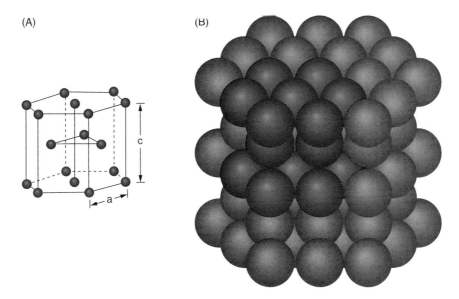

(A)   (B)

■ **FIGURE 3.4**   The hexagonal close-packed structure. (A) The unit cell; note the two lattice parameters, *a* and *c*. (B) Fraction of a crystal showing atoms touching each other. The basal plane of one unit cell is shown in color.

3.2.1 **The Hexagonal Close-Packed Structure (HCP)**

The **hexagonal close-packed structure** is obtained by placing a third layer so that its atoms are directly above those of the first as shown in Figures 3.3A and 3.4. It is convenient to label this stacking as ABAB. The structure shown, of course, extends in all directions throughout the crystal. In this structure, we discern a **unit cell**, (Figure 3.4) shown in color, that repeats itself in all three directions. This unit cell possesses two **lattice parameters**, *a* and *c*. *a* is the repeat unit of the crystal in the hexagonal plane, called the **basal plane**, and *c* is the periodicity of the crystal perpendicular to this plane. If the atoms were perfect spheres, the ideal ratio of the two lattice parameters would be $c/a = 1.633$. The real c/a ratio of most metals deviates from this value, as shown in Table 3.1. Hexagonal close-packed crystals have the atomic packing factor 0.74, which means that the volume of the spheres in the structure occupies 74% of the space. Metals such as Zn, Mg, Be, Ti, Co, Zr, and others crystallize in the hexagonal close-packed structure.

3.2.2 **The Face Centered Cubic Structure (FCC)**

When the close-packed hexagonal planes are stacked in such a way that three layers occupy different vertical positions as shown in Figure 3.3B, and the fourth is on top of the first, the fifth on top of the second, and so on, one obtains the face centered cubic structure. This structure corresponds to ABCABC stacking of the hexagonal planes. Like the HCP structure, it also constitutes a closest possible packing of spheres with the atomic packing factor of 0.74.

Table 3.1 Lattice Constants of Selected Metals at Room Temperature.

Body centered cubic

Metal	a, nm
Chromium	0.2885
Iron	0.2867
Molybdenum	0.3147
Potassium	0.5247
Sodium	0.4291
Tantalum	0.3298
Tungsten	0.3165
Vanadium	0.3023

Face centered cubic

Metal	a, nm
Aluminum	0.4050
Copper	0.3615
Gold	0.4079
Lead	0.4950
Nickel	0.3524
Platinum	0.3924
Silver	0.4086

Hexagonal close packed

Metal	a, nm	c, nm	c/a ratio
Cadmium	0.2979	0.5617	1.890
Zinc	0.2665	0.4947	1.856
Ideal HCP			1.633
Beryllium	0.2286	0.3584	1.568
Cobalt	0.2507	0.4069	1.623
Magnesium	0.3210	0.5211	1.623

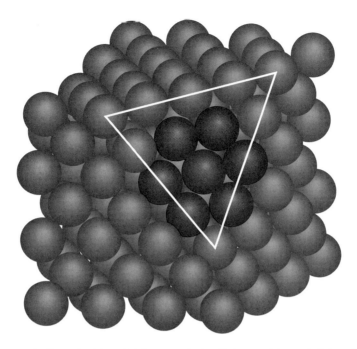

■ **FIGURE 3.5** The face centered cubic structure is truncated at one corner to show the hexagonal close-packed planes shown in color.

When seen from a different direction, this structure has cubic symmetry where atoms are placed at the corners and in the centers of cubic faces shown in Figure 3.5. The metals Ca, Sr, Al, Ni, Cu, Rh, Pd, Ag, Ir, Pt, and Au crystallize in the FCC structure.

Figure 3.6 shows the unit cell of this structure. It repeats itself in all three directions to form the crystal. It is a cube with atoms at its corners and in the centers of its faces, hence its name: **Face Centered Cubic (FCC)** structure. It has only one lattice parameter a, namely, the length of an edge of the cube. It is convenient to draw the unit cell through the centers of the atoms at its edges. Therefore, the atoms at the plane centers are cut in half: the other half belongs to the neighboring cell. For the same reason, only one eighth of the corner atoms belong to the unit cell. The FCC unit cell contains four atoms: $6 \times 1/2 = 3$ in the faces and $8 \times 1/8 = 1$ at the corners.

3.2.3 **The Body Centered Cubic Structure (BCC)**

A large number of metals, notably iron at room temperature, crystallize in a structure that has cubic symmetry and does not represent the closest possible packing of spheres. This structure is shown in Figure 3.7. Its unit cell has atoms at the corners and one in the center of the cell, hence its name, **body centered cubic structure**.

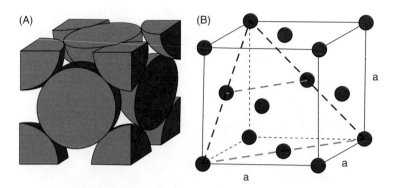

■ **FIGURE 3.6** The unit cell of the face centered cubic (FCC) structure. It is drawn through the centers of the atoms at its edges. (A) Representation of the atoms in their real size, as they touch each other. (B) Representation of the centers of the atoms, showing their positions. The dotted lines show half of a hexagon in the dense plane of atoms.

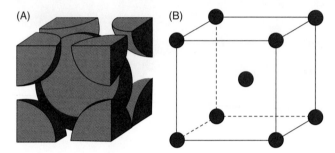

■ **FIGURE 3.7** The body centered cubic (BCC) structure. (A) The unit cell showing the atoms in their real size and touching each other. (B) Representation of the unit cell, showing the positions of the centers of the atoms.

The BCC structure contains two atoms per unit cell, one in its center and one (i.e., $8 \times 1/8 = 1$) at its corners. Other metals possessing BCC structure are Cr, Mo, W, Ta, Na, K, and so on. The length of the unit cell is the lattice parameter a; it has the same size in all three directions and represents the periodicity of the crystal.

An abbreviated list of the crystal structures of metals and their lattice parameters is found in Table 3.1. These structures have been measured by X-ray diffraction which gives us the symmetry of the structures and the lattice parameters (see Chapter 19). These limited crystallographic data enable calculation of a considerable amount of information about the material when we assume that the atoms are hard spheres.

3.2.4 **Atomic Radii**

Knowing the lattice parameters from diffraction measurements, one **defines** the radii of the atoms by considering them as rigid spheres that touch each other. In FCC structures, atoms touch along the cube face diagonal (see Figure 3.6) so that $r + 2r + r = a\sqrt{2}$, and

$$r = \frac{a\sqrt{2}}{4}$$

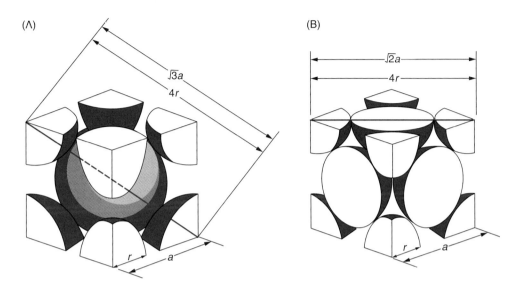

(A) (B)

■ **FIGURE 3.8** Calculation of the atomic radius in BCC and FCC structures.

For Al where $a = 0.4045\,\text{nm}$, $r = 0.143\,\text{nm}$.

In BCC structures, atoms of radius r touch along the cube body diagonal. Thus, as shown in Figure 3.8, the space diagonal of the cell is equal to four atomic radii: $4r = a\sqrt{3}$, and

$$r = \frac{a\sqrt{3}}{4}$$

Since $a = 0.2866\,\text{nm}$, the calculated radius of an iron atom is $0.1241\,\text{nm}$.

3.2.5 Atomic Packing Factor

The atomic packing factor (APF) is defined as the ratio of the volume of atoms, assumed to be spheres, to the volume of the unit cell.

For BCC structures there are two atoms per cell. Therefore, $\text{APF} = 2 \times (4/3\pi r^3)/a^3$. But $r = \sqrt{3}/4a$; after substitution $\text{APF} = \sqrt{3} \times \pi/8$ or 0.680, a number that is independent of atomic size. In the case of FCC structures, $\text{APF} = 4 \times (4/3 \times \pi r^3)/a^3$. Since $r = a\sqrt{2}/4$, $\text{APF} = 0.740$. This demonstrates that of the two structures, FCC is more densely packed.

3.2.6 The Density of the Material

The crystal structure allows us to calculate the density of a material. Since the FCC structure has four atoms per cubic cell of volume a^3, the number of atoms per unit volume is $N = 4/a^3$. In aluminum, with the lattice parameter $a = 0.4045\,\text{nm}$,

$$N = 4/(0.4045)^3 = 60.44 \text{ atoms/nm}^3 = 6.044.10^{28} \text{ atoms/m}^3 = 6.044.10^{22} \text{ atoms/cm}^3$$

The atomic weight of aluminum is 26.98; this is the mass of 1 mole or $N_A = 6.023.10^{23}$ atoms per gram mole. The density of aluminum is then calculated as

$$\rho = (6.044.10^{22}/6.023.10^{23}) \times 26.98 = 2.70 \text{ g/cm}^3.$$

(The universal unit for density is 2.70 Mg/m^3 but it is seldom used in practice as it does not speak to the senses.) This is a **theoretical density**, namely, the density of a perfect aluminum crystal. Materials contain defects such as vacancies and impurities and their actual density may differ from their theoretical value. A comparison of theoretical and actual densities is often used as a measure of the crystalline perfection of a material.

3.2.7 **Allotropy or Polymorphism**

Allotropy, or polymorphism, is the ability of certain materials to exist in different crystal structures. The most important polymorphism of metals is that of iron. At low and moderate temperature, iron has the BCC crystal structure shown in Figure 3.7. When iron is heated above $913\,°C$ (1,186K), its structure changes to FCC, Figure 3.6. Its structure reverts to BCC at still higher temperatures. We shall see in Chapter 6 that this property of iron is a magnificent present nature gave us; it is what makes steel possible. Another well-known polymorphism is that of carbon, which exists as diamond, graphite, pyrolytic graphite, glassy carbon, and more recently discovered fullerenes, carbon nanotubes, and graphene layers. These will be discussed further in later chapters.

EXAMPLE 3.1   *(a) What is the fractional volume change in iron as it transforms from BCC to FCC at 910°C? Assume that a(BCC) = 0.2910 nm and a(FCC) = 0.3647 nm. (b) What is the volume change if Fe atoms of fixed radius pack as hard spheres?*
ANSWER   The volume (V) per atom is equal to volume per unit cell/atoms per unit cell. For BCC Fe, $V_{(BCC)}$ per atom $= a^3_{(BCC)}/2$. Similarly for FCC Fe, $V_{(FCC)}$ per atom $= a^3_{(FCC)}/4$. Substituting, $V_{(BCC)}/\text{atom} = (0.2910)^3/2 = 0.01232 \text{ nm}^3$; and

$V_{(FCC)}/\text{atom} = (0.3647)^3/4 = 0.01213 \text{ nm}^3$

Therefore, $\Delta V/V = (12.13 - 12.32)/12.32 = -0.0154$ or -1.54%, a result consistent with the abrupt contraction observed when BCC Fe transforms to FCC Fe.

If atoms packed as hard spheres, $a_{(BCC)} = 4r/3^{1/2}$ and $a_{(FCC)} = 4r/2^{1/2}$. Therefore,

$\Delta V/V = \{[4r/2^{1/2}]^3/4 - [4r/3^{1/2}]^3/2\}/[4r/3^{1/2}]^3/2 = (5.657 - 6.158)/6.158 = -0.08135$ or -8.135%

The two values for $\Delta V/V$ differ considerably. The results for part b are independent of the lattice parameter of the involved unit cells and therefore, of the particular metal. Obviously, atoms in real metals and iron, in particular, are not hard spheres of fixed radius. Other metals like manganese and plutonium also exist in both BCC and FCC crystal forms of different lattice parameters.

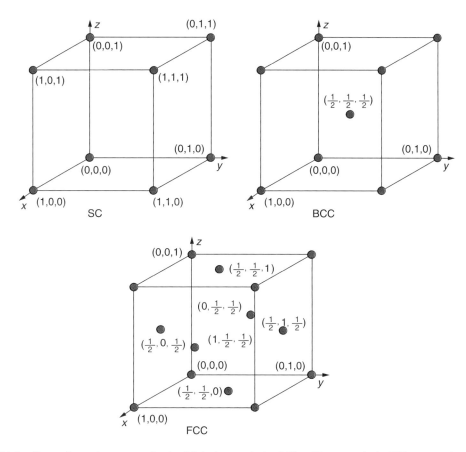

■ **FIGURE 3.9** The coordinates of atoms in simple cubic (SC), body centered cubic (BCC), and face centered cubic (FCC) structures. (No elements crystallize in the SC structure.)

3.3 **COORDINATES OF ATOMIC POSITIONS, DIRECTIONS, AND PLANES**

For the cubic crystals, one utilizes Cartesian coordinates and the rules of analytical geometry. In order to make the coordinates independent of a particular material, one uses dimensionless coordinates in which the length of the unit cell is unity; this is shown in Figure 3.9.

3.3.1 **Atomic Positions**

Atomic positions at the edges of the cube are expressed by the integers $(0,0,0)$, $(1,0,0)$, $(0,1,0)$, $(0,0,1)$, $(1,1,0)$, $(0,1,1)$, $(1,0,1)$, $(1,1,1)$. Positions inside the cube have fractional numbers. The positions of the atoms in the BCC structure and the FCC structure are shown in Figure 3.9.

Note that the directions of the x, y, and z axes are always shown as in Figure 3.9: the x-axis toward the reader, the y-axis to the right, and the z-axis toward the top. To avoid confusion, do not draw the axes in directions different from the standard presentation.

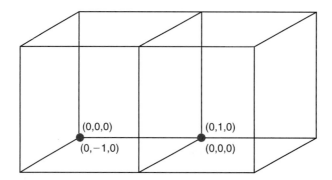

■ **FIGURE 3.10**   Two unit cells in a periodic structure. The point (0,1,0) of the unit cell on the left (upper coordinates) is also the origin (0,0,0) of the cell on the right (lower coordinates). To draw points or directions with a negative y-coordinate in the unit cell, it is convenient to move the origin to the right. Similarly, one moves the origin to the front for negative x-coordinates and to the top for negative z-coordinates. One, two, or three such moves can be used.

The crystal is a three-dimensional periodic stacking of such unit cells. Therefore, the corner points are shared with neighboring unit cells. The (1,0,0) point, for instance, is also the origin (0,0,0) point of the cell in front of the one shown; the (0,1,0) point is the (0,0,0) of the unit cell to the right, as shown in Figure 3.10. It is therefore permissible to position the origin (0,0,0) at any convenient corner when negative coordinates are used.

The simple unit cell thus contains only one atom at (0,0,0); the atoms at (1,0,0), (0,1,0), (1,1,0), and so on belong to neighboring cells. The BCC cells contain two atoms, at (0,0,0) and at (½,½,½). Similarly, the FCC structure has four atoms per unit cell, at (0,0,0), (½,½,0), (½,0,½), and (0,½,½); all atomic positions that contain a 1 belong to neighboring cells. This periodic description is equivalent to the one given in Figures 3.6 and 3.7, but avoids cutting atoms into halves or eighths.

3.3.2 Directions

In analytical geometry, the coordinates of a vector are the coordinates of its end point minus the coordinates of its initial point. Directions are indicated by coordinates in brackets [] and not separated by a comma. These are the **Miller indices** of a direction. They are always written as integers by multiplying all numbers by the largest common denominator. Thus, the direction [301] in Figure 3.11 starts at (0,0,0) and ends at the point (1,0,1/3), the vector joining these points is [1 0 1/3]. Multiplying all numbers by 3 makes it 3 times longer; this is the **same direction** [301]. The main directions in cubic systems are shown in Figures 3.11 to 3.13. Their end points are marked as dots. Note that the [110] direction is shown twice in the figure: parallel vectors represent the same direction and have the same coordinates.

Some directions require negative coordinates. These are written with a minus sign on top of the number. Figure 3.12 shows the $[01\bar{1}]$ direction. Its starting point is (0,0,1) and its end point is (0,1,0). With the rule governing the coordinates of directions, we obtain

$$(0,1,0) - (0,0,1) = [01\bar{1}]$$

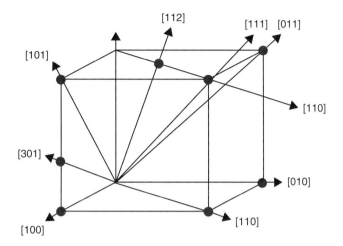

■ **FIGURE 3.11**   Some directions. The end points are marked as dots.

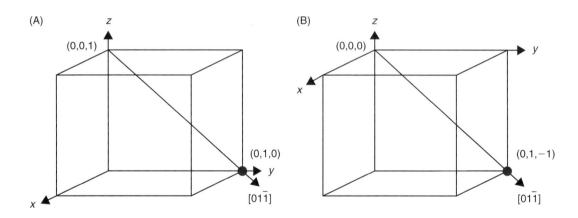

■ **FIGURE 3.12**   Two equivalent drawings of the [01 $\bar{1}$] direction. (A) [01 $\bar{1}$] = (0,1,0) − (0,0,1). (B) The origin is moved in order to contain the vector inside the unit cell.

An alternative way of obtaining the coordinates of this direction is shown in Figure 3.12B. We displace the origin to the top of the cell so that it becomes the new (0,0,0) point. In these coordinates the initial point is (0,0,0) and the end point is (0,1, −1), yielding the same direction coordinate.

When given a direction to draw, one divides all indices by the largest one to ensure the vector remains inside the unit cell. Then, selecting a proper origin, one draws the vector to the end point. Let us draw the direction [4 $\bar{1}$ 2]. This is the same direction as [1 ¼ ½]. Since the *y*-scale is negative, we move the

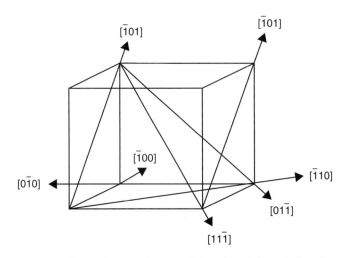

■ FIGURE 3.13   Directions with negative indices. Displacement of the origin (different for each direction) allows drawing them inside the unit cell.

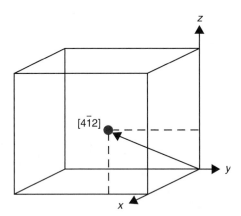

■ FIGURE 3.14   Drawing the direction $[4\,\bar{1}2]$. The origin is moved to the right; the end point has coordinates $(1, -\frac{1}{4}, \frac{1}{2})$.

origin to the right on the y scale; this is the starting point of the vector. Then we note the point $(1, -\frac{1}{4}, \frac{1}{2})$ and draw the vector joining these two points. This is shown in Figure 3.14.

The edges of the cube have coordinates [100], [010], and [001]. These cube edges form a **family** denoted by <100>. In the same way, all face diagonals [110], [011], and [101] form the <110> family, and the space diagonals [111], $[\bar{1}11]$, $[1\bar{1}1]$, and $[11\bar{1}]$ form the <111> family.

Generally, all the permutations of [uvw] with their negatives form the <uvw> family.

3.3.3 **Planes**

Figures 3.4, 3.5, and 3.7 show quite clearly that, in a crystal, the positions of atoms define various planes. Like points and directions, crystallographic planes are identified by Miller indices.

Any three points define a plane. The following simple recipe can be used to identify a given plane in cubic crystals:

1. Note the coordinates of the intercepts of the plane with each of the three axes.

2. Take reciprocals of these numbers.

3. Reduce the reciprocals to the smallest integers by clearing fractions.

The resulting triad of numbers placed in parentheses without commas, (hkl), represents the Miller indices of the plane in question. Any time a plane passes through the origin, the above recipe will not work. In such a case it must be remembered that the origin can be shifted to any other corner. Another alternative is to translate the plane parallel to itself until intercepts are available. By either of these means it is always possible to have the involved plane slice through the unit cell. In the case of cube faces one intercept (e.g., x) is 1 and the other two intercepts extend to infinity. Therefore, $1/1 = 1$, $1/\infty = 0$, $1/\infty = 0$, and the planar indices are (100). Figure 3.15 shows a number of typical planes identified with the above rule.

(100) type planes that are equivalent have (010), (001), $(\overline{1}00)$, $(0\overline{1}0)$, and $(00\overline{1})$ indices. These six planes constitute the $\{100\}$ family. Generally all permutations of (hkl) with their negatives constitute the $\{hkl\}$ family. (Note that they are designated with $\{\ \}$).

In order to draw the (*hkl*) plane, take the inverse, $1/h$ and place it on the x-scale, place $1/k$ on the y scale and $1/l$ on the z-scale. These three points define the plane that can be drawn by joining them. If one of the indices is zero, the plane is parallel to the corresponding axis (it joins it at ∞). For every negative Miller index, move the origin to the other extremity of the corresponding cell axis.

3.4 DENSE PLANES AND DIRECTIONS

The planes with densest atomic packing are the chemically most stable planes; they are the planes that slide on each other in plastic deformation and the planes along which brittle materials cleave. The densest directions, along which neighboring atoms touch each other, are the directions along which planes slide on each other in plastic deformation. Figures 3.5 and 3.7 show crystals bounded by their $\{100\}$ faces. In FCC crystals, Figure 3.5 shows that the atoms touch each other along the <110> family of directions. In the BCC structure, we have seen in Figure 3.7 that atoms touch each other along the space diagonals or <111> family of directions.

The $\{111\}$ planes are the densest planes in the FCC structure. They are in fact the close-packed hexagonal planes we described earlier. This is easily verified by drawing a (111) plane and placing the atoms on the FCC positions. Figure 3.7 will show, as well as Figures 3.5 and 3.6, that the densest directions are <110>. In the BCC structure, the densest planes are $\{110\}$ and the densest directions, in which atoms touch, are the <111> family. This is illustrated in Figure 3.16.

3.5 DEFECTS IN CRYSTALLINE SOLIDS

No solid piece of material is a perfect crystal. Most solid objects are **polycrystalline**; they consist of numerous small single crystals or **grains** of random orientation. In these grains, errors abound in the

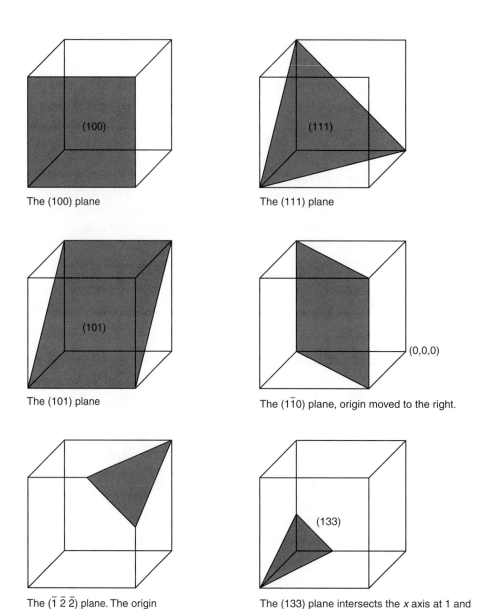

The (100) plane

The (111) plane

The (101) plane

The (1$\bar{1}$0) plane, origin moved to the right.

The ($\bar{1}$ $\bar{2}$ $\bar{2}$) plane. The origin is shifted on all three axes.

The (133) plane intersects the x axis at 1 and the y and z axes at 1/3.

■ **FIGURE 3.15**   Some representative planes in cubic crystals.

placement of atoms and the stacking of planes. At the outset a distinction should be made between such crystallographic defects, and gross manufacturing defects and flaws such as cracks and porosity. Here our concern is with **lattice defects** having atomic size dimensions. It will be found throughout this book, and may be surprising at first, that lattice defects are responsible for many desirable

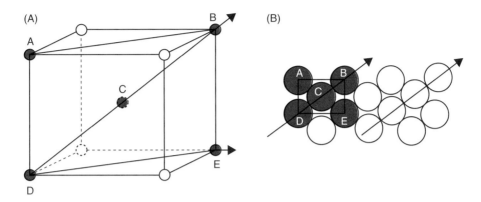

■ **FIGURE 3.16** The (110) plane of the BCC structure. (A) Inside the unit cell. (B) Plane drawing. The dense directions A-C-E and D-C-B are space diagonals of the <111> family; these are shown as arrows.

properties of materials; lattice defects are introduced to strengthen metals, and defects are essential in the creation of semiconductor devices. With few exceptions, one can state that a perfect crystal is perfectly useless. To a great extent, materials engineering consists of the introduction and management of the right defects.

3.5.1 Point Defects

A good way to recognize **point defects** is through Figure 3.17. All of the defects involve one or, at most, two atoms.

Vacancies A vacancy is a missing atom at a lattice site (Figure 3.17). Vacancies exist in all classes of crystalline materials.

Lattice vacancies are essential for phenomena involving atomic motion or diffusion in solids. Atoms that are completely surrounded by nearest neighbor lattice atoms are not mobile. But if there is an adjacent vacancy, then the two can exchange places and atomic motion is possible.

Interstitial Atoms These are atoms that do not occupy regular crystalline positions but sit in interstices between regular atoms. They can be host atoms or impurities.

Impurities These are foreign atoms in the material. Some impurities are undesirable and must be removed by purification. Silicon, used to produce integrated circuits, for instance, can tolerate less than one impurity per 10 million Si atoms. Other foreign atoms are introduced for their beneficial effects. Examples are the doping of semiconductors to make them n- or p-type conductors. Impurities are also introduced to increase the electrical resistance and the mechanical strength of metals.

Impurities are **substitutional** when they occupy a regular crystal site otherwise occupied by a host atom; they are **interstitial** when they fit in the interstices between the host atoms. Host atoms can also be displaced and occupy interstitial sites as shown in Figure 3.17.

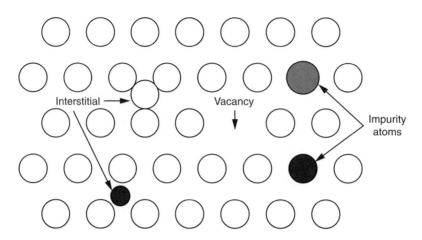

■ **FIGURE 3.17**   Point defects in metals.

3.5.2 **Dislocations**

Dislocations play a central role in plastic deformation. They extend along a **line** of atoms in a crystalline matrix. There are two types of dislocations.

The **edge dislocation** can be imagined to arise by inserting an extra half plane of atoms. The resulting edge dislocation defect, denoted by the symbol ⊥, is the line of atoms at the bottom of the inserted plane. It is shown in Figure 3.18A together with one of its chief attributes—the **Burgers vector**. If a closed loop (e.g., four atoms up, three atoms to the right, then four atoms down and three atoms to the left) is made about a perfect lattice, then the end point coincides with the starting point. A similar traverse around a region containing the core of an edge dislocation will not close, and the vector connecting the end point to the initial point is known as the Burgers vector b (Figure 3.18A). Its magnitude is one lattice spacing. Figure 3.19 shows an edge dislocation observed in the transmission electron microscope at very high resolution.

The second type of dislocation can be imagined to arise by first making a cut halfway into the lattice. Then one half is sheared up, the other down until a total relative displacement of one atomic spacing occurs. The resulting **screw dislocation** is shown in Figure 3.18B together with the Burgers vector that defines it. Making a clockwise circuit about the axis of the dislocation is like going down a spiral staircase. The closure error or Burgers vector is parallel to the screw axis. In this case b is **parallel** to the dislocation line.

It frequently happens that a single continuous dislocation line acquires mixed edge and screw character merely by turning a 90° angle corner in the crystal. A plan view of such a mixed dislocation is depicted in Figure 3.20.

3.5.3 **Grain Boundaries**

When the ordered atomic stacking extends over the whole piece of material, it forms a **single crystal**. All sorts of precious and semiprecious gems found in nature are single crystals. Single crystals are

(A)

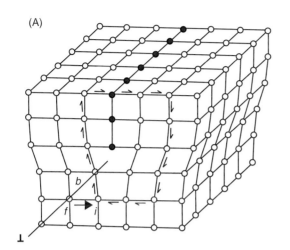

(B)

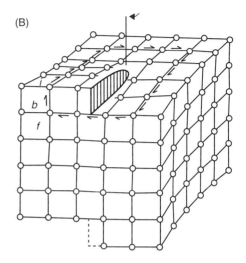

■ **FIGURE 3.18** The dislocation. (A) Edge dislocation. (B) Screw dislocation.

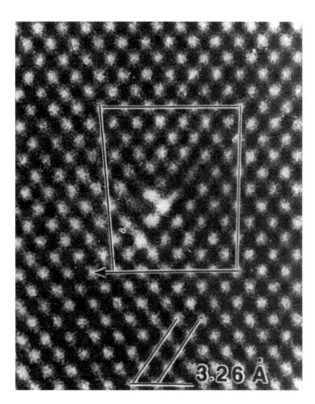

■ **FIGURE 3.19** Transmission electron micrograph of an edge dislocation in germanium. From A. Bourret and J. Desseaux, *Journal de Physique* C 6, 7 (1979).

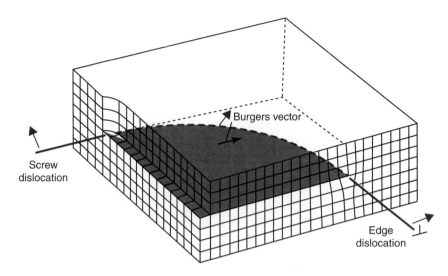

■ **FIGURE 3.20**   Curved dislocation with edge and screw components that emerge at mutually perpendicular surfaces. The colored arrow shows the direction of dislocation motion.

■ **FIGURE 3.21**   The grain structure of α brass. Individual grains show in different shades of gray because of their different orientations. Courtesy of G. F. Vander Voort, Buehler LTD

grown for electronic (silicon, quartz), magnetic (garnets), optical (ruby), and even metal turbine blade applications. They can weigh 50 kg and be 30 cm in diameter. Most all materials used in engineering, however, are polycrystalline. They consist of a large number of small crystals, or grains, that differ from each other by their crystalline orientation. This is illustrated in Figure 3.21 for the case of brass.

The grains are usually a few micrometers in size, but their size can vary from several millimeters down to a small fraction of a micrometer. In some materials, different grains can also be composed of different substances.

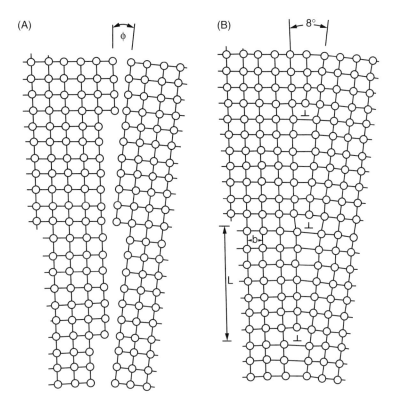

■ FIGURE 3.22   Small angle grain boundary. When the difference in orientation between two grains is a small angle, the grain boundary takes the form of an array of dislocations.

Grain boundaries are the interfaces between the grains. The simplest model of a grain boundary involves tilting two adjacent single crystal grains relative to each other by a small angle ϕ as illustrated in Figure 3.22A. When the crystals are welded together at the interface (Figure 3.22B), the grain boundary consists of isolated edge dislocations stacked vertically a distance L apart, where L is essentially b/ϕ.

Since grain boundaries are relatively open structurally, atoms located in grain boundaries have disordered, nonequilibrium environments (see Figure 3.23); they are more energetic than those within the bulk. Therefore, grain boundaries are the preferred location for diffusion, atomic segregation, phase transformation, precipitation, or chemical reactions (e.g., etching, corrosion). Furthermore, grain boundaries help resist the processes of deformation under stress (Chapter 4). But they are weaker than the bulk, and when the material breaks, the fracture often propagates along a grain boundary; this is intergranular fracture (Chapter 2).

3.5.4 **Microstructure and Crystal Structure**

The size and shapes of the grains in the solid constitute its **microstructure**. This is distinct from the crystal structure, which describes the relative positions of atoms in a grain.

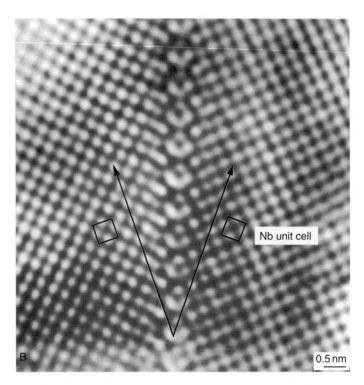

■ **FIGURE 3.23** Electron microscope picture of a grain boundary in niobium. It is apparent that grain boundaries contain many defects and voids in which impurities can segregate. Courtesy of G. H. Campbell, Lawrence Livermore Laboratory.

3.6 MECHANISMS OF PLASTIC DEFORMATION

We have seen already in Chapter 1 that plastic deformation occurs by the sliding of atoms over each other. We know also that plastic deformation is restricted to crystals. Amorphous metals are as hard as glass and, like the latter, break when the stress is excessive. We now examine the role that crystal structure and defects play in the plastic deformation of metals.

3.6.1 Slip Systems

If we observe the plastic deformation of a single crystal, we find that, after the deformation, the material is still a crystal with the same structure. Entire planes glide over each other so that the atoms wind up at equilibrium crystalline positions again. Crystallographic planes that are smooth and dense require a smaller shear stress to slide than planes that contain steps or in which the atoms are further apart. Plastic deformation takes place only by the sliding of these dense planes, which are known as **slip planes**. The sliding of these slip planes occurs preferentially along the directions that offer the smallest resistance; these are the **slip directions**. The slip directions and the slip planes form the **slip systems** of the crystal structure. These are illustrated in Table 3.2.

Table 3.2 The Slip Systems of Crystals. Note that the Minimum Shear Stress for FCC Metals Is Much Smaller than for BCC Metals. The HCP Metals Have Fewer Slip Systems.

Structure and material	Slip plane	Slip direction	Number of slip systems	τ_{CRSS} (MPa)	Slip geometry
FCC	{111}	$\langle 1\bar{1}0 \rangle$	{4} × $\langle 3 \rangle$ = 12		
Ag (99.99%)				0.58	
Cu (99.999%)				0.65	
Ni (99.8%)				5.7	
Diamond cubic	{111}	$\langle 1\bar{1}0 \rangle$	{4} × $\langle 3 \rangle$ = 12		
Si, Ge					
BCC					
Fe (99.96%)	{110}	$\langle 1\bar{1}1 \rangle$	{6} × $\langle 2 \rangle$ = 12	27.5	
Mo	{110}	$\langle 1\bar{1}1 \rangle$		49.0	
HCP					
Zn (99.999%)	(0001)	[11$\bar{2}$0]	{1} × $\langle 3 \rangle$ = 3	0.18	
Cd (99.996%)	(0001)	[11$\bar{2}$0]		0.58	
Mg (99.996%)	(0001)	[11$\bar{2}$0]		0.77	
Al$_2$O$_3$, BeO	(0001)	[11$\bar{2}$0]			
Ti (99.99%)	(10$\bar{1}$0)	[11$\bar{2}$0]	{3} × $\langle 1 \rangle$ = 3	13.7	
Rocksalt	(110)	[1$\bar{1}$0]	{6} × $\langle 1 \rangle$ = 6		
LiF, MgO					

In face centered crystals (FCC), the {111} planes are the densest and the slip directions are along the dense lines of atoms, which are in the <110> family of directions. There are four planes in the {111} family, and each has three directions in the <110> family. The FCC crystal can slide in 12 different directions.

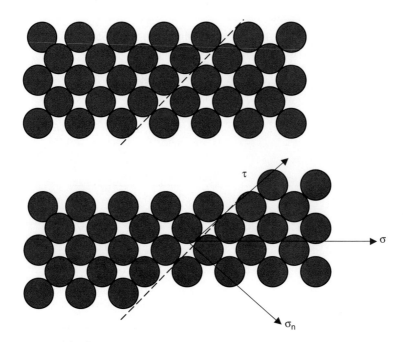

■ FIGURE 3.24   Model of plastic deformation in a metal. (Top) The undeformed crystal. The dotted line represents a glide direction in a glide plane. (Bottom) One plane slips over the other causing permanent, plastic, deformation. σ is the applied tensile stress; τ is the resolved shear stress, which is responsible for the deformation; σ_n is the stress normal to the shear direction.

Since the {111} planes in the FCC structure are very dense, the minimum shear stress necessary for slip in these materials is quite low. FCC metals are ductile and are preferred where extensive plastic deformation is required in manufacture.

The densest planes of body-centered crystals (BCC) are the {110} family of planes, and the slip directions are the <111> family. There are six {110}-type planes and in each one, there are two <111> directions. Thus there are 12 slip systems in the BCC crystal. The {110} planes in the BCC structure are not as dense and smooth as the {111} planes of the FCC structure; the minimum shear stress for plastic deformation in BCC metals is therefore higher than in FCC materials.

Hexagonal close-packed crystals (HCP) have fewer slip systems than the cubic metals. In Zn, for instance, the slip plane is the basal plane (0001), which has three slip directions belonging to the <11$\bar{2}$0> family. In titanium, the slip systems consist of the three prismatic planes of the {1010} family, which have only one slip direction each. Thus HCP crystals have only three slip systems. Consequently they do not deform as easily, especially when polycrystalline. HCP metals are stronger and more brittle than cubic metals. For this reason, HCP cobalt is often alloyed into metals to increase their resistance to wear.

3.6.2 Deformation of a Crystal in a Tensile Test

If plastic deformation is always a shear deformation responding to a shear stress, how does a bar deform in a tensile test? As illustrated in Figures 3.24 and 3.25, deformation takes place along a slip direction that lies in a slip plane at an angle with the tensile force. This is demonstrated in Figure 3.1 by

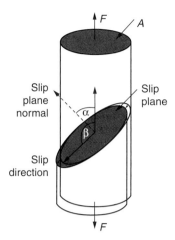

■ **FIGURE 3.25** The slip plane and slip direction in a tensile test of a single crystal.

the plastic deformation of a zinc crystal which, according to Table 3.2, has only one slip plane. α is the angle between the normal to the slip plane and the tensile force direction, and β the angle between the slip direction and the tensile force. The tensile force F can be decomposed into a slip force $F_s = F \cos \beta$ and a force normal to the slip plane $F_n = F \sin \beta$. The area of the slip plane is $A_s = A/\cos\alpha$ so that the tensile force F generates the resolved shear stress

$$\tau = (F/A) \cos\alpha \cos\beta = \sigma \cos\alpha \cos\beta \tag{3.1}$$

Plastic deformation will occur on the plane and along the slip directions that provide the largest value of $\cos\alpha\cos\beta$. Table 3.2 indicates the critical resolved shear stress τ_{CRSS} necessary for plastic deformation of several metals. Obviously the minimum tensile stress is then

$$\sigma = \tau_{CRSS}/(\cos\alpha \cos\beta) \tag{3.2}$$

3.6.3 **The Role of Dislocations in Plastic Deformation**

Look at Table 3.2. The critical resolved shear stress necessary to induce plastic deformation in perfect metals is very small. How could entire planes slide over each other with such small stresses, given the strength of the metallic bond? In reality, the crystallographic planes do not slide rigidly, all at once; they slide atom by atom through the movement of dislocations. This is shown in Figure 3.26 for the case of an edge dislocation.

Under the effect of the applied shear stress τ, atom A moves slightly to the right and atom 2 moves to the left. The result is that the A-2 distance becomes smaller than the distance B-2, the bond A-2 becomes stronger than bond B-2, and the dislocation moves to the right. The process continues: the dislocation then moves to atom C and further until it emerges on the surface of the crystal and forms a step. The result is that the plane containing the atoms A,B,C has slipped over the plane containing

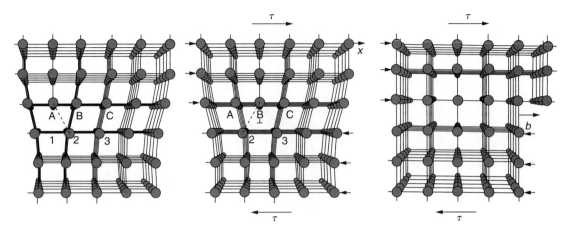

■ **FIGURE 3.26** Plastic deformation associated with edge dislocation movement across a stressed crystal. When the dislocation emerges at the surface, a step is produced.

■ **FIGURE 3.27** Movement of a dislocation with a kink. As the kink moves (color arrow), the dislocation moves from position A to position B.

atoms 1,2,3. In addition, the whole dislocation does not move as a straight line but one atom at a time. A kink is formed in the dislocation; this kink moves along until the whole dislocation has advanced by one Burgers vector (i.e., atom length). This is illustrated in Figure 3.27.

It is now easy to see why plastic deformation requires such a small shear stress. In a dislocation with a kink, the stress must shift the position of only one atom at a time; also, the bonds A-2 and B-2 are only slightly different in strength and, finally, positions A and B of the dislocation possess the same energy.

In amorphous solids, dislocations cannot be defined, and this mechanism for easy plastic deformation does not exist: Metallic glasses do not deform plastically but break under sufficient stress.

It is important to note that this easy movement of dislocations is possible because of the crystal structure of the material. If the periodicity of the lattice is disturbed by any lattice defect, the propagation of the dislocation requires a larger stress and the yield stress of the metal is increased. In other words, **lattice defects increase the strength of a metal**. The methods for strengthening metals will be examined in Chapter 4. In a disordered (amorphous) structure, dislocations do not exist, and plastic deformation is practically impossible except at high temperatures where the material flows like a very viscous liquid.

The easy movement of dislocations is restricted to metals. In covalent ceramics, the directions of the chemical bond are rigidly prescribed by those of the valence orbitals (usually sp^3 hybrids). In ionic ceramics, the presence of positive and negative ions renders the formation and movement of dislocations difficult because of electrostatic forces.

■ SUMMARY

1. Most solid materials are crystalline: their atoms or molecules are arranged in symmetrical structures that are periodic over extended space. Most materials applied in structural applications are polycrystalline; they are composed of a large number of small crystals called grains; many functional materials are single crystals. Glasses and most polymers are amorphous, like liquids: their atoms or molecules assume random positions.

2. Most elements crystallize in the face centered cubic (FCC), the body centered cubic (BCC) or the hexagonal close-packed (HCP) structure. Compound materials, especially ceramics, present complex crystal structures.

3. Polymorphism is the ability of a material to crystallize in different structures. It is also called allotropy.

4. The unit cell of a structure is the volume that repeats itself in three dimensions throughout the crystal. Atoms in the corners, edges, and faces of a unit cell are shared with the neighboring cells. The length of the edges of the unit cells is the lattice parameter. Cubic crystals have only one lattice parameter, a, by reason of symmetry; hexagonal crystals have two lattice parameters, a and c, and more complex structures have three lattice parameters.

5. Positions of atoms, directions, and planes in cubic crystals are described in Cartesian coordinates where the length of the cell is unity. The coordinates of directions and planes are the Miller indices. Directions are written in whole integers in brackets [ ]. Planes are designated in whole numbers in parentheses ( ). All directions and planes with the same character (e.g., edges of the cube or face diagonals, faces of the cube) have Miller indices that are permutations of the same coordinates, including their negatives (e.g., [100], [010], [001] for edges) and form a family of directions (e.g., <100>) or a family of planes (e.g., {111}).

6. The unit cell of the FCC structure contains four atoms: one at the corners and three in the centers of the faces. Its atomic packing factor is the largest possible with spheres. Its densest planes are {111} and its densest directions are <110>.

7. The unit cell of the BCC structure contains two atoms, one at the corners and one in the center of the cube. Its atomic packing factor is lower than that of FCC. Its densest planes are {110} and its densest directions <111> .

8. The HCP structure is a stacking of close-packed, hexagonal, planes of atoms. Its atomic packing factor is the same as for FCC, the largest possible with spheres.

9. A crystallographic direction is a vector, independent of the latter's length, [120] = [240] = [½ 1 0]. It is written with the set of smallest integers. It extends from the origin to the atom with the same coordinates, that is, from the origin to the point (1/2,1,0) in our example. For directions with negative indices, it is convenient to shift the origin to the other edge of the corresponding axis.

10. A plane is defined by the three points at which it intersects the coordinate axes. One notes the point of intersection (one or a fraction); its inverse is the Miller index. To draw the (hkl) plane, one writes the inverse $1/h$ and draws it on the x-axis; one draws $1/k$ on the y-axis and $1/l$ on the z-axis and joins the three points to draw the plane. When $h = 0$, $1/h = \infty$ and the plane is parallel to the x-axis.

11. Crystal defects are departures from the periodic array of atoms. They are responsible for many useful properties of materials.

12. Point defects are vacancies, interstitials, and impurities. The latter can be substitutional or interstitial.

13. Dislocations are line defects. The edge dislocation is the edge of an additional plane of atoms. The screw dislocation is a line along which the crystal is sheared by one unit cell.

14. Grain boundaries are the interfaces between grains.

15. Plastic deformation occurs by the sliding of planes over each other.

16. This sliding takes place only on the smoothest (i.e., densest) planes in the easy glide directions.

17. The sliding of planes occurs by the movement of dislocations, and this takes place by the motion of kinks in the dislocation.

18. Dislocation motion requires small stresses because the metallic bond is not directional.

■ KEY TERMS

■ PROBLEMS AND QUESTIONS

3.1. Cesium metal has a BCC structure with a lattice parameter of 0.6080 nm. What is the atomic radius?

3.2. Using Table 3.1, compute the radius of nickel, gold, chromium, and tungsten atoms.

3.3. Compute the atomic densities of nickel, gold, chromium, and tungsten.

3.4. Compute the theoretical specific mass (mass density) of nickel, gold, chromium, and tungsten.

3.5. Draw the (111), (110), and (100) planes of copper and place the atoms on each so that they touch. If one builds these planes with solid spheres, and places a similar sphere on top of them, on which plane could one move the sphere with least resistance? In what direction would it move with least resistance?

3.6. Draw the (111), (110), and (100) planes of chromium and place the atoms on each so that they touch. If one were to build these planes with solid spheres, and one were to put a similar sphere onto them, on which plane could one move the sphere with least resistance? In what direction would it move with least resistance?

3.7. Based on atomic weights and structural information show that gold and tungsten essentially have the same density. Calculate the density of each.

3.8. Demonstrate that the densities of FCC and ideal HCP structures are identical if sites are populated by atoms of the same size and weight.

3.9. Cobalt exists in an FCC structure with a = 0.3544 nm and in the HCP structure with the same atomic radius. What is its theoretical density in the FCC structure? What is the theoretical density of HCP Co?

3.10. Draw the $(\bar{1}10)$, $(1\,\bar{1}0)$, $(01\,\bar{1})$, $(\bar{1}\,\bar{1}1)$, $(22\,\bar{1})$ and $(\bar{2}1\,\bar{1})$ planes in a BCC structure and draw the centers of the atoms that sit on them.

3.11. Draw all the directions of the $<111>$ family and give the Miller indices of each.

3.12. Draw the (100) and the (200) planes of the BCC and FCC structures and draw the centers of the atoms that sit on them.

3.13. What are the Miller indices of directions that form the intersections of the colored planes of Figure 3.15 with the cubes?

3.14. The linear thermal expansion coefficient of iron is $\alpha = 12.10^{-6}$ K^{-1} and its density at room temperature is 7.8 g/cm^3. Compute the lattice constant a and the density of BCC iron at 914°C. Assuming that the radius of the iron atoms does not change, compute the lattice constant and density of FCC iron at 914°C.

3.15. A vacancy moves from the surface of a crystal toward the interior: what do the atoms of the crystal do when this happens? Discuss how vacancies increase the diffusion rate in solids.

3.16. Look at the structure of an edge dislocation and justify that "pipe diffusion," which is the diffusion of atoms along a dislocation core, is much faster than in the bulk.

3.17. Observe a grain boundary and justify why diffusion along a grain boundary is much faster than through the grains.

3.18. Explain why surface diffusion is faster than grain boundary diffusion, pipe diffusion, or bulk diffusion.

3.19. The surface energy is caused by the unsatisfied bonds of surface atoms that lack the neighbors they would have inside the crystal. Study Figure 3.22 and give reasons why (1) the grain boundaries have a surface energy (i.e., a grain boundary energy) and (2) why the grain boundary energy is smaller than the surface energy.

3.20. What do the atoms do in elastic deformation? What do atoms do in plastic deformation?

3.21. Describe the arrangement of atoms in a crystal and in an amorphous material.

3.22. What is the unit cell in the structure of a material?

3.23. Describe the BCC structure. How many atoms does it have in a unit cell?

3.24. Describe the FCC structure. How many atoms does it have in a unit cell?

3.25. Describe the HCP structure. Compare its density with that of an FCC crystal if the radius of their atoms is the same.

3.26. What structure(s) represent(s) the densest possible packing of atoms?

3.27. What are the coordinates of atoms in a BCC crystal?

3.28. What are the coordinates of the atoms in an FCC crystal?

3.29. What is the densest plane in the FCC structure?

3.30. What is the densest plane in the BCC structure?

3.31. What is the significance of the densest crystal plane in terms of mechanical properties?

3.32. Describe the polymorphism of iron. Define polymorphism.

3.33. Draw the glide planes and glide directions in FCC and BCC structures.

3.34. Name and define point defects.

3.35. Define or draw an edge dislocation.

3.36. Define or draw a screw dislocation.

3.37. What is the microstructure of a material?

Chapter 4

Strengthening and Forming Metals

The strength of metals depends on obstacles that impede dislocation motion. Any type of lattice defect is an effective obstacle: foreign atoms, misplaced atoms, precipitates, internal surfaces, and even other dislocations strengthen a metal. In this chapter we examine various methods by which one can control the strength and hardness of metals and how one can make them more resistant to fracture, to fatigue, and to creep. The yield stress σ_y, the tensile strength σ_T, the ductility (elongation to fracture ε_F), and the resistance to fracture (the toughness) can be controlled by processing.

LEARNING OBJECTIVES

After studying this chapter, the student will be able to:

1. Explain the role of crystal perfection and crystal defects in the strength of metals.

2. Describe the strain and stress fields of a dislocation.

3. Perform solution strengthening.

4. Perform precipitation hardening.

5. Utilize grain size to increase the strength and toughness of metals.

6. Perform strain hardening.

7. Describe annealing and its effect on the microstructure and mechanical properties of metals.

8. Prescribe ways to increase the fracture resistance of metals.

9. Prescribe how to increase the fatigue life of metals.

10. Explain why modern jet engine blades are fabricated as single crystals.

4.1 STRENGTHENING A METAL

Our objective is to obtain a metal with high yield strength. This requires increasing the stress necessary to move a dislocation. In Chapter 3, we saw that dislocation motion is easy when the metal is a perfect single crystal. Practically any departure from crystalline perfection will be effective in increasing the stress necessary to move a dislocation. Crystal defects used for the industrial strengthening of metals are

- Foreign atoms. These are atoms of a different element that are introduced by alloying. The foreign atoms can be dissolved at random in the solid and result in **solution strengthening** or solution hardening. They can be entire aggregates of matter that is not soluble in the host crystal; they are precipitates in alloys and result in **precipitation hardening**.

- Grain boundaries. Grain boundaries are effective obstacles to dislocation motion. The deliberate introduction of a large density of grain boundaries leads to **strengthening by grain refinement**.

- Dislocations. Dislocations impede each other's movement. Their deliberate introduction causes **strain hardening**, also called work hardening or cold working.

It is worth noting that an increase in strength is accompanied by a decrease in ductility. This was examined in Chapter 2: when the material is harder, the blunting of the crack tip by plastic deformation is reduced and the stress concentration at the crack is increased.

4.1.1 Solution Strengthening

In the cases examined here, the "foreign" atoms are dissolved in the host metal; they occupy lattice positions at random; therefore the strengthening obtained in this way is called solution strengthening or solution hardening.

A look at Figure 4.1 shows that dislocations introduce large elastic strains in their vicinity. In an edge dislocation (Figure 4.1A), the extra plane of atoms crowds the lattice; it introduces large compressive strains above the dislocation line and equally large tensile strains below. In the screw dislocation (Figure 4.1B), large shear strains exist around the dislocation line. Impurities also introduce elastic deformation in their vicinity. A large foreign atom puts its environment in compression and a small atom, attracting its neighbors to itself, and creates a region of tensile deformation.

A large impurity atom in the compressed region exacerbates the compression and increases the resulting stress energy: the impurity repels dislocation. If the large impurity is in the tensile strain region, it decreases the total strain and stress energy: it attracts the dislocation but traps it in its vicinity. Similarly, a small impurity decreases the stresses in the compression region and increases them in the tensile region. In all cases, the impurity is an obstacle to the movement of a dislocation, by attracting or repelling it. Although it is not as easily visualized, impurities hamper the movement of screw dislocations as well.

The difference in bond strength between foreign and host atoms also disrupts the perfect periodicity responsible for easy dislocation motion and increases yield strength, and so do vacancies and interstitials. It is not difficult to see, for instance in Figure 3.26, that a very weak bond between atoms A and 2 or a strong bond between B and 2 would impede the motion of the dislocation to the right.

Figure 4.2 shows the changes in mechanical properties of copper when nickel is alloyed with it. We note that an increase in strength causes a decrease in ductility as we have discussed in Chapter 2.

(A)   (B)

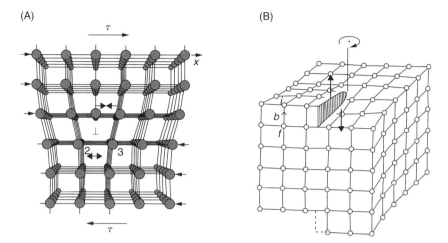

■ **FIGURE 4.1**   Elastic strains in the vicinity of a dislocation. (A) In the edge dislocation, compressive strains above and tensile strains below the dislocation. (B) Large shear stresses exist a long the edge dislocation line.

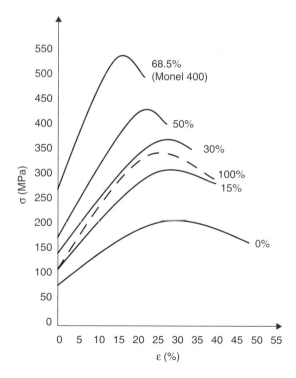

■ **FIGURE 4.2**   Stress-strain curves of copper-nickel alloys as a function of nickel content. The curve for pure nickel is shown dashed.

4.1.2 **Precipitation Strengthening or Precipitation Hardening**

Precipitates are small particles with different composition, structure, and often bonding type from the host metal; they are more powerful obstacles to the propagation of dislocations than dissolved alloy atoms. An example is aluminum alloyed with as little as 4.4% copper. A heat treatment that will be discussed in Chapter 7 produces microscopic precipitates of the compound $CuAl_2$. As Figure 4.3 shows, these precipitates increase the yield strength of aluminum from 35 MPa to 345 MPa, a factor of 10. In copper, the addition of 1.7% beryllium and 0.2% carbon increases the yield strength from 69 MPa to about 200 MPa when the beryllium and carbon are dissolved (solution strengthening) and to 1,070 MPa when they form precipitates!

It is not difficult to understand why precipitates are more effective than solid solutions in restricting dislocation motion. To begin with, precipitates are much larger than single atoms. Their crystal structure is different from that of the host metal; this mismatch can create large elastic strains and stresses in the particles and at their interface with the host. Usually, also, precipitates are compounds that form because of a strong chemical bond between their constituents. Often they are ceramic particles with directional and ionic bonds that preclude any plastic deformation; this is the case of WC and MoC precipitates in tool steels.

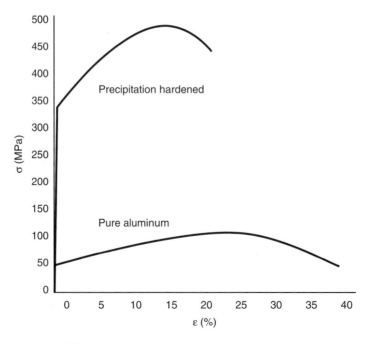

■ **FIGURE 4.3**   Precipitation hardening of aluminum.

It is important to know what alloys form solid solutions and which ones form precipitates and under what conditions. This is an important aspect of metallurgy that we will treat in Chapters 5 and 6.

4.1.3 The Strength of Polycrystalline Materials

Most metals in practical use are polycrystalline: they consist of a large number of small grains, which are crystals oriented in random directions. Grain boundaries are regions of severe lattice disorder; they are strong obstacles to the movement of the dislocations. The random orientation of the grains also contributes to the strength of metals. When a polycrystalline material is subjected to a tensile stress, in each grain the densest planes glide over each other in the easy glide directions; each grain tends to deform in a different direction but remains in contact with its neighbors. This restricts the possible slip directions of the individual grains and requires larger stresses than in a single crystal. Thus **polycrystalline materials have higher yield stresses than single crystals**.

Note that in the examples of Sections 4.1 and 4.2 that showed the properties of polycrystalline copper and aluminum, the yield strengths of 35 MPa for Al and 69 MPa for copper already reflect the grain boundary strengthening. They would be much weaker in the single crystal state.

4.1.4 Strengthening by Grain Refinement

A metal with a fine grain structure contains more grain boundaries that impede dislocation motion; it is stronger than a coarse-grained metal. The Hall-Petch relation

$$\sigma_y = \sigma_o + \frac{k}{\sqrt{d}} \tag{4.1}$$

describes how the yield strength σ_y of metals increases as the average size d of the grains decreases. Here σ_o is the yield strength of a metal with very large grains and k is a constant. Figure 4.4 illustrates the Hall-Petch behavior for brass.

4.1.5 Nanostructured Materials

The Hall-Petch relation has encouraged researchers to produce materials with grain sizes in the tens of nanometers. A yield strength of 400 MPa has been obtained in pure copper with 100 nm grains. (If it were possible to prepare brass with an average grain size of 50 nm, an extrapolation of Figure 4.4 would predict a yield strength of 1,775 MPa.) The Hall-Petch relation is valid for grains as small as about 50 nm; as the grains become still smaller, the metal becomes weaker. Some nanostructured materials will be described in Chapter 18.

4.1.6 Highly Strained and Amorphous Metals

It is possible to obtain highly disordered metals through which the propagation of dislocation is all but impossible. The most important case is that of steel, which is obtained in a very hard and brittle form, called Martensite, through rapid cooling from a high temperature. Steel will be treated in detail in Chapters 6 and 7.

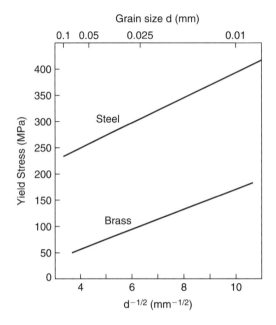

■ **FIGURE 4.4**   Hall-Petch Relation: The dependence of the yield strength of brass on the average grain size d. The upper scale denotes the grain diameter.

Metals can be made amorphous by very rapid cooling from the melt. Valve seats in engines are hardened by irradiation with short, high-power, laser pulses which melt a thin layer at the surface of the engine block. This layer cools so rapidly, in contact with the cold substrate that its atoms are "frozen" in place before crystallization and are unable to regain their crystal positions. The result is an amorphous layer in which dislocations cannot exist. Amorphous metals (Metglas®) can also be obtained by the very rapid cooling where the molten metal is poured between two chilled rolls that form a thin sheet as shown in Figure 4.5. Metglas is very hard and brittle.

4.1.7 Strain Hardening

In various stress-strain curves we have observed that the stress needed for plastic deformation increases with the strain; the ultimate tensile strength is larger than the yield strength. This phenomenon is called **strain hardening, strain strengthening or work hardening, or also cold working** when the deformation is performed at room temperature. Figure 4.6 illustrates the effect of strain hardening on cartridge brass, and Figure 4.7 shows the microstructure of the material at different stages of cold working.

Not only does plastic deformation proceed by the movement of existing dislocations, it also generates large numbers of dislocations by mechanisms that are beyond the scope of this text. Thus, the larger the deformation, the higher the dislocation density. These dislocations impede each other's movement

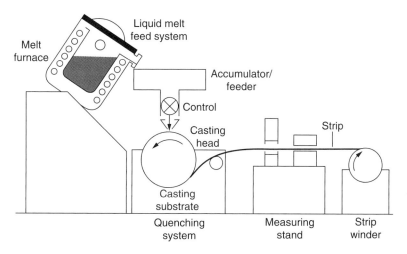

■ **FIGURE 4.5**   Melt quenching used to produce amorphous metals. The liquid metal is poured onto a cooled metal drum and cools rapidly. Courtesy of Metglas Inc.

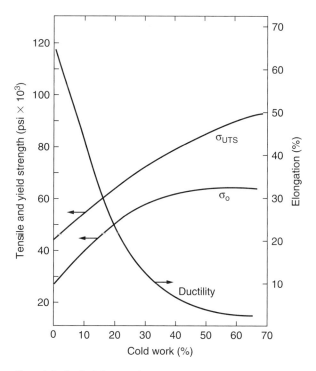

■ **FIGURE 4.6**   Strain hardening of brass (a Cu-Zn alloy). Stress-strain curve.

■ **FIGURE 4.7** Microstructures of cold-rolled cartridge brass (70%Cu-30%Zn). (A) Before rolling. (B) 30% reduction. (C) 50% reduction. (D) 75% reduction. 80X. Courtesy of G.F. Vander Voort, Buehler Ltd.

as follows. When two dislocations approach each other, they form a deformation field that is the vectorial sum of the two individual strain fields. When the dislocations have the same sign (i.e., the extra plane of atoms is on the same side), the resultant strain increases as the dislocations approach and so does the total elastic stress energy: the two dislocations repel each other. When the dislocations have opposite signs, the resultant strain and the elastic stress energy decrease and the dislocations attract each other. In either case, they impede each other's motion and require a higher stress to move. The result is strain hardening.

Strain hardening is a widespread method for strengthening metals. It is commonly accomplished by rolling, where a piece of metal is deformed between two rolls; or by forging, where the metal is deformed by hammering. Rolling and forging allow larger strains than tensile deformation because they do not produce necking.

4.2 **INCREASING THE DUCTILITY BY ANNEALING**

Strain hardening presents problems in manufacturing. Most metallic objects are shaped by plastic deformation during which the metal becomes hard and brittle. It is desirable to undo the effects of strain hardening and recover the soft and ductile material. This is achieved by annealing.

Annealing is a processing method in which a material is kept at elevated temperatures for a certain time and cooled slowly. Annealing removes residual stresses and reverses the effects of strain hardening. Slow cooling maintains equal temperatures throughout the piece and avoids the creation of residual stresses. The strong and brittle brass of Figure 4.6 obtained after cold working recovers the lower strength of 43 ksi and high ductility of 65% after annealing. Its microstructure reverts to that of Figure 4.7A.

Annealing modifies the mechanical properties of the material because the thermal energy at elevated temperatures facilitates the movement of atoms and dislocations to a configuration closer to equilibrium.

Annealing proceeds in three successive stages: **recovery**, **recrystallization**, and **grain growth**.

Recovery removes the internal stresses that resulted form uneven cooling or extensive deformation. Elevated temperature facilitates the movement of the dislocations under the influence of residual stresses until the latter disappear.

In the recrystallization phase, dislocations move and rearrange themselves in a way that reduces the total elastic strain energy that surrounds them. By aligning themselves as shown in Figure 4.8, the dislocations create boundaries between new grains that are tilted with respect to each other and contain fewer dislocations. When strain hardening has been extensive, this phenomenon creates a large number of small grains.

The final phase of annealing is grain growth. At elevated temperatures, the size of the larger grains increases and the small grains disappear. This is driven by the decreases in total grain boundary energy: The disappearance of the small grains and the growth of the larger grains decreases the total grain boundary area of the solid and, with it, the grain boundary energy.

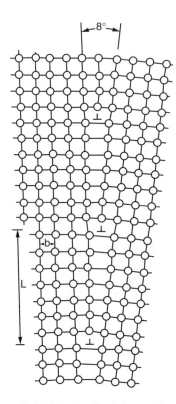

■ **FIGURE 4.8** Grain boundary formed by the alignment of edge dislocations. The tilt between the two grains accommodates the half planes of atoms without compressive and tensile stresses.

In sum, annealing removes residual stresses, decreases the density of dislocations, and lowers the grain boundary area. The result is a lowering of the yield strength and an increase in ductility.

Figure 4.9 illustrates schematically the changes in a material as it is cold worked and subsequently annealed, and Figure 4.10 shows the microstructure of pure iron after the various stages of cold working and annealing.

4.3 INCREASING FRACTURE RESISTANCE

In many applications of metals, the resistance to fracture is more important than its strength. Think of a metal bridge, for instance, or the frame of a bicycle. There are a number of ways the fracture toughness of a metal can be increased.

4.3.1 Materials Selection

In many applications, safety requires the selection of a metal that is fracture resistant. In these cases, one selects a metal with sufficient toughness rather than one with high strength.

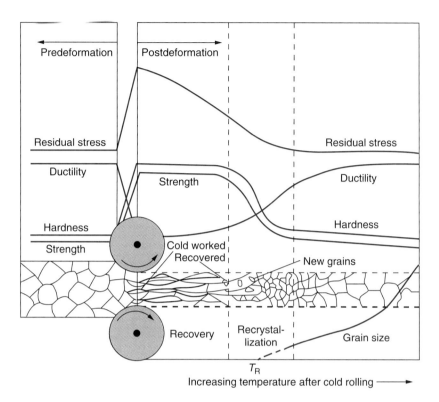

■ **FIGURE 4.9**   Schematic trends in structure and mechanical properties of a metal that is cold worked and annealed. From Z.D. Jastrzebski, *The Nature and Properties of Engineering Materials*, 2nd ed., Wiley, New York (1976), reprinted with permission.

4.3.2 **Annealing**

It is obvious from Section 4.2 that annealing increases the ductility and therefore the toughness of metals. It also removes unknown residual stresses that can be tensile and add themselves to the ones resulting from applied loads.

4.3.3 **Introduce Compressive Residual Stresses**

The fracture resistance can be increased by introducing compressive residual stresses at the surface of a piece. Such stresses tend to close cracks and counteract tensile stresses at the surface. This can be achieved by the diffusion of elements that crowd the crystal lattice or by mechanical means such as shot peening. In the latter process, the surface is bombarded by steel or glass beads that introduce a compressive residual stress which counteracts the applied tensile stress.

4.3.4 **Fine Grain Structure**

Contrary to other strengthening methods, the reduction in grain size is capable of increasing the yield strength without decreasing ductility. Grain boundaries are effective barriers that stop the

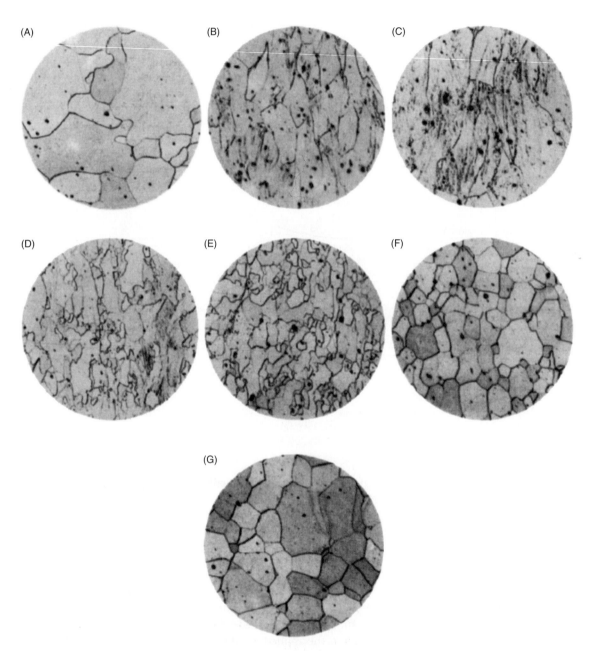

■ **FIGURE 4.10** Cold working and annealing of 99.97% pure iron. (A) Before deformation. (B) After 60% cold working. (C) After annealing at 510℃ for 1 hr. (D) After annealing at 510℃ for 10 hr. (E) After annealing at 510℃ for 100 hr. (F) After annealing at 675℃ for 1 hr. (G) After annealing at 900℃ for 1 hr.

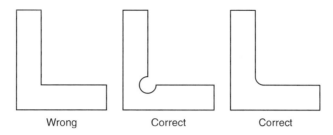

Wrong   Correct   Correct

■ **FIGURE 4.11**   Designing to avoid stress concentrations.

advance of cracks. Fracture can only occur by repeated crack initiation and change of crack direction at each grain boundary. Such crack extension absorbs mechanical energy and raises the toughness of the material.

4.3.5 **Polished Surfaces**

A rough surface contains notches that are similar to cracks in producing local stress concentrations as discussed in Chapter 2. A polished surface retards the initiation of surface cracks.

4.3.6 **Design for Fracture Resistance: Avoid Stress Concentrations**

We have seen in Chapter 2 that large stresses exist at the tip of a crack. Design faults such as sharp interior corners produce similar stress concentrations (also called stress raisers) and must be avoided. Blueprints always specify a radius of curvature at interior corners. Figure 4.11 illustrates designs that promote or avoid stress concentrations.

4.4 **INCREASING FATIGUE LIFE**

We have seen in Chapter 2 that fatigue failure is a fracture phenomenon; a crack advances by a small amount at each application of the cyclic stress. Therefore the methods that increase fatigue life are similar to the ones for increasing fracture strength.

4.4.1 **Avoid Stress Concentrators**

Keyways on shafts, holes, abrupt changes in cross sections, sharp corners, and so on, all stress raisers, are design features that should be avoided in components subjected to repetitive loading. Stress concentrators mean operation at higher stress levels on the *S-N* curve. This in turn reduces the number of stress cycles to failure.

4.4.2 **Polished Surfaces**

Crack nucleation is facilitated at micro-crevices and grooves on rough surfaces. Grinding, honing, and polishing of surfaces remove these sources of potential cracks.

4.4.3 **Residual Stresses**

Residual compressive stresses in surface layers, introduced by shot peening or diffusion, allow higher levels of applied tension to be tolerated. As a result, fatigue strength and life can be increased.

4.4.4 **High Strength Surfaces**

Strengthening surfaces can enhance fatigue resistance; carburization and deposition of hard coatings are ways to achieve this. Shot peening not only introduces compressive stresses but strengthens the surface by cold working.

4.4.5 **Homogeneous Material**

Since fatigue starts by the nucleation of a crack in a region of increased stress, it is important to avoid inclusions of material with different mechanical properties such as large particles of oxides or carbides. Steels for the production of ball bearings, which are exposed to large numbers of stress cycles, are produced by vacuum metallurgy that extracts oxygen and other gases.

4.4.6 **Avoid Corrosion and Environmental Attack**

Localized chemical attack will always reduce fatigue life. Metal removed from exposed grains and grain boundaries leaves a generally rough surface, pits, and corrosion products behind, that serve as incipient cracks.

4.5 **CREEP RESISTANCE**

Let us review the observations (Chapter 2). Creep takes place at stresses below the yield stress; therefore it does not operate by dislocation motion. It happens at temperatures above half the melting point (in degrees Kelvin); these are the same temperatures at which diffusion occurs. Clearly diffusion, that is, the random motion of atoms caused by thermal agitation, is the major mechanism of creep. Observations of creep failure reveal the dominant role grain boundaries play in the phenomenon; by diffusion of atoms in the grain boundaries, the grains slide along each other.

For this reason, modern jet engine turbine blades are grown as **single crystals** by a method of directional solidification.

Creep resistance is also sought through the development of refractory alloys with high melting temperatures. These alloys contain carbide particle precipitates, some of which reside in grain boundaries and restrict the sliding of grains. Stainless steels and superalloys containing nickel and cobalt are notoriously creep resistant.

4.6 **MECHANICAL FORMING OF METALS**

The largest number of metallic objects are made by plastic deformation, illustrated in Figure 4.12. In all cases, the metal is first cast into a simple billet. Casting will be discussed in Chapter 6. The shaping of metals can be done at room temperature or when the metal is hot. Cold forming is preferred for

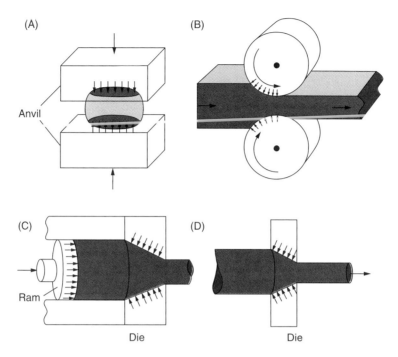

■ FIGURE 4.12 Typical mechanical forming methods. (A) Open die forging. (B) Rolling. (C) Extrusion. (D) Drawing. Arrows show forces on workpiece.

rolling and stamping of sheet metal; it is often done with the purpose of hardening the material by strain hardening. Hot forming is done during rolling and forging of large pieces; the elevated temperature facilitates the plastic deformation and provides some simultaneous annealing that prevents excessive hardening and brittleness.

Sheet metal, bars, rails, and I-beams, for example, are manufactured by **rolling**, in which the billet is introduced between two large rolls. The familiar aluminum foil roll is obtained by repeated rolling from a large ingot.

In **forging**, the metal is hammered until it acquires the desired shape. Complex shapes, for instance the connecting rods or the crankshaft of an engine, hand tools, turbine disks, railroad wheels, and gears, are shaped by **die forging**, in which the "hammer and anvil" consist of dies that are shaped accordingly. Figure 4.13 illustrates the forging of a crankshaft.

Bars of complex shape are produced by **extrusion** where the metal is pushed through a die of the desired shape by a ram. Wires are manufactured in a similar way, except that the wire is drawn through the die.

Sheet metal forming is a large class of fabrication techniques in which objects are manufactured from metal sheets. **Stamping** between dies is used in the fabrication of car bodies and many other objects. **Deep drawing** is the plastic deformation of a thin sheet to fabricate soda cans, cooking pans, or other deep hollow objects. See Figure 4.14.

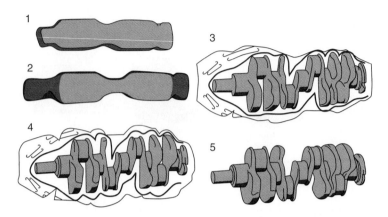

■ **FIGURE 4.13**   Stages in the closed die forging of a crankshaft. (1,2) Preliminary roll forging. (3) Blocking in closed dies. (4) Finishing in closed dies. (5) Flash trimming. From *Metals Handbook*, 8th ed., Vol. 5, American Society of Metals, Metals Park, OH (1970).

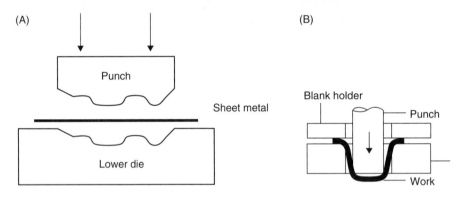

■ **FIGURE 4.14**   (A) Stamping. (B) Deep drawing of a metal can.

4.7 **CUTTING AND MACHINING**

Metals can be shaped by different modes of cutting. A sharp cutting tool made of a hard material penetrates the metal and separates a "chip" by plastic deformation. This is illustrated in Figure 4.15.

The different types of metal cutting are **drilling** of cylindrical holes, **turning** of cylindrical objects in a lathe, **milling** of flat surfaces by cutting tools attached to a wheel, and **sawing** by a serrated hard saw blade. Most metals can be machined by any of the above methods by a sufficiently hard cutting tool. The latter is generally a hardened steel, a ceramic (SiC or diamond) or a metal-ceramic composite (WC/Co also called cermet or hardmetal). **Grinding** is the removal of material from surfaces by the penetration of small very hard grains of ceramic, often diamond. The hardest metals, especially hardened steels, can only be shaped by grinding. The strength and machinability of metals make them the material of choice for high precision equipment and complex machinery.

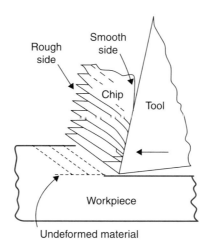

■ FIGURE 4.15   Cutting a metallic workpiece in a lathe or milling machine. The metal is removed by plastic deformation.

■ SUMMARY

1. In general, crystal defects increase the strength of a metal because they impede the movement of dislocations.

2. Solution strengthening is the increase in strength due to foreign atoms dissolved in the host metal.

3. In precipitation hardening grains (precipitates) of a foreign substance impede dislocation movement.

4. Grain refining increases the strength of a metal because grain boundaries are obstacles to dislocation movement.

5. Work hardening is the increase in the strength of a metal caused by deformation. The deformation increases the densities of dislocations which impede each other's movement.

6. The effects of work hardening can be reversed by annealing. At elevated temperatures, atoms and dislocations move more easily. The dislocations move in a way to reduce residual stresses. this is recovery. Later, the dislocations rearrange themselves and form new grain boundaries: this is recrystallization. Finally grain growth reduces the number of grain boundaries. The result is a lower yield stress and larger ductility.

7. Fracture resistance of metals is increased by selection of ductile metals, by fine grain structures, by annealing, by introducing compressive stresses at the surface (shot peening), by polishing the surfaces, and by design that avoids stress concentrations (no sharp interior corners)

8. Fatigue resistance is increased by all methods increasing fracture resistance and by avoiding inclusions and corrosion.

9. Creep resistance is obtained by the choice of metals with high melting point. The high creep resistance of turbine blades is obtained by growing them as single crystals.

10. Most metallic objects are fabricated by plastic deformation. Bars, rails, and sheets are made by rolling an ingot. Bars and rubes with complex shapes are made by extrusion where the metal is pushed through a die. Wires are fabricated by pulling the metal through a die.

11. Forging consists of hammering the metal to the proper shape. Complex shapes are obtained by die forging.

12. Sheet metal objects are made by stamping or deep drawing.

13. Cutting and machining occurs by plastic deformation of the metal. Machining operations include drilling, turning in a lathe, milling, sawing, and grinding. In all these methods a hard tool penetrates the metal.

■ KEY TERMS

A
amorphous metal, 111

C
cartridge brass, 112
cold working, 112
copper, 108
corrosion, 120
creep resistance, 120
cutting, 122

D
deep drawing, 121
die forging, 121
dislocation, 108
drilling, 122
ductility, 115

E
extrusion, 121

F
fatigue life, 119
forging, 121
fracture resistance, 116
   annealing, 116
   grain refinement, 117
   residual stresses, 117

G
grain growth, 115
grain refinement, 108
grinding, 122

H
Hall-Petch, 111

I
inclusions, 120

M
machining, 122
materials selection, 116
mechanical forming of
   metals, 120
Metglas, 122
milling, 122

N
nanostructured materials, 111

P
polished surfaces, 119
polycrystalline materials, 111
precipitates, 110
precipitation hardening, 108, 110
precipitation strengthening, 110

R
recovery, 115
recrystallization, 115
residual stresses, 117
rolling, 121

S
sawing, 122
sheet metal forming, 121
single crystals, 120
solution strengthening, 108
stamping, 121
strain hardening, 108, 112
strengthening, 107
strengthening by grain
   refinement, 111
stress concentrations, 119
stress concentrators, 119

T
turning, 122

W
work hardening, 112

■ REFERENCES FOR FURTHER READING

[1]   Ashby, M.F. *Materials Selection in Mechanical Design*, 3rd ed., Butterworth-Heinemann, Woburn, MA (2005).

[2]   Ashby, M.F. Jones, D.R.H. *Engineering Materials, An Introduction to Their Properties and Applications*, 3rd ed., Butterworth-Heinemann, Woburn, MA (2005).

[3]   Handbook, A.S.M. Vol. 8, *Mechanical Testing and Evaluation*, ASM International, Materials Park, OH (2000).

[4]   Courtney, T.H. *Mechanical Behavior of Materials*, 2nd ed., Waveland Press, Long Grove, IL (2005).

[5]   Dieter, G.E. Bacon, D. *Mechanical Metallurgy* (Materials Science and Engineering), 3rd ed., McGraw-Hill, New York (1989).

[6]   Dowling, N.E. *Mechanical Behavior of Materials*, 3rd ed., Prentice Hall, Englewood Cliffs, NJ (2005).

■ PROBLEMS AND QUESTIONS

4.1.   How have dislocations helped to explain the following phenomena:

  a.   The yield stress of mild steel.

  b.   The structure of small angle grain boundaries.

  c.   Work hardening of metals.

  d.   Surface slip offsets on deformed crystals.

4.2.   Look at a drawing of an edge dislocation and discuss the elastic strains of the crystal around the dislocation. Can you see that two dislocations repel each other? State why.

4.3.   From Figure 4.2, make a graph showing the yield stress and the tensile strength of Cu-Ni alloys as a function of nickel content. Continue the graph qualitatively all the way to 100% nickel. Explain the shape of the curve.

4.4.   Explain in your own words the statement: "A perfect material, namely a perfectly pure crystal without defects in structure is perfectly useless." Apply this to the strength of metals. (You will later apply it to electric resistance, color, the function of semiconductors, and hard magnets.)

4.5.   From Figure 4.4 estimate the strength of the material if it is nanostructured with a grain size of 50 nm.

4.6.   Use the tables in Chapter 7. Compare the mechanical properties obtained by solution strengthening, strain hardening, and precipitation strengthening, and explain the difference.

4.7.   Is the strength of a nanostructured metal maintained at high temperatures? Explain.

4.8.   Of solution hardening, precipitation hardening, and strain hardening,

  a.   Which one is reversed most easily by annealing and which one is not?

  b.   Which one remains effective at high temperatures?

4.9.   What problem do you foresee during deep drawing of metal cups

  a.   If the punch used has too small a radius of curvature at the edge?

  b.   If there is too much clearance between punch and die?

  c.   If the sheet is held too loosely due to insufficient hold down forces?

  d.   If the necking strain is exceeded?

4.10.   Define solution strengthening and explain how it works.

4.11.   Define precipitation hardening. Describe what happens in the material and how it strengthens it.

4.12.   Define work hardening and describe how it occurs at the atomic level.

4.13. Describe the effect of grain size on the mechanical properties of metals.

4.14. Define annealing. Describe what the engineer does and what effect it has on the mechanical properties of metals.

4.15. Describe the effect of annealing on the microstructure of a metal.

4.16. Which strengthening methods used in metals can be reversed by annealing and which ones cannot? Explain your answer.

4.17. Describe recrystallization: its effect on the microstructure and why it occurs.

4.18. When being rolled into sheets, metals must be annealed; otherwise they will tear. Explain why annealing is necessary and how it modifies the metal to allow further rolling.

4.19. Name two methods by which the fracture resistance of a metal can be increased by processing and explain why they are effective.

4.20. Name two methods by which the fatigue resistance of a metal can be increased by processing and explain why they are effective.

4.21. How are turbine blades processed to provide them with maximum creep resistance?

4.22. Describe the fabrication of a rail.

4.23. Describe the fabrication of aluminum foil.

4.24. Describe the fabrication of a wire.

4.25. Describe the fabrication of an aluminum soda can.

4.26. How is the metallic body of a car shaped?

4.27. Describe how cutting in a lathe or a milling machine occurs by plastic deformation.

4.28. How does one shape a very hard material?

Phase Diagrams

When two metals are mixed together to form an alloy, three things can happen: the two elements mix together and they form a solid solution; the two elements do not mix and separate into different phases and they form a phase mixture; or the two elements bond strongly to each other and form a compound. Phase diagrams represent the behavior of binary alloys as they solidify from the melt: they indicate what phases form for each composition and temperature. Phase diagrams offer guidance on how to process alloys in order to obtain the desired mechanical properties described in Chapters 2 and 4.

LEARNING OBJECTIVES

After studying this chapter, the student will be able to:

1. Sketch the phase diagram of a solid solution and analyze this diagram to determine the composition and amount of each phase for a given overall composition and temperature.

2. Sketch the phase diagram of a simple eutectic alloy and define the eutectic point, liquidus, solidus, and solvus lines. Analyze this diagram in terms of number, composition, and amount of the phases.

3. Sketch a simple phase diagram with an intermediate compound and analyze the diagram in terms of number, composition, and amount of the phases.

4. Analyze the phase diagram of the iron-carbon system that underlies steel.

5.1 BEHAVIOR OF BINARY ALLOYS

In Chapter 4, we saw that metals are strengthened by introducing foreign atoms, either randomly dispersed in the host metal, or in the form of precipitates. For this reason practically all structural metals utilized today are alloys that are obtained by melting together two or more different metals. In order

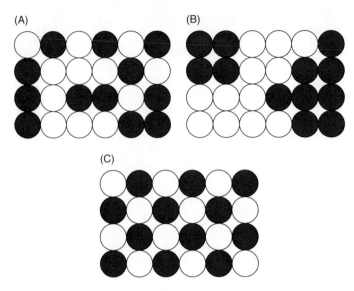

■ **FIGURE 5.1**   Schematic illustration of differing configurations of A and B atoms in binary alloys. (A) Solid solution. (B) Segregation in a eutectic. (C) Compound formation.

to process the alloys or utilize alloys effectively, it is important to know what one obtains when the molten mixture cools to form a solid alloy.

Most commercial alloys contain a number of elements that have been added in various quantities to the host metal, often through empirical development. It would be impossible to understand all the interactions between the different alloying elements in an introductory treatment, but it is possible to understand the principles of alloy formation in the simple case of **binary alloys,** namely alloys composed of two elements. When two metals are molten together and solidified, three things can happen, depending on the strength of the bond between similar and different atoms: the alloy can form a solid solution, a mixture of distinct phases, or a compound. This is illustrated in Figure 5.1.

In a **solid solution,** the atoms of both elements A and B occupy the lattice positions in the crystal **at random**. The classic example is the copper-nickel alloy. The solid solution is obtained because the strength of the Cu-Ni bond is intermediate between those of the Cu-Cu and the Ni-Ni bonds.

A **multi-phase mixture** is formed when the A-B bond is weaker than the A-A and the B-B bond; each component has a low solubility in the other. (It is the solid equivalent of a water-oil mixture.) In this case, the alloy solidifies by segregating into two distinct phases, one rich in element A and one rich in element B. This will be illustrated by the lead–tin and bismuth–tin alloys, which are the materials used as solder. Materials that form mixtures of immiscible phases are commonly used for precipitation strengthening.

A **compound** is obtained when the A-B bond is stronger than the A-A and the B-B bonds. This will be illustrated by the semiconductor gallium arsenide.

The behavior of such alloys is codified in the **phase diagrams**. A phase diagram shows what phases an alloy will form at all temperatures and compositions in **thermodynamic equilibrium**, which is the state that nature aspires to and is obtained **if sufficient time is allowed**. This is the topic of this chapter.

In practice, the materials predicted by the phase diagrams are obtained only **with very slow cooling**. When the alloy is cooled rapidly, the equilibrium may not be reached and the structure of the alloy will depend on the kinetics of the phase transition. Such rapid cooling is an important tool of the engineer and will be treated in Chapter 6.

5.2 **PHASES, COMPONENTS, AND PHASE DIAGRAMS**

Let us take a glass of water and pour some sugar into it and add ice cubes. We now have two components: water and sugar. (Ice and water are the same component, H_2O.) We also have three phases: water, ice, and solid sugar. Now let us stir until the sugar is dissolved in the water. We now have two components: water and sugar and two phases: the water-sugar solution and the ice cubes. If we wait until all the ice melts, we obtain one phase and two components.

A **component** is an ingredient in the alloy. Note that, in our example, we have used water and sugar as the components. Water is H_2O and sugar is a complex hydrocarbon molecule. Why did we not use hydrogen, oxygen, and carbon as components? Because that is not useful: neither the water nor the sugar reacts chemically in this system.

A **phase** is a homogeneous, physically distinct, portion of matter that is present in a nonhomogeneous system. It may be a single component, a solution, or a compound. In our example, we start with a liquid phase (pure water) and two solid phases (ice and sugar); as the sugar is dissolved, we have a liquid phase (sugar solution in water) and one solid phase (ice); we end with one phase (sugar solution in water).

The **composition** of a system represents the relative amounts of the components it contains. One defines the overall composition C_O of the system (how much sugar and how much H_2O in liquid or ice form are in the glass) and the compositions of the phases (how much sugar in the ice C_S, how much sugar in the liquid C_L).

We will also be concerned about the **amount of a phase**, which is the fraction of the mixture that is in the particular phase. (How many percent ice (f_S) and how many percent liquid (f_L) are in the glass?)

In a binary phase diagram (illustrated in Figures 5.2, 5.3, and 5.8) we plot the composition of the alloy on the horizontal scale and the temperature on the vertical scale. The overall composition C_O remains of course constant, but the compositions of individual phases vary. These diagrams are valid for atmospheric pressure, at which most practical casting and thermal processing are performed.

5.3 **SOLID SOLUTIONS**

Copper-nickel alloys form solid solutions. No matter how much nickel we add to the copper, we obtain a solid in which the copper and nickel atoms occupy crystalline positions at random. This is

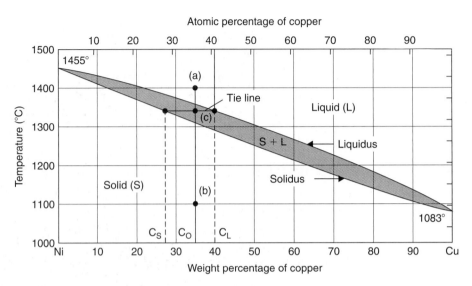

not surprising; copper and nickel are neighbors in the periodic table so their electronegativies are similar, their atomic radii are comparable ($r = 0.128$ nm for Cu and $r = 0.125$ for Ni), and both crystallize in the FCC structure.

The phase diagram of copper-nickel alloys is shown in Figure 5.2. As in all phase diagrams, the x-scale represents the composition C of the alloy. In the present example, it shows the percentage of nickel contained in the alloy. We have a choice of representing the composition in atomic percent or weight percent. The weight fractions are more convenient for engineering purposes; these are usually represented in the bottom scale. Atomic percentages, which are of scientific interest, are usually shown in the top scale. The two scales are, of course, different because copper and nickel have different atomic weights.

Pure nickel has a well-defined melting point of 1,455°C; it is liquid above this temperature and solid below. When nickel is heated to 1,600°C and allowed to cool, its temperature decreases to 1,455°C, where it remains constant until all the nickel is solidified. Once all the nickel is solid, its temperature decreases again. The same is true for pure copper, at 1,083°C. (See Figure 5.3.) This is not the case for the alloys. Take the alloy containing 35% nickel shown as the vertical line (ab) in Figure 5.2. When we mix 35 kg of copper and 65 kg of nickel, we form the **overall composition C_O** (35 wt%). When this alloy is cooled from point (a), namely 1,400°C, it starts solidifying at 1,360°C; it continues to cool, but more slowly until it is completely solid at 1,310°C. Between these 1,360°C and 1,310°C, we obtain two phases, one liquid and one solid. At 1,340°C, for instance, the liquid phase contains 40 wt% Cu and the composition of the solid phase is 27 wt% Cu, no matter how long we wait at this temperature. This phenomenon is described in the diagrams at all compositions by the **liquidus line** above which the

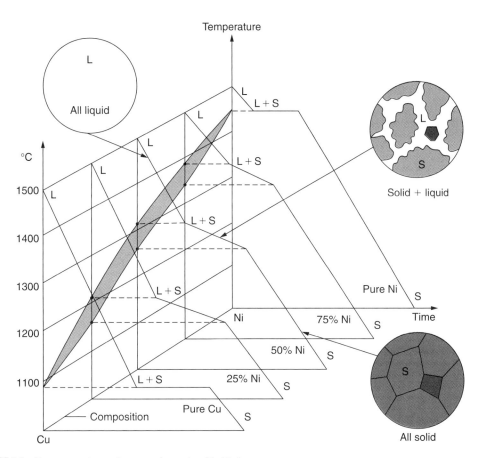

■ **FIGURE 5.3**   Temperature-time cooling curves for a series of Cu-Ni alloys.

alloy is completely liquid and the **solidus line** below which it is completely solid. The region between these two lines, (shaded in the figure) is a **two-phase** region: it contains a solid and a liquid phase that have different compositions. The evolution of the temperature for the copper-nickel alloys is shown in Figure 5.3. We recall that Figure 5.3 is an equilibrium diagram, which means that the times are very long.

5.4 ANALYSIS OF BINARY PHASE DIAGRAMS

5.4.1 Single Phase

In a single phase field (in this or any other phase diagram) the **chemical** and **physical** analysis of what is present in thermodynamic equilibrium is simple. The composition of the material is equal to the overall composition. Let us consider the 35 wt% Cu alloy (close in composition to Monel metal, which is used in petroleum refineries) and heat it to 1,400°C (state a, Figure 5.2). The molten alloy would contain 35 wt% Cu and 65 wt% Ni; and 100% of what is present is liquid. Similarly, if the original alloy is cooled to 1,100°C (state b), the phase diagram tells us that the material is all solid and has a composition equal to the overall composition (i.e., 35 wt% Cu and 65 wt% Ni).

In any single phase field of any binary equilibrium phase diagram the rules are:

1. The chemical composition of the phase is the same as the overall composition C_O

2. Physically, the alloy is either 100% homogeneous liquid or 100% homogeneous solid.

3. Single phase alloy liquids and solids are generally stable over some range of composition and temperature.

5.4.2 Two Phase Mixture, Tie Line Construction, and Chemical Composition

If the system is in a two phase (L + S) field, a **physical mixture** of the two phases exists. The chemical and physical analyses follow different rules from those of the single phase field. Both analyses center on the construction of a **tie line** or **isotherm**; this is a horizontal line (constant temperature) drawn through the state point (c) so that it extends to the left and right until it ends at the single phase boundaries on either side of the overall composition. Such a tie line is shown at 1,340°C for the initial 35 wt% Cu-65 wt% Ni alloy.

Within the L + S field, the composition C_S of the solid is given by the intersection of the tie line with the solidus curve (read off on the horizontal axis). The composition C_L of the liquid is given by the intersection of the tie line with the liquidus curve. For example, in the 35 wt% Cu-65 wt% Ni alloy at 1,340°C the solid has a composition C_S = 27 wt% Cu, while the liquid has a composition C_L = 40 wt% Cu. See Figure 5.4.

5.4.3 Two Phase Mixture, Lever Rule, and Relative Phase Amounts

We are only halfway through our analysis of the two phase mixture. Now that we know the **chemical composition** of each phase let us determine the **physical composition, namely,** the relative proportions or weights of each **phase** present. The problem is to determine the fraction f_L that is liquid and the fraction $f_S = (1 - f_L)$ that is solid. If the total alloy weighs W grams, the weight of the liquid phase is $W \times f_L$ and the amount of copper it contains is $W \times f_L \times C_L$. Because no copper disappears, the total amount of copper in the alloy is the amount in the solid + the amount in the liquid

$$WC_O = WC_1 f_L + WC_S(1 - f_L) \qquad (5.1)$$

The term on the left represents the total weight of copper in the alloy, and the terms on the right the weight of copper in the liquid and the solid phases. Solving for f_L yields the **Lever Rule**

$$f_L = (C_S - C_O)/(C_S - C_L) \text{ and } f_S = (C_O - C_L)/(C_S - C_L) \qquad (5.2)$$

Note that, in order to obtain the amount of **liquid** f_L, one measures the distance from the overall composition C_O to the composition of the **solid** and divides the result by the length of the tie line.

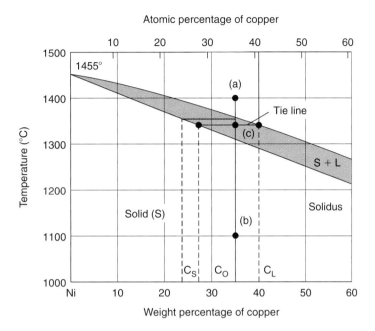

Atomic percentage of copper

The phase whose boundary is closest to the initial composition C_O is present in the greatest amount. When applied to our 65 wt% Ni-35 wt% Cu alloy at 1,340°C, the fraction of solid is $(40 - 35)/(40 - 27) = 0.385$ and that of liquid is $(35 - 27)/(40 - 27) = 0.615$.

The following analogy could help us remember the rule. The moon lies: it has the shape of a D when it is growing and the shape of a G when it is diminishing. The same applies to the tie line: go from C_O to C_S to obtain f_L and from C_O to C_L to obtain f_S.

In summary, for any two phase field in any binary equilibrium phase diagram the rules are:

1. Draw a horizontal tie line corresponding to the chosen temperature, extending to the left and right phase limits from the overall composition C_O. (Tie lines are only drawn in two phase fields. They make no sense in a one phase field.)

2. The chemical composition of the two phases is given by the ends of the tie line, extended vertically down to, and read off, the horizontal axis.

3. The weight fraction of each phase within the two phase mixture is given by the Lever Rule, Equation (5.2).

5.4.4 **Coring**

We have just seen that the composition C_S of the solid phase changes as the system cools down. As solidification starts at 1,355°C, the incipient solid phase contains $C_S = 23\%$ Cu (color in Figure 5.4); this changes to 27% at 1,340°C and finally to $C_S = C_O = 35\%$ Cu when solidification is complete at 1,310°C. The change in composition requires the diffusion of copper and nickel atoms, which takes

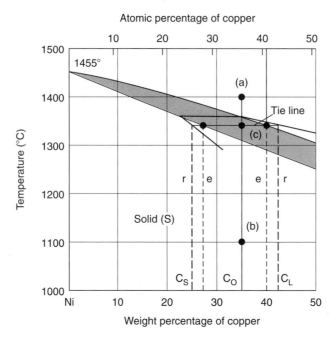

■ FIGURE 5.5   Coring. The black lines represent equilibrium (e). The color lines show the solidification with rapid cooling (r): the solid is richer in Ni and the liquid poorer in Ni than at equilibrium

time. Thus Figure 5.2 is an **equilibrium** phase diagram that represents the alloy only if sufficient time is allowed for the atoms in the solid phase to diffuse (i.e., theoretically infinitely slow cooling).

During industrial casting of metals, such slow cooling is neither feasible nor desirable. Alloys are cooled rather quickly in the foundry. What happens under these conditions is illustrated in Figure 5.5. The nickel does not have time to diffuse out of the solid phase, and the first solid that forms is richer in Ni than the overall alloy. This nickel is missing in the liquid, which is enriched in copper as indicated in Figure 5.5. The final result is **a cored structure** in which the centers of the grains are enriched and the outsides are depleted of Ni. Such nonequilibrium structures are sometimes undesirable and are corrected by subsequent thermal treatment, such as a lengthy heating in the solid state which is called **soaking**.

5.5 **THE BINARY EUTECTIC PHASE DIAGRAM**

Many pairs of elements (e.g., Bi-Cd, Sn-Zn, Ag-Cu, Al-Si, etc.) do not mix. The bond between the two elements A-B is weaker than the A-A and the B-B bond. Each metal can dissolve a limited amount of the other. These pairs solidify in a manner similar to Pb-Sn whose **eutectic** phase diagram is shown in Figure 5.6. The eutectic is so named for the critical **eutectic point** corresponding to the **eutectic temperature** (183°C in Sn-Pb) and **eutectic composition** (61.9% Sn in Sn-Pb).

On the left, the diagram possesses a solid solution designated as α **phase**. The **solidus** and **solvus lines** specify the maximum amount of tin that can be dissolved in the lead at each temperature.

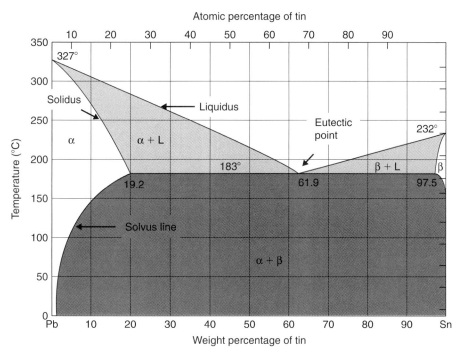

■ **FIGURE 5.6** Pb-Sn phase diagram showing eutectic reaction. Two-phase regions are colored.

Maximum solubility occurs at 183°C, where the alloy can contain 19.2 wt% tin. As the temperature is lowered, the amount of tin soluble in the lead decreases, reaching about 1% at room temperature. On the right side of the diagram, the tin-rich β phase can dissolve 2.5 wt% lead at 183°C. This solubility decreases rapidly to near zero as the temperature is lowered.

Let us see how an alloy of the eutectic composition of 61.9 wt% tin solidifies. The diagram shows that the liquid alloy will not solidify even below the melting point of tin. At the eutectic temperature of 183°C, the liquid alloy undergoes the eutectic reaction; it **decomposes** into a mixture of two solid phases: a Pb-rich solid solution α containing 19.2% Sn, and a Sn-rich solid solution β (with 97.5% Sn). The reaction is invariant: the temperature and the phase compositions remain constant during the transformation of eutectic liquid to α + β. This is followed by cooling to room temperature of the two phase solid. During the cooling, tin is expelled from the α phase and lead is expelled from the β phase; at room temperature, the eutectic consists of almost pure lead and tin. See Figure 5.7.

The eutectic microstructure that develops is a lamellar array of alternating plates of the two phases shown in Figure 5.8C. Most eutectic alloys form such a lamellar array, but sometimes they solidify as a dispersion of one phase in a matrix of the other.

The rules of phase analysis in single and two phase fields given previously are equally applicable to eutectic systems and will be applied in the example below.

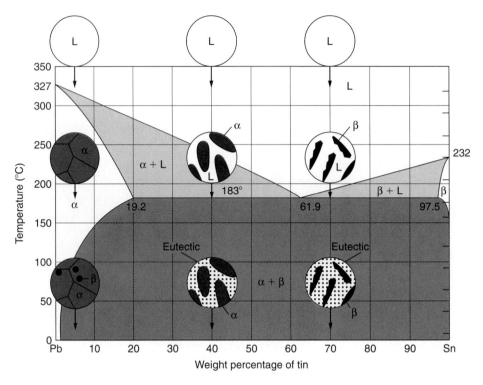

■ FIGURE 5.7   Schematic microstructures of three alloys in the Pb-Sn system at temperatures of 330°C, 184°C, and 25°C. The eutectic microstructure is composed of alternate α and β layers.

Let us cool an alloy with a composition between 61.9 and 19.2%; we choose 50 wt% Sn. Below 230°C this alloy forms a two-phase system of solid α phase and liquid.

At 200°C, for example, the α phase contains about 18% Sn and the liquid phase 57% Sn. As the temperature approaches 183°C, the composition of the liquid approaches the eutectic point and that of the solid changes to 19.2% Sn. Just below 183°C, the alloy consists of the **proeutectic** α phase that was formed at higher temperatures and a solid eutectic (consisting of α with 19.2% Sn and β with 97.5% Sn). As the alloy cools, the proeutectic α expels tin, following the solvus curve, and becomes almost pure lead at room temperature. The two phases in the eutectic also become purer lead and tin, as in the above case. A polished and etched section of the alloy is shown in Figure 5.8B.

An alloy with less than 19.2 wt% tin forms α phase and liquid and finally all α phase until it cools. Below 183°C, the solid solution rejects tin according to the solvus curve; no eutectic is formed. The same holds for alloys with more than 97.5 wt% tin.

(A)

(B)

(C)

(D)

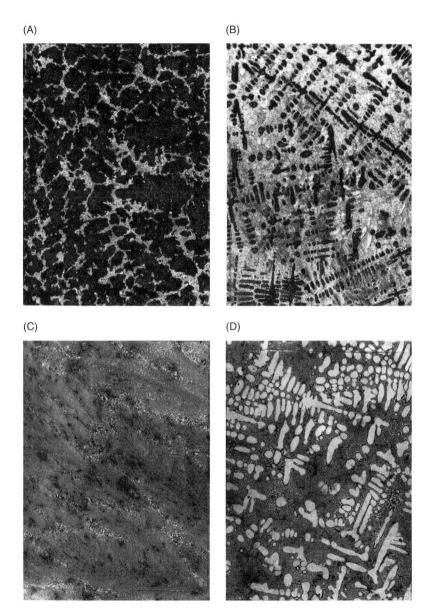

■ **FIGURE 5.8** Microstructures of Pb–Sn alloys at room temperature. (A) 30 wt% Sn-70 wt% Pb. Dark proeutectic α phase surrounded by eutectic. (B) 50 wt% Sn-50 wt% Pb. Dark proeutectic α phase within the light eutectic. (C) 63 wt% Sn-37 wt% Pb. Two-phase eutectic microstructure. (D) 70 wt% Sn-30 wt% Pb. Light proeutectic β phase surrounded by eutectic.

EXAMPLE 5.1 *For the alloys, (1) 5 wt% Sn-95 wt% Pb, (2) 40 wt% Sn-60 wt% Pb, and (3) 70 wt% Sn-30 wt% Pb, identify the phases present and determine their compositions and relative amounts at 330°C, 184°C, and at 25°C.*
ANSWER The answers are tabulated below. They are read off Figure 5.7.

Temp (°C)	Phases	Chemical Composition (wt% Sn)	Phase amounts (wt. fraction)	Comment
1. 5 wt% Sn-95 wt% Pb				
330	L	5	1.0	all liquid
184	α	5	1.0	all solid
25	α	~2	$(100 - 5)/(100 - 2) = 0.97$	no eutectic
	β	~100	$(5 - 2)/(100 - 2) = 0.03$	in grain boundaries
2. 40 wt% Sn-60 wt% Pb				
330	L	40	1.0	all liquid
184	α	19.2	$(61.9 - 40)/(61.9 - 19.2) = 0.51$	proeutectic
	L	61.9	$(40 - 19.2)/(61.9 - 19.2) = 0.49$	liquid
182	α	19.2	$(61.9 - 40)/(61.9 - 19.2) = 0.51$	proeutectic
	$\alpha + \beta$	61.9	$(40 - 19.2)/(61.9 - 19.2) = 0.49$	eutectic
Eutectic	α	19.2	$(98 - 61.0)/(97.5 - 19.2) = 0.46$	
Eutectic	β	98	$(61.9 - 19.2)/(97.5 - 19.2) = 0.54$	
25	α	~2	$(100 - 40)/(100 - 2) = 0.61$	
	β	~100	$(40 - 2)/(100 - 2) = 0.39$	
3. 70 wt% Sn-30 wt% Pb				
330	L	70	1.0	all liquid
184	β	97.5	$(70 - 61.9)/(97.5 - 61.9) = 0.23$	proeutectic
	L	61.9	$(97.5 - 70)/(97.5 - 61.9) = 0.77$	liquid
182	β	97.5	$(70 - 61.9)/(97.5 - 61.9) = 0.23$	proeutectic
	$\alpha + \beta$	61.9	$(97.5 - 70)/(97.5 - 61.9) = 0.77$	eutectic
25	α	~2	$(100 - 70)/(100 - 2) = 0.31$	eutectic
	β	~100	$(70 - 2)/(100 - 2) = 0.69$	eutectic and proeutectic

EXAMPLE 5.2   *For a 1 kg specimen of the 70 wt% Sn-30 wt% Pb alloy from Example 5.1, what weight of Pb is contained within the eutectic microstructure at 182°C?*

ANSWER   In 1 kg of the 70 wt% Sn-30 wt% Pb alloy, the eutectic mixture weighs 0.79 kg (from Example 5.1, the amount of eutectic does not change). Of this, 61.9% is Sn and 38.1% is Pb. Therefore, the amount of Pb $=$ (0.79)(0.381) $=$ 0.30 kg.

Schematic microstructures of these alloys are sketched in Figure 5.7. They should be compared with actual room temperature microstructures of several alloys in the Pb-Sn system (Figure 5.8).

A noteworthy feature of eutectic systems is the low melting temperature at the eutectic point; it is lower than the melting points of either component. For this reason, the alloy containing 63 wt% Sn-37 wt% Pb is common electrical solder, while the 50 wt% Sn-50 wt% Pb alloy was used in plumbing. (Recently, the use of Pb in plumbing has been outlawed. Some lead-free solders include 95% tin and 5% antimony. The latter has a melting point very close to that of Sn.)

5.5.1 **The Formation of Precipitates**

The decrease in solubility with decreasing temperatures is utilized in precipitation strengthening of metallic alloys. For any Pb-Sn alloy containing less than 19.2 wt% Sn, the material is a solid solution (α phase) at 183°C. As the temperature is lowered, the composition C_O reaches the solubility limit (it crosses the solvus line) and β phase is precipitated. This is the case in Example 4.1, for 5% Sn at 25°C, as the microstructure sketch shows. These precipitates are obstacles to dislocation motion. We shall visit this

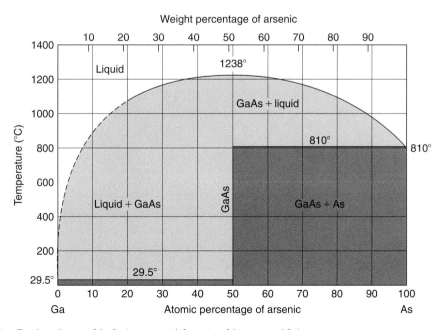

■ **FIGURE 5.9**   The phase diagram of the Ga-As system with formation of the compound GaAs.

concept again in Chapter 6 where we will examine the use of cooling rates to control the size of the precipitates.

5.6 **INTERMEDIATE COMPOUNDS AND INTERMEDIATE PHASES**

Consider the Ga-As phase diagram depicted in Figure 5.9. A dominant feature in the phase diagram is the central **compound** GaAs, containing equiatomic amounts of Ga and As. The Ga-As chemical bond is stronger than the Ga-Ga and the As-As bonds; therefore the alloy forms the compound GaAs. Compounds often have well-defined stoichiometries, in this case 50 at% Ga-50 at% As (at% stands for atom percent), and are denoted by vertical lines in phase diagrams. The strong bond between gallium and arsenic atoms is reflected in the melting temperature of the compound, which is higher than that of either component.

The interpretation of this phase diagram follows the rules we have examined before. Let us assume that we have a melt that is slightly rich in Ga, say 40 at% As and 60% Ga. Solidification will start around 1,230°C. At 1,200°C, the system contains solid GaAs and a liquid phase containing 32% As. There will be $(40 - 32)/(50 - 32) = 44.5\%$ GaAs and 55.5% liquid. At 30°C we have almost pure Ga and GaAs. The Lever Rule yields $(50 - 40)/(50 - 0) = 20\%$ liquid Ga and 80% solid GaAs. Below 29.5°C, which is the melting point of gallium, we find 20% solid gallium and 80% GaAs. On the As-rich side of the diagram, the analysis is similar; the arsenic solidifies at 810°C.

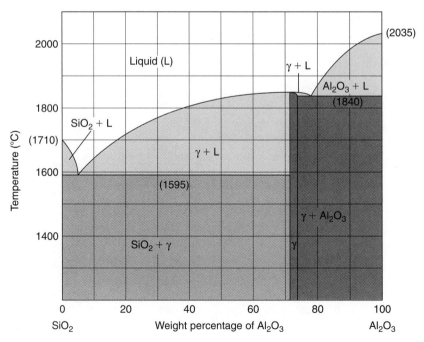

■ **FIGURE 5.10**   The SiO_2-Al_2O_3 phase diagram.

The GaAs phase diagram is important in semiconductor technology where compound diodes and lasers are fabricated by **liquid phase epitaxy**. Consider a Ga-rich melt containing 10 at% As that is cooled slowly just below 930°C so that it enters the two phase L + GaAs field. A small amount of solid GaAs separates from the melt. If a single crystal GaAs wafer is introduced into the melt, it provides a template for the rejected GaAs to deposit on. The deposited layer continues the crystal structure of the substrate. In the process, thin single-crystal epitaxial layers are grown.

The SiO_2-Al_2O_3 system of Figure 5.10 is instructive. The components are not elements, but the compounds SiO_2 (silica) and Al_2O_3 (alumina). These are considered components because neither substance decomposes at the temperatures shown. The system forms the intermediate compound Mullite with nominal composition $3Al_2O_3 \cdot 2SiO_2$. This compound, marked as γ phase, exists over a broader stoichiometric range than does GaAs, namely from 71 to 73 wt% alumina. Alloy mixtures of these oxides are used to make refractory bricks that line high temperature industrial furnaces.

Mullite can be considered as the end compound of the SiO_2-Mullite eutectic system to its left and of the Mullite-alumina eutectic system to its right. The analysis of the phase diagram on either side of Mullite is identical to that of a eutectic system. Remember that the tie lines in a phase analysis never extend beyond a given two phase region. The tie lines in the SiO_2-Mullite system (left half of the diagram) extend from pure SiO_2 to the γ phase; likewise, the tie lines of the Mullite-Al_2O_3 system extend from 73% alumina to pure alumina.

5.7 **PERITECTIC SOLIDIFICATION**

Not uncommon in the catalog of solidification sequences is the invariant **peritectic** reaction, which can be written in the generic form

$$L + \beta \rightarrow \alpha \tag{5.3}$$

Peritectic reactions often occur when there is a large difference in melting point between the two components. The liquid reacts with the higher-temperature solid to form a new solid phase. On the phase diagram the peritectic reaction appears as an upside down eutectic. There are few examples of alloy systems that exhibit only peritectic behavior over the complete temperature range. One is the platinum-rhenium system, whose phase diagram is shown in Figure 5.11. Above the peritectic temperature of 2,450°C, solid solution solidification (β) prevails in Re-rich alloys. Below this temperature down to the melting point of Pt, liquid exists with Pt-rich compositions, which also solidify as solid solutions α.

Only alloys with compositions between ~43 wt% Re and ~54 wt% Re participate in the peritectic reaction. For any alloy in this range a liquid containing ~43 wt% Re and a solid β containing ~54 wt% Re coexist just above 2,450°C. The relative amounts depend on the initial composition and can be calculated from the Lever Rule. Given sufficient time, the β phase will react with the liquid and produce the α phase as Equation (5.3) suggests. But, as the temperature drops, β will be rejected from α and both phases will persist down to lower temperatures. Unlike eutectics, peritectics have no distinguishing microstructural features.

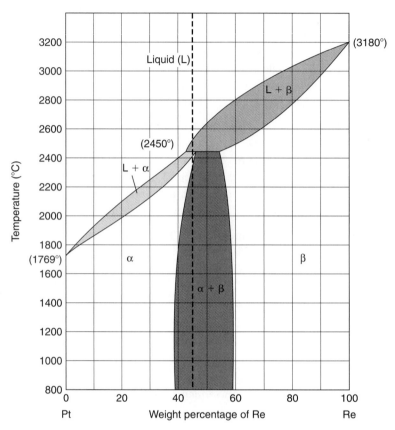

■ **FIGURE 5.11**   Peritectic phase diagram of Pt-Re.

The issue of equilibrium is an important one in peritectic solidification. Since peritectic reactions are slow, they do not readily proceed to completion unless maintained at high temperature for long times. During practical cooling rates the peritectic reaction may be bypassed, yielding nonequilibrium phases that may not appear on the phase diagram.

5.7.1 **The Cu-Zn Phase Diagram**

The Cu-Zn phase diagram shown in Figure 5.12 is an example of multiple intermediate phases. The α, β, γ, ε, and η phases are actually solid solutions with different crystal structures. The transition from liquid to solid takes place by peritectic reactions. The system includes several commercially important coppers and brasses. For example, α brass, or cartridge brass, contains 30 wt% zinc, while Muntz metal or β brass contains 40 wt% zinc.

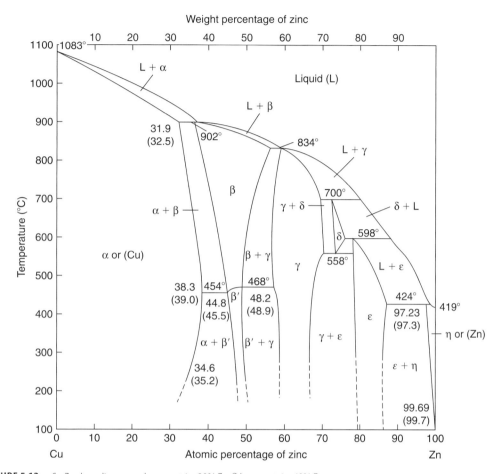

■ FIGURE 5.12 Cu–Zn phase diagram. α brass contains 30% Zn, β brass contains 40% Zn.

5.8 **THE IRON-CARBON SYSTEM AND STEELS**

The best known and arguably most important phase diagram in metallurgy is the iron-carbon diagram shown in Figure 5.13. From this diagram, much can be learned about plain carbon steels and cast irons, two of the most widely utilized classes of structural metals. Note that the diagram does not extend to pure carbon because all reactions of interest occur at concentrations below 6.67 wt% carbon, which corresponds to the compound Fe_3C. Thus the vertical line at the right corresponds to Fe_3C which is called cementite.

At the extreme left of the diagram, pure iron is polymorphic. It has the BCC crystal structure at temperatures up to 914°C; then it transforms to the FCC structure above 914°C and back to BCC at 1,394°C. This polymorphism is an extraordinary present nature has offered us: it is responsible for the existence

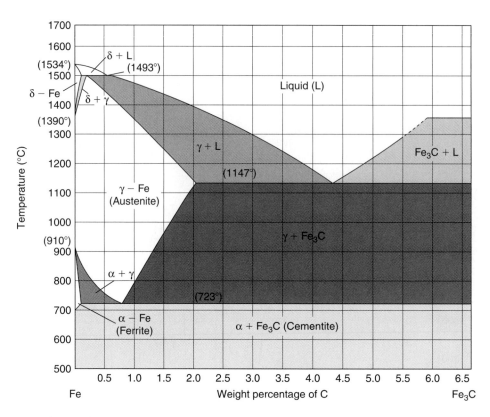

■ **FIGURE 5.13** The phase diagram of the iron-carbon system. The diagram is shown only up to 6.67% carbon, which corresponds to the compound Fe₃C. All colored areas are two-phase regions.

of steel, from the ductile steels used in buildings and bridges to the hard tool steels used in cutting tools and ball bearings.

The two crystal structures of iron are shown in Figure 5.14. FCC iron is shown at left; it can dissolve carbon in the interstices between the atoms. A dissolved carbon atom is shown in black in the figure. Iron in the BCC structure, shown on the right, does not have interstices large enough to accommodate carbon atoms. Thus FCC iron (the γ phase in Figure 5.13) can dissolve appreciable amounts of carbon (up to 2.14 wt% = 8 at% at 1,147°C); but the BCC structure (the α phase) cannot dissolve more than 0.022 wt% (= 0.08 at%) carbon at 723°C and practically none at room temperature. The α phase is called **ferrite**, and the γ phase is referred to as austenite.

When the austenite solid solution containing 0.76 wt% carbon is cooled below 727°C, it undergoes a **eutectoid** reaction into almost pure α iron (ferrite) and Fe₃C. The eutectoid that is formed is called **pearlite**; it consists of alternating layers of ferrite and Fe₃C and is shown in Figure 5.15. Note that ferrite, being pure α iron, is very soft and cementite is a hard ceramic. This combination provides steel

(A)

(B)

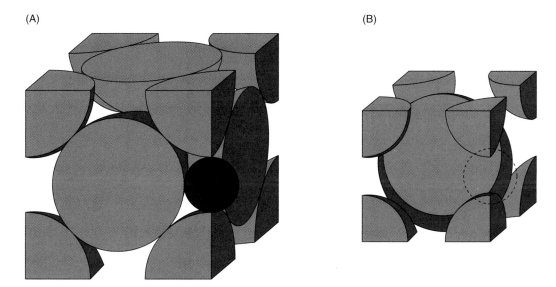

■ **FIGURE 5.14** The crystal structures of iron in steel. Left: the FCC structure with a dissolved carbon atom in black; it exists at $T > 914°C$ under the name of γ phase or austenite. Right: the low-temperature BCC structure. The latter does not have interstices large enough to accommodate carbon atoms. It forms the α phase or ferrite.

■ **FIGURE 5.15** The microstructure of pearlite. The dark lines are cementite layers, the bright areas represent ferrite. The eutectoid is called pearlite because under the microscope it shimmers like mother of pearl. Courtesy of G. F. Vander Voort, Buehler Ltd.

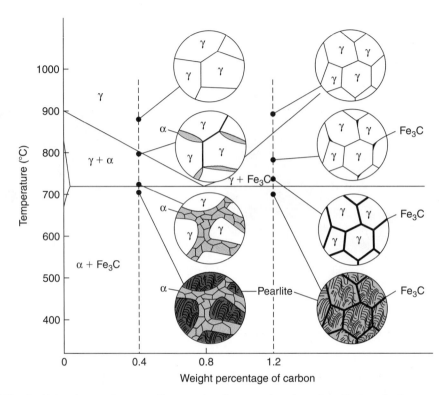

■ **FIGURE 5.16**   Equilibrium phase transformations of hypoeutectic and hypereutectic steels together with schematic microstructures.

with strength and ductility. The reaction and its product are called eutectoid instead of eutectic because a solid phase (austenite), not a liquid, is transformed into a two-phase system. (We also note in the phase diagram, a eutectic point at 4.3 wt% carbon and 1,147°C, but this is of minor technical importance.)

Let us analyze the phases produced from melts containing three different amounts of carbon, namely, 0.4, 0.76, and 1.2 wt% carbon. A look at Figure 5.13 reveals that, at 900°C, say, all three alloys are austenite solid solutions.

We start with the steel that contains 0.76 wt% C; this is a **eutectoid steel**. Cooling from the melt, it will form an austenitic solid solution all the way down to 727°C. At this temperature, it undergoes the eutectoid reaction and forms pearlite (Figure 5.15).

Now let us examine a **hypoeutectoid steel**, namely, one that contains less than 0.76 wt% C. For this purpose, we enlarge in Figure 5.16 the iron-carbon phase diagram in the vicinity of the eutectoid point (0.76 wt% C and 727°C).

At 900°C, the alloy containing 0.4 wt% carbon is austenite. The microstructure, sketched in a circle, consists of γ-phase grains. As the temperature passes the solvus line at 800°C, the γ phase is enriched in carbon and rejects **proeutectic α iron**. This segregation usually occurs at grain boundaries as sketched in the circle. As we cool the alloy to just above the eutectoid temperature, more α iron is rejected and the carbon concentration in the γ phase approaches 0.76 wt%. We can estimate the amount of proeutectoid α by means of the Lever Rule:

$$f_\alpha = (0.4 - 0.022)/(0.76 - 0.022) = 51\%$$

Now we cool to just below the eutectoid temperature; the proeutectoid α is not affected, but the remaining austenite (γ) undergoes the eutectoid transformation into pearlite (which is α + Fe_3C). There is now 51% proeutectic α and 49% pearlite. The microstructure of the steel is sketched in the circle. Since the α iron is soft, hypeutectoid steels are softer and more ductile than eutectoid steels.

The **hypereutectoid steel** at 1.2% C is analyzed in the same fashion. At 900°C, it is austenite (single phase solid solution). As we cool to 800°C, the solubility limit is exceeded, and cementite (Fe_3C) is precipitated to the grain boundaries; the amount of the latter increases as we cool to, say, 728°C. The Lever Rule can again compute the amount of proeutectoid Fe_3C (7.5%). When we cool through the eutectoid point, the remaining austenite (containing 0.76 wt% C) undergoes the eutectoid transformation and becomes pearlite. The microstructure is sketched in Figure 5.16.

With all three materials, cooling from 722°C to room temperature does not change the microstructure, but the ferrite rejects the little carbon it contained and becomes pure BCC iron. Since cementite is a hard material, hypereutectoid steels are harder and less ductile than pearlite. We see that, in steels, the higher the carbon concentration, the harder the material.

This, however, is not the whole story. So far, we have examined **equilibrium** phase diagrams, namely, materials that are cooled very slowly. All three steels we have examined can be made much harder by rapid cooling. These will be examined in the next chapter where we will study the kinetics of phase transformations.

■ GLOSSARY

A **phase** is a homogeneous, physically distinct, portion of matter that is present in a nonhomogeneous system. It may be a single component or a mixture.

A **component** is an ingredient of a chemical system.

The **composition of a phase** (C) is the relative amount of the components it contains.

The **amount of a phase** (f), is the fraction of the mixture that is in the particular phase.

In a **solid solution**, the components occupy random positions in the crystal; they can be substitutional or interstitial.

The **liquidus line** represents the lowest temperatures of a liquid as a function of composition. It forms the boundary between the liquid phase and the two phase system.

The **solidus line** represents the highest temperature of solids as a function of composition. It is the border between the solid phase and the two phase region.

The **solvus line** represents the limit of solubility on one component in the other. It is generally the limit between a solid solution phase and a two phase eutectic region.

Precipitates are formed when the concentration of a component exceeds the solubility limit.

A **eutectic** is an alloy of immiscible components. At the **eutectic point** the liquid transforms into the **eutectic**, which consists of two distinct phases arranged in alternating layers.

A **proeutectic** phase is formed above the eutectic temperature.

A **eutectoid** reaction is similar to a eutectic reaction; the difference is that a solid phase (instead of a liquid) reacts to produce two phases. The result is called a **eutectoid**.

A **eutectoid steel** contains the eutectoid concentration (0.76 wt%) of carbon.

A **hypoeutectoid steel** contains less than 0.76 wt% carbon.

A **hypereutectoid steel** contains more than 0.76% carbon.

Austenite is the FCC solid solution in steel; its other name is γ **phase**.

Ferrite is the BCC form of iron in steel. It contains very little carbon. It is the α **phase**.

Cementite is the compound Fe_3C containing 6.67 wt% carbon in steel.

Pearlite is the eutectoid phase composed of ferrite and cementite in steel.

■ SUMMARY

1. Phase diagrams show the phases one obtains in an alloy system for every composition and temperature. They represent the thermodynamic equilibrium situation which may require long times to be reached, especially at the lower temperatures.

2. Binary alloys can form solid solutions, eutectics, or compounds, depending on the relative strengths of the bonds between the atoms.

3. In alloys, a two-phase region separates the liquid phase from the solid in the phase diagram. In this region, the liquid and the solid have different compositions. The composition of the liquid is given by the intersection of the liquidus curve, with a isothermal tie line representing the temperature of interest; the composition of the solid is the intersection of the tie line with the solidus curve.

4. The amounts of phases can be computed by the Lever Rule. The amount of solid is the ratio of the length of tie line between the overall composition and the liquidus curve to the length of the tie line between the liquidus and solidus. The amount of liquid phase is the ratio of the line between the solidus and the overall composition divided by the length between liquidus and solidus. (Note: for the liquid phase, go to the solidus line and vice versa!).

5. Solid solutions are one phase systems in which both constituents are randomly distributed. They occur over the whole composition range for solid solution solids; they occur at both ends of the diagram for eutectics.

6. Eutectics form when the two components are poorly soluble in each other. At the eutectic point (a specific composition and temperature), the liquid separates into two solid phases (which form the eutectic). Solids formed above the eutectic point are called proeutectic.

7. Compounds form when the A-B bond is stronger than the A-A or B-B bond. They can appear as the end phases of eutectic regions.

8. Steel is an alloy of iron with 0.2 to 1.2 wt% carbon.

9. The phase diagram of steel has a eutectoid point at which the high-temperature FCC solid solution (austenite) decomposes into a eutectoid (pearlite) formed of BCC iron (ferrite) and the compound Fe_3C (cementite).

10. At hypoeutectoid compositions, a proeutectoid ferrite forms in addition to the pearlite. Such steels are softer than eutectoid steels.

11. At hypereutectoid compositions, a proeutectoid cementite forms in addition to the pearlite, resulting in a harder steel than pearlite.

■ KEY TERMS

■ REFERENCES FOR FURTHER READING

[1]   ASM Handbook, Vol. 3. *Phase Diagrams*. ASM International, Materials Park, OH (2000).

[2]   M. Hillert, *Phase Diagrams and Phase Transformations: Their Thermodynamic Basis*, 2nd ed., Cambidge University Press, Cambridge, UK (2007).

[3]   P. Haasen, B.L. Mordike, *Physical Metallurgy (Paperback)*, 3rd ed., Cambridge University Press, Cambridge, UK (1996).

■ PROBLEMS AND QUESTIONS

5.1.   A solder manufacturer wishes to make a batch of solder having the eutectic composition. On hand is a supply of 250 kg of electrical solder scrap containing 60 wt% Sn-40 wt% Pb, and 1,250 kg of plumbing solder scrap containing 60 wt% Pb-40 wt% Sn. If all of the scrap is to be melted, how much of which metal must be added to achieve the desired 61.9 wt% Sn composition?

5.2.   Gallium arsenide crystals are grown from Ga rich melts. Can you mention one reason why As-rich melts are not used for this purpose?

5.3.   Do a complete phase analysis (phases present, their amounts, and their compositions) for Cu-Ni with 70% copper at 1,300°C, 1,200°C, 1,100°C, and 22°C.

5.4.   Do a complete phase analysis (phases present, their amounts, and their compositions) for the lead-tin alloy:

   a. With 10 wt% Sn at 350°C, 275°C, 225°C, 150°C, and 50°C.

   b. With 30 wt% Sn at 350°C, 225°C, 184°C, 182°C, and 50°C.

   c. With 61.9 wt% Sn at 350°C, 184°C, 182°C, and 50°C.

   d. With 90 wt% Sn at 250°C, 200°C, and 50°C.

5.5.   Do a complete phase analysis (name of phases present, their amounts, and their compositions) for steel at:

   a. 0.5 wt% C, 1,600°C, 1,300°C, 724°C, 700°C, and 22°C.

   b. 0.76 wt% C, 1,600°C, 1,300°C, 724°C, 700°C, and 22°C.

   c. 1.5 wt% C, 1,600°C, 1,400°C, 1,100°C, 800°C, 724°C, and 22°C.

5.6.   A 90 wt% Sn-10 wt% Pb alloy is cooled from 300°C to 0°C.

   a. What is the composition of the alloy in terms of atomic or molar percent?

   b. Draw a cooling curve for this alloy.

   c. In what ways does the equilibrium microstructure of this alloy differ from that of the 90 wt% Pb-10 wt% Sn alloy?

5.7.   The Al_2O_3-Cr_2O_3 phase diagram resembles that of Cu-Ni. Al_2O_3 melts at 2,040°C and Cr_2O_3 melts at 2,275°C.

   a. Sketch the phase diagram.

   b. How many components are there in this system?

   c. It is desired to grow a ruby single crystal with the composition 22 wt% Cr_2O_3. On your phase diagram, indicate the melt temperature and temperature for growth.

5.8.   Suppose you have a large quantity of silver amalgam (Ag-Hg alloy) scrap. The alloy is liquid at low temperatures and has much less value than the isolated pure metals. Design a physical process to separate these metals.

5.9.   In the Ga-As binary system, perform total chemical composition and phase amount analyses for an 80 at% Ga-20 at% As alloy at 1,200°C, 1,000°C, 200°C, and 29°C. Why is it preferable to use a slightly Ga-rich alloy compared to using an As-rich alloy when making GaAs crystals?

5.10. It is desired to pull an alloy single crystal of composition 78 at% Ge-22 at% Si from a binary Ge-Si melt (Figure 5.17).

    a. At what temperature should the crystal be pulled?

    b. What melt composition would you recommend?

    c. As the crystal is pulled, what must be done to ensure that its stoichiometry is maintained constant?

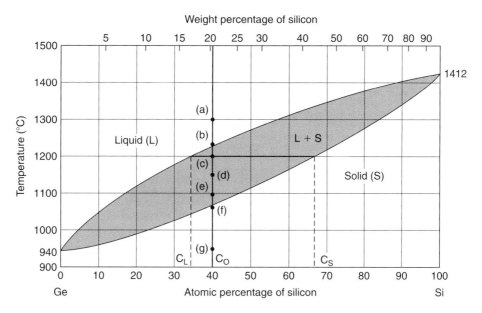

■ **FIGURE 5.17** The Ge-Si phase diagram.

5.11. Consider the 50 wt% Pt-50 wt% Re alloy.

    a. This alloy is heated to 2,800°C and cooled to 800°C. Sketch an equilibrium cooling curve for this alloy.

    b. Perform complete equilibrium chemical and physical phase analyses at 2,800°C, 2,452°C, 2,448°C, and 1,000°C.

5.12. Make enlarged sketches of the phase regions that surround the two highest isotherms in the Cu-Zn phase diagram (Figure 5.12).

    a. What is the name of the solidification behavior displayed in both cases?

    b. Which line(s) on this phase diagram represent the solubility of Zn in the α phase?

    c. Which line(s) on this phase diagram represent the solubility of Zn in the β phase?

    d. Which line(s) on this phase diagram represent the solubility of Cu in the γ phase?

    e. Is it possible for the solubility of a component to drop as the temperature is raised? If possible, give an example in this phase diagram.

5.13. Perform quantitative phase analyses at temperatures (C and f) corresponding to the four states indicated for the 0.4% C and 1.2% C steels in the $Fe-Fe_3C$ system of Figure 5.13. Do your analyses correspond to the sketched microstructures?

5.14. Up until the Middle Ages, iron was made by reducing iron ore (e.g., Fe_2O_3) in a bed of charcoal (carbon). Steel objects were shaped by hammering the resulting **solid** sponge iron that was reheated to elevated temperatures using the same fuel. As the demand for iron and steel increased, furnaces were made taller and larger. They contained more iron ore and charcoal that were heated for longer times. An interesting and extremely important thing occurred then! **Liquid** iron was produced. Why? (Molten iron issuing from the bottom of the furnace could be readily cast, a fact that helped to usher in the Machine Age.)

5.15. Distinguish between intermediate phases and compounds.

5.16. a. A 40 wt% Al_2O_3-60 wt % SiO_2 melt is slowly cooled to 1,700°C. What is the proeutectic phase, and how much of it is there?

   b. Furnace bricks are made from this composition. Is the melting point of these bricks higher or lower than that of pure silica bricks?

   c. What is the highest melting temperature attainable for bricks containing both SiO_2 and Al_2O_3?

5.17. A 10 g sphere is made of steel containing 1.1 wt% C. Suppose the ball is austenized and slowly cooled to room temperature.

   a. What is the weight of pearlite present?

   b. What is the total weight of ferrite present?

   c. What is the total weight of cementite present?

   d. What is the weight of cementite present in the pearlite?

   e. What is the weight of cementite present as the proeutectoid phase?

5.18. What information is contained in a phase diagram? Does it represent material obtained after fast cooling?

5.19. Sketch the phase diagram of a hypothetical solid solution. Indicate on it the solidus line, the liquidus line, the solvus lines, and the eutectic point. Omit the ones that are not relevant, if any.

5.20. What kind of phase diagram is relevant for an alloy of which the bond between different atoms is much weaker than the bonds between same atoms?

5.21. What type of alloy does one make when a low melting point is desirable (as in solder)?

5.22. Sketch the phase diagram of a hypothetical eutectic alloy. Indicate on it the solidus line, the liquidus line, the solvus lines, and the eutectic point. Omit the ones that are not relevant, if any.

5.23. Indicate the relative bond strengths in an A-B alloy for the different types of phase diagrams. Pair the correct letters and numbers.

Bond strengths

A. (A-A) < (A-B) < (B-B)
B. (A-A) and (B-B) > (A-B)
C. (A-A) and (B-B) < (A-B)

Alloy type

1. Eutectic.
2. Solid solution
3. Intermediate compound.

5.24. Describe the thinking in the derivation of the Lever Rule.

5.25. How many phases are present in a solid solution at room temperature?

5.26. What is a proeutectic alloy? What is a hypoeutectoid steel?

5.27. What defines a steel?

5.28. What is austenite? Ferrite? Cementite? Pearlite?

5.29. What type of alloy is formed when the two components are poorly soluble in each other?

5.30. Describe the benefits of the polymorphism of iron.

5.31. What is a hypereutectoid steel?

5.32. How do the mechanical properties of steel vary with the amount of carbon it contains?

Chapter 6

Reaction Kinetics and the Thermal Processing of Metals

In this chapter we explore what materials can be obtained when processing is too rapid for thermodynamic equilibrium. Old blacksmiths knew that steel can be made hard by quenching the hot steel in water. We first learn how nature proceeds to form a new phase (e.g., how it solidifies from the melt): very small nuclei of the new phase form and then grow by diffusion of atoms. Knowing the temperature-dependence of nucleation and diffusion rates, we establish the technologies of steel making and precipitation-hardening.

LEARNING OBJECTIVES

After studying this chapter, the student will be able to:

1. Describe martensite and how it is formed by rapid cooling.

2. Describe the kinetics of phase transitions at various temperatures in terms of nucleation and diffusion.

3. Use the temperature-dependence of reaction rates in the processing of materials.

4. Explain the formation of pearlite and martensite in terms of reaction kinetics.

5. Use the time-temperature-transformation (TTT) diagram of steel to prescribe the proper treatment for obtaining desirable mechanical properties of steel.

6. Describe how the mechanical properties of plain carbon steels depend on composition and heat treatment.

7. Define hardenability and describe how to increase it.

8. Explain the effect of alloying elements on the crystal structure of iron and its consequences for hardenability and the different forms of stainless steel.

9. Distinguish wrought and heat treatable aluminum and describe their mechanical properties.

10. Design the proper ageing process for aluminum.

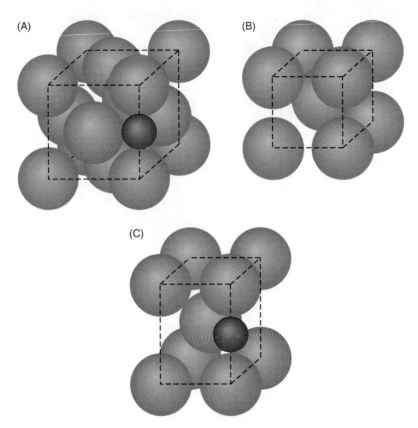

■ **FIGURE 6.1** (A) FCC iron with interstitial carbon atom (austenite). (B) BCC iron (ferrite); its interstices are too small to accommodate carbon atoms. (C) BCT (distorted BCC) cell with "frozen-in" carbon atom (martensite).

6.1 QUENCHED AND TEMPERED STEEL

It is generally known that very hard steel can be obtained by quenching hot steel in water or in oil. In Chapter 5, we saw that slow cooling of steel forms a ductile and relatively soft pearlite by a eutectoid reaction at $727\,°C$ because the low-temperature BCC structure of iron does not have interstices large enough to accommodate the carbon (Figure 6.1A and B). The eutectoid reaction takes time because the carbon atoms dissolved in austenite must travel in order to form Fe_3C particles. When austenitic steel is cooled rapidly by quenching in water or in oil, the carbon atoms do not have time to move into cementite lamellae but are **"frozen" in place**. Because of the presence of carbon, the low-temperature structure is distorted into a **body centered tetragonal** (BCT) structure (Figure 6.1C); this new material is **martensite**. Carbon can also be inserted on the x- or the y-axis and produce deformation in those directions. The martensitic structure consists of very thin needles that are distorted in different directions (Figure 6.2).

This structure is so stressed and has so many interfaces that it is practically impossible for a dislocation to move through it. The structure is very hard and so brittle that it is of limited use. One recovers

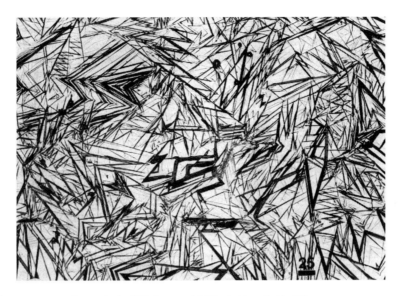

■ **FIGURE 6.2**   The microstructure of martensite (×200). Courtesy of G. F. Vander Voort. Buehler LTD.

a measure of ductility but gives up some hardness by **tempering** the martensite, that is, by reheating at temperatures below 727°C. Tempered martensite is the material of choice for cutting tools, ball bearings, and other applications requiring a very hard metal.

This example shows us a new method for producing materials: by adjusting the heating or cooling rates we can obtain materials that are not in thermal equilibrium and present interesting novel properties. In steel making, the question arises: "How fast does one have to cool the steel to obtain martensite and what materials does one obtain with different cooling rates?" In order to answer this question and design effective processing methods, it is necessary to understand how a phase transformation proceeds with time. This is the subject of phase transformation kinetics.

6.2 THE KINETICS OF PHASE TRANSFORMATIONS

When the temperature of steel decreases below the eutectoid point, the transformation from austenite to pearlite proceeds in two steps. In the first step, carbon atoms expelled from the austenite coalesce into very small particles of cementite called **nuclei**. Carbon atoms then **diffuse** through the steel toward these nuclei, which **grow** in size. The transformation thus occurs through **nucleation and growth of the new phase**. If nucleation is rapid and diffusion slow, one obtains a large number of small cementite particles; when nucleation is slow and diffusion rapid, one obtains a coarse pearlite that consists of a smaller number of larger cementite particles. If cooling is so rapid that neither nucleation nor diffusion has time to proceed, the steel becomes martensite. The term **martensitic transformation** is used for any phase transformation in which no diffusion takes place. We note that martensite is not included in the iron-carbon phase diagram (Figure 5.13) because it is not an equilibrium phase.

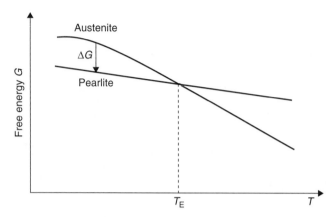

■ FIGURE 6.3   Free energy change in a phase transition. The phases coexist at T_E.

6.2.1 **Nucleation**

Why does a phase transformation occur? It does so because the two phases have different free energy $G = U - TS$ as illustrated in Figure 6.3. The free energy of austenite decreases more rapidly with increasing temperature because the disordered solid solution has a higher entropy S than the separated pure iron and Fe_3C; therefore the two curves cross at the temperature T_E. Above T_E, austenite has the lower free energy G and is stable. Below T_E the free energy G of pearlite is lower than that of austenite; the drop in free energy ΔG is the driving force for the transition from austenite to pearlite. Obviously, T_E is the eutectoid temperature. At $T < T_E$, carbon atoms move around and coalesce into very small particles of Fe_3C called **embryos**.

As Figure 6.3 shows, the driving force for the transition increases linearly as the temperature drops below T_E.

$$\Delta G_{A \to p} = -A(T_E - T) \tag{6.1}$$

The total energy gain ΔG is proportional to the volume V of the embryo. If we assume that the embryos are spheres with radius r and ΔG_V is the change in free energy per unit volume

$$\Delta G = -V \times |\Delta G_V| = -4\pi/3 r^3 \times |\Delta G_V| \tag{6.2}$$

(Note that ΔG_V is negative; we use the absolute value $|\Delta G_V|$ in order to remove any doubt that ΔG is a decrease in energy and therefore a driving force for the phase transformation.)

Each embryo has an interface with the original material, and this interface costs surface energy

$$\Delta E_\gamma = 4\pi r^2 \gamma \tag{6.3}$$

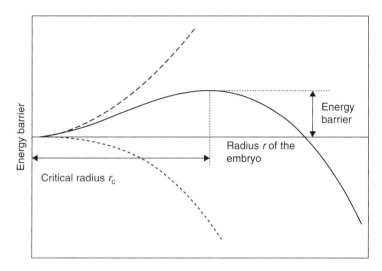

■ FIGURE 6.4   Energy barrier to nucleation. Dotted line, free energy change, Equation (6.1); dashed line, surface energy, Equation (6.2); solid line, total energy barrier, Equation (6.3).

which is proportional to the area $4\pi r^2$ of the interface. The specific surface energy γ is on the order of 0.1 to 1 J/m². ΔE_γ is an increase in energy, and therefore an obstacle to the transformation.

The total change in energy is the sum of the change in free energy and the surface energy:

$$\Delta E = -4\pi/3 r^3|\Delta G_V| + 4\pi r^2 \gamma \tag{6.4}$$

This is shown in Figure 6.4. The volume increases or decreases faster than the area when the radius increases or decreases. (When the radius doubles, the volume increases by a factor of eight and the area by a factor of four.) At a low radius of the new particle, the surface energy is larger than the bulk energy, but the bulk energy wins out when the particle grows beyond a certain radius r_c.

The energy barrier has a maximum ΔE_N at r_c. This maximum is an energy barrier for nucleation. Embryos with a radius smaller than r_c form spontaneously but lower their energy by dissolving again. Embryos with $r > r_c$ lower their total energy by growing: these are the nuclei for the phase transformation. r_c is the **critical radius** of the nucleus; it is obtained when the derivative of Equation (6.4) is zero:

$$r_c = 2\gamma/|\Delta G_V| \tag{6.5}$$

By inserting r_c into Equation (6.4) we find that the energy barrier for nucleation is

$$\Delta E_N = \frac{16\pi\gamma^3}{3(\Delta G_V)^2} \tag{6.6}$$

The carbon and iron atoms in the solid all possess thermal vibration energy. Of all the embryos that are constantly formed at $T < T_E$, only the fraction that possesses a thermal energy larger than ΔE_N can overcome the barrier and grow. According to Boltzmann statistics, this fraction is

$$f = \exp\left(-\frac{E_N}{kT}\right) \tag{6.7}$$

Thus, the lower the energy barrier, the higher the nucleation rate. $\Delta G_V = 0$, the barrier ΔE_N is infinite and no transformation takes place at the transition temperature T_E. According to Equations (6.6) and (6.1), the energy barrier decreases and the nucleation rate increases as the temperature drops below T_E; the material must be **undercooled** ($T < T_E$) for nucleation of the new phase to start. The larger the temperature difference ($T_E - T$), the higher the nucleation rate.

There is an alternative method for increasing the nucleation rate: it consists in decreasing the interfacial energy γ. This will be discussed in Chapter 7 where the use of grain refiners will be described as a method for the control of grain size.

The above discussion was focused on the formation of pearlite, but the concepts have general validity for all phase transitions such as melting, solidification, precipitation-hardening, and so on.

6.2.2 Diffusion Rates

After nucleation, the phase transformation proceeds by diffusion. In the formation of pearlite, carbon atoms diffuse toward the cementite nuclei, which grow in the process. The movement of an atom from one equilibrium position to the next requires a certain amount of energy, which constitutes the **activation energy Q_D for diffusion**.

The probability P that an atom will acquire the required energy Q_D follows the Boltzmann statistics,

$$P(Q_D) = \exp\left(-\frac{Q_D}{RT}\right) \tag{6.8}$$

The diffusion rate $D(T)$ follows the same law:

$$D(T) = D_O \exp\left(-\frac{Q_D}{RT}\right) \tag{6.9}$$

The value for the diffusion coefficient of carbon in α iron is

$$D_\alpha(C) = 0.004 \exp(-19.2 \text{ kcal}/RT) \text{ cm}^2/\text{s}$$

In γ iron the diffusion coefficient of carbon is

$$D_\gamma(C) = 0.12 \exp(-32.0 \text{ kcal}/RT) \text{ cm}^2/\text{s}$$

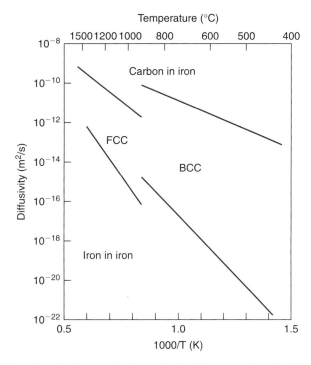

■ **FIGURE 6.5**   Diffusion coefficient of iron and carbon atoms in iron as a function of temperature. Note that the inverse temperature $1/T$ is plotted on the x-scale. High temperatures are on the left as shown in the upper scale.

Because of the exponential expression in Equation (6.9), the diffusion rate is a rapidly increasing function of temperature that is commonly represented in an **Arrhenius plot** as shown in Figure 6.5. Note the logarithmic scale for the diffusion coefficients. The diffusion rate of carbon in α iron decreases by about a factor of 100 as the temperature decreases from 727°C to 400°C. The growth rate of the particles of new phase thus decreases rapidly as the temperature decreases.

> The Arrhenius plot and equations similar to Equation (6.9) describe the temperature dependence of all thermally activated processes, in which thermal agitation provides the energy to overcome an energy barrier. It applies, among others, to diffusion, viscous flow, evaporation, and chemical reactions. The rate of these processes increases rapidly with temperature.

When a phase transformation occurs by cooling, the nucleation rate is high and diffusion is slow at low temperatures. The nucleation rate is low and diffusion is rapid closer to the transition temperature. Rapid cooling results in fine microstructure; slow cooling produces a coarse microstructure.

By combining these two concepts, we can analyze the formation of pearlite or martensite in the processing of steel. This is represented in the TTT or time-temperature-transformation diagram of Figure 6.6.

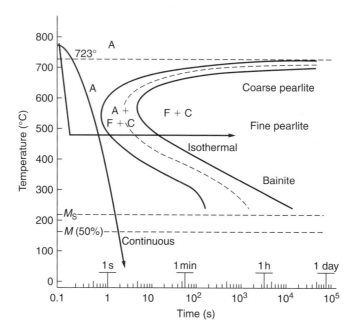

■ FIGURE 6.6   Isothermal TTT diagram for the transformation of austenite to pearlite and martensite.

6.3 **THERMAL PROCESSING OF STEEL**

6.3.1 **The TTT Diagram (Time-Temperature-Transformation Curves)**

The TTT diagram, Figure 6.6, shows how much time is necessary for the transformation of austenite to pearlite at any given constant temperature. Figure 6.6 is valid for plain carbon eutectoid steel (containing 0.76 wt% C), a model alloy because there are no proeutectoid phases (αFe or Fe$_3$C) to contend with and the austenite transforms only to pearlite.

It is obtained in the following way. A large number of thin specimens, each the size of a dime, is first austenitized at a temperature of, say, 885°C. Then one by one they are quickly brought to a chosen transformation temperature T_t below the eutectoid temperature of 727°C. Suppose T_t is 400°C. The first specimen is kept at 400°C for 1 s and then rapidly quenched in cold water. For the second specimen the time at 400°C might be 4 s prior to quenching. This procedure is repeated for the remaining specimens with times suitably chosen to track the transformation of γFe to αFe + Fe$_3$C until it is certain that austenite is no longer present. Now another T_t is chosen, for example, 700°C, and a new set of austenitized specimens are sequentially transformed **isothermally** and quenched. This procedure is repeated until a sufficient number of temperature-time treatments have been carried out. All of the specimens are then prepared for optical microscopy in order to estimate the relative amounts of phases present. In the days prior to computerized image analysis, one can imagine how tedious this chore was.

The starting time is arbitrarily associated with the appearance of 1% pearlite and the finishing time with 99% pearlite. After plotting the pair of times for each transformation temperature, the so called **time-temperature-transformation** or **TTT** curve displayed in Figure 6.6 emerges.

Why it takes a long time to form pearlite at both high and low temperatures, but a short time in between, is no longer a mystery. Close to the transition temperature, the time to nucleate the transformation is long because ΔG is small and ΔE_N is large. At 700°C, about 2 min are needed to start the reaction. At 660°C, nucleation is faster and the reaction starts after 10 s and is completed in about 100 s. At 500°C, the reaction starts in 1 s and is complete in 8 s. These increasing rates reflect the increase in nucleation rate. As the temperature is lowered further, the reaction is slower because the diffusion rate decreases rapidly. At 280°C, the reaction takes 2 min to start and 1 hr to complete.

The microstructure of the transformed steel reflects the temperature dependence of nucleation and diffusion. At high temperatures, nucleation is slow and diffusion is fast: few nuclei are formed but the cementite plates grow rapidly: the result is a coarse pearlite formed of relatively thick platelets. At lower temperatures, where nucleation is fast but growth is slow, one obtains a fine pearlite formed of closely spaced fine platelets of ferrite and cementite. Below 550°C, diffusion is so slow and nucleation so rapid that the steel is no longer composed of cementite plates but contains needles and plates of cementite so fine and spaced less than one micrometer so that they can only be observed in the electron microscope. This material is strong and tough and bears the name of **Bainite**.

When the material is quenched so rapidly that it avoids the nose of the TTT diagram, one obtains **martensite**. The martensitic transformation starts at 220°C and is 90% complete at 130°C as shown in Figure 6.6.

6.3.2 **Continuous Cooling Transformation (CCT)**

Aside from a few special treatments, steel is rarely transformed **isothermally**. Rather, it is cooled **continuously** from austenitizing temperatures to room temperature and transforms incrementally in a complex way dependent on the cooling rate. We would like to map the kinetics of austenite decomposition along continuous cooling curves as we did previously along the horizontal isotherms but it is more difficult to generate a **continuous cooling transformation** (CCT) curve than a TTT curve. Nevertheless, it is found that transformation during continuous cooling is delayed. This displaces the CCT curve down and to the right of the TTT curve as shown in Figure 6.7 where the two behaviors are superimposed. The left curve indicates the beginning of the transformation and the right curve its completion as before. The novel feature is the line joining the two just below 500°C. At that temperature transformation to pearlite stops; the remaining austenite will transform to martensite when the relevant temperatures ($T < 220$°C) are reached. No Bainite is produced in continuous cooling of plain carbon steels because all the austenite has been transformed to pearlite at higher temperatures.

According to the figure, a cooling rate lower than 5°C/s produces coarse pearlite; a cooling rate between 5°C/s and 40°C/s results in fine pearlite. More rapid cooling results in a fine pearlite and slower cooling in coarse pearlite. Cooling rates between 40°C/s and 140°C/s generate partial transformation to pearlite and transformation to martensite for the remainder, and more rapid cooling than 140°C results in martensite.

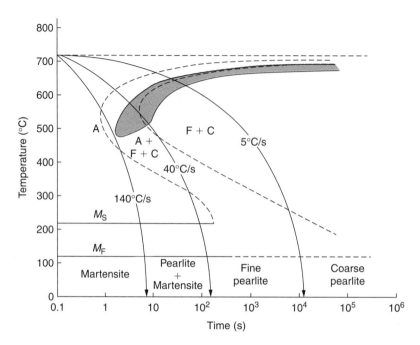

■ **FIGURE 6.7**   TTT (segmented) and CCT (shaded) curves for eutectoid steel.

In practice, martensite is fabricated by quenching the austenite in water or in oil. Water quenching results in a harder and more brittle steel but in large residual stresses that can cause fracture of the piece if the steel is not tempered (see below). Oil-quenched martensite is softer and less brittle, and it also contains smaller internal stresses.

6.3.3 Hardenability of Steel

When a piece of steel is quenched in oil or water, only a relatively thin layer below the surface is cooled rapidly enough to form martensite. **Hardenability** is a measure of the **thickness of steel that can be hardened by quenching**. (It is not the hardness.) The Jominy end-quench test rates the hardenability of steels. All that is needed is a fixture with a hole in it to support a hot test bar and a source of water placed a fixed distance beneath it (Figure 6.8A). A round Jominy bar of standard dimensions is austenized at elevated temperature and then quickly transferred to the fixture where it is end quenched in a specified way. After cooling, a flat is ground on the bar length, and hardness measurements are recorded along it at regular intervals. Jominy test results for a number of steels are plotted in Figure 6.8B, with approximate cooling rates indicated on the upper abscissa.

Alloying has a dramatic impact on the response of steel to heat treatment and particularly of their hardenability. It does this by altering the shape and relative position of the TTT curve on the time axis. The most important alloying element in steel is carbon, which shifts the TTT curve to the right. With

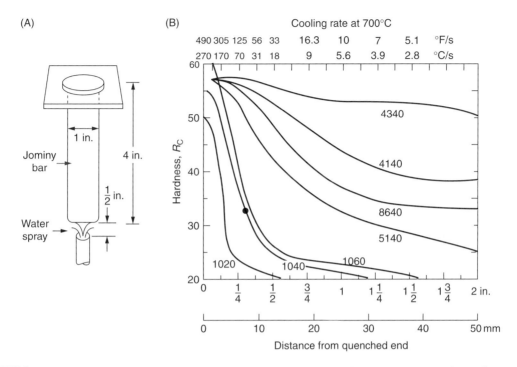

■ **FIGURE 6.8** Jominy test of hardenability. (A) The test: a water spray quenches the extremity of a steel bar that was heated to transform it to austenite. (B) Hardness as a function of distance from the quenched end for various alloy steels. (1020, 1040, and 1060 are plain carbon steels containing 0.2, 0.4, and 0.6 wt% C).

plain carbon steels that contain less than about 0.2 wt% C, parts thicker than a knife blade cannot be cooled fast enough to miss the TTT curve which is shifted far to the left. At higher carbon levels the hardness of the martensite and the ease with which it forms increase.

Other common alloying elements in steel, for example, Mn, Ni, Si, Cr., Mo, V, and W, also cause the TTT curve to translate to the right. In differing degrees and for different reasons these elements slow down the austenite to pearlite transformation. They are homogeneously distributed on substitutional sites within the austenite, but when this phase is no longer stable they partition either to the ferrite or the carbide phase. Manganese, nickel, and silicon segregate to the ferrite; the other elements listed tend to form carbides. Pearlite formation relies on solid state diffusion processes in the austenite to redistribute these atoms to the appropriate phase boundaries. But compared to carbon which diffuses interstitially, and hence rapidly, these substitutional atoms diffuse more slowly through the lattice, thus decelerating the overall transformation rate. Boron has an interesting effect in inhibiting the austenite to pearlite transformation that should be mentioned. Very small B additions have a profound effect in displacing the TTT curve to the right, a fact capitalized upon to conserve strategic metals. Apparently boron segregates to and stabilizes the prior austenite grain boundaries, making them less accommodating sites for pearlite nucleation.

The benefits of alloying are apparent in the hardenability curves of Figure 6.8B. Relative to eutectoid steel, the TTT curve for 4340 steel is shifted toward longer times. The critical cooling rate required to miss the knee of the curve is considerably reduced, for example, by more than a factor of 10 relative to carbon steels. This means that it is possible to produce a martensitic structure even if the cooling rate is quite low and less severe quenches can be used to produce martensite. What may be only attainable by a water quench in carbon steel could be easily realized by oil quenching an alloy steel.

These same concepts have wider application: in solidification from the melt, rapid cooling produces a fine grain structure and slow cooling a coarse structure. Extremely fast cooling prevents crystallization and produces amorphous solids. In metals, the cooling rates necessary for obtaining amorphous materials are extremely high (10^6 deg/s). Metglass can only be produced as thin foils by quenching a thin flow of liquid metal between two water-cooled copper rolls. Very hard amorphous metal films are generated on the surface of valve seats in automobile engines by a short and intense pulse of laser heat: a thin film on the surface is melted and cools very rapidly. In glass, long times are necessary for crystallization; normal processing results in amorphous glass. The latter can be crystallized by prolonged heating.

6.3.4 Standard Types of Steel

Let us now examine the types of steel that are obtained by the various heat treatment schedules.

Annealed steel is obtained by furnace cooling from the austenite phase. It is a relatively soft and very ductile material that consists of pearlite and proeutectic ferrite.

Rolled steel is an annealed or air-cooled pearlite with proeutectic phases that has been strain hardened by rolling.

Tempered steel is tempered martensite. When the steel is quenched from the austenite phase, the resulting martensite is so brittle that it is unfit for use. The steel is further processed by **tempering** where the martensite is reheated to temperatures below the eutectoid temperature (i.e., to 400°C or 600°C). At these temperatures residual stresses are relieved and carbon diffuses to form fine rods of cementite Fe_3C. Upon heating to temperatures near 700°C, the cementite particles assume a spherical shape forming a steel called **spheroidite**. This transformation decreases the hardness of the material somewhat but restores the necessary ductility.

The mechanical properties of these steels are given in Table 7.1.

6.4 HEAT TREATMENT OF ALUMINUM HARDENING BY PRECIPITATION

Aluminum is a widely used metal because of its low density and its resistance to corrosion. In its pure and annealed state, it has a low yield strength of 24 MPa. This strength can be increased to about 160 MPa by cold working and to 345 MPa by precipitation-hardening. In order of form precipitates, the aluminum must be alloyed with an element that is soluble at high temperatures and precipitates at

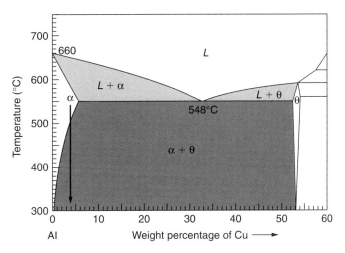

■ **FIGURE 6.9**   Phase diagram of the Al–Cu system.

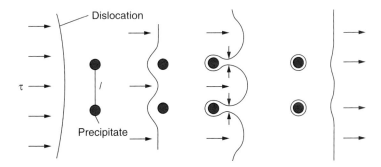

■ **FIGURE 6.10**   Movement of dislocation through widely spaced precipitate particles. The dislocation moves to the right under stress. Precipitate particles impede its movement, but it bows, passes the particles, and reforms.

room temperature. Fortunately, many binary systems display this behavior; the classic example is the Al-Cu system shown in Figure 6.9.

When an alloy containing 4 wt% Cu is slowly cooled in the α phase from 550°C, a coarse $\theta(CuAl_2)$ phase precipitates from the solid solution. The resulting two-phase structure would be moderately harder and stronger than pure Al. When the particles are spaced far apart, they barely inhibit the motion of dislocations. The latter can easily balloon outward past the particles and recombine later as shown in Figure 6.10.

Optimum strengthening requires a high density of small precipitate particles. When the particles are closely spaced, dislocations cannot bow sufficiently and the particles behave as effective obstacles; more stress is required for further dislocation motion.

■ **FIGURE 6.11** Transmission electron microscopy image of an Al-3.8% Cu alloy aged at 240°C for 2 hr, showing θ′ precipitates. Specimen offered by JEOL.

6.4.1 The Precipitation-Hardening Treatment

The task is to obtain a large number of very fine precipitates. This means a high precipitation rate and slow diffusion, which occur at temperatures below 200°C. Precipitation is obtained by the following process:

1. Maintain the alloy at 550°C for some time to ensure homogeneous distribution of the copper solid solution in the α field.

2. **Quench** to a low temperature (e.g., room temperature or lower). Quenching prevents the achievement of the equilibrium condition and the solid solution of Cu is "frozen in."

3. **Age** at temperature T_p (e.g., 150°C). The low temperature ensures a very high nucleation rate and precipitates a large number of very small particles. The microstructure of an aged Al-4% Cu alloy is shown in Figure 6.11.

The resultant strength and ductility changes are shown in Figure 6.12 for an Al-4% Cu alloy at several temperatures. Initially, the hardness of the material increases because of the formation of very fine particles. Optimum strengthening is achieved after 1 day at 190°C and 60 days at 130°C, The strength of the material is decreased by **overageing**, that is, by heating for longer times. Diffusion of the Cu atoms leads to particle coarsening: the smaller precipitate particles disappear and the structure contains fewer, larger θ particles that are less effective in stopping the dislocations.

6.4.2 Some Precipitation-Hardened Alloys

There are commercial precipitation-hardening alloys based on virtually all of the nonferrous metals. The Al-4 wt% Cu alloy was the first one developed and remains popular. Other commercially important Al precipitation-hardening alloys contain zinc and magnesium either singly or in combination with other elements. In Al-Mg alloys, for example, a quadrupling of the yield stress relative to annealed Al is attainable with a 5 wt% Mg addition.

Copper can undergo extraordinary hardening with a couple of percent beryllium. A tensile strength approaching 1.4 GPA (200,000 psi), some three times that of very heavily cold worked pure copper, is

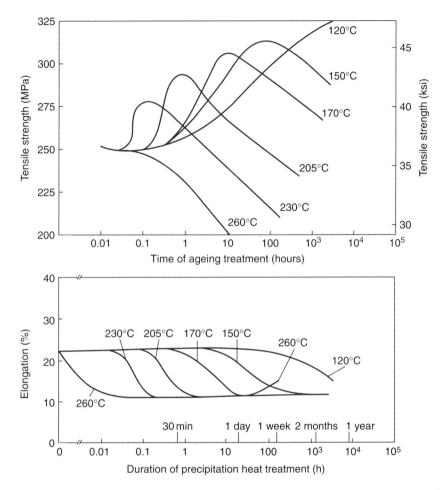

■ **FIGURE 6.12**   Variation of the tensile strength and ductility of an Al-4% Cu alloy with ageing time. Strength increases because of the formation of precipitates. It decreases with overageing because of particle coarsening. From Metals Handbook, 9th ed. Vol. 4, American Society for Metals, Metals Park, OH (1981).

possible. The resulting alloy is widely used for spring contacts. Other notable precipitation-hardening Cu base alloys contain Ti, Zr, and Cr. The latter are added in small amounts in order to avoid lowering the electrical conductivity.

The popular titanium alloy, Ti-6Al-4 V, also precipitation-hardens. It combines high tensile strength and good workability, qualities that are important in aircraft engine components.

■ SUMMARY

1. Interesting properties can be obtained by preventing thermal equilibrium and exploiting the kinetics of phase transitions.

2. In a phase transition, the new phase is first nucleated; it then grows by diffusion of atoms in the material.

3. Nucleation is a competition between the driving force (a decrease in free energy) and an energy barrier (the interfacial energy between new and old phase).

4. Diffusion is the random motion of atoms or molecules. Diffusion rates increase rapidly with temperature.

5. Slow cooling causes slow nucleation and rapid diffusion; it produces coarse structures.

6. Rapid cooling causes rapid nucleation and slow diffusion; it produces fine structures.

7. Quenching of steels causes a cooling rate that is too high for nucleation of pearlite and produces martensite.

8. The TTT diagram shows the times required for isothermal transition from austenite to pearlite.

9. CCT (Continuous Cooling Transformation) curves show the types of material obtained when austenite is cooled continuously at various rates.

10. Martensite is a very disordered structure that is extremely hard and brittle.

11. Tempered steel is martensite that has been reheated to restore some ductility.

12. Hardenability measures the thickness of steel that can be transformed into martensite. It can be increased by alloying with nickel.

13. Aluminum containing 5% Cu can be hardened by a heat treatment because of the limited solubility of Cu. Al is first stabilized at the eutectic temperature, then quenched and aged around 150°C. Overageing produces coarse precipitate structure and weakens the material.

14. Alloys based on most nonferrous metals can be hardened by a precipitation heat treatment.

■ KEY TERMS

A
ageing, 166
ageing of aluminum, 166
Al-4 wt% Cu, 166
Al-Mg, 166
annealed steel, 164
Arrhenius plot, 159

B
Bainite, 161
BCT, 154
body centered tetragonal
    structure, 154
Boltzmann statistics, 158

C
CCT, 161
CCT curve, 161

continuous cooling
    transformation, 161
critical radius, 157

D
diffusion rates, 158

E
embryos, 156
entropy, 156

F
free energy, 156

G
growth, 155

H
hardenability, 162

heat treatment of
    aluminum, 164

J
Jominy Test, 162

K
kinetics of phase
    transformations, 155

M
martensite, 154
martensitic transformation, 155

N
nucleation, 155, 156

O
oil-quenching, 162

■ REFERENCES FOR FURTHER READING

[1] R.J. Borg, G.J. Dienes, *Solid State Diffusion*, Academic Press, San Diego, CA (1988).

[2] E.M. Dobkin, M.K. Zuraw, *Principles of Chemical Vapor Deposition*, Springer, New York (2003).

[3] M.P. Groover, *Fundamentals of Modern Manufacturing: Materials, Processes, and Systems*, Wiley, New York (2006).

[4] S. Kalpakjian, S. Schmid, *Manufacturing Processes for Engineering Materials*, 5th. ed, Addison-Wesley (1991), Prentice Hall, Englewood Cliffs, NJ (2007).

[5] D.A. Porter, K.E. Easterling, *Phase Transformations in Metals and Alloys* (Paperback), Van Nostrand Reinhold, UK (1992).

[6] F. Reidenbach, *ASM Handbook, Vol. 5: Surface Engineering*, ASM International, Materials Park OH (1994).

[7] D.L. Smith, *Thin-Film Deposition: Principles and Practice*, McGraw-Hill, New York (1995).

■ PROBLEMS AND QUESTIONS

6.1. Devise heat treatments for 1080 steel that would convert the following. Refer your treatments to the TTT curve.

   a. Pearlite to martensite.

   b. Martensite to coarse pearlite.

   c. Martensite to fine pearlite.

   d. Pearlite to bainite.

6.2. Devise heat treatments for 1080 steel that would convert the following. Refer your treatments to the CCT curve.

   a. 100 % pearlite to a mixture of 50% pearlite + 50% martensite.

   b. A mixture of 75% pearlite + 25% martensite to 100% tempered martensite.

   c. Austenite to a mixture of 50% martensite + 50% tempered martensite.

6.3. Since time immemorial, blacksmiths have forged stone-carving tools starting with essentially pure iron, containing perhaps 0.05% carbon. Assuming you have such round bar blanks and a blacksmith's hearth, detail the steps you would follow to forge a flat-edged stone-carving chisel. Note that chisels must be very hard and tough at the cutting edge, but very soft where the hand grasps it in order to absorb mechanical impact and prevent a ringing sensation.

6.4. Heat conduction follows the same kinetics as mass diffusion. Suppose a cylindrical Jominy bar, initially at 800°C, is suddenly quenched at one end to 20°C. In snapshot fashion, schematically sketch temperature-distance profiles that evolve in the bar at a sequence of increasing times during cooling.

6.5. An austenitized Jominy bar is end-quenched with oil rather than water. How would this change the resulting hardness-distance profile of the Jominy curve?

6.6. Upon viewing microstructures along a water-quenched Jominy bar of 1060 steel, the following phases were observed starting from the quenched end: martensite; martensite plus pearlite; fine pearlite; coarse pearlite. At what distances along the 4 in Jominy bar would each of these structures be observed?

6.7. Vanadium, a common alloying element in tool steels, refines the grain size by reducing the prior austenitic grain size. Consider a steel with and without V. Both steels are austenitized at high temperatures and end-quenched. Sketch the expected microstructures along the bar for each steel. Note: Pearlite nucleates at austenite grain boundaries.

6.8. The heat treatments noted below are carried out in 1080 steel. In each case what phases will be present at the end of the treatment?

    a. Austenitize at 750°C; quench to 0°C in 1 s.

    b. Austenitize at 750°C; rapid quench to 650°C; isothermally anneal for 1000 s; quench to 0°C in 1 s.

    c. Austenitize at 750°C; very rapid quench to 250°C; heat to 350°C and isothermally anneal for 500 s; quench to 0°C in 1 s.

    d. Austenitize at 800°C; rapid quench to 550°C; isothermally anneal for 1000 s; heat to 750°C; quench to 0°C in 1 s.

    e. Austenitize at 800°C; rapid quench to 350°C; isothermally anneal for 700 s; heat to 550°C; quench to 0°C in 1 s.

6.9. Why does the hardness of martensite and pearlite increase with carbon content?

6.10. A gear made of 5140 steel has a hardness of R_c 30 when quenched in oil. What probable hardness would a 4340 steel gear of identical dimensions exhibit if heat-treated in the same way?

6.11. In order to capitalize on weight savings, a steel is quenched and tempered, strengthening it from 1,100 MPa where K_{IC} is 76 MPa$\sqrt{}$m to 2,000 MPa where $K_{IC} = 26$ MPa$\sqrt{}$m. What change in critical flaw size occurs for a design stress of 800 MPa?

6.12. Design an ageing heat treatment (temperature and time) for a 6061 aluminum alloy that would yield a tensile strength of 275 MPa coupled with an elongation of 20%. (The treatment should be economical to perform.)

6.13. Define a phase transition. Define and describe the two phenomena that succeed each other in a phase transition.

6.14. Describe the two energies that compete in nucleation.

6.15. Describe diffusion. What causes diffusion?

6.16. How do diffusion rates change with temperature? Sketch a temperature dependence of diffusion or write the equation describing it.

6.17. Describe the rates of nucleation and diffusion as one cools down from the equilibrium point. Namely, near the equilibrium point, which phenomenon is fast and which is slow? Far below the equilibrium temperature, which phenomenon is fast and which is slow?

6.18. Describe the effect of the cooling rate on the microstructure.

6.19. What is quenching of steel, and what effect does it have on the microstructure and mechanical properties of steel?

6.20. Does quenching have the same effect on Cu-Ni alloys as it does on steel?

6.21. Draw the TTT diagram for steel and sketch in it a cooling schedule that produces soft pearlite and a schedule that produces martensite.

6.22. What is martensite? What is its crystal structure? What are its remarkable mechanical properties (desirable and undesirable)?

6.23. What is tempered steel? How is it obtained and why?

6.24. Define hardenability. How is it measured? Give one means of increasing it.

6.25. What is heat-treatable aluminum? What is the heat treatment it describes?

6.26. Describe ageing of aluminum. Describe overageing and its consequence.

7

Metallic Alloys and Their Use

This chapter reviews the most important metallic alloys available commercially. Metals are divided into ferrous and nonferrous alloys. Ferrous alloys include the various kinds of steel and cast iron. Steels are generally available as annealed, rolled, and tempered. Plain carbon steels contain carbon and manganese as alloying elements and are the least expensive. By the addition of small amounts of other elements that provide additional precipitation strengthening, one produces high-strength low-alloy (HSLA) steels, tool steels, and stainless steels. Cast irons contain more carbon and silicon than steels. Gray cast iron is corrosion-resistant, absorbs noise, and is self lubricating; it is very brittle. White iron contains large amounts of cementite and is abrasion resistant. Cast irons exist in more ductile varieties. The nonferrous alloys include copper and its alloys, most notably brass and bronze; aluminum that is precipitation strengthened (heat-treatable alloys) or work hardened (wrought alloys); magnesium, titanium alloys, and superalloys. The latter have been developed for jet engines and operate at high temperatures.

Most metals are cast from the molten stage. Continuous casting produces metals for further mechanical processing. Sand casting, die casting, and lost wax casting produce complex near final shapes.

Metal surfaces are modified to increase the hardness (for instance in gears), to provide corrosion resistance, as thermal barriers in turbines, or for appearance. Surface modification is done by diffusion of certain elements (carbon, nitrogen) into the surface layer or by shot peening. Coatings are deposited by physical vapor deposition (evaporation of a source or sputtering of the source by a plasma discharge), by chemical vapor deposition, welding, thermal spray methods, electroplating, or electroless plating.

LEARNING OBJECTIVES

After studying this chapter, the student will be able to:

1. Name the principal classes of metallic alloys used in technology.

2. Name the main characteristics of these classes of alloys.

3. Read a table of properties and make a selection of a metal for a given application.

4. Select an appropriate casting method.

5. Describe the effect and action of grain refiners.

6. Describe the processing of single crystals by the Czochralski and directional solidification methods.

7. Describe the surface hardening of materials by diffusion.

8. Describe the various techniques used for the deposition of coatings.

7.1 **TYPES OF ALLOYS**

In this chapter we consider metals that are employed in mechanical and structural engineering applications. Materials that achieve commercial status have necessarily passed a tough development that includes: (1) meeting stringent mechanical property specifications (tensile strength, elongation, toughness, etc.); (2) developing reproducible processing and fabrication methods (casting, mechanical forming, joining, etc.); (3) extensive testing to ensure reliable operation in components and structures; and (4) competitive pricing.

It is customary to divide metallic alloys into ferrous alloys, which are based on iron, and nonferrous alloys, which represent all others.

7.2 **FERROUS ALLOYS**

Ferrous alloys are produced in tonnages 10 times those of aluminum- and copper-base alloys combined. Iron is not only abundant, but easily and cheaply extracted. After alloying and fabrication, steels are significantly cheaper than either Al or Cu alloys. But the major reason that accounts for the use of iron and steel alloys is their superior strength and their extraordinary versatility.

7.2.1 **Steels**

Plain Carbon Steels   For structural applications (e.g., skyscrapers, bridges) as well as for sheet metal enclosures of all kinds (auto bodies, appliances, etc.), inexpensive tools and wire, unalloyed **plain carbon steels** have an adequate combination of strength and ductility.

These steels contain between 0.20–1.2% carbon and 0.45–0.8% manganese. Manganese is added to protect the steel against the harmful effects of sulfur and phosphorus which are always present in iron ore and cannot be completely removed. Sulfur and phosphorous tend to segregate to the grain boundaries of the metal where they decrease the chemical bond between the grains and cause fracture of the metal at intolerably low stresses. Manganese has a high chemical affinity for sulfur and phosphorus and "scavenges" these elements by forming MnS, MnP, and Mn_3P_2 particles that are not only harmless but provide additional precipitation-strengthening. Plain carbon steels are designated by a 10 followed by two digits that indicate the amount of carbon. Thus 1020 steel contains 0.2% C and 1080 steel contains 0.8% C. Steels are commercially available in three forms: annealed, as rolled, and tempered (which actually means "quenched and tempered"). We have already discussed the consequences of these treatments: annealing produces the softest and most ductile steels. Rolled steels are strengthened by deformation strengthening, and quenched and tempered steels are strengthened by the martensitic transformation during rapid cooling. Note that 1020 steel does not contain enough carbon to be hardened by quenching and exists only as rolled and annealed.

7.2.2   High-Strength Low-Alloy Steels (HSLA)

Higher strengths can be achieved in steels by the addition of small amounts, usually less than 1%, of chromium, nickel, tungsten, vanadium, titanium, niobium (called columbium in metallurgy), and aluminum. Cr, W, V, Nb, and Ti form carbide particles that further reinforce the steel by precipitation hardening. Silicon and manganese scavenge dissolved oxygen and form strengthening SiO_2 or MnO particles. The alloying elements also increase the hardenability of quenched and tempered steels as discussed in Chapter 6. See Table 7.1.

Stainless Steels   Stainless steels are resistant to oxidation and other forms of corrosion. This property is achieved by alloying them with at least 12 wt% chromium. The latter forms a stronger bond with oxygen than with iron and develops a hard protective surface layer of Cr_2O_3 when the steel is subjected to an oxidizing environment.

Chromium crystallizes in the BCC structure. When added to iron, it stabilizes the BCC (ferritic) structure of steel. The addition of nickel, which has the FCC structure, or carbon to the steel favors the FCC (austenitic) phase. When 17% Cr and very little carbon are added to the iron, as in 430 **ferritic stainless steel**, the austenitic FCC structure is entirely suppressed and the steel maintains the BCC structure at all temperatures. The addition of sufficient amounts of nickel lowers the free energy of the FCC structure enough for the alloy to form **austenitic stainless steel** that crystallizes in the FCC structure at all temperatures. Because plastic deformation is easier in FCC crystals than in BCC, austenitic stainless steels possess a large ductility. They are used where extensive deformation is required. With proper amounts of chromium and carbon, one obtains **martensitic stainless steels** that possess FCC structure at high temperature and BCC at low temperature. These steels can be hardened by quenching and tempering. Note that a higher hardness can be obtained with martensitic stainless steels than with plain carbon steels (Table 7.1).

Tool Steels   These are high-carbon steels, containing between 0.85 and 1.5 wt% carbon. The hardness, strength, and ductility needed for tools are obtained with larger additions of chromium, nickel, molybdenum, tungsten, and vanadium to the iron. Manganese content ranges from 0.1–1.4%, depending on the alloy.

Table 7.1 Mechanical Properties and Applications of Some Steels.

Composition (wt%)	Condition	Strength		Ductility % elongation	Applications
		Tensile (MPa)	Yield (MPa)		
Plain carbon steels					
1020	Rolled	450	330	36	Steel plate,
(0.20 C, 0.45 Mn)	Annealed	390	300	36	structural sections
1040	Rolled	620	415	25	Shafts, studs
(0.40 C, 0.45 Mn)	Annealed	520	350	30	
	Tempered	800	595	20	Gears
1060	Rolled	815	485	17	Spring wire, forging
(0.60 C, 0.65 Mn)					dies, railroad wheels
	Annealed	630	485	22	
	Tempered	1,100	780	13	
1080	Rolled	970	585	12	Music wire helical springs, chisels, die blocks
(0.80C, 0.75 Mn)	Annealed	615	375	25	
	Tempered	1,300	980	12	
High strength low alloy steels					
4340 (0.40 C,	Annealed	745	470	22	Truck, auto, aircraft
0.90 Mn, 0.80 Cr,	Tempered	1,725	1,590	10	parts, landing gear
0.2 Mo, 1.83Ni)					
5160 (0.60 C,	Annealed	725	275	17	Automobile coil and leaf springs
0.80 Cr, 0.90 Mn)	Tempered	2,000	1,775	9	
8650 (0.50 C,	Annealed	710	385	22	Small machine axles
0.55 Ni, 0.50 Cr	Tempered	1,725	1,550	10	
0.8 Mn)					
Stainless steels					
Ferritic					
430	Annealed	550	375	25	Decorative trim, acid tanks Combustion chambers
(17Cr, 0.012 C max)					

(Continues)

Table 7.1 — *Continued*

Composition (wt%)	Condition	Strength		Ductility % elongation	Applications
		Tensile (MPa)	Yield (MPa)		
Austenitic **304** (0.08 C,19 Cr, 9 Ni, 2.0 Mn)	Annealed	580	290	55	Chemical and food processing equipment
316 Li, (0.03 C, 17 Cr,13 N 02.0 MN,2.5 Mo)	Annealed Cold Worked	490 863	190 690	40 12	Surgical implants
Martensitic					
440A (17 Cr, 0.70 C, 0.75 Mo,1.0 Mn)	Tempered	1,830	1,690	5	Cutlery, bearings surgical tools
440 C (17 Cr, 1.1 C)	Tempered	1,970	1,900	2	Ball bearings, valve parts
Tool steels		Hardness (R_C)			
T1 (18 W,4 Cr,1 V, 0.75 C)		60 at 260°C, 50 at 600°C			Lathe and milling machine tools, drills
M2 (6 W,5 Mo, 4 Cr, 2 V, 0.9 C)		60 at 260°C, 50 at 600°C			Taps, reamers, dies

7.2.3 Cast Irons

Engine blocks, machine tool bodies, piston rings, manhole covers, and many other objects with complex shapes are cast. Cast irons usually contain between 3 and 4.5% carbon in addition to silicon and manganese. Silicon is added to improve the fluidity of the molten iron; it also stimulates the formation of graphite and provides corrosion resistance.

By examining the Fe-C phase diagram, Figure 5.13, we convince ourselves easily that the melting temperature of these alloys is between 1150 and 1300°C, near the eutectic point. Some cast irons contain small additions of nickel and molybdenum. Depending on the heat treatment and composition, we obtain four types of cast iron.

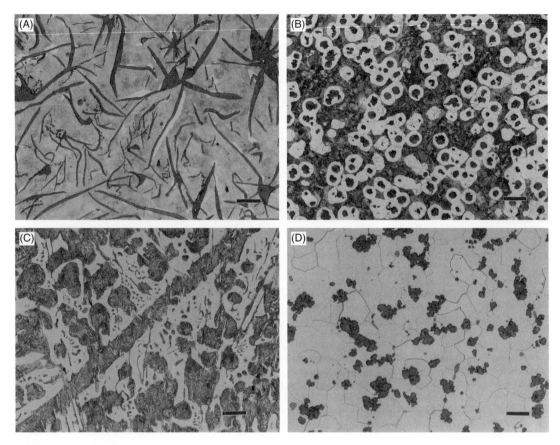

■ **FIGURE 7.1** The microstructure of cast irons. (A) Gray cast iron showing the graphite flakes in black. (B) Nodular cast iron. The graphite is present in spherical nodules. (C) White cast iron. The white areas are cementite. (D) Malleable cast iron. The carbon is present in black rosettes surrounded by ferrite. Courtesy G. F. Vander Voort. Buehler LTD.

Gray Cast Iron Gray cast iron possesses interesting properties that make it ideal for engine blocks, gear boxes, machine tools, pistons, cylinders, piston rings, and manhole covers. Its use in engine blocks and gear boxes exploits the fact that **gray cast iron absorbs noise and vibrations**. Pistons and piston rings use the fact that **gray cast irons are self lubricating** and manhole covers benefit from its **excellent corrosion resistance**. What gives the material these quite unique properties? Gray cast iron contains around 3.5% carbon and 2–2.5% silicon. At moderate cooling rates, it solidifies into pearlite and the excess carbon forms graphite in the shape of flakes. This microstructure is shown in Figure 7.1A. At low cooling rates, gray cast iron solidifies into ferrite and graphite. There are two reasons for the particular shape of the graphite: (1) graphite crystallizes in the form of atomic layers that are weakly bonded to each other (Figure 8.16), so that flakes are a natural low-energy shape; and (2) sulfur that is dissolved in the iron segregates to the graphite-iron interfaces and lowers their surface energy (like soap in water). When gray cast iron is fractured, its fracture surface appears gray because of the graphite, hence the name.

The graphite flakes do not transmit vibrations in the material and are responsible for the sound absorption of the material. When the surface of gray cast iron is rubbed, the graphite is smeared over its surface and provides lubrication. Graphite flakes have one drawback that limits the applications of this material: they have the shape of cracks (see Chapter 2) and render the material very brittle.

Ductile Cast Iron Ductile cast iron overcomes the brittleness of gray cast iron by shaping the graphite as spherical nodules (see Figure 7.1B). To this end, the iron is processed to reduce the sulfur and phosphorus to very small levels (<0.03% S and <0.1% P), and a small amount of magnesium is added just before casting (0.05% is enough). The magnesium reacts with sulfur and phosphorus and prevents them from segregating to the graphite interface. This increases the surface energy of the graphite, which gives it a spherical shape that avoids the stress concentrations caused by the flakes. The result is a strong and ductile cast iron that still possesses the valuable qualities provided by the graphite. Ductile cast iron is also called "nodular cast iron." Ductile cast iron can be further processed to modify its properties. When the material is annealed, carbon diffuses from the pearlite and joins the graphite, leaving a **ferritic** matrix. The iron can also be quenched and tempered by austenitizing above 727°C and quenching and tempering.

White Cast Iron White cast iron contains much less silicon (~1.2%); it solidifies into large amounts of cementite surrounded by pearlite (see Figure 7.1C). Polished surfaces of the material appear white because of the cementite. This material is extremely brittle but very hard; it is used where abrasion resistance is important and risk of fracture is reduced. A prolonged heat treatment at 800–900°C decomposes this cementite into ferrite and irregular nodules of graphite, which produces **malleable cast iron** that is strong and ductile, as its name implies (see Figure 7.1D). Composition, strength, and applications of cast irons are shown in Table 7.2.

7.3 **NONFERROUS ALLOYS**

This term designates metals that do not contain iron. Nonferrous alloys comprise a large number of materials, since the majority of elements in the periodic table are metals. We limit ourselves here to the most important ones that are utilized in structural applications. These are the copper alloys, namely copper, brass, and bronze, which are used mostly for their corrosion resistance in homes, statuary, and water containers; the aluminum alloys, which are used for their light weight; the titanium alloys, which are recently finding increased use despite the difficulty and cost of their processing; the ultra light magnesium alloys; and the superalloys, which have been developed to withstand the high temperatures and stresses of jet engines.

7.3.1 **Aluminum Alloys**

The first thoughts that come to mind when thinking of aluminum are "airplanes" and "aluminum foil." These embody two main characteristics of this metal. Aluminum has a low density of $2.7\,\mathrm{g/cm^3}$

Table 7.2 Properties and Uses of Cast Irons.

Composition (wt%)	Condition	Strength		Ductility % elongation	Applications
		Tensile (MPa)	Yield (MPa)		
Gray cast irons					
G 3500 Ferritic	As cast	250			Cylinder blocks, gear boxes
(3.2 C, 2.0 Si, 0.7 Mn)					
G4000 Pearlitic	As cast	280			
(3.3 C, 2.2 Si, 0.7 Mn)					
Ductile cast irons					
60-40-18 Ferritic	Annealed	410	270	18	Valve and pump bodies
(3.5 C, 2.2 Si, 0.05 Mg, 0–1.0 Ni, <0.03 S)					
80-55-06 Pearlitic	As cast	550	380	6	Crankshafts, gears, rollers
(3.3 C, 2.4 Si, 0.05 Mg, 0–1.0 Ni, <0.03 S)					
120-90-02 Martensitic	Quench. + Temp.	830	620	2	Gears, rollers
(3.4 C, 2.2 Si, 0.05 Mg, 0–1.0 Ni, <1.0 Mo, < 0.03 S)					
Malleable cast irons					
32510 Ferritic	Annealed	340	220	10	General engineering
(2.2 C, 1.2 Si, 0.04 Mn)					
45008 Pearlitic	As cast	440	310	8	General engineering
(2.4 C, 1.4 Si, 0.75 Mn)					
M7002 Martensitic	Quench. + Temp.	620	430	2	Connecting rods
(2.4 C, 1.4 Si, 0.75 Mn)					

and is very ductile because of its FCC crystal structure. This permits the rolling of aluminum into the thin foils we are familiar with and also permits the fabrication of metal sheets used in airplane construction.

Aluminum readily forms a surface layer of very strong Al_2O_3 that provides it with excellent corrosion resistance in most environments, but makes it difficult to weld. See Table 7.3.

Table 7.3 Aluminum Alloys.

Composition	Condition	Strength		Ductility % elongation	Applications
		Tensile (MPa)	Yield (MPa)		
Wrought alloys non-heat-treatable					
1100 (99.0 min Al, 0.12 Cu)	Annealed half-hard	90 125	25 100	25 4	Sheet metal
3003 (0.12 Cu, 1.2 Mn, 0.1 Zn)	Annealed	110	40	30–40	Cooking utensils pressure vessels, piping
5052 (2.5 Mg. 0.25 Cr)	Strain hardened	230	195	15	Aircraft fuel lines fuel tanks, rivets
Wrought alloys, heat treatable					
2024 (4.4 Cu, 1.5 Mg, 0.6 Mn)	Annealed Heat treated (T6)	220 (max) 440 (min)	100 (max) 345 (min)	12 5	Aircraft structural parts
6061 (1.0 Mg, 0.3 Cu, 0.2 Cr)	Annealed Heat treated (T6)	150 (max) 290 (min)	82 (max) 240 (min)	16 10	Truck and marine structures
7075 (5.6 Zn, 2.5 Mg, 1.6 Cu 0.23 Cr)	Annealed Heat treated (T-6)	275 (max) 504 (min)	145 (max) 428 (min)	10 8	Aircraft and other structures
Cast alloys					
356 (7 Si, 0.3 Mg)	Sand cast Heat treated (T-6)	10 (min) 30 (min)	140 (min) 150 (min)	3 3	Transmission and axle housings, truck wheels, pump housings
413 (12 Si, 2 Fe)	Die cast	297	145	2.5	Intricate die castings
T-6 = solution heat treated, then aged.					

Aluminum alloys are designated as **wrought** or **cast**. The wrought alloys are intended for shaping by rolling, extrusion, or forging. These procedures also strengthen them by strain hardening. Certain aluminum alloys are strengthened by precipitation hardening (Chapter 6). These are the **heat treatable alloys**. Alloy 2024 is strengthened by precipitation of a θ phase as described in Chapter 6. Alloy 7075 is strengthened by precipitation of $MgZn_2$ particles.

■ **FIGURE 7.2**   Copper cladding on a roof.

7.3.2 **Copper Alloys**

Copper is the first metal that has been used, long before the bronze age. It is very soft and ductile and therefore easily shaped. Like all FCC metals, its strength increases remarkably by cold working and can be as high as 345 MPa. (The Ice Man, who lived in 3000 BC and was found in a glacier in the Alps, was carrying a copper axe). An attractive property of copper and its alloys is their corrosion resistance in air and in sea water.

Oxygen-Free, High-Conductivity (OFHC) copper undergoes a vacuum treatment to remove dissolved oxygen and possesses high electrical conductivity, making it suitable for electrical applications. Tough pitch copper, which is less pure, finds application in roofing, gutters, and other architectural elements. It acquires an attractive green color when weathered (Figure 7.2).

Brass is an alloy of copper and zinc. It is somewhat stronger than copper, is easy to cast and to machine, and finds many applications in radiators, electrical fixtures, faucets and water fixtures, musical instruments, and so on. It is strengthened by solution hardening and cold working. Figure 7.3 shows an example of a brass trumpet.

Bronze is an alloy of copper with tin, lead, or any element other than Zn or Ni. Its high strength relative to copper and its low melting point allowed the use of tin bronze in early antiquity, defining the bronze age. Bronze has excellent corrosion resistance and acquires an attractive patina through weathering, therefore this material is much used in sculpture (Figure 7.4). Leaded bronze is used in sliding bearings because it does not weld to steel.

Beryllium copper is a precipitation hardened alloy that attains the strength of hardened steel. It is used in bellows, spring contacts, and diaphragms. See Table 7.4.

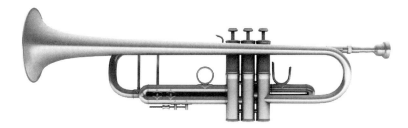

■ **FIGURE 7.3** Brass trumpet.

■ **FIGURE 7.4** Bronze statue: The Thinker by Auguste Rodin.

7.3.3 **Magnesium Alloys**

Magnesium is a very light metal, with a density of $1.45\,\text{g/cm}^3$, lighter than aluminum ($2.7\,\text{g/cm}^3$). It is difficult to cast because it burns in air and cover fluxes must be used in casting. (The old photographic "flashes" used the bright flame of burning magnesium.) Magnesium has the HCP structure and therefore is difficult to shape by deformation. Therefore, magnesium alloys are used mostly as castings, where low weight is paramount.

Table 7.4 Properties and Applications of Copper and Some of Its Alloys.

Composition	Condition	Tensile strength	Yield stress	Ductility (% elong)	Applications
Wrought alloys					
OFHC (99.99 Cu)	Annealed	220	70	45	Bus conductors, electrical
	Cold worked	345	310	6	wiring, vacuum seals
Electrolytic tough pitch	Annealed	220	70	45	Auto radiators, gutters,
	Cold worked	345	310	6	roofing,
Brass (70 Cu, 30 Zn)	Annealed	325	105	62	Radiators, lamp fixtures,
	Cold worked	525	435	8	musical instruments, ammunition, architectural hardware
Beryllium copper (1.7 Be, 0.20 Co)	Solution treated	410	190	60	Bellows, spring contacts,
	Cold worked	1,240	1,070	4	diaphragms, fasteners, switch parts, welding equipment
Casting alloys					
Leaded bronze (5.5Sn, 5Pb,5Zn)	As cast	255	120	10	Flanges, pipe fittings, water pump parts, sculpture, ornamental fixtures

7.3.4 Titanium Alloys

Titanium possesses a relatively low density of $4.5\,g/cm^3$, very high strength even in the pure state (660 MPa), good creep resistance, and extraordinary resistance to corrosion. For these reasons, titanium alloys are finding increased application despite the difficulty of its processing. Titanium alloys (tensile strength 1,100 MPa) are used extensively in the construction of aircraft, for instance in compressors and other components of jet engines, mainframes, and landing gear. With its corrosion resistance, titanium finds applications in propeller shafts of ships, in desalinization plants, and in the paper industry. Titanium alloys are also utilized as cladding for modern buildings (Figure 7.5) and as frames for eyeglasses. They are increasingly used in sports equipment because of their high strength-to-weight ratio. Finally, titanium alloys are among the very few metals that are approved for use in human prostheses (see Chapter 17). Tooth implants are now generally made of titanium. See Table 7.5.

Titanium is very reactive to oxygen, carbon, hydrogen, and iron at elevated temperatures so that special techniques must be used in its processing and casting. For these reasons, titanium is about five times more expensive than aluminum.

■ **FIGURE 7.5**   Guggenheim Museum in Bilbao (Spain) with titanium sheet cladding. (Frank Gehry, architect.)

Table 7.5 Composition, Mechanical Properties, and Applications of Titanium Alloys.

Denomination	Composition	Condition	Tensile Strength (MPa)	Yield Strength (MPa)	Ductility % elong	Applications
Commercially Pure	99% Ti	Annealed	660	590	20	Jet engine parts, airframes, chemical processing, and marine uses
Ti-6Al-4V		Annealed	950	870	14	Prosthetic implants, chemical equipment
		Heat treated and aged	1100	1000	8	Aircraft engine components, processing equipment, airframe
Ti-8Al-1Mo-1V		Annealed	950	890	15	Jet engine components

7.3.5 **Superalloys**

The particular demands of high performance aircraft engines (Figure 7.6) have done far more to accelerate the development of strengthened advanced metals and superalloys than any other set of applications.

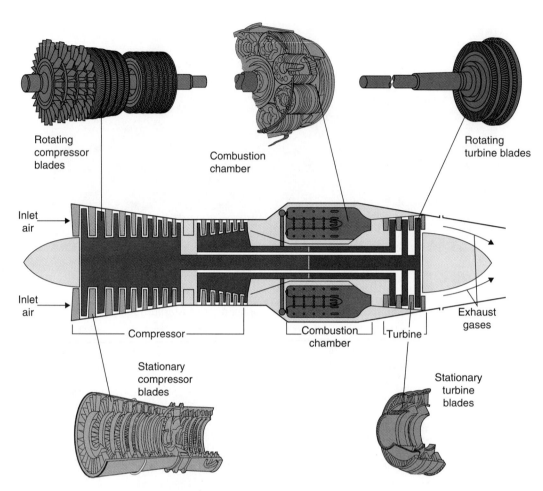

■ FIGURE 7.6 Turbojet engine components. Air is pulled into the compressor, composed of a series of fan blades attached to rotating disks; surrounding these are stationary compressor blades that redirect the airflow between rotors. Titanium alloys are used in the compressor. In the combustion chamber, the compressed air is mixed with fuel and ignited. The expanding, hot exhaust gases move with great velocity through the turbines. Nickel-base and cobalt-base superalloys are employed in these latter engines sections. Adapted from "Advanced Metals" by B.H. Kear, Copyright © 1996 Scientific American Inc.

The harsh thermal, stress, and oxidation environment faced by components in the tail portion of the compressor, the combustion chamber, and the turbine have only been satisfactorily countered through the use of superalloys. Properties of several popular nickel and cobalt base superalloys are listed in Table 7.6. The nickel-base alloys derive their strength from the presence of block-like Ni_3Al or gamma prime (γ') precipitates distributed within the matrix of the gamma (γ) phase as shown in Figure 7.7. Dislocations propagate easily through γ, a relatively ductile FCC austenite-like phase, but upon encountering γ' their motion is severely impeded. This is due to the ordered nature of this

Table 7.6 Composition and Properties of Selected Superalloys.

Alloy	Composition	Tensile Strength (MPa)	Yield Strength (MPa)	E (GPa)	Elongation (%)
Hastelloy X	50 Ni, 21 Cr, 9 Mo 18 Fe, Co+ W~2	755	385	210	57
Inconel 718	53 Ni, 19 Cr, 19 Fe, 5 Nb, 3 Mo, 1 Ti	1385	1295	205	28
Udimet	55 Ni, 18 Cr, 15 Co, 3 Mo, 5 Ti	1575	1240	230	16
Haynes 25	54 Co, 20 Cr, 10 Ni 15 W, 1 Fe	1020	470	230	64

■ **FIGURE 7.7**   Electron micrograph of the block structure of nickel-aluminum γ' phase in a nickel-base superalloy (40,000✕). Courtesy of N.D. Persons, United Technologies Research Center.

compound which requires Al atoms to occupy only cube corners and Ni atoms to populate only face centered sites of their FCC unit cell. The motion of a dislocation would disrupt this order and result in high-energy Ni-Ni or Al-Al pairings; it is therefore strongly impeded. See Table 7.6.

With so many atomic components, it is not surprising that a number of strengthening mechanisms interact synergistically in these alloys as the temperature increases. Ni_3Al and Ni_3Ti precipitates confer matrix hardening, while carbides that are often present tend to stabilize grain boundaries. All of these particles are set within a tough γ solid solution matrix.

Their higher temperature strength ($\sim$1,100°C vs. $\sim$1,000°C) earmarks cobalt-base superalloys for use in the combustion chamber and in neighboring stationary turbine vanes. Strengthening occurs by the presence of MoC and WC particles that collect at grain boundaries, restraining them and the matrix from deforming under stress.

7.4 **SOLIDIFICATION OF METALS**

As a rule, metal alloys are processed in the liquid state. Solid metals are obtained by pouring or **casting** the melt into a mold where it cools and solidifies. The liquid metal is either cast as simple blocks called **ingots**, bars and plates that are further shaped by rolling, extrusion, or forging as described in Chapter 5, or they are cast into a mold that approximates the final desired shape of the piece. The term **casting** usually refers to the latter case; the alloys used in this method, called **cast iron**, **cast aluminum**, and so on, are specially formulated for this process, and the solid pieces obtained are called **castings**. Engine blocks, sewing machines, valve bodies, manhole covers, statues, and other objects with complex shapes are manufactured by casting.

7.4.1 **Casting**

Liquid metals have very low viscosities (e.g., $\eta \approx 10^{-4}$ Pa-s) and therefore flow readily. Coupled with high densities, liquid metals easily fill molds while less dense nonmetallic fluxes, foreign matter (drosses), and oxide slag float to the surface where they can be skimmed off. High heat transfer rates make it possible to complete solidification rapidly, making casting production rates economical. A wonderfully diverse collection of casting processes has been developed over thousands of years. New processes as well as modifications of older ones continue to appear.

Sand casting is the preferred method for large pieces. It is illustrated in Figure 7.8A. Wooden or plastic patterns of the desired shapes, together with attached gates and risers, are embedded in special casting sand contained within the two halves of a split mold, called **drag** and **cope**. This allows the pattern and its attached gates and risers to be easily removed to create the cavity, which is then filled by molten metal after the halves have been rejoined. The gates are cavities in the sand into which the metal is poured, and the risers are cavities that allow trapped air to escape the mold; the length of the gates and risers also produces a sufficient pressure to force the liquid metal to fill the mold.

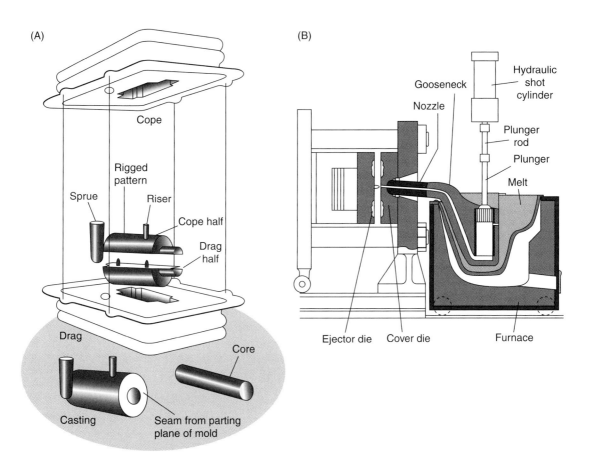

(A)

Cope

Rigged pattern

Sprue   Riser

Cope half

Drag half

Drag

Core

Casting

Seam from parting plane of mold

(B)

Gooseneck

Nozzle

Hydraulic shot cylinder

Plunger rod

Plunger

Melt

Ejector die   Cover die

Furnace

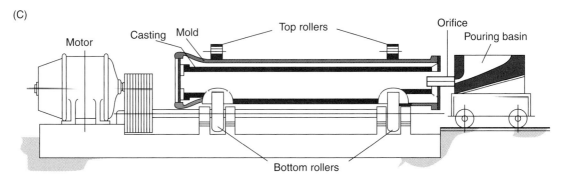

(C)

Motor

Casting   Mold

Top rollers

Orifice

Pouring basin

Bottom rollers

■ **FIGURE 7.8** Casting. (A) Exploded view of sand casting. The drag is the lower half of the mold into which the "drag half" of the pattern is inserted. The cope is the upper half of the mold. The metal is poured into the sprue, fills the cavity in the sand, and rises through the riser to generate enough pressure to fill the mold. (B) Die casting: a pump on the right injects the metal into the die. (C) Centrifugal casting of a pipe.

A modern variant of sand casting is **lost foam casting** where the pattern is made of styrofoam instead of wood. The pattern is embedded in casting sand, and the molten metal is poured directly into the styrofoam. The latter is molten and burned by the hot metal, which replaces it and fills the cavity. Lost foam casting is cost-effective because it reduces the amount of labor.

In **die casting**, Figure 7.8B, the molten metal is injected under pressure into a mold that is made of steel, called a die. The fabrication of the steel die is expensive, so that the technique is used for the manufacture of large numbers of small pieces that require good precision and a smooth surface. It is obviously restricted to metals with relatively low melting points such as zinc, aluminum, and magnesium.

Figure 7.8C shows the **centrifugal casting** of a pipe in which the mold rotates.

Statues and the blades of jet engines are cast by the **lost wax** method: a copy of the statue is made of wax, to which a type of wax plumbing has been added that directs the flow of the molten metal (usually bronze). The assembly is then immersed in a plaster-like slurry that is allowed to solidify. The whole is then inserted upside-down into a furnace that melts the wax and leaves the mold with a cavity, into which the molten metal is poured.

Continuous casting is an efficient way to cast steel bars intended for further shaping. In this method, the molten metal is poured into a cooled mold under which it is withdrawn in semi-solid condition. The mold can give the bars the desired profile. The latter can be cut to desired length or fed into a rolling machine where it is shaped further. The continuous casting method is more energy efficient than the pouring of ingots because it avoids the need for reheating the material before rolling. Figure 7.9. shows a schematic of continuous casting of steel. The method is also used to cast aluminum ingots ready for rolling into bars or plates. Because of the formation of hard aluminum oxide on the surface of the billet, the latter must be scalped before further processing.

7.4.2 Control of Grain Size

We know that fast cooling results in fine grain structure and slow cooling produces coarse grains. Rapid cooling is not always possible because the cooling rate is limited by the size of the piece or slow cooling may be required to avoid internal stresses. Equiaxed, fine-grained structure can be produced with the use of small additions of **grain refiners** to the melt. Commercially, titanium and boron are added to aluminum, and zirconium is added to magnesium.

Grain refiners act by **heterogeneous nucleation**. As the molten aluminum, for instance, approaches the solidification temperature, the grain refiner addition solidifies first and forms a large number of small particles in the melt. The interfacial energy γ_{gr} between the grain refiner and solid aluminum is smaller than the interfacial energy γ_{ls} between liquid and solid aluminum (see Equations (6.3) and (6.4)). Therefore, nucleation of the solid aluminum occurs at each grain refiner particle.

In the absence of grain refiners, nuclei are formed on the cooler mold surfaces. From there, they grow in the shape of columns and form the so-called columnar grain structure. Figure 7.10 is a schematic of the columnar structure grown without grain refiners.

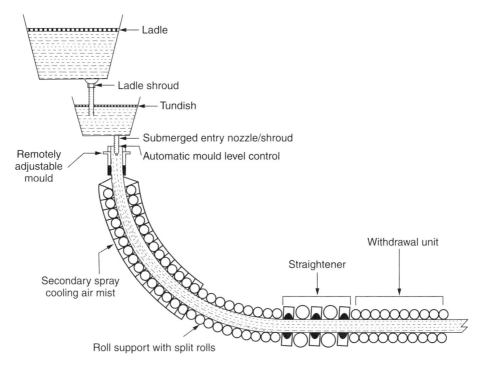

■ **FIGURE 7.9**   Continuous casting of steel. The molten steel is poured from the ladle into a tundish that controls the flow into a water-cooled copper mold in which a surface layer solidifies. The strand is cooled as it travels along the water-cooled rolls. It can then be cut into desired lengths or fed into a hot-roll machine that takes advantage of the high temperature of the steel.

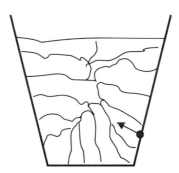

■ **FIGURE 7.10**   Formation of columnar structure. Nuclei form on the wall of the mold and grains grow in the shape of columns.

Heterogeneous nucleation is used in cloud seeding to make rain. Small particles of potassium bromide are distributed in the clouds with an aircraft. These particles serve as nucleation sites for the condensation of water vapor into raindrops.

7.4.3 Making Single Crystals

In order to obtain a single crystal, one must allow the formation of a single nucleus from which the material grows and suppress any further nucleation. The two most important methods are Czochralski crystal pulling used in the production of silicon boules in the semiconductor industry (see Chapter 11) and directional solidification used, for instance, in the production of single-crystal turbine blades for jet engines.

7.4.4 The Czochralski Single Crystal Pulling Method

A heated **crucible** maintains the molten material just at the melting temperature, and a **seed crystal** of the desired orientation is lowered into the melt (Figure 7.11, left). The material crystallizes around the seed, continuing its crystal structure. As more material is solidified, the resulting single crystal is slowly pulled upward (Figure 7.11, right). The crystal, called a **boule**, is rotated during growth in order to obtain a regular shape. For the solidification to take place heat must be removed from the system. This heat is conducted upward through the solid crystal, the seed, and its support.

7.4.5 Directional Solidification of Single Crystal Turbine Blades

Single crystal turbine blades are grown by the lost wax technique. The wax model of the blade is prepared in a modified shape: at its bottom is a **"pigtail"** that will be discussed shortly. The superalloy

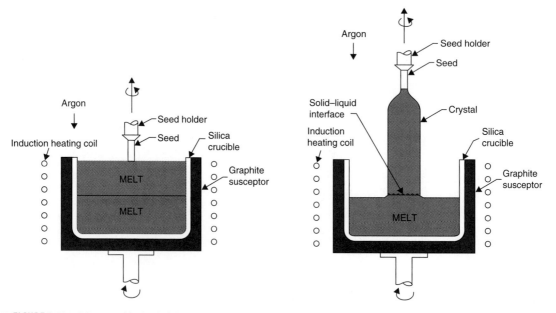

■ **FIGURE 7.11**   Schematic of the Czochralski single crystal pulling method. Left: initiation; a seed is put in contact with the melt. Right: the single crystal (color) grows from the seed and is pulled away from the melt while being rotated.

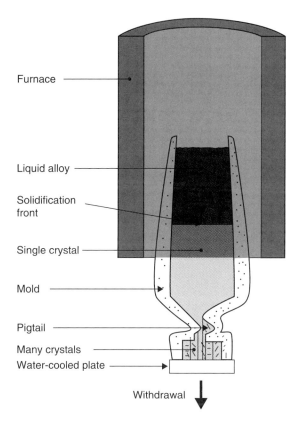

Furnace

Liquid alloy

Solidification front

Single crystal

Mold

Pigtail

Many crystals

Water-cooled plate

Withdrawal

■ **FIGURE 7.12**   Fabrication of a single crystal turbine blade by directional solidification. The mold filled with molten alloy is slowly pulled out of the furnace to allow directional solidification. Molten metal is shown dark; solid is light in color. As the solid-liquid interface moves up, most crystals terminate at the walls of the pigtail and one grain continues into the blade where it grows.

melt is then poured into the mold, and the latter is slowly pulled down from the furnace (Figure 7.12). Solidification starts below the pigtail, and several grains are nucleated and grow. As the piece is pulled out of the furnace, the solid region grows upward. Most "grains" terminate at the walls of the pigtail; only one grain (or crystal) extends into the blade region and grows by directional solidification as the structure continues to be pulled out of the furnace.

In commercial turbine blades, the wax model contains a hollow ceramic armature that provides for the passage of air that enters at the bottom of the blade and exits at its tail. This provides cooling of the blade in service.

7.5 **SURFACE PROCESSING OF STRUCTURAL MATERIALS**

There are many reasons to modify the surface of a material. One may wish to make it harder than the bulk to increase its resistance to wear or scratching; one may wish to increase its resistance to corrosion or to protect it from the hot gases of a jet engine; or one may simply desire an esthetically pleasing appearance. We will describe the most important techniques for surface modification together with their purposes.

7.5.1 Diffusion Treatments

Consider the fabrication of a gear. The surfaces of its teeth slide and rub against other teeth at high pressure; in such a situation it is important to avoid wear and abrasion of the teeth surfaces. The wear resistance of a metal is proportional to its hardness. However, making the gear out of a hard metal is not permissible: the material must be tough so that the teeth do not break under the load, and it must be soft enough so the gear can be machined economically; so gears are typically made of steel with about 0.2% carbon. The solution is to harden the surface by a surface treatment. The treatment of choice is carburization.

Carburization of Steel   Carburization of steel is the most widely used diffusion surface treatment. The intent is to boost the carbon concentration at the surface so that it can be made hard by subsequent quenching and tempering. The result is a case hardened layer at the surface. In the carburization process, carbon-rich gases, such as methane and CH_4-CO-H_2 mixtures, are made to flow over low- or medium-carbon steels (0.1–0.4 wt% C) maintained at temperatures of ~900°C. Pyrolysis of the gas at the metal surface releases elemental carbon that **diffuses** into the austenite, which can dissolve about 1.25 wt% C at this temperature. A surface layer, typically 1 mm thick, is enriched to about 1 wt% C. After quenching and tempering, a wear-resistant **case** consisting of a hard **martensite** surrounds the softer but tougher interior. Many automotive parts, machine components, and tools such as gears, camshafts, and chisels require this combination of properties and therefore are carburized.

Even harder steel surfaces can be produced by nitriding or carbo-nitriding. In nitriding, pyrolysis of ammonia at 525°C provides nitrogen that penetrates the steel. After two days, case layers extending 300 μm deep can be expected. Typical components that are nitrided include extrusion-, deep drawing-, and metal-forming tools and dies. Through diffusion processing at temperatures much below the melting point, such machined parts avoid thermal distortion. The resulting surfaces retain high hardness and are resistant to wear and to corrosion without requiring subsequent heat treatment.

Other commercial diffusion processes involve the introduction of the elements aluminum, boron, silicon, and chromium into the surface layers of metals. The corresponding aluminizing, boronizing, siliciding, and chromizing treatments yield surfaces that are considerably harder or more resistant to environmental attack than the original base metal. For example, coatings based on Al have been used for decades to enhance the resistance of materials to high temperature oxidation, hot corrosion, particle erosion, and wear. Metals undergoing aluminizing treatments include nickel and iron-base superalloys, heat-resistant alloys, and a variety of stainless steels. Aluminized components find use in diverse applications – nuclear reactors, aircraft, and chemical processing and coal gasification equipment.

The Mathematics of Diffusion   Diffusion may be defined as the migration of atoms (or molecules) by random movement. A current of atoms in one direction results from a gradient in concentration.

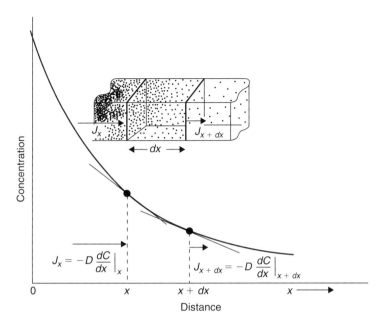

■ **FIGURE 7.13**   Diffusion of atoms from a surface into the bulk of a material.

The atoms move from a region of high concentration to one of low concentration. Consider the volume element extending from distance x to $x + dx$ in Figure 7.13.

At x, the concentration is higher than at $x + dx$. Atoms at distance x move at random, that is, from left to right and from right to left. Atoms at position $x + dx$ do the same. There are more atoms moving from point x than there are atoms moving from point $x + dx$. Therefore there is a net current of atoms from left to right, and this current is proportional to the difference in concentrations at the two ends of the volume element, namely, to the concentration gradient.

$$J = -D \, dC/dx \qquad (7.1)$$

The negative sign in the equation expresses the fact that a positive current flows along a negative gradient (decreasing concentration). Irrespective of concentration units, values of D, the diffusion coefficient or diffusivity, are expressed in units of m^2/s or cm^2/s.

Because of the concentration gradient, more atoms diffuse into our volume element through plane x than leave through plane $x + dx$. Therefore, the concentration of A atoms in the element increases.

$$dC/dt = -dJ/dx \qquad (7.2)$$

With Equation (7.1), we can write

$$dC/dt = D \, d^2C/dx^2$$

We recall that the concentration changes with space and time and rewrite the equation as

$$dC(x,t)/dt = D \, d^2C(x,t)/dx^2 \tag{7.3}$$

This partial differential equation describes **non-steady state** diffusion under conditions where D is assumed to be constant. The objective is to find $C(x,t)$ and the consequences that flow from it.

Let us apply this equation to diffusion into a semi-infinite solid (one that extends from $x = 0$ to $x = \infty$). If atoms are supplied at the boundary from an inexhaustible source so that the **concentration at the surface** remains a constant C_s, and it is assumed that the **initial concentration** in the matrix is C_o, the solution to Equation (7.3) gives the concentration of diffusing atoms $C(x,t)$ as a function of position and time as

$$[C(x,t) - C_o]/[C_s - C_o] = Erfc(z) \tag{7.4}$$

where z is

$$z = x/2(Dt)^{1/2} \tag{7.5}$$

Like trigonometric functions, values for the complementary error function are tabulated as a function of the argument z. Table 7.7 gives some values of the error function.

Thus a given concentration $C(x,t)$ is associated with a given value of z. The position where the concentration has that particular value advances in the solid as

$$z = x/2(Dt)^{1/2} \quad \text{or} \quad x = 2z\sqrt{Dt}$$

Table 7.7 Values of the Complementary Error Function Erfc.

Z	Erfc	Z	Erfc	Z	Erfc	Z	Erfc
0	1	0.40	0.5716	0.85	0.2293	1.6	0.0237
0.025	0.9718	0.45	0.5245	0.90	0.2030	1.7	0.0262
0.05	0.9436	0.50	0.4795	0.95	0.1791	1.8	0.0109
0.10	0.8875	0.55	0.4367	1.0	0.1573	1.9	0.0072
0.15	0.8320	0.60	0.3961	1.1	0.1198	2.0	0.0047
0.20	0.7773	0.65	0.3580	1.2	0.0897	2.2	0.0019
0.25	0.7237	0.70	0.3222	1.3	0.0660	2.4	0.0007
0.30	0.6714	0.75	0.2888	1.4	0.0477	2.6	0.0002
0.35	0.6206	0.80	0.2579	1.5	0.0339	2.7	0.0001

An example will illustrate the way to use the function: to simplify matters, let us assume that $C_o = 0$: There is none of the diffusing material in the body to start with. In this case, $Erfc = C(x,t)/C_S$. When $Erfc = 0.4795$, namely, when the diffused concentration is about 48% of C_S, the argument is $z = 0.5 = x/2(Dt)^{1/2}$. In other words, the front where $C(x,t)$ is about half surface concentration advances with time as $x = \sqrt{Dt}$. In a similar vein, the concentration $C(x,t) = 0.7237C_S$ advances with time t at the depth $z = 0.25 = x/2\sqrt{Dt}$ or $x = 0.5\sqrt{Dt}$.

Note that all concentrations move in the solid as $x = 2z\sqrt{Dt}$, namely, as the square root of time. (They move rapidly at very short times and more slowly as time goes on.) The diffusion distance is proportional to the square root of the diffusion coefficient D, and the latter is a rapid function of temperature: the higher the temperature, the more rapid is diffusion. We have already discussed the temperature-dependence of the diffusion coefficient in Equation (6.7) and Figure 6.5. Figure 7.14 illustrates the rapid temperature dependence of the diffusion coefficient of nitrogen in iron.

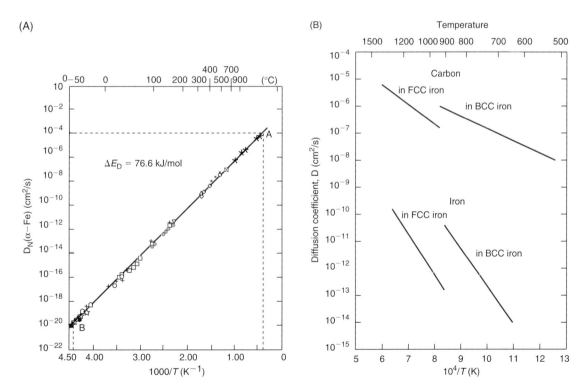

■ **FIGURE 7.14**   (A) Arrhenius plot of the diffusion coefficient of nitrogen in iron. Note that the sense of the abscissa is reversed from that of Figure 6.5 and that the diffusion coefficient is given in cm²/s rather than m²/s. (B) Arrhenius plot of the diffusion coefficients of carbon and iron in iron. After D. N. Beshers, *Diffusion*, American Society of Metals, Metals Park, OH (1973).

EXAMPLE 7.1   Steel gears containing 0.20 wt% C are exposed to a carbon-containing gas atmosphere that maintains a surface concentration of carbon equal to 1.0 wt%. How long will it take the carbon concentration to reach a level of 0.5 wt% at a distance 0.1 cm below the surface? Assume $D = 5.1 \times 10^{-6}$ cm^2/sec.

ANSWER   Substitution in Equation (7.4) yields $(0.5 - 0.2)/(1.0 - 0.2) = 0.375$ or $Erfc[(x/2\,(Dt)^{1/2}] = 0.375$. From Table 6.2. the argument is estimated to be $z = 0.63$ by interpolation. Therefore, $x/2(Dt)^{1/2} = 0.63$. Substituting for x and D,

$$t = (x/0.63)^2\, 4D = (0.1 \text{ cm})^2/(0.63)^2 \times 4 \times 5.1 \times 10^{-6} \text{ cm}^2/s = 1,235 \text{ s or } 20.6 \text{ min.}$$

Because the diffusion zone is small compared to typical gear dimensions, the assumption of an "infinite" solid is reasonable, validating the use of Equation (7.4).

7.5.2 Laser Hardening

Laser hardening transforms a thin layer of steel into martensite. A piece of steel is irradiated with a short pulse of high-power laser light that heats a surface layer, approximately 1 mm thick, into the austenite region. Because of the cold metallic substrate, this heated layer is cooled very rapidly and forms martensite. This technique has a number of desirable properties.

1. The technique is local and permits the treatment of selected areas. No subsequent machining of the surface is required.

2. Minimum heat input is required so that the treatment avoids thermal warping.

3. The treatment is self-quenching; no further thermal treatment is required.

Laser hardening is effective also on cast iron. The valve seats on a cast iron engine block are selectively hardened by this method.

7.5.3 Coatings

The resistance of metals to corrosion and to wear can also be increased by coating with an appropriate material. Often, coatings are applied to replace material that has been removed by wear. **Thermal Barrier Coatings** (TBCs) are used in turbines and jet engines to protect the metal turbine blades from the high temperature of the gases. The TBCs consist of partially stabilized zirconium oxide, a ceramic material that possesses a low thermal conductivity and a thermal expansion coefficient close to that of metals. The latter property is important because it avoids large stresses between the coating and the substrate when the temperature of the blade increases at start-up and decreases again at the end of operation. Coatings are applied with a variety of techniques, which we examine briefly.

Physical Vapor Deposition   The physical vapor deposition (PVD) techniques of **evaporation** and **sputtering** are the means employed to deposit thin films. Evaporation involves heating the source metal until it evaporates at appreciable rates. The source is heated by an electron beam that is accelerated to a high voltage and bent by a magnetic field so that it impinges on the surface of the charge to be evaporated (Figure 7.15). This technique assures the purity of the deposited material because the hot filament, which also evaporates tungsten, is hidden below the charge and the crucible can be cooled.

The process is carried out in a chamber (Figure 7.16) operating at vacuum pressures that are typically 10^{-10} atm. The substrates, carefully positioned relative to the evaporation source, are rotated to yield

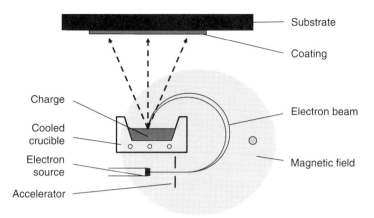

■ **FIGURE 7.15**   e-beam evaporation source. The hot filament is hidden below the crucible, and the beam is bent in order to impinge on the charge.

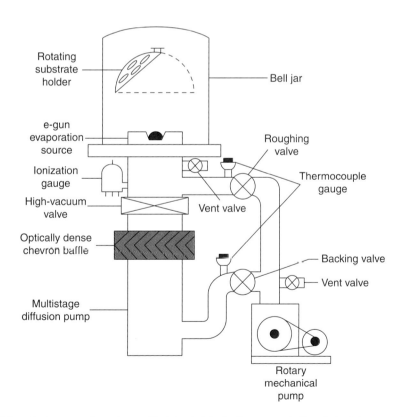

■ **FIGURE 7.16**   Schematic of an evaporation system used for integrated circuits. Atoms evaporated from the heated source condense on the rotating Si wafers.

uniform metal coverage. The thermal barrier coatings of turbine blades are commercially deposited by electron beam evaporation.

Vapor deposition of alloys presents a problem. During evaporation of alloys the constituent elements evaporate more or less independently. Therefore the composition of the coating is different from that of the source.

Examples of the use of evaporation include the deposition of thermal barrier coatings of ZrO_2 on turbine blades operating at high temperatures (in jet engines or power plants); metallization of polymer films and the deposition of decorative coatings or of optical coatings on eyeglasses and windows.

Sputtering   In this technique the metal or alloy to be deposited is fashioned into a plate and placed parallel to the substrate within a chamber. An inert gas (Ar) at low pressure (~0.01 atm) is introduced, and a potential difference of a few thousand volts DC is established between source and substrate. Under these conditions an electrical **glow discharge** is sustained between electrodes, much like what occurs in the neon light tube. The gas atoms become ionized and together with electrons and neutral atoms create a plasma. Energetic positive ions are drawn to the negative source where they impact and dislodge atoms. These fly through the sputtering system (Figure 7.17), deposit on the substrate, and build the film. Insulating material can be deposited by applying an AC voltage between source and substrate. Sputtering has the disadvantage of low film deposition rates, but it has the great advantage of maintaining stoichiometry in deposited alloy films. Deposition of magnetic alloy coatings on discs for data storage applications is a growing application.

Sputtering and evaporation are line-of-sight methods: the atoms or molecules travel in straight lines toward the material to be coated. When the latter has a complex geometry, with holes or overhangs,

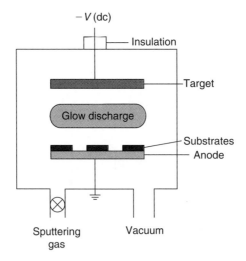

■ **FIGURE 7.17**   Schematic of sputtering system. Ionized sputtering gas ions impact the cathode target surface, ejecting metal atoms which deposit on the substrate.

part of the surface may not be reached by the straight-moving atoms. For such geometries, one employs Chemical Vapor Deposition.

Chemical Vapor Deposition (CVD)   Many important coatings are synthesized from high temperature gas mixtures. As an example consider the TiN, TiC, and Al_2O_3 compound coatings deposited on tungsten carbide metal machining and cutting tools. (The gold color commonly seen on cutting tools is due to a TiN coating). Each coating has a hardness double that of hardened tool steel, a hard material in its own right. Only 5–10 µm (5,000–10,000 nm) thick layers are deposited; this coating thickness is enough to bestow considerable wear resistance to the underlying tool so that it can last 5 to 10 times longer than uncoated tools before requiring sharpening. A CVD reactor is shown schematically in Figure 7.18.

Input gases react at the substrate surface and form the compound deposit while gaseous products are swept out of the reactor. Some chemical reactions used in CVD are:

$$2TiCl_{4(g)} + N_{2(g)} + 4H_{2(g)} = 2TiN(s) + 8HCl_{(g)} \text{ at } 1,000°C \tag{7.6a}$$

$$TiCl_{4(g)} + CH_{4(g)} = TiC(s) + 4HCl_{(g)} \text{ at } 1,000°C \tag{7.6b}$$

$$2AlCl_{3(g)} + 3CO_{2(g)} + 3H_{2(g)} = Al_2O_3(s) + 3CO_{(g)} + 6HCl_{(g)} \text{ at } 1,000°C \tag{7.6c}$$

Each reaction takes place at roughly $1,000°C$ under approximate equilibrium conditions. All of the reactants and products, with the exception of the coatings denoted by (s), are gaseous; this is a general feature of CVD reactions. Several coated tools are reproduced in Figure 7.19 together with the structure of a modern multilayer coating on a WC tool substrate.

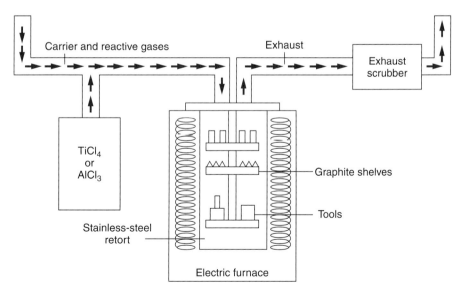

■ **FIGURE 7.18**   Schematic of a chemical vapor deposition reactor used to deposit hard coatings of TiN, TiC, or Al_2O_3.

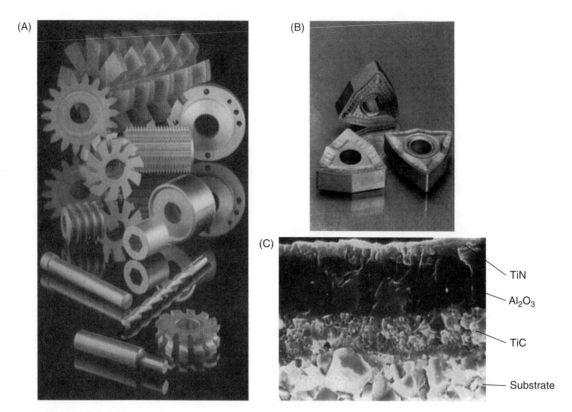

■ **FIGURE 7.19** (A) Assorted cutting tools coated with TiN. (B) Multilayer coated lathe cutting tool inserts. (C): Scanning electron microscope image of a trilayer coating, deposited by CVD on a WC cutting tool substrate ($\times$3,500). The TiC bonds well to the WC. Protection against wear is provided by the hard Al_2O_3 layer. The TiN layer is chemically stable and provides low friction. A: Courtesy of Multi-Arc Scientific Coatings. C: Courtesy S. Wertheimer, ISCAR.

Halogen Lamps

Halogen lamps can be considered as CVD reactors in your home. These lamps are brighter than ordinary light bulbs because the tungsten filament is operated at much higher temperatures. The reason they do not fail by evaporation is a chemical reaction with a halogen gas (iodine and bromine) that is added to the argon. At low temperatures the reaction runs $W + 6I \rightarrow WI_6$; the latter is a gas. At the high temperatures of the filament, the WI_6 decomposes. Thus the halogen picks up the evaporated tungsten deposited on the walls and brings it back to the filament. Hotter spots on the filament receive more tungsten, becoming thicker and less resistant. Thus the halogen not only conserves the filament but equalizes its temperature.

Welded Coatings Coatings can be applied by arc welding. A rod of the material to be deposited strikes an electric arc with the substrate. This arc produces a very high local temperature that melts the material from the rod and deposits it on the surface. This technique deposits thick coatings (several

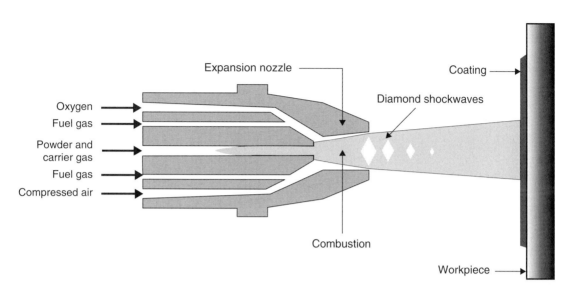

■ FIGURE 7.20 Schematic of an HVOF spray gun.

millimeters) and is most appropriate for repairs, for instance after excessive wear. Welded coatings are rough and need post-deposition machining when a smooth surface is required.

Thermal Spray Coatings In thermal spray coating techniques, the material to be deposited is partially molten in a spray gun and projected against the substrate at very high velocities, approaching 900 m/s (three times the speed of sound). The result is a coating approaching theoretical density with bond strengths to the substrate above 10,000 psi = 70 MPa. The hot gas jet can be produced by a plasma discharge in the torch (plasma spray) or by a flame (HVOF or High-Velocity Oxy-Fuel deposition). In HVOF, fuel gas (propane, propylene, or hydrogen) and oxygen are supplied at high pressure; combustion occurs outside the nozzle but within an air cap supplied with compressed air (Figure 7.20). The compressed air pinches and accelerates the flame and acts as a coolant for the HVOF gun. Powder is fed at high pressure axially from the center of the nozzle.

HVOF has become the quality standard for carbide and cermet materials. Ceramics require the higher temperatures of plasma spray.

Spray techniques produce relatively rough surfaces that may require post-deposition machining of polishing.

Electrochemical Deposition In this technique, salts of the metal to be deposited are dissolved in water and placed in an electrochemical cell (Figure 7.21). An electric potential is established between an inert electrode and the substrate. The material to be coated constitutes the negative electrode (cathode) of the system. The positive metallic ions are driven toward the negative electrode on which they are neutralized and deposited. The anode can consist of the metal to be deposited and is progressively

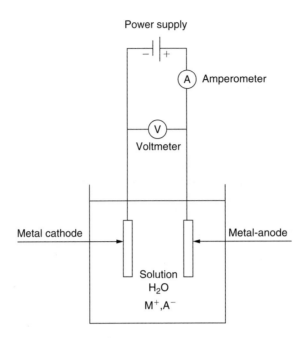

dissolved in the process. The science and technique of electrodeposition are described in more detail in Chapter 15.

Electroless deposition is related to electroplating in that the ions of the metal to be deposited are dissolved in an electrolyte. In electroless plating, the electrons for the neutralization (reduction) of the ions are not provided by an outside circuit, but by a chemical contained in the solution. In the case of electroless nickel plating, the reducing chemical is sodium hypophosphite. As a consequence, the nickel contains between 2% and 12% phosphorus. The low-phosphorous nickel is harder, and the high-phosphorous nickel is more corrosion resistant.

■ SUMMARY

1. Ferrous alloys comprise plain carbon steels; HSLA (high-strength low alloy steels); stainless steels; tool steels; and cast irons.

2. Plain carbon steels contain up to 0.9 wt% carbon and a small amount of Mn. The latter serves to scavenge the impurities sulfur and phosphorus. They are available as annealed, rolled, and tempered steels.

3. LAHS steels contain less than 2 wt% Cr, Mo, or Ni. These additions provide a strength about 50% above that of plain carbon steels.

4. Stainless steels are made corrosion resistant by the addition of about 17 wt% chromium. The latter forms a protective Cr_2O_3 layer on the surface.

5. Stainless steels exist in ferritic, austenitic, and martensitic form. Ferritic stainless steels contain Cr and little carbon; they exist in BCC only. Austenitic steels exist in FCC structure because of the addition of Ni; they are very ductile. Martensitic steels contain Cr and 0.75–1% C. They are quenched and tempered and are very hard.

6. Tool steels are very hard (R_C 60). They contain appreciable amounts of W, Mo, and V. They are used as cutting tools in lathes, milling machines, saws, and so on.

7. Gray cast irons contain up to 3.5% carbon and 2.5% Si. They solidify into pearlite or ferrite; the excess carbon becomes graphite in flake form. Graphite flakes make gray iron brittle, self lubricating, and sound absorbing. Gray cast irons are corrosion resistant.

8. White iron contains about 2% carbon and 1.2% silicon. It solidifies into ferrite and cementite. It is abrasion resistant but very brittle.

9. Cast irons can be made ductile or malleable by treatment that gives the graphite a spherical shape (nodular).

10. Aluminum alloys exist as wrought and heat treatable alloys. Wrought aluminum is strengthened by cold working. Heat treatable aluminum is strengthened by precipitation hardening. The alloy is first homogenized at high temperature, then quenched and aged.

11. Copper is available as oxygen-free, high-conductivity (OFHC) copper used for electric conductors and as tough pitch copper of lower purity used mostly for architectural elements. It acquires relatively high strength by cold working. It has excellent corrosion resistance and acquires an attractive green color by weathering.

12. Brass is a copper-zinc alloy that is easy to cast and machine and possesses good corrosion resistance.

13. Bronze is copper alloyed with tin or other metals except zinc and nickel. Its beautiful appearance in weathering is responsible for its wide use in statuary. It is used in sliding bearing material because it does not weld to steel.

14. Copper beryllium is a strong precipitation-hardened material.

15. Manganese alloys are used for their low specific weight.

16. Titanium alloys combine high strength with low weight. Their superior corrosion resistance make them suitable for orthopedic and architectural applications.

17. Superalloys consist mainly of nickel, chromium, and molybdenum. They were developed for jet engines and maintain their strength and corrosion resistance at high temperature.

18. Casting is the industrial solidification of materials from the melt. A fine grain structure can be obtained in certain metals (Al) by the introduction of grain refiners. These accelerate nucleation by presenting a surface with low interfacial energy.

19. Single crystals are obtained by suppressing nucleation except at one point. In the Czochralski method, nucleation occurs at one seed that is the only material below the melting point. Turbine blades are grown as single crystals by directional cooling: nucleation occurs in a small space and the material grows from there.

20. The surface hardness, the fatigue resistance, the thermal conductivity, the resistance to corrosion, and the appearance of a metal can be improved by modifying its surface.

21. Surface hardening can be achieved by diffusion of carbon or nitrogen, by rapid heating and quenching with a laser beam, or by shot peening.

22. Coating of a surface is achieved by physical vapor deposition (evaporation or sputtering), by chemical vapor deposition (where a reactive gas decomposes on the surface and deposits the desired species), or by electroplating.

■ KEY TERMS

A
aluminum
    heat treatable, 179, 203
    wrought, 179, 203, 206
aluminum alloys, **177**
austenitic stainless steel, 173

B
beryllium copper, 180
boule, 190
brass, 180
bronze, 180

C
carbo-nitriding, 192
carburization, 192
cast iron, **175**
casting, 186
centrifugal casting, 188
chemical vapor deposition
    (CVD), 199
coatings, 196
columnar grain structure, 188
continuous casting, 188
control of grain size, 188
cope, 186
copper, 180
copper alloys, **180**

crucible, 190
Czochralski, 190
Czochralski crystal pulling, 190

D
die casting, 188
diffusion treatments, 192
directional solidification, 190
drag, 186
dross, 186

E
electroless deposition, 202
evaporation, 196

F
ferritic stainless steel, 173
ferrous alloys, 172

G
galvanic deposition, 201
gate, 186
glow discharge, 198
grain refiners, 188
graphite, 176
gray cast iron, **176**

H
hardenability, 173
heterogeneous nucleation, 188

high-strength low alloy steels
    (HSLA), 173

I
ingots, 186
initial concentration, 194
interfacial energy, 188

L
Laser hardening, 196
lost foam casting, 188
lost wax casting, 188

M
magnesium, 181
malleable cast iron, **177**
manganese, 173
martensitic stainless steels,
    173
mathematics of diffusion, 192

N
Ni_3Al, 184
nitriding, 192
nodular cast iron, **177**
nonferrous alloys, **177**

O
OFHC, 180

■ REFERENCES FOR FURTHER READING

[1] ASM Handbook, Vol. 1, *Properties and Solution: Irons, Steels, and High Performance Alloys*, ASM International Materials Park, OH (2004).

[2] ASM Handbook, Vol. 2, *Properties and Solution: Nonferrous Alloys and Special-Purpose Materials*, ASM International Materials Park, OH (2004).

■ PROBLEMS AND QUESTIONS

7.1. What is an automobile made of? From the web or an automotive magazine, what metals are used for important parts of an automobile? Choose at least 4 parts, name the metal used, and justify the choice by the match between the requirements and the properties of the material.

7.2. If you need a tempered shaft with yield strength higher than 950 MPa, why would you prefer 4340 steel to a 1080 in tempered form? What is the role of the nickel addition?

7.3. Use the information in Table 7.1. Draw a bar graph of the mechanical properties of plain carbon steels as a function of carbon content. For each composition, draw the tensile and yield strengths of the three forms vertically and the ductility vertically below the abscissa. Explain the effect of carbon.

7.4. What properties make 304 steel preferred for cooking pots? Why not 430 steel?

7.5. Why would you use 1080 tempered steel for a chisel rather than 440°C or T1?

7.6. Sketch the stress-strain curves for 304, 430, and 440A steels and explain the role of the various alloying elements. (Draw the elastic portion to proper scale!)

7.7. Can you harden 304 steel by quenching and tempering? Explain your answer.

7.8. Transform the wt% composition of T1 tool steel into at% and explain the role of carbon in this alloy.

7.9. Discuss the influence of carbon, nickel, chromium, and tungsten additions on the allotropy of iron.

7.10. How would you make the enclosure of a gearbox, and what material would you choose? (Gears are noisy.)

7.11. Select a material for piston rings and justify your choice.

7.12. Among objects you encounter today, describe one made of copper, one of bronze, and one of brass and describe the performance requirement that justifies the choice.

7.13. What properties of bronze are responsible for making the Bronze Age possible? (Cite more than mechanical properties; consider the state of the technology at the time.)

7.14. What element makes steel stainless?

7.15. What is the role of nickel in stainless steel?

7.16. What is the role of Mn additions in steels?

7.17. Compare the processing and properties of annealed, rolled, and tempered steel.

7.18. Cite an application for each of annealed, rolled, and tempered steel.

7.19. What are the alloying elements and their concentrations in low alloy high strength steels? How do they strengthen the steel?

7.20. Name the three classes of stainless steel, describe their defining properties, and cite an application for each.

7.21. Chromium is BCC. What is its effect on the phase diagram of stainless steels?

7.22. Nickel and carbon are FCC. What is their effect on the phase diagram of stainless steels?

7.23. What is the outstanding property of tool steels?

7.24. Cite one alloying element in tool steels and its concentration.

7.25. Compare the carbon content of steels and cast irons. Looking at an iron-carbon phase diagram, explain the benefit of the high carbon concentration in cast irons.

7.26. What is the role of silicon in cast iron?

7.27. What is the microstructure (phase content) of gray cast iron? Describe two advantages and a major disadvantage of this microstructure.

7.28. Considering the mechanical weakness of gray cast iron, what is the strategy for making it ductile?

7.29. What is the microstructure (i.e., phase content) of white cast iron? What is the benefit of this microstructure?

7.30. What is wrought aluminum, and how is it strengthened?

7.31. What is heat treatable aluminum, and how is it strengthened?

7.32. Describe the processing of heat treatable aluminum. Why is quenching not sufficient?

7.33. Explain how quenching and ageing provides the aluminum with very fine precipitates. Use the kinetics of phase transformation.

7.34. Describe what overageing aluminum does to its mechanical properties and why.

7.35. Describe the properties of copper that justify its architectural uses.

7.36. What is OFHC copper, how is it processed, and what is it used for?

7.37. What is the composition of brass? Cite an attractive property and a use.

7.38. What is the composition of bronze? Cite a mechanical use of this alloy.

7.39. Why has bronze been used for statuary since very early times? Give two possible reasons, one for use and one for processing.

7.40. Describe the desirable properties and a use of copper beryllium alloys.

7.41. Why do automobile enthusiasts use "mag wheels"?

7.42. Compare the properties of titanium to those of steel.

7.43. What makes titanium expensive?

7.44. What property motivated Frank Gehry to clad the Bilbao Guggenheim Museum in titanium?

7.45. What property justifies the use of titanium alloys as a biomaterial?

7.46. What casting process would you recommend to make the following items?

   a. A large bronze ship propeller.

   b. Steel railroad wheels.

c. Aluminum sole plates for electric steam irons.

d. Iron base for a lathe.

e. Aluminum engine block for a lawn mower.

f. Steel pressure vessel tubes.

g. Nickel base alloy turbine blades.

h. Large brass plumbing valves.

7.47. a. From Figure 7.15, estimate the diffusion coefficient of nitrogen in iron at 0°C, 200°C, and at 700°C.

b. At what distance from the surface is the nitrogen concentration about half the surface concentration C_S in 1 s, in 16 s, in 1 min, and in 1 hr at 700°C?

c. At what distance is the nitrogen concentration 23% of CS at the same times and temperature?

d. What are the distances corresponding to the concentrations and times of part (b) when the temperature is 200°C?

7.48. Make an Arrhenius plot of the diffusion coefficients

T(°C)	0	100	200	300	400	500
D(m²/s) =	$7 \cdot 10^{-22}$	$4 \cdot 10^{-18}$	$1.5 \cdot 10^{-15}$	$2.5 \cdot 10^{-14}$	$5 \cdot 10^{-13}$	$3.5 \cdot 10^{-12}$

7.49. A 1010 steel is carburized by a source that maintains a surface carbon content of 2.0 wt% C. It is desired to produce an 0.80 wt% C concentration 0.05 cm below the steel surface after a 1 hr treatment. At what temperature should carburization be carried out?

7.50. In your own words, describe the entire sequence of processes in the fabrication of a gear.

7.51. You wish to fabricate sunglasses that consist of plastic (see Chapter 9) covered with a semitransparent metal film. (One can see through the metal because it is very thin.) Select a method for depositing the metal film and describe the manufacturing process.

7.52. Compare physical and chemical vapor deposition. Describe the advantages and disadvantages of each for the following applications:

a. Deposition of metal onto a polymer.

b. Deposition of coatings onto a piece with a complex shape.

c. Deposition of ceramic coatings with very high melting point.

d. Rapidity of deposition.

7.53. Search the web or the literature and name an object that has been coated by

a. Evaporation.

b. Sputtering.

c. Chemical vapor deposition.

d. Electroplating.

e. HVOF.

f. Plasma spray.

g. Welding.

For each, state why the particular method is the most appropriate. Simplicity and cost are important considerations.

7.54. Describe or sketch the casting process.

7.55. How does one obtain a fine-grained cast aluminum?

7.56. What is the principle used in making single crystals?

7.57. Sketch the Czochralski method for growing single crystals.

7.58. Sketch or describe the directional solidification method of growing single crystal turbine blades.

7.59. Describe one method for obtaining a hard surface in a metal.

7.60. Describe physical vapor deposition. (A sketch is optional but may help.)

7.61. Describe chemical vapor deposition.

7.62. Compare the respective advantages of physical and chemical vapor deposition.

7.63. For what application were superalloys developed? What are their outstanding properties?

7.64. What, approximately, is the composition of superalloys?

Chapter 8

Ceramics

Ceramics include natural stones; clays and porcelains; electric insulators; abrasives; glass; and cement. These materials are hard and brittle; they do not conduct electricity and are often transparent. These properties are the consequences of a filled valence band. The chemical bond is covalent for elemental solids and ionic or mixed covalent-ionic for compounds. The nature of the chemical bond controls the crystal structure of the materials. These bonds are generally stronger than in metals and give the solid a high melting temperature that prevents casting from a melt. Ceramics cannot be machined the way metals are. Ceramic pieces are fabricated by forming a paste, consisting of the ceramic powder and water, into a near-final shape and solidifying it by firing. Firing causes sintering of the ceramic particles through diffusion of atoms or molecules. Glass is an amorphous silicon oxide with additions of sodium, magnesium, or boron. These additions form positive ions that neutralize the oxygen atoms and allow a disordered structure. Glass does not have a melting point but increases its viscosity, upon cooling, to values so high that the glass cannot be deformed at room temperature. The fabrication of glass objects makes use of its high viscosity. Cement is a ceramic that hardens by a chemical reaction with water. Hardening increases with time and reaches its final value in more than a year.

LEARNING OBJECTIVES

After studying this chapter, the student will be able to:

1. Name the distinguishing mechanical, electrical, and optical properties of ceramics and relate them to their electronic band structure.

2. Name the principal groups of traditional ceramics and name their most important constituents.

3. Name some modern high-performance ceramics and some of their applications.

4. Describe the geometry of structures based on the sp^3 bonding of silicon- and carbon-based ceramics.

5. Name and describe the several allotropes of carbon.

6. Describe the fundamental elements of the structure of glass and the role of structure modifiers.

7. Distinguish the solidification mechanisms of crystalline and amorphous solids.

8. Define the glass transition temperature.

9. Design the processing of crystalline ceramics at room temperature and the final solidification.

10. Explain how sintering provides the ceramic with its final solidity.

11. Distinguish the solidification phenomena of crystalline ceramics, glass, and cement.

12. Describe the main fabrication techniques of glassy objects.

13. Process tempered glass and explain its mechanical properties.

14. Name the principal components of cement and explain how it solidifies.

8.1 THE TYPES OF CERAMICS AND THEIR DEFINING PROPERTIES

The class of ceramics includes all inorganic solids that are not metallic. It is convenient to distinguish four main groups of structural ceramics:

1. Traditional ceramics such as stones, clay products, refractories, and abrasives.

2. Synthetic high-performance ceramics (including semiconductors).

3. Glass.

4. Cement and concrete.

Stones, clay, and porcelain are natural ceramics. Clay and porcelain have been used for many centuries for the fabrication of dishes, vases, and sanitary equipment. Porcelain is also used as an electric insulator in furnaces, switches, and in high-voltage power transmission lines. The synthetic ceramics have recently assumed critical importance in applications where extreme hardness, wear resistance or strength, and chemical stability at high temperatures are required. The low weight of the silicon nitride is also exploited in turbocharger turbines for their low inertia (Figure 1.11) and in high-speed ball bearings, for instance in dentist's drills.

In this chapter we are concerned with structural ceramics. Ceramics for functional applications will be examined in Chapters 11 to 14.

We saw in Chapter 1 that the valence band of ceramics is entirely filled with electrons. According to Pauli's exclusion principle, no change is possible in the shape, position, motion, or energy of the valence electrons. As a consequence, ceramics present the following properties.

- Ceramics cannot deform plastically: they are very hard and brittle.

- The chemical bond of ceramics is generally stronger than that of metals; consequently, ceramics have a higher melting point than other materials.

These features of ceramics have consequences for their utilization, which we review briefly here.

8.1.1 Mechanical Performance of Ceramics

Chapters 1 and 2 discuss the mechanical properties of ceramics. We know that these materials are unreliable under tensile stresses because the length of microcracks in the material is unknown. The designer must take care to subject them to compressive stresses only.

Many ceramics are porous. For certain applications, notably in bricks, porosity is introduced intentionally and is quite large. A small amount of porosity is present in most ceramics as a result of their processing, as we shall see below. Porosity decreases the elastic modulus as well as the fracture strength of the material. In high-performance applications such as ceramic ball bearings, care is taken to eliminate the porosity.

8.1.2 Thermal Properties of Ceramics

The covalent-ionic bonds of ceramics are usually stronger than the metallic bond and most ceramics are chemically and mechanically stable at higher temperatures than metals; therefore ceramics are used as **refractory linings** in furnaces and in crucibles for the melting of metals. Table 8.1 shows the thermal properties of a few representative ceramics and those of some metals and dry air for comparison.

Ceramics are also used as **thermal insulators**. The thermal conductivity of ceramics is generally lower than that of metals but varies widely. Diamond has the highest thermal conductivity known. Silicon nitride and alumina have a thermal conductivity similar to that of stainless steel. Zirconium oxide is the best thermal insulator and has a thermal expansion coefficient similar to that of metals. For this reason it is used as a thermal barrier coating on turbine blades; its low thermal conductivity provides protection to the blade, and its thermal expansion coefficient allows it to expand and contract approximately together with the metallic blade, minimizing the risk of spalling off. Dry air has lower thermal conductivity than any material, and vacuum transmits heat only as radiation. Construction bricks are intentionally porous in order to reduce heat conduction through the solid material. Thermal insulation of homes is provided by glass fibers; glass itself has a low thermal conductivity, but most of the insulation is provided by the air trapped between the fibers. The Space Shuttle orbiter shown in Figure 8.1 also relies on ceramics for thermal insulation tiles composed largely of SiO_2 and Al_2O_3 fibers. These tiles are often coated with high-thermal-emissivity borosilicate glass layers to reject heat by radiation. Some 30,000 tiles protect the underlying aluminum frame from heating appreciably. The orbiter nose must withstand reentry temperatures as high as 1,450°C and is therefore made of a very

Table 8.1 Thermal Properties of Ceramics (Representative Values).

	Thermal conductivity k W/mK (at 20°C)	Thermal expansion α 10^{-6} (°C)$^{-1}$ (at 20°C)	Melting point °C
Diamond	900–3000	1	4,350
Graphite	100–190	2–5	
Alumina	25	7.4	2,050
Silicon nitride	30	3.1	dissociates
Magnesium oxide	36	10	2,800
TiO$_2$	6.5	7.1	1,870
Zirconia	2	9.6	2,715
Fused quartz	1.5	0.4	
Glass soda lime	1.1	9	
Glass pyrex	1.1	3.3	
Dry brick	0.04		
Cement mortar	0.5	10	
Cu	400	17	1,083
Stainless steel	15	10–17	~1,500
Air	0.025		

special ceramic, a carbon/carbon composite (see Section 9.8.3). In these and many other applications, there are no viable substitutes for ceramics.

Thermal shock is a problem with many ceramics, especially glass and zirconium oxide. If one pours cold water into a container made of soda-lime glass, the latter breaks. The cold surface of the glass shrinks because of thermal expansion. This establishes a strain ε between the surface and the hot interior and creates a stress $\sigma = \varepsilon E$, where E is Young's modulus. If the thermal conductivity k of the ceramic is small, the temperature gradient and the strain ε are large; the stress exceeds the fracture strength σ_f of the material, which breaks. This is thermal shock. If the material has a high thermal conductivity k, heat flows from the bulk to the surface; the gradient of temperature, the strain, and the stress are relatively low, and the material does not break. From the above, it is intuitive that the thermal shock resistance TSR can be expressed as

$$\text{TSR} = \frac{k\sigma_f}{\alpha E} \tag{8.1}$$

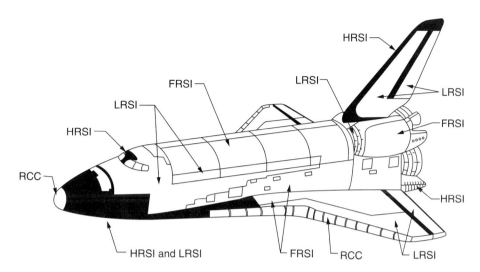

■ **FIGURE 8.1**   Thermal protection system for the Space Shuttle. HRSI: high-temperature reusable surface insulation (1,260°C); FRSI: felt reusable surface insulation (430°C); RCC: reinforced carbon/carbon composite (1,450°C). Temperatures that must be withstood are indicated. Reprinted from S. Musikant, *What Every Engineer Should Know about Ceramics*, p. 149. Courtesy of Marcel Dekker, Inc.

The best known example of thermal shock resistance is that of Pyrex, a glass with a low thermal expansion coefficient α, which will be discussed in Section 8.5.1. It is easily apparent from the data of Table 8.1 that zirconia has low resistance to thermal shock. This prevents its use as a wear-resistant coating; it fractures because of frictional heat.

8.2 TRADITIONAL CERAMICS

8.2.1 Stone

Stone and rock are ceramics that have always been important building materials. The popular ones are granite, limestone, and marble. Granite is a mixture of feldspar (aluminosilicate containing sodium, potassium, or calcium), quartz (SiO_2), and mica (aluminosilicate). Stones are agglomerates of crystallites held together by a glassy network. Their brittle nature facilitates quarrying and polishing; their low coefficients of thermal expansion ($\sim 8 \times 10^{-6}$°C^{-1}), relatively low density (~ 2.5–2.7 g/cm^3), and low tendency to absorb water are beneficial attributes in structural applications.

8.2.2 Clay Products

The term *ceramics* has traditionally meant fired **whiteware** and **structural clay** products (Figure 8.2). Their common ingredient is clay, which varies widely in chemical, mineralogical, and physical characteristics. Clay basically consists of aluminosilicate layers in the form of tiny crystalline platelets

■ **FIGURE 8.2**   American Indian earthenware bowl. Earthenware contains iron ions. When fired in an oxidizing atmosphere, the product contains Fe_2O_3 and has the familiar red color of bricks. When fired in a reducing atmosphere, the product contains Fe_3O_4 and is black as shown here. Maria Martinez (1997–1980) Black on Black Avanyu Bowl, Courtesy Mark Sublette Medicine Man Gallery, Tucson, AZ.

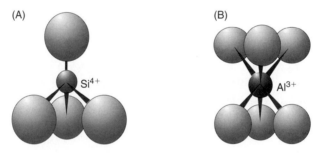

■ **FIGURE 8.3**   (A) SiO_4 tetrahedron. (B) AlO_6 octahedron.

that readily slide over each other. These layers are formed by a sheet structure of SiO_4^{-4} tetrahedra (Figure 8.3A) attached to sheets of AlO_6 octahedra (Figure 8.3B); electrical neutrality is maintained by the sharing of oxygen atoms and by hydrogen, sodium, and potassium or magnesium ions.

Figure 8.4 shows the structure of pure kaolinite; an electron microscope picture of kaolinite crystals is shown in Figure 8.5. In combination with water, clays form plastic masses that are easily shaped.

Whiteware includes sanitary ware, electrical insulators, dishes, and decorative porcelains. These materials are made from mixtures of clay, feldspar, and flint. Porcelain, an important member of this group, is a complex mixture of K_2O-SiO_2-Al_2O_3 that is fired at very high temperature ($\sim$1,400°C), yielding a translucent product free of porosity. For a long time, China was the only producer of porcelain; the Chinese possessed mines of a very fine clay at Kaoling (kaolinite) and were already able to obtain temperatures of 1,300°C in the Han dynasty (220 BC to 200 AD).

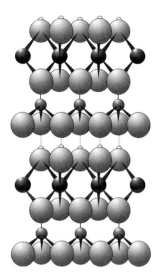

■ **FIGURE 8.4**   Structure of kaolinite. Large grey spheres: oxygen; black spheres: aluminum ions; small grey: silicon ions; small white: hydrogen ions. The dotted lines show the hydrogen bonds between the layers. The kaolinite platelets have a thickness of 0.7 nm and extend to 0.5 μm.

■ **FIGURE 8.5**   Scanning electron microscope image of kaolinite platelets. Courtesy of R. Anderhalt, Philips Electronics North America Corp.

Structural clay products include bricks, tiles, and sewer pipes. Bricks contain a high degree of porosity to provide thermal insulation. Other examples of similar compositions include stoneware and earthenware such as terra cotta. Clays of different composition are used in these products, and firing temperatures are lower than for whiteware. Surface glazes (glasses) are applied for decorative purposes and to make fired clay impervious to water penetration.

8.2.3 **Refractories**

Refractories are used for thermal insulation, crucibles, and hardware in all kinds of casting operations as well as in high-temperature processing and heat treatment furnaces. Whether they line the immense blast furnaces used in steel making or line molds for casting turbine blades, they must withstand direct contact with molten metals and glasses (e.g., slags) without decomposing or cracking. Refractory oxides used for these general purposes most commonly include fireclays containing alumina-silicates, magnesium oxide, relatively pure silica, and zirconia (ZrO_2).

8.2.4 **Abrasives**

Abrasives consist of natural silicon oxide (flint); corundum, a natural form of aluminum oxide; synthetic silicon carbide (carborundum) or diamond dust. Abrasive ceramic materials are critical in the grinding, lapping, and polishing of parts requiring high dimensional tolerances. The sharp abrasive grains must be hard enough to penetrate the material being cut. They are commonly bonded to grinding wheels or to paper and cloth (e.g., sandpaper), but are also used in loose form or embedded in pastes and waxes.

8.3 **SYNTHETIC HIGH-PERFORMANCE CERAMICS**

In the past several decades there has been a rapidly growing interest in what are called **fine**, **advanced**, or **engineering** ceramic materials. The functions, properties, and applications of these ceramics are conveniently displayed in a condensed format in Figure 8.6.

Modern, high-performance ceramics are synthesized; they include alumina (Al_2O_3); silica (SiO_2); other oxides, such as TiO_2, ZrO_2, Na_2O, Li_2O; carbides WC, TiC, SiC, BC; nitrides: Si_3N_4, TiN, BN; and borides TiB_2. Since they are compounds of metals with the lighter elements N, C, and O, their density is usually lower than that of metals.

A glance at the applications reveals the economic potential of these high-tech materials. It is instructive to compare the conventional and new ceramics with the aid of Figure 8.7 which draws distinctions in cartoon form. It is in the applications that remarkable differences between conventional and new ceramics are evident. A sampling of these together with the materials involved is listed in Table 8.2.

8.4 **THE CRYSTAL STRUCTURES OF CERAMICS**

The crystal structure of ceramics is determined by their composition and the type of their chemical bonds. In covalent ceramics, the directions of the chemical bonds shape the crystal structure. In ionic materials, the structure is largely controlled by the sizes of the ions and the electric charges they carry. Since ceramics are compounds that sometimes have a large number of atoms per formula, we must expect their structures to be rather complex. In fact they are not all well established yet.

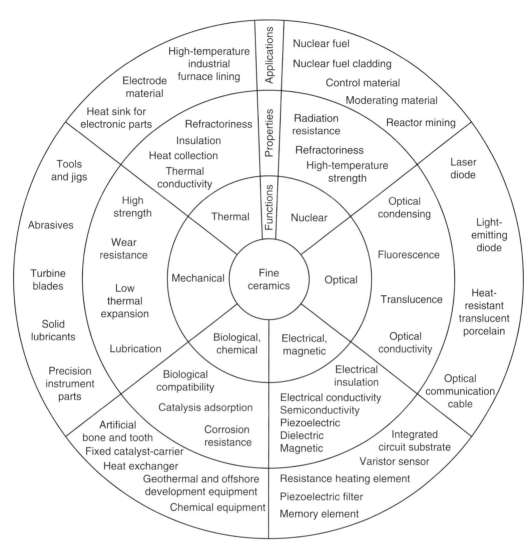

■ **FIGURE 8.6** Functions, properties, and applications of high-performance ceramics. From the Fine Ceramics Office, Ministry of International Trade and Industry, Tokyo.

8.4.1 The Diamond, Zincblende, and Wurtzite Structures

The crystal structure of diamond, silicon, and germanium is imposed by the geometry of the sp^3 hybrid electron orbitals that form the covalent bond. As a consequence, the structure of these elements is not the densest possible. When these solids melt, they contract.

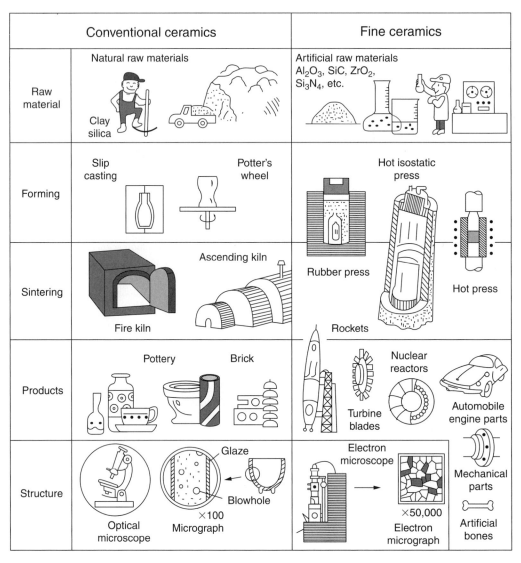

■ FIGURE 8.7 Contrast between synthetic and conventional ceramics. From the Fine Ceramics Office, Ministry of International Trade and Industry, Tokyo.

A familiar solid that has a structure dictated by the direction of the bonds is ice: when ice melts, the water molecules assume random positions that are more compact. Ice contracts upon melting and water expands when it freezes. This makes ice float on water and causes the rupture of water pipes in unheated houses in winter.

Table 8.2 Synthetic Ceramics and Their Applications.

Structural applications	Materials
1. Machinery	
Heat-resistant structural materials	Si3N4, SiC,
	Cordierite (2MgO · 2Al$_2$O$_3$ · 5SiO$_2$)
	ZrO$_2$ (transformation toughened)
	Mullite (Al$_2$O$_3$ · 2SiO$_2$)
Heat engine components	Si$_3$N$_4$
Turbine engine	
High-temperature bearings	
High-temperature gas and molten metal transport pipes	
Ceramic heat exchanger	Cordierite
Wear-resistant materials	SiC, Al$_2$O$_3$, ZrO$_2$, TiC, DLC[a]
Low-density materials (turbochargers)	Si$_3$N$_4$
High-rigidity bearings	Si$_3$N$_4$
2. Cutting Tools	WC, Al$_2$O$_3$, TiC, TiN, diamond
Coatings	TiC, TiN Cubic BN, Diamond-like Carbon (DLC)

Functional applications	Materials
3. Nuclear fuel	UO$_2$
Reactor components	SiC, B$_4$C, Al$_2$O$_3$
Dosimeters	CaF$_2$, K$_2$SO$_4$, LiF
4. Electronics	
Integrated circuit substrates	Al$_2$O$_3$, AlN, BeO, SiC
Capacitors	BaTiO$_3$, SrTiO$_3$
Thermistor (temperature sensor)	(NiMn)$_3$O$_4$, KTaNbO$_3$
Piezoelectrics	PZT (lead zirconate titanate), LiNbO$_3$
Ferroelectrics	BaTiO$_3$, LiTaO$_3$, Pb(TiZr)O$_3$
Magnetics	Ferrites (NiZn)Fe$_2$O$_4$ garnets
Batteries	Graphite
5. Sensors	
Oxygen	Y-doped ZrO$_2$
Humidity	Ti-doped MqCr$_2$O$_4$
Hydrocarbon gas	Doped SnO$_2$
6. Optical	
Lasers	Ruby (Cr doped Al$_2$O$_3$)
	Nd doped Yttrium iron garnet
Transparent windows	Al$_2$O$_3$, MgF$_2$, ZnSe, Glass
Electro-optical materials	LiNbO$_3$, KH$_2$PO$_4$
7. Medical and bioengineering	
Bone, artificial joints,	Apatite hydroxide (for bone meal)
Tooth replacements	Porcelain, Al$_2$O$_3$ for implantation

From S. Musikant, What Every Engineer Should Know about Ceramics, *Marcel Dekker, New York (1991) .*
[a]DLC is diamond-like carbon, a hydrogenated form of amorphous carbon.

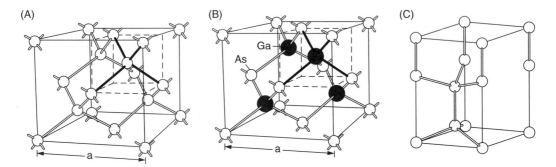

■ **FIGURE 8.8** (A) The diamond structure. (B) The zincblende structure where one half of the atom sites are occupied by Ga and the other by As atoms. (C) The hexagonal wurtzite structure.

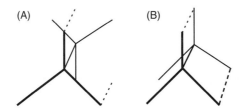

■ **FIGURE 8.9** Rotations of the structure around an sp^3 bond, (A) leading to the diamond and zincblende structures, and (B) leading to the wurtzite structure.

Figure 8.8A shows the diamond structure. Figure 8.8B shows the structure of the III-V semiconductor compound gallium arsenide. This compound has a low degree of ionicity, and its structure is governed by the symmetry of the sp^3 bonds. While the angles between the sp^3 hybrids are fixed at 109°, the rotation around such a bond is easy; therefore two different structures are possible. Rotation A in Figure 8.9 produces the zincblende structure shown in Figure 8.8B, and rotation B generates the hexagonal wurtzite structure shown in Figure 8.8C. Therefore the ceramics based on silicon and carbon, such as SiO_2 and SiC, can have different crystal structures. Silicon carbide, for instance, has a cubic structure similar to Figure 8.8B and a hexagonal wurtzite structure as in Figure 8.8C.

8.4.2 The Structures of Compounds

The crystal structures of compounds, which are usually ceramic and possess a mixed covalent-ionic structure, can be very complex.

Linus Pauling, twice a Nobel laureate, has formulated useful rules to predict structural coordination in terms of ionic size:

- Positive and negative ions must alternate for electrostatic cohesion;

- The ions, which have different sizes, must touch each other so that the structure does not collapse.

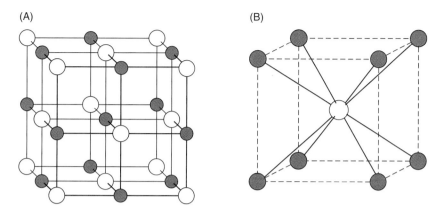

■ **FIGURE 8.10**   (A) The NaCl structure. (B) The CsCl structure. White atoms are Cl.

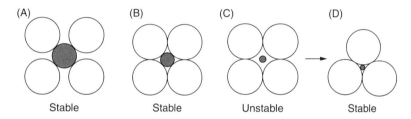

■ **FIGURE 8.11**   Stable and unstable coordination between ions of different size. When the cation becomes smaller than in (B), it no longer touches the anions and the structure is unstable (C); the anions arrange themselves so that they touch the cation (D).

These principles are responsible for the two structures of sodium chloride and cesium chloride, shown in Figure 8.10.

The Coordination of Ions   Why are eight Cl^- ions coordinated to each Cs^+ ion in CsCl, but only six Cl^- ions to each Na^+ ion in NaCl? The coordination number N_c, defined as the number of anions surrounding the cation, is dependent on the ratio of the cation to anion radii (i.e., r_c/r_a). Cations, having lost an electron charge, are usually smaller than anions that have gained electronic charge. Assuming a hard-sphere ionic model, the central cation cannot remain in contact with all the surrounding anions if it is too small. Under these conditions, the structure would tend to fly apart because of electrostatic repulsion between anions; a smaller N_c would be favored. This is illustrated in Figure 8.11 where stable and unstable ionic coordination configurations are indicated.

The stability criterion dictates that the cation touch all anions when anions touch each other. Structures that are stable as a function of r_c/r_a are shown in Figure 8.12. Three-dimensional **tetrahedra** ($N_c = 4$) are predicted for $r_c/r_a > 0.255$; **octahedra** ($N_c = 6$) for $r_c/r_a > 0.414$, **cubes** ($N_c = 8$) for $r_c/r_a > 0.732$, and **close-packed face centered cubes** or **cuboctahedra** when $r_c/r_a > 1$.

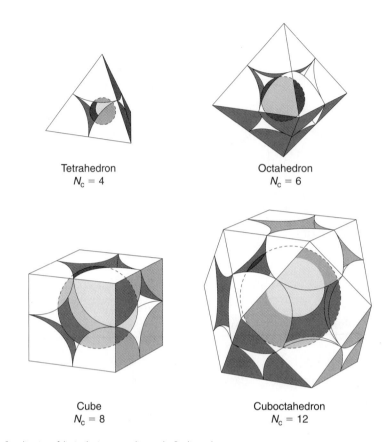

Tetrahedron
$N_c = 4$

Octahedron
$N_c = 6$

Cube
$N_c = 8$

Cuboctahedron
$N_c = 12$

■ **FIGURE 8.12** Coordination of dissimilar ions according to the Pauling scheme.

EXAMPLE 8.1 *Calculate the critical values of r_c/r_a for octahedral and cubic coordination.*

ANSWER In octahedral coordination, the ions assume the planar geometry shown in Figure 8.12B at the octahedron midsection. Therefore, $2r_a\sqrt{2} = 2r_a + 2r_c$. Solving, $r_c/r_a = \sqrt{2} - 1 = 0.414$. For r_c/r_a values greater than this critical value, octahedral coordination is stable.

In the case of cubic coordination ions touch along the cube diagonal. Therefore, $2r_a\sqrt{3} = 2r_a + 2r_c$. Solving $r_c/r_a = \sqrt{3} - 1 = 0.732$. For r_c/r_a values greater than this critical value cubic coordination is stable.

EXAMPLE 8.2 *Predict values of N_c for NaCl and CsCl.*

ANSWER Values for r in nm are $r(Na^+) = 0.098$, $r(Cs^+) = 0.165$ and $r(Cl^-) = 0.181$.

For NaCl, $r(Na^+)/r(Cl^-) = 0.098/0.181 = 0.544$. Therefore, $N_c = 6$.

For CsCl, $r(Cs^+)/r(Cl^-) = 0.165/0.181 = 0.912$. Therefore, $N_c = 8$.

These values are consistent with the observed structures in Figure 8.10.

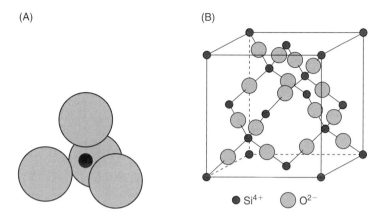

■ **FIGURE 8.13**   (A) Silicon oxide SiO_4^- pyramid based on the sp^3 orbitals of silicon. (B) Structure of cristobalite; the silicon atoms occupy the same positions as in the diamond structure; each oxygen ion is shared by two silicon ions.

The structure of compounds that contain silicon or carbon is governed by the geometry of the sp^3 hybrid bonds in addition to the two principles stated above. Silicon nitride, Si_3N_4, for example, possesses seven atoms per formula. Four nitrogen atoms in the sp^3 configuration must surround each Si atom, and every N atom touches three Si atoms arranged in a plane. The result is a complex hexagonal structure.

There exist three distinct structures (polymorphs) of SiO_2: quartz, cristobalite, and trydimite; they consist of different ways of stacking the SiO_4^- pyramids shown in Figure 8.13. The pyramids are stacked in such a way that every oxygen atom belongs to two neighboring pyramids (i.e., is bonded to two silicon atoms). It is easy to verify that this sharing results in an electrically neutral SiO_2. The crystalline structure of cristobalite is sketched in Figure 8.13B.

With the rules for coordination based on ionic size, the geometric demands of molecular orbitals, and the relatively large molecular formulae, it is not surprising that ceramics can form rather complex crystal structures. The crystal lattices corresponding to the structures observed in ceramics are shown in Table 8.3 and in Figure 8.14.

8.4.3 **Polymorphism or Allotropy**

We have already seen that certain materials crystallize in several different structures. Such is the case of silicon carbide and silicon oxide. Polymorphs sometimes present differences in their properties. Two important cases of polymorphism we shall discuss here are those of carbon and zirconia.

The valence orbitals responsible for chemical bonds of **carbon** can be sp^3 hybrids or sp^2 hybrids. The sp^3 **hybrid** bonds form **diamond**, which has the structure shown in Figure 8.15A. Diamond is the hardest material known. A chemical bond with sp^2 **hybrids** forms **graphite**, which has the hexagonal layered structure shown in Figure 8.15B. The chemical bond within the layers is strong, but the bond between the layers is weak. Graphite is a soft material. Its hexagonal layers glide over each other easily, making graphite a solid lubricant. (It is worth emphasizing that graphite is a lubricant only in humid

Table 8.3 Space Lattices and Crystal Geometries.

Crystal system (Bravais lattice)	Axial lengths and interaxial angles	Examples
Cubic (Simple cubic, body centered cubic, face-centered cubic)	Three equal axes, Three right angles $a = b = c, \alpha = \beta = \gamma = 90°$	Au, Cu, NaCl, Si, GaAs
Tetragonal (Simple tetragonal, body centered tetragonal)	Two of the three axes equal, three right angles $a = b \neq c, \alpha = \beta = \gamma = 90°$	In, TiO_2
Orthorhombic (Simple orthorhombic, body centered orthorhombic, base-centered orthorhombic, face-centered orthorhombic)	Three unequal axes, three right angles $a \neq b \neq c, \alpha = \beta = \gamma = 90°$	Ga, Fe_3C
Rhombohedral (Simple rhombohedral)	Three equal axes equally inclined, three angles $\neq 90°$ $a = b = c, \alpha = \beta = \gamma$	Hg, Bi
Hexagonal (Simple hexagonal) (Hexagonal close-packed)	Two equal axes at 120°, $a = b \neq c, \alpha = \beta = 90°$, $\gamma = 120°$	Zn, Mg
Monoclinic (Simple monoclinic, base-centered monoclinic)	Three unequal axes, one pair of axes not at 90° $a \neq b \neq c, \alpha = \gamma = 90°, \beta \neq 90°$	$KClO_3$
Triclinic	Three unequal axes, three unequal angles $a \neq b \neq c, \alpha \neq \beta \neq \gamma$	Al_2SiO_5

atmospheres; it is unusable in space.) In 1985, scientists were able to produce small carbon balls consisting of 60 carbon atoms with sp^2 bonding, named C_{60} or **Fullerenes** (often also **Bucky balls**) after Buckminster Fuller who popularized this structure in buildings. Another variant that has been synthesized more recently is the **carbon nanotube**. These new polymorphs of carbon are presently the object of much research and possess interesting chemical and mechanical properties; they will be discussed further in Chapter 18. These structures are also shown in Figure 8.15.

Pyrolytic carbon is a disordered form of graphite that is obtained by the pyrolysis (i.e., decomposition by heat) of hydrocarbons. It is a hard and very wear-resistant material. It acquires these properties from its turbostratic structure, which is similar to that of graphite, but with many more defects. The stacking of the sheets is not ordered as in graphite, and the layers are curved and kinked; this prevents them from sliding over each other and provides for the mechanical resistance of the material.

Zirconium oxide ZrO_2 possesses three crystal structures, depending on temperature. It is simple cubic between 2,370°C and its melting point of 2,700°C, it is tetragonal between 1,170°C and 2,370°C, and it is monoclinic below 1,170°C. The transition from tetragonal to monoclinic structure is associated with a ~9% volume expansion that makes fabrication difficult because of stress-induced cracking. Zirconia

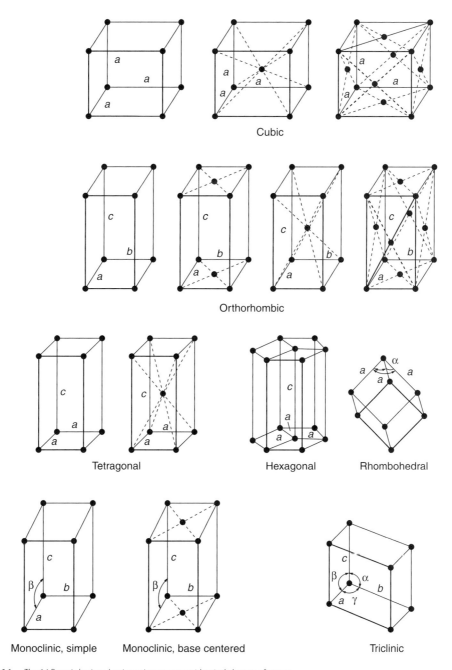

Cubic

Orthorhombic

Tetragonal   Hexagonal   Rhombohedral

Monoclinic, simple   Monoclinic, base centered   Triclinic

■ **FIGURE 8.14**   The 14 Bravais lattices. Lattice points represent identical clusters of atoms.

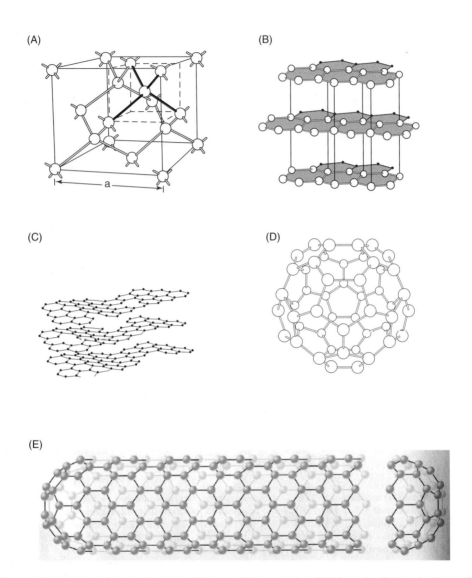

(A)

(B)

(C)

(D)

(E)

■ **FIGURE 8.15** The polymorphs of carbon. (A) Diamond. (B) Graphite. (C) Pyrolytic carbon. (D) Fullerene C_{60}. (E) Nanotube. Diamond bonds are sp^3 hybrids, and the others are sp^2 hybrid orbitals.

can be toughened, however, through the addition of 8 wt% Y_2O_3. Different amounts of CaO and MgO will also work. In such alloys, tetragonal ZrO_2 is stable at about 1,100°C, and this phase can be retained at room temperature by rapid cooling. The transformation from the tetragonal to monoclinic phase is sluggish but is triggered by the tensile stresses surrounding a crack (see Section 2.7.3). The **increase in volume** caused by the transformation produces a compressive stress field around transformed particles. The compressive stresses seal cracks shut, inhibit their further advance, and toughen the material.

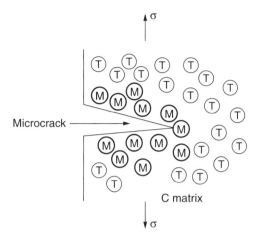

■ **FIGURE 8.16**   Toughening of Partially Stabilized Zirconia (PZT). The tensile stresses around the crack tip trigger the transformation of tetragonal to monoclinic zirconia. The resultant increase in volume cancels out the tensile stresses and retards crack propagation.

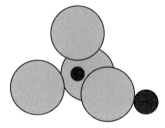

■ **FIGURE 8.17**   Silica tetrahedron with positive ion attached. This ion provides one electron to the oxygen and obviates the need to share this oxygen with another tetrahedron.

This is schematically shown in Figure 8.16. This alloy is **partially stabilized zirconia** (PZT); its fracture toughness approaches 12 Pa√m, which is several times higher than that of other ceramics.

8.5 **GLASS**

Glass is one of the most versatile materials and also one of the oldest. Obsidian, a common volcanic glass, was widely used for arrowheads, spearheads, and knives in the Stone Age. The earliest example of an object completely crafted of glass dates to ~7000 BC. Today 700 different glass compositions are in commercial use. These are fabricated into tens of thousands of products.

8.5.1 **Structure and Composition**

The basic component of glass is silicon oxide, and its basic structural element is the SiO_4^{-4} tetrahedron shown in Figure 8.17. Besides SiO_2, glass contains elements such as sodium, potassium, boron or lead that are dissolved as ions. These atoms are **structure modifiers**. They attach themselves to oxygen atoms at the corners of the tetrahedra; forming an ionic bond, they transfer an electron to the oxygen and

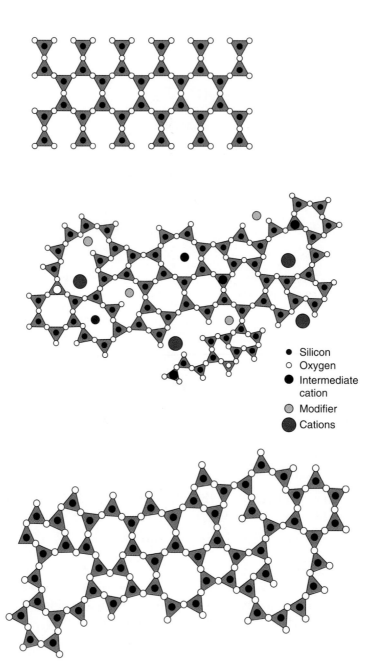

■ FIGURE 8.18 Structures of crystalline silica (top), glass (middle), and amorphous silica or fused quartz (bottom).

remove the need for the corner to be shared with another tetrahedron. The result is that the rigid crystalline structure is no longer necessary; the tetrahedra form an irregular, amorphous structure shown in Figure 8.18. Thus, a **silica glass is a disordered arrangement of SiO_4 tetrahedra**.

Glass often also contains **intermediate oxides** such as CaO, PbO, and Al_2O_3; these assume both network and modifier sites in the glass. Their role is to achieve certain desirable properties that will be discussed below. A schematic representation of a glass with modifiers and intermediates is shown in Figure 8.18.

Another classification of additives emphasizes their functional nature. In this scheme the common oxides are divided into **glass formers** (SiO_2), **fluxes**, and **stabilizers**. Fluxes are added to lower the melting and working temperatures by decreasing the viscosity; they are the structure modifiers. Common fluxes are Na_2O, K_2O, and B_2O_3. Stabilizers (e.g., CaO, MgO, and Al_2O_3) are added to prevent crystallization and improve chemical durability.

Silicate glass compositions are listed in Table 8.4. Soda-lime glasses, typically containing (by weight) 72% SiO_2, 15% Na_2O + K_2O, 10% CaO + MgO, 2% Al_2O_3, plus 1% miscellaneous oxides, account for 90% of the tonnage of all glass produced. With minor composition variations, they are used to fabricate plate glass, containers, and light bulbs.

Borosilicate glasses such as Pyrex contain ~13% B_2O_3. Replacing the Na_2O and K_2O structure modifiers with boron oxide provides Pyrex with a low thermal expansion coefficient $\alpha = 3 \times 10^{-6}(°C)^{-1}$ and

Table 8.4 Composition (wt%) and Uses of Silicate Glasses.

Glass[a]	SiO_2	Al_2O_3	B_2O_3	Na_2O	K_2O	CaO	PbO	MgO	Applications
0010	63	1		7	7		22		Lamp tubing
0080	73	1		17		5		4	Lamp bulbs
1720	62	17	5	1		8		7	Ignition tubes
7070	71	1	26	1	1				Low-loss electrical
7740	81	2	13	4					General
7900	96	0.3	3						High temperature
8363	5	3	10				82		Radiation shielding
Window	72	1		14		10		3	
Fiberglass	54.5	14.5	8.5	0.5		22			Composites
Borosilicate	81	2.5	12	4.5					Low expansion chemical industry

[a] *The four-digit codes refer to Corning glasses. From J. R. Hutchins and R. V. Harrington, Glass: Encyclopedia of Chemical Technology, 2nd ed., Wiley, New York (1966).*

a high resistance to thermal shock, as shown in Equation (8.1). Pyrex is widely used in laboratory glassware and cooking ware. Aluminosilicate glasses are chemically durable and resistant to crystallization and degradation at elevated temperature. The addition of lead to glass increases its index of refraction, giving it optical properties close to those of diamond. High-lead silica glasses are used for radiation shielding, optical applications, and decorative art (lead crystal glass) applications.

An amorphous form of pure silica exists as **fused quartz**. When quartz or sand (which consists of quartz crystals) is molten and cooled naturally, the SiO_4^- tetrahedra of which it is composed are too large and sluggish to order themselves into a crystalline structure; they form an amorphous network shown in Figure 8.18.

8.5.2 **Solidification of Glassy Melts**

In contrast to crystalline solids, **amorphous materials do not have an exact melting temperature**. As they cool from the melt, they do not undergo a phase transition; their viscosity increases until it is so large that no deformation can be obtained at ordinary cooling rates. This is shown in Figure 8.19.

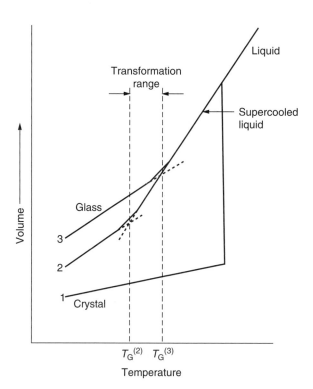

■ **FIGURE 8.19**   Volume-temperature relationship for (1) a crystal, (2) a slowly cooled glass, and (3) a rapidly cooled glass. Other properties of glass like refractive index and heat content show similar changes as a function of temperature.

Below the **glass transformation temperature** T_G, the volume decreases less rapidly with temperature because the molecules are frozen in place. The viscosity becomes so large that structural change cannot be detected in the time scale of measurement. Melts cooled at a lower rate have more time to relax and form a denser glass. More rapid cooling raises T_G.

8.6 PROCESSING OF CERAMICS

Many crystalline ceramics cannot be cast because of their high melting temperature; often, they dissociate chemically before they melt. Because of their high hardness and lack of ductility, they cannot be shaped by plastic deformation. The starting material in the fabrication of ceramic objects is a paste composed of small solid particles, water, and often a binder. The paste is easily shaped at room temperature by various methods and acquires a modest strength after drying. A ceramic piece formed at room temperature is called a **green**. Its final solidification and hardness are acquired by firing. We describe the formation of the green body in Section 8.6.1 and the final solidification by firing in Section 8.6.2.

8.6.1 Forming the Green Body

Green bodies of both traditional and new ceramics are formed by a number of methods that are enumerated below.

Hand Forming This is the traditional technique used by artisans and artists to produce pottery and statues. Pots are usually **thrown** on a turning potter's wheel and shaped by hand. Other pottery, such as that shown in Figure 8.2, is hand-formed by coiling ropes of clay.

Slip Casting In this technique, used for industrial production of pots, a slip is first prepared; it consists of a suspension of clay or ceramic powder in water. This is cast into an absorbent mold as illustrated in Figure 8.20A. After sufficient water loss, the liquid slurry remaining in the middle liquid is poured out and the partially wet solid body is removed and dried further prior to firing. Vases as well as large and complex parts like Si_3N_4 turbine rotors have been produced this way.

Pressing High-performance ceramics require as little porosity as possible In this case the specially prepared dry powder is placed in a die and pressed under high uniaxial forces to make a green compact (Figure 8.20B). Only relatively small parts can be made this way. An improvement over applying uniaxial compressive loads is **isostatic** compaction of the powder. Here the part is enclosed in a flexible airtight rubber bag and immersed in a chamber filled with hydraulic fluid. Hydrostatic pressure (uniform in all directions) is applied during (cold) compaction, eliminating the complex pressure distribution throughout the die due to mold-wall friction. Electrical insulators, bushings, magnetic components, and spark plug bodies are produced this way for subsequent sintering. **Hot pressing** combining compaction and sintering operations is also practiced to make a variety of highly dense ceramic components; this technique is used to produce silicon nitride bearing balls.

(A)

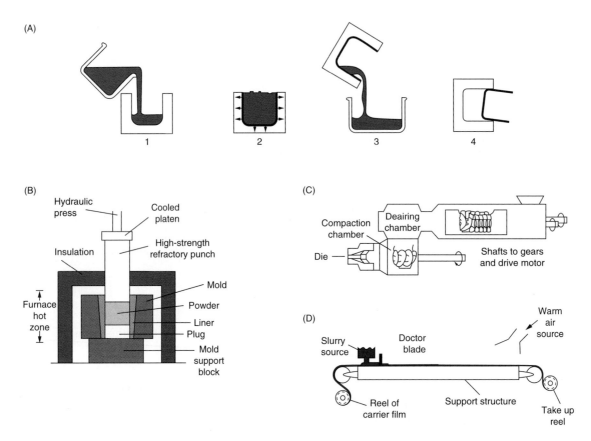

■ **FIGURE 8.20** Forming the green (low temperature processes) in ceramic processing. (A) Slip casting. (B) Pressing. (C) Injection molding; this machine can also be used for extrusion when the paste does not enter a mold. (D) Tape forming.

Extrusion The slurry is extruded through a die as shown schematically in Figure 8.20C. Long tubes, rods, and honeycomb structures used to shield thermocouples or for heat exchanger applications are made this way. The technique is similar to the extrusion of metals except that an auger is sufficient to push the soft material through the die.

Tape Casting This process, shown in Figure 8.20D, is used to make thin ceramic parts like alumina substrates for integrated circuit chips and special capacitors. A thin layer of slip is laid on a flat carrier that can be a paper sheet or polymer film. A doctor blade controls the slip thickness resulting in a tape that can be stamped into small shapes. These can be stacked or coiled for subsequent firing.

Injection Molding In this hybrid process, ceramic powder is mixed with thermosetting polymer binder and injected under pressure into a die or mold. The bodies are then ejected and new ones are replicated. Complex parts with excellent dimensional control can be made this way.

8.6.2 **Densification**

Firing   The dried green ware is fired in a furnace to densify it into a hard body. Firing temperatures depend on the composition of the ceramic as Table 8.5 indicates.

Firing proceeds in three stages. At the lower temperatures (<450°C), water is expelled and the binder is burned off. As the temperature is increased, **vitrification** takes place: certain components of natural ceramics form a liquid glass that infiltrates the pores between the grains. Above 1,000°C, sintering fuses the ceramic crystals together.

Sintering   The final stage of solidification consists of **sintering** in which the grains of ceramic fuse together; this is shown in Figures 8.21 and 8.22. Two natural phenomena are crucial to sintering; these are the diffusion of molecules at high temperature and the decrease of surface energy. Diffusion is sufficiently fast for sintering at half the melting temperature in degrees Kelvin. Diffusion is the random movement of molecules made possible by their thermal vibration energy. Surface diffusion is much more rapid than diffusion through the bulk because the diffusing molecules need not displace their neighbors in order to change place. Consider two grains that are in contact, as shown in Figure 8.21.

Table 8.5 Firing Temperatures of Ceramics.	
Ceramics	**Approximate firing temperature**
1. Porcelain enamels (on cast irons and steels)	650–950°C
2. Clay products (bricks, sewer pipes, earthenware, pottery)	1,000–1,300°C
3. Whitewares (porcelain, china, sanitary ware)	1,000–1,300°C
4. Refractories (alumina, silica and magnesia brick, silicon carbide)	1,300–170°C
5. Synthetic high-performance ceramics (aluminas, nitrides, carbides, titanates)	1,300–1,800°C

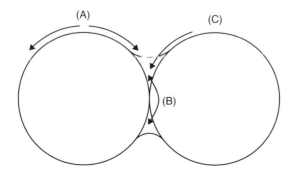

■ **FIGURE 8.21**   Sintering of two particles. (A) Random surface diffusion of molecules. Bulk diffusion (B) and surface diffusion (C) into the crevice between the grains reduce the total surface area and therefore the total energy of the two grains. Surface diffusion is faster than bulk diffusion.

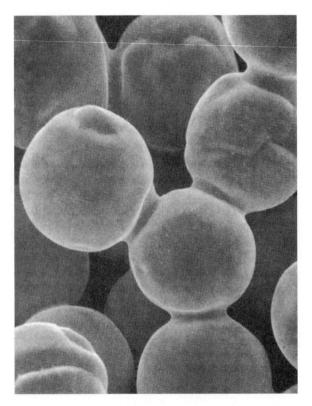

■ **FIGURE 8.22**   Partially sintered particles. As sintering proceeds further, the contact areas grow, the particles move closer together, and porosity decreases. The material shrinks in the process. Courtesy of Professor M. German, Mississippi State University.

Their total surface area is that of two spheres. At elevated temperatures, molecules at the surface of the grains move in random directions. This is shown as A.

The molecules that diffuse to the contact area fill the crevice between the two grains and reduce the surface area, and therefore the total surface energy. This lowering of energy favors the diffusion into the crevice shown as B. The result is that the two grains fuse together. Sintering of the grains is rarely complete: pores are left in the junction between three grains. This porosity weakens the material and may be undesirable. For this reason, sintered ceramics are specified with an indication of density: 95% theoretical density means that the volume of the material includes 5% porosity. Two methods are used to minimize porosity where mechanical strength is important. One is **Hot Isostatic Pressing** (HIP) in which the ceramic is encapsulated in a glass skin and subjected to high pressure gas during sintering: the particles are thus pressed together; the other is **liquid phase sintering,** in which a sintering aid that liquefies at the sintering temperature is added to the powder and penetrates the pores.

Since the particles fuse together, the material shrinks during sintering. Allowance must be made for shrinkage in the design of ceramic pieces.

Table 8.6 Operation and Viscosity.

Operation	Viscosity (Pa · S)
1. Glass melting and fining (bubble elimination)	5–50
2. Pressing	50–700
3. Gathering or gobbing for forming	50–1,300
4. Drawing	2,000–10,000
5. Blowing	1,000–3,000
5. Removal from molds	10^3–10^6
6. Annealing	10^{12}–10^{15}
7. Use	10^{13}–10^{15}

After sintering, ceramics can only be machined by abrasion. Abrasion is the "scratching" of the material by small grains of a hard material: ceramics are machined by diamond powder bonded to a polishing pad or a metal slicing wheel. The machining of ceramics is slow and expensive.

8.6.3 Fabrication of Glass Objects

The fabrication of glass objects exploits the increasing viscosity of glass as its temperature is lowered. Viscosity determines virtually all of the melting, forming, annealing, sealing, and high-temperature heat treatments of glass. Typical viscosity ranges for the various operations are shown in Table 8.6.

The temperatures at which these operations are carried out differ with glass composition; they are depicted in Figure 8.23 for a number of commercial silica-based glasses. Each glass has a specific viscosity-temperature dependence and an individual set of strain, annealing, softening, and working-point temperatures. These are defined in the following ways.

Strain Point The temperature at which internal stresses are reduced significantly in a matter of hours. At this temperature the viscosity is taken as $10^{13.5}$ Pa · s.

Annealing Point The temperature at which the viscosity is about 10^{12} Pa · s and the internal stresses are reduced to acceptable commercial limits in a matter of minutes.

Softening Point At this temperature glass will rapidly deform under its own weight. For soda-lime glass this occurs at a viscosity of 10^7 Pa · s.

Working Point At this temperature glass is soft enough for most of the common hot working processes. The working point corresponds to a viscosity of 10^3 Pa · s.

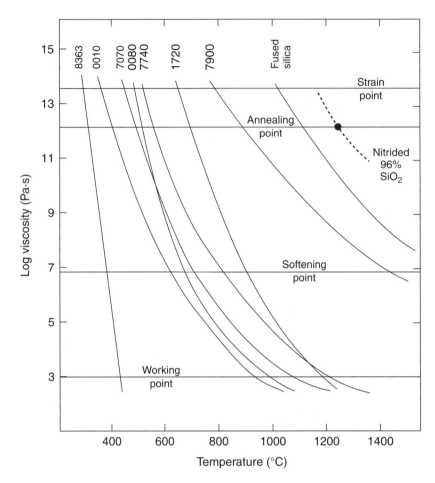

■ **FIGURE 8.23**   Variation of the viscosity of various glasses with temperature. The four-digit codes refer to Corning glasses whose compositions are given in Table 8.3.

From J.R. Hutchins and R.V. Harrington, *Encyclopedia of Chemical Technology*, 2nd ed., Vol. 10, Wiley, New York (1966).

Figure 8.24 illustrates the various methods that are in use in the fabrication of glass objects. These are:

1. Pressing.

2. Blowing.

3. Casting.

4. Rolling and float molding.

Pressing   An example of pressing involves the forming of glassware by compressing it in a mold with a plunger. The process (Figure 8.24A) resembles closed-die forging of metals and compression molding of polymers. Parts with weights up to 15 kg are pressed using pressures of 0.7 MPa (100 psi).

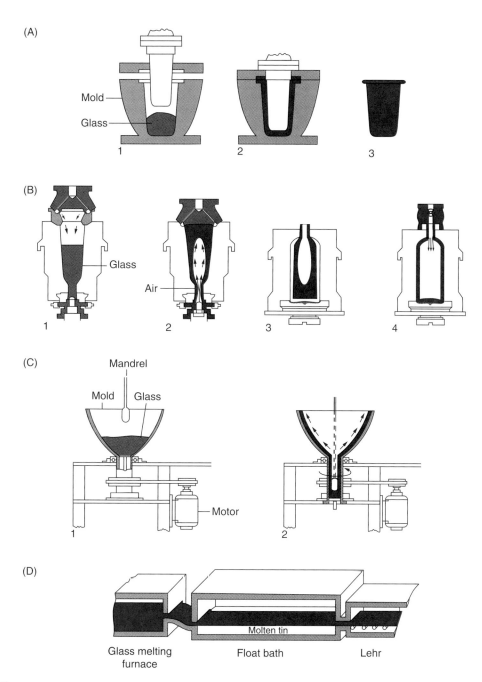

■ **FIGURE 8.24** Processes to form and shape glass. (A) Pressing a glass container. (B) Blowing a bottle. (C) Centrifugal casting of a tube. (D) Plate glass manufacture by the Pilkington method. From J.R. Hutchins and R.V. Harrington, *Encyclopedia of Chemical Technology,* 2nd ed., Vol. 10, Wiley, New York (1966).

Blowing Hand blowing of glass is practiced today much as it was in antiquity. Glass is gathered on the end of an iron blowpipe and the glass blower shapes the "gathers" with lungpower or the use of compressed air. Artistic glassware is hand-blown while continually rotating the glass. The largest pieces reach weights of 15 kg, lengths of almost 2 m, and diameters of 1 m.

Glass containers, light bulb envelopes, and many other mass-produced objects are machine-blown. Gobs of glass are delivered to the **blank mold** where they are preformed either by blowing or pressing. The blank is blown in the **blow mold** to finish the operation (Figure 8.24B).

Casting In this process glass is poured into or on molds, tables, or rolls. The largest piece ever cast was the Mount Palomar reflector, a cylinder measuring 5.08 m (200 in) in diameter and 0.457 m thickness. Other products routinely cast are radiation-shielding window blocks and television and cathode ray tubes. The latter are centrifugally cast, and as the molten gobs are spun the glass flows up to create a uniform wall thickness (Figure 8.24C).

Rolling and Float Molding Both of these methods produce plate glass. In continuous processes, shown in Figure 8.25, raw materials are fed in at one end of very large horizontal furnaces (with capacities of 1,000 tons) where successive melting and refining of glass occurs. At the other end the glass is fed into a pair of cooled rollers, and the emerging ribbon of solid glass is conveyed on rollers through an annealing lehr. Plate glass nearly 4 m wide and 1 cm thick can be rolled at rates of over 6 m/min. The surfaces of rolled plate glass must be further ground and polished.

The Pilkington float molding process is now widely used to make plate glass. In this process soft glass is floated on the flat surface of a molten tin alloy that controls the lower glass surface temperature. The upper surface of the glass develops by gravity. As the temperature is reduced, the glass stiffens and is conveyed to be annealed. This process produces a distortion-free plate glass of flatness comparable to that produced by rolling. A drawback of the process is incorporation of Sn in the glass.

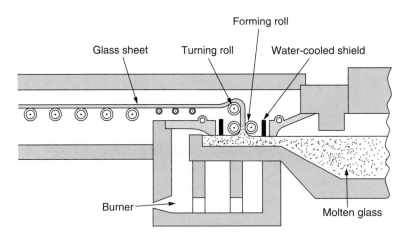

■ FIGURE 8.25 Continuous rolling of plate glass.

All one has to do is look through plate glass windows of the last century to appreciate how far we have come in glass plate manufacture.

8.6.4 **Tempered Glass**

The temperature-dependence of the viscosity (Figure 8.23) is utilized for the manufacture of tempered glass that resists fracture and shatters into small pieces when it finally breaks. Tempered glass equips, for instance, the side windows of cars. The hot glass sheet leaving the furnace is rapidly cooled by cold air blown on it. Its surface cools, shrinks, and hardens first. As the body of the glass sheet cools, its shrinkage is resisted by the hard surface. As a result, the glass acquires compressive residual stresses on its surface and tensile stresses in its center. The surface stresses resist fracture and, if they are overwhelmed, the tensile stresses in the bulk cause the glass to break into small pieces.

8.6.5 **Glass Ceramic (Vitroceram or Pyroceram)**

The amorphous state of glass does not represent thermodynamic equilibrium but results from the inability of the large SiO_4 units to position themselves into regular crystalline positions. Partial crystallization of glass can be obtained by annealing the glass. One obtains a pore-free ceramic that is extremely resistant to thermal shock. The crystalline phase, in fact, has a negative thermal expansion coefficient: it contracts with increasing temperature. This contraction compensates for the expansion of the glassy phase, which results in zero thermal expansion. This avoids the stresses caused by uneven temperatures. Glass ceramics are widely used in cooktops of electric ranges.

8.7 **CEMENT AND CONCRETE**

In tonnage consumed, cement and concrete far exceed that of steel, wood, and polymers combined. They are the essential ingredient in some of the largest structures built in this century such as high-rise buildings, airport runways, and dams.

Concrete is composed of cement, aggregates (sand, gravel, crushed rock), and water. Cement, a generic term for concrete binder, is the key ingredient. The most important binder for concrete is Portland cement, which is produced from an initial mixture of 75% limestone ($CaCO_3$), 25% clay, assorted aluminosilicates, and iron oxides and alkali oxides. This mixture is ground and fed into a rotary kiln (a large cylindrical rotating furnace) together with powdered coal. Progressive reactions at temperatures extending to $1,800°K$ break down clays, decompose the limestone to yield quicklime (CaO), and fuse them to produce clinker (pellets) 5–10 mm in size. After cooling, the latter is mixed with 3–5 wt% gypsum ($CaSO_4$) and ground to the powder that is Portland cement.

Portland cement is composed of the following identifiable compounds:

Calcium oxide (lime)	CaO usually denoted by	C
Silicon oxide (silica)	SiO_2 usually denoted by	S
Aluminum oxide (alumina)	Al_2O_3 usually denoted by	A

Some three quarters by weight is composed of

Tricalcium silicate (C_3S) (i.e., $3CaO \cdot SiO_3$ or Ca_3SiO_5)

Dicalcium silicate (C_2S) (i.e., $2CaO \cdot SiO_2$).

Tricalcium aluminate (C_3A) (i.e., $3CaO \cdot Al_2O_3$).

Different proportions of these and other ingredients are blended to produce cements that either set slowly or more rapidly, liberate less or more heat of hydration, or are intended to resist degradation by water containing sulfates.

Water is added to the cement, which solidifies through chemical reactions that are complex and not entirely understood. Cement does **not** harden by drying, but the water is incorporated in the solid by a chemical reaction. Cement does not shrink when hardening.

Several important hydration reactions occur at different rates and can be written as:

$$C_3A + 6H \rightarrow C_3AH_6 + \text{heat (hours)}$$

$$2C_3S + 6H \rightarrow C_3S_2H_3 + 3CH + \text{heat (days)}$$

$$2C_2S + 4H \rightarrow C_3S_2H_3 + CH + \text{heat (months)}$$

Here, H stands for water (H_2O). Thus C_3AH_6 is $3Ca(OH)_2 \cdot 2Al(OH)_3$ and CH is $Ca(OH)_2$ in ordinary chemical nomenclature.

The first reaction causes the cement to set; the second and third reactions cause hardening. Note that all these reactions are exothermic: the cement heats up as it sets. In constructions where a large volume of concrete is needed, such as in water dams, one must avoid high temperatures that could result from rapid exothermic reactions. These temperatures cause internal stresses that result in fracture. In this case, one uses a cement with a larger proportion of dicalcium silicate C_2S, which takes months to complete its reaction. Fast hardening cements, used for small construction, contain a larger proportion of C_3S that completes its reaction in a matter of days. One often observes water sprayed on top of construction sites the purpose of this spray is to cool the concrete while it sets.

Wet cement is only workable as long as it sets, and this period establishes the pot life of the mix. A typical time-dependence of heat evolution and hardening of cement is shown in Figure 8.26 After a large initial burst of heat there are much smaller secondary peaks that coincide with the start of solidification. During the setting period cement has no strength, but with solidification, strength rises. The setting time of cement is used in building an attractive stone wall. When you build a wall with cement between the stones, let the cement set for a few hours. (The calcium aluminate hardens.) It is then possible to remove any excess cement with a brush and water and obtain a neat wall (see Figure 8.27). It is advisable to build the wall in the morning and clean it in the evening. If one waits overnight, the cement is too hard and the excess can no longer be removed. Several days are necessary for sufficient solidity.

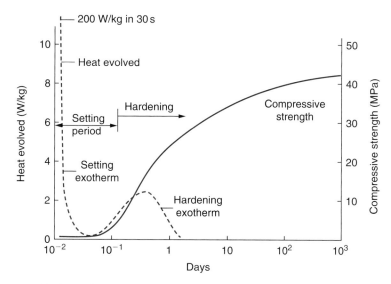

■ FIGURE 8.26   The setting and hardening of Portland cement. Dotted line and left scale: evolution of heat during the reaction. Solid line and right scale: increase in the compressive strength of cement with time. After G. Weidman, P. Lewis, and N. Reid, *Materials in Action Series: Structural Materials*, Butterworths, London (1990)

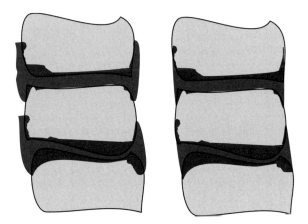

■ FIGURE 8.27   Brushing of cement after a few hours of setting. Left: stone wall with cement as applied. Right: stone wall with excess cement removed by simple brushing and rinsing with water.

■ SUMMARY

1. Ceramics are hard and brittle, electrically insulating, and often transparent. They have a fully occupied band of valence electrons.

2. Ceramics include stones and clay products, synthetic high-performance ceramics, glass, and cement.

3. Ceramics are chemical compounds, most often oxides or carbides, with the exception of diamond, silicon, and germanium. Their chemical bond is either covalent or covalent-ionic.

4. The crystal structure of ceramics is determined by the geometry of the chemical bond for covalent materials, and by local electric neutrality and the relative sizes of ions in ionic compounds.

5. The basic structural element of silicon oxides and glass is the SiO_4^{4+} tetrahedron.

6. Diamond, graphite, fullerenes, nanotubes, and glassy carbon are five polymorphs of carbon.

7. Glass is an amorphous assembly of SiO_4^{4-} tetrahedra made possible by the addition of structural modifiers (e.g., Na^+) that obviate the necessity of sharing corners of the tetrahedra.

8. Since glass does not crystallize, it does not possess a precise melting temperature but increases in viscosity as the temperature decreases.

9. The manufacturing techniques of glassy objects are based on the high viscosity of glass.

10. Tempered glass is obtained by rapid cooling of the glass surface. This introduces a compressive residual stress in the surface that increases the fracture resistance and a tensile stress in the bulk that causes shattering into small pieces when fracture occurs.

11. The fabrication of ceramic pieces proceeds in two major steps: forming of a green (at room temperature) and firing.

12. In the preparation of a green, one prepares a paste or a slurry consisting of the ceramic powders, water, and, often, a binder. The green is formed by any method appropriate for a paste. The green is allowed to dry and acquires a modest strength. In that condition, it can be machined.

13. Ceramics are solidified by firing. In the first stage, the green is dried and the binder is burned off; at higher temperatures, oxide components of natural ceramics form a liquid glass (vitrification); finally the ceramic acquires its final strength by sintering.

14. During sintering, the grains fuse together by diffusion of the molecules. In the process, the material shrinks. Complete fusion of the grains is difficult; sintered ceramics remain porous.

15. The porosity of sintered ceramics is reduced or eliminated by Hot Isostatic Pressing (HIP) or by liquid-phase sintering.

16. Cement is a mixture of calcium and silicon oxides (with possible additions of alumina and iron oxides) and water. It solidifies by a chemical reaction that incorporates the water in the structure. This process takes several months to completion. Cement does not dry and does not shrink on setting.

■ KEY TERMS

A
abrasion, 235
abrasives, 216
aggregate, 239
allotropy, 223

AlO_6 octahedra, 214
alumina, 211
aluminosilicate, 239
aluminosilicate glasses, 230
aluminum oxide (alumina), 239

annealing point, 235

B
BC, 216
blank mold, 238

■ REFERENCES FOR FURTHER READING

[1] M.W. Barsoum, *Fundamentals of Ceramics* (Series in Materials Science and Engineering) (Paperback), Taylor and Francis, London (2002).

[2] D. Kingery, H.K. Bowen and D.R. Uhlmann, *Introduction to Ceramics*, 2nd ed, Wiley-Interscience, New York (1976).

[3] D.W. Richerson, *Modern Ceramic Engineering*, 2nd ed, Marcel Dekker, New York (1992).

■ PROBLEMS AND QUESTIONS

8.1. Why is fused silica more difficult to shape than soda-lime glass?

8.2. Select a suitable abrasive to polish wood, steel, silicon nitride, or diamond.

8.3. Refractories are usually very light. Explain the cause and the purpose of this.

8.4. Stone walls of houses are much thicker than brick walls. Explain why.

8.5. Why can most flowerpots not be left outdoors in winter?

8.6. The insulating tiles of the Space Shuttle are soft and lightweight. What is their structure?

8.7. A scientist wanted to find out whether molten silicon is a semiconductor or a metal. The scientist had great difficulties because the container of the silicon broke when cycling between liquid and solid state. What in the properties of silicon caused this difficulty?

8.8. Figure 8.8 shows that zincblende has the cubic structure. Show that wurtzite is hexagonal.

8.9. Sketch, to scale, a crystal that would have the NaCl structure but cesium instead of sodium ions. Why would this structure not be stable?

8.10. Verify that cristobalite is in fact similar to the structure of silicon where the Si-Si bond is replaced by Si-O-Si.

8.11. Explain how it is possible that diamond is the hardest material known and graphite is so soft that it is used as a lubricant.

8.12. In fact a TTT diagram, similar to the one for the processing of steel, can be drawn for all phase transitions. Sketch a TTT diagram for the crystallization of iron and the crystallization of glass. Explain the difference in terms of the structure and size of the structural units that would have to arrange themselves into a crystal.

8.13. By which method would you fabricate the green of

a. An artistic ceramic pot.

b. A ceramic statue.

c. The copy of a statue.

d. Large numbers of porcelain figurines.

e. Ceramic tiles.

f. Ceramic tubes.

g. Roof tiles (the Romans formed their roof tiles by bending clay over their thigh).

8.14. Why is it desirable to use glassy sintering aids?

8.15. Describe two methods to decrease the porosity of a ceramic.

8.16. Select a cement composition for the construction of a house, the construction of a large dam, and for small repairs.

8.17. Molten glass is compounded to flow easily during shaping or mold filling operations. Glazes and enamels are also glasses, but they must adhere to the fired ceramic or metal surface. Would you expect the compositions to differ in glasses as opposed to glazes and enamels? Mention one required property difference.

8.18. Crucibles used for melting 50 kg charges of silicon for single-crystal growth are made of the purest grade of fused silica and have the shape of a rounded cup with a flat bottom. The wall thickness is ,6 mm and the maximum crucible diameter is ,25 cm. Suggest a way to make these crucibles.

8.19. Describe the processing of a tempered glass plate, starting with the molten glass, and explain how the residual stress develops.

8.20. One way to strengthen surface layers of glass is to chemically diffuse in oversized alkali ions (e.g., K^+) that replace the original Na^+ ions. What state of stress develops at the surface and interior of a piece of soda lime glass so treated?

8.21. Contrast structural and chemical differences in Al_2O_3 used for the following applications:

a. Substrates for microelectronic applications.

b. Transparent tubes for sodium vapor lamps.

c. Furnace brick.

Note any special processing requirements for these applications.

8.22. Why do ceramic bodies undergo much larger shrinkages than powder metal parts during high-temperature processing?

8.23. Porosity in ceramics is usually undesirable, but there are applications where pores are desired. Contrast these applications.

8.24. Why are thick ceramic wares more prone to crack than thin ones during their processing?

8.25. What is the probable method for fabricating the following ceramic objects?

a. Ming dynasty vase.

b. Bathroom wall tiles.

c. Cups for a tea set.

d. Spark plug insulation.

e. A ball for a bearing.

8.26. Use Figure 8.26 to determine the best time to clean the cement of a stone wall as in Figure 8.27.

8.27. Explain, in terms of Pauli's exclusion principle, why ceramics are hard and brittle.

8.28. Describe the two principles that determine the crystal structure of ceramics.

8.29. Glass is an amorphous material. Does this mean that all its atoms are positioned at random? Describe the structure of glass.

8.30. Carbon has five polymorphs. Cite four of them.

8.31. Describe the chemical composition of glass.

8.32. What is the function of structure modifiers in glass? Name one important structure modifier.

8.33. What is Pyrex glass? What is its outstanding property, and how is it achieved?

8.34. Describe or name the phenomenon by which the following materials solidify: glass, porcelain, silicon nitride, cement.

8.35. What property of glass is responsible for the particular fabrication techniques used?

8.36. What is tempered glass? What are its outstanding properties, and how are these achieved?

8.37. How are the side windows of a car manufactured? How do they increase the safety of the passengers? How is this safety achieved?

8.38. What are the major steps in the fabrication of ceramic objects?

8.39. Describe the fabrication of a ceramic ball for a ball bearing, starting with the powder.

8.40. What ceramic are ball bearing balls made of? What property of that material is determinant?

8.41. Describe the different stages in the firing of a ceramic. What physical phenomenon is responsible for the final solidification?

8.42. What are the two phenomena involved in sintering?

8.43. Cement is a mixture of ceramics and water. Does cement dry when it solidifies?

8.44. Why is it possible to build the structures of a seaport under water with cement?

8.45. One sometimes sees water sprayed over a cement (concrete) construction. What is the purpose of this water?

Chapter 9

Polymers

Polymers, namely plastics, rubbers, and resins, are light, relatively soft materials that can easily be formed into films, sheets, bottles, or complex shapes. Most polymers are organic materials, consisting mostly of carbon and hydrogen. Thermoplastic polymers consist of very long molecules (chains) that are bound to each other by weak secondary bonds. The materials are amorphous or partly crystalline: they do not have a definite melting point; when they are cooled from the liquid state, their viscosity increases until they are solid. This property allows one to blow the material like glass and form thin films or bottles. Partial crystallization increases the strength of the material. Thermosets solidify by a chemical reaction that creates primary bonds between the chains. They tend to be more solid and stable to higher temperatures. Rubbers (elastomers) are polymers with a distinct molecular structure; their response to stresses is viscoelastic (sluggish), a property that is used in the design of tires.

LEARNING OBJECTIVES

After studying this chapter, the student will be able to:

1. Differentiate polymers from other classes of materials on the basis of their molecular structure and chemical bonds.

2. Distinguish between a thermoplastic polymer, a thermoset polymer, and a rubber.

3. Describe the role of primary and secondary bonds in polymeric solids and how secondary bonding accounts for the mechanical and thermal properties of plastics.

4. List the major properties of polymers that make them attractive and unattractive relative to other engineering materials for a particular application.

5. Sketch the monomer units of polyethylene, PVC, and PTFE.

6. Relate molecular weight and degree of polymerization.

7. Describe the temperature-dependence of the viscosity and the importance of the glass transition temperature in amorphous thermoplastics.

8. Describe the crystallization of polymers.

9. Describe the major methods of melt-processing of polymers.

10. Explain why epoxies and other thermosetting polymers are formed by using two or more components that must be mixed.

11. Define the term *crosslink* and describe the role of crosslinking in rubbers.

It is difficult to imagine that polymers were unknown 100 years ago. Since then, polymeric materials, which are commonly known as plastics, have replaced glass and ceramic materials in food containers, dinnerware, and assorted kitchen and toilet accessories; they are increasingly replacing metal in automobiles and aircraft, and they take the place of leather and natural fibers in clothing and furniture. Due largely to their ease of processing, high strength-to-weight ratios, low density (typically $\sim 1.5\,\mathrm{Mg/m^3}$ compared to $\sim 7.5\,\mathrm{Mg/\,m^3}$ for metals), and competitive costs, polymers and polymer-based composite materials continue to find expanded usage practically everywhere.

The origin of synthetic polymers can be traced to the American Wild West. In an effort to provide saloon patrons with enough billiard balls, John Hyatt sought a substitute for the scarce and expensive ivory spheres in use at the time. His early efforts resulted in balls consisting of a core of ivory dust bonded with shellac and an exterior coating of the somewhat unstable collodion. The latter frequently exploded when the balls collided, prompting every gunman in the saloon instinctively to draw his six-shooter. In 1868, Hyatt successfully adapted a material developed some 14 years earlier by Alexander Parkes in England. It consisted of a mixture of cellulose nitrate and camphor. This first manmade plastic, known as celluloid, was still widely used until recently in making toys, particularly dolls. Its dangerous flammability spawned substitutes, and when Baekeland patented the process for making Bakelite in 1907 the Age of Plastics formally arrived.* It only required an additional 80 years or so until the total volume of synthetic polymer products manufactured in the world exceeded that of metals produced.

9.1 **DEFINITION OF A POLYMER**

A **polymer** consists of very large molecules that are made up of small molecular units or **mers** linked together into chainlike or network structures. The simplest polymer is polyethylene. Look at its name: polyethylene means "many ethylene molecules". A polyethylene molecule can consist of many thousands of mers and can have a length approaching a micrometer. Such a molecule is referred to as a **macromolecule**.

---

*R.A. Higgins, *The Properties of Engineering Materials*, R. E. Krieger Publishing Co., Huntington, NY (1977).

Polymers are practically divided into plastics, rubbers, adhesives, and resins. Plastics are defined as materials that can be molded into shape; rubbers are characterized by large elastic deformations; adhesives are used to join materials (glues); resins serve as the matrix in composites, which will be treated in the next chapter. From a scientific viewpoint, one divides polymers into thermoplastics and thermosets. **Thermoplastics** soften and liquefy on heating and are processed in the liquid state by a variety of extrusion and molding processes similar to those used for glass. Polyethylene, polyvinyl chloride, polypropylene, and polystyrene, in that order, are the four most widely produced thermoplastics. **Thermosets** solidify by a chemical reaction; this reaction is accelerated by heating. Thermosets include phenolic resins, epoxies, unsaturated polyesters, and polyurethanes—substances that cannot be melted like thermoplastics and therefore cannot be recycled simply by heating. Their most important usage is in adhesives and in composites. Thermoplastic production is approximately six times that of thermosets in tonnage and value. **Rubbers**, or **elastomers**, form another group of polymeric materials. The natural rubber industry based on latex predated the introduction of plastics. During World War II, Japanese occupation of South Asian countries cut off the supply of natural rubber. English and American scientists realized that latex is a polymer with a distinctive molecular structure, which they replicated in synthetic elastomers such as polybutadiene, styrene-butadiene, polyisoprene, and silicone rubber.

Although our concern here is with synthetic organic polymers, the definition is broad enough to include inorganic polymers based on silicon. The word "synthetic" is important because polymers are also contained within natural animal (wool, leather), insect (silk), and vegetable (wood, cotton) substances, but they do not concern us here. Cellulose and lignin are natural polymers found in vast quantities in wood. Some natural polymers are so well designed by nature that they have not yet been replaced by the synthetic variety.

9.2 SYNTHESIS OF POLYMERS

Polymers are synthesized by a chemical reaction that binds the mers together into a macromolecule. For this reason, the starting materials must be chemically reactive. In polyethylene, reactivity is provided by the double bond between its two carbon atoms.

$$
\begin{array}{cc}
\text{H} \quad \text{H} & \text{H} \quad \text{H} \\
\text{C}=\text{C} & -\text{C}-\text{C}- \\
\text{H} \quad \text{H} & \text{H} \quad \text{H}
\end{array}
$$

This double bond can be opened, making the valence available for bonding to other ethylene molecules, forming polyethylene. This type of reaction leads to **addition polymerization** or **chain polymerization**.

In another type of synthesis, reactions occur between an organic acid

$$
\begin{array}{c}
\text{O} \\
\parallel \\
\text{R}-\text{C}-\text{OH}
\end{array}
$$

$$\overset{\displaystyle H}{\underset{\displaystyle |}{}}$$

and an amine R—NH or an alcohol R—OH.

As usual in organic chemistry, R denotes an organic radical. For instance, in acetic acid, R is a methyl radical CH_3, forming the acetic acid CH_3COOH.

This reaction attaches the two reagents together and liberates water; it is a **condensation polymerization**.

9.2.1 Addition Polymerization

Polyethylene is synthesized from ethylene gas by a chemical reaction involving a starter or catalyst. This catalyst is a radical, namely, a molecule with an unsatisfied valence; it is often a peroxide

$$H_2O_2 \rightarrow 2\ HO\text{—}\ \text{ or } ROOR \rightarrow 2\ RO\text{—}$$

Benzoyl peroxide is a frequently used catalyst for polymerization.

The peroxide attacks the double bond of an ethylene molecule and bonds to it. The resulting molecule is again a radical (which means that it has an unsatisfied, chemically active, bond).

$$
RO\text{—}+ \overset{\displaystyle H\ \ H}{\underset{\displaystyle H\ \ H}{C{=}C}} \rightarrow RO\text{—}\overset{\displaystyle H\ \ H}{\underset{\displaystyle H\ \ H}{C\text{—}C}}\text{—}
$$

The new molecule attacks other ethylene molecules in a similar way and bonds with it. This is shown in Figure 9.1.

The process continues until a long molecule is formed, consisting of many thousands of ethylene molecules; this is polyethylene. It will end when the molecule encounters another growing chain or a peroxide catalyst.

$$
RO\text{—}\overset{\displaystyle H\ H\ H\ H\ H\ H}{\underset{\displaystyle H\ H\ H\ H\ H\ H}{C\text{—}C\text{—}C\text{—}C\text{—}C\text{—}C}}\text{—}+\text{—}\overset{\displaystyle H\ H\ H\ H\ H\ H\ H\ H}{\underset{\displaystyle H\ H\ H\ H\ H\ H\ H\ H}{C\text{—}C\text{—}C\text{—}C\text{—}C\text{—}C\text{—}C\text{—}C}}\text{—}OR
$$

The polymerization process just described is called **addition polymerization**.

Activated ethylene + Ethylene monomer ⟶ Polymerized ethylene

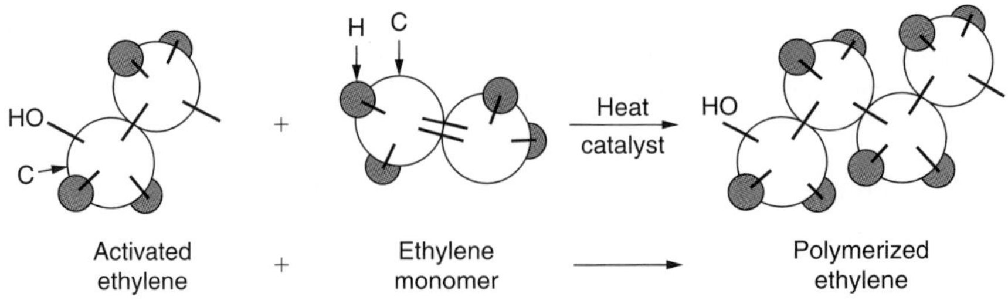

■ **FIGURE 9.1**   The polymerization of ethylene.

9.2.2 **Condensation or Step Polymerization**

Condensation polymerization is another important type of reaction that leads to the formation of thermoplastics. The important polymer nylon 6,6 is produced via the following chemical reaction:

$$
\underset{\text{Hexamethylene diamine}}{\overset{\text{H}\quad\quad\text{H}}{\text{HN -(CH}_2)_6\text{ -N-H}}} + \underset{\text{Adipic acid}}{\overset{}{\text{HO -}\underset{O}{\overset{||}{\text{C}}}\text{ -(CH}_2)_4\text{ -}\underset{O}{\overset{||}{\text{C}}}\text{ -OH}}} \rightarrow \underset{\text{Nylon (monoamide)}}{\overset{\text{H}}{\text{H}_2\text{N -(CH}_2)_6\text{ - N- }\underset{O}{\overset{||}{\text{C}}}\text{-(CH}_2)_4\text{ -}\underset{O}{\overset{||}{\text{C}}}\text{ -OH}}} + \underset{\text{water}}{\overset{}{\text{H}_2\text{O}}}
$$

Here two different organic molecules, an amine and an acid, combine to form the molecule of interest through elimination of water (H + HO). Once formed, each nylon molecule still has reactive groups at each end like those of the original precursors, and they can undergo repeated reactions to extend the polymer chain of Nylon 6,6:

$$
-\left[\overset{\overset{\text{H}}{|}}{\text{N}}-(\text{CH}_2)_6-\overset{\overset{\text{H}}{|}}{\text{N}}-\underset{\underset{\text{O}}{||}}{\text{C}}-(\text{CH}_2)_4-\underset{\underset{\text{O}}{||}}{\text{C}}\right]_N-
$$

Note that no initiators are required; condensation monomers are intrinsically reactive, and each molecule has an equal probability of reacting.

Similarly, when organic acids react with alcohols, an organic **ester** $-\overset{\overset{\text{O}}{||}}{\text{C}}\text{O}-$ linkage binds the original molecule remnants, and water is rejected. In this way, commercially important thermoplastic **polyesters**, like polyethylene terephthalate (PET), polybutylene terephthalate (PBT), Dacron, and Mylar are produced.

If the chain consists of N polyethylene molecules; we symbolize it as

$$
-\left[\overset{\overset{\text{H}\quad\text{H}}{|\quad|}}{\underset{\underset{\text{H}\quad\text{H}}{|\quad|}}{\text{C}-\text{C}}}\right]_N-
$$

The repeat unit $-\overset{\overset{\text{H}\quad\text{H}}{|\quad|}}{\underset{\underset{\text{H}\quad\text{H}}{|\quad|}}{\text{C}-\text{C}}}-$ is a mer.

9.2.1 **Molecular Weight**

The number N of mers is the **degree of polymerization**; this can be very large, often reaching $N = 10,000$, and results in molecules larger than one micrometer. Such molecules are called macromolecules.

The **molecular weight** of the polymer is the sum of all atomic weights, in other words it is N times the **molecular weight of the mer**. The molecular weight of ethylene is $M = 2 \times 12 + 4 \times 1 = 28$. With $N = 10,000$, for example, the molecular weight of the polyethylene chain is 280,000. In the processing of polymers, it is not possible to obtain all chains of the same length. Thus, in practice, the degree of polarization and the molecular weight given for a polymer are **average values**. Figure 9.2 shows a typical distribution of molecular weights. In the particular example, the average molecular weight is 25,850 g/mole.

With the geometry of the sp^3 hybrid molecular orbitals that characterize the bonding of carbon, a section of polyethylene takes the shape shown in Figure 9.3. The real shape of a polyethylene molecule is not a straight bar as suggested by Figure 9.3. While the sp^3 hybrid rigidly imposes the angle of 109.5°

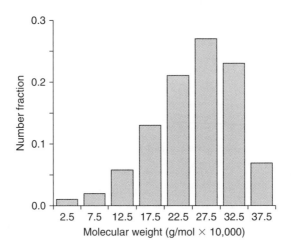

■ **FIGURE 9.2**   Distribution of molecular weights in a polyethylene sample.

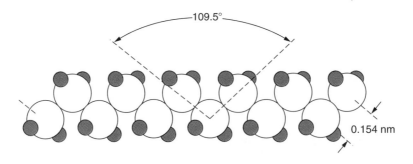

■ **FIGURE 9.3**   Portion of a polyethylene molecule. The white atoms are carbon; the dark atoms are hydrogen.

■ **FIGURE 9.4** (A) Possible rotation of the bond around its axis. As a consequence, the molecule can have the straight shape (B) or the irregular shape where the angle between bonds remains at 109.5°.

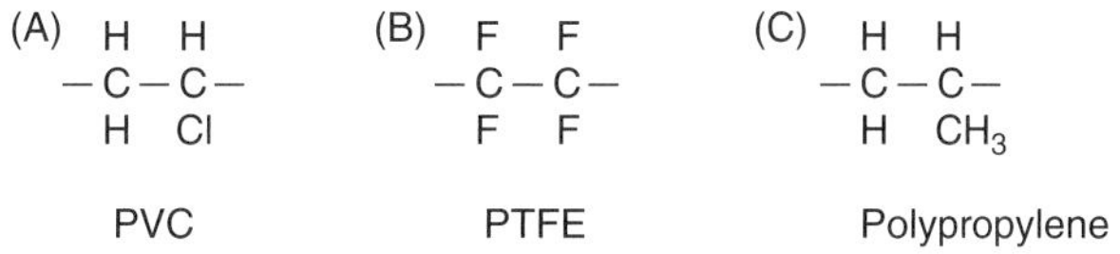

(A)
H H
— C — C —
H Cl

PVC

(B)
F F
— C — C —
F F

PTFE

(C)
H H
— C — C —
H CH_3

Polypropylene

■ **FIGURE 9.5** The mers of (A) PVC; (B) PTFE; and (C) polypropylene.

between bonds, a bond can rotate easily around its axis as illustrated in Figure 9.4. Consequently, the polymer chains are flexible and have an irregular shape.

9.3 POLYMERS AND SECONDARY BONDS; THERMOPLASTICS

We have already examined the chemical bonding of polymers in Chapter 1; we briefly recapitulate it for convenience. The bonds between the carbon atoms and between carbon and hydrogen inside the chain are covalent. The valence electrons completely fill the valence band; therefore polymers are insulators and are often transparent.

Polyethylene chains are held together by a weak van der Waals bond. The **van der Waals bond** is about 50 times weaker than the C—C and the C—H bonds inside the chain. **The weakness of the van der Waals bond is responsible for the softness of polyethylene.**

Other polymers, such as polyvinyl chloride (PVC), polytetrafluoroethylene (PTFE, which is commonly known by its trade name Teflon©), and polypropylene are built on the same principle as polyethylene. Their mers have a similar structure as that of polyethylene except that one hydrogen is replaced by chlorine in PVC or by a methyl group CH_3 in polypropylene and all four hydrogen atoms are replaced by fluorine in PTFE. These are shown in Figure 9.5.

PVC is a stronger polymer than polyethylene, and the bonding between its chains will explain this (Figure 9.6). Chlorine has a high electronegativity and attracts electron charge, creating a mixed covalent-ionic C-Cl bond and a **permanent dipole**. The attraction between these dipoles forms a **secondary bond** that is about five times stronger than the van der Waals bond, but still 10 times weaker than the

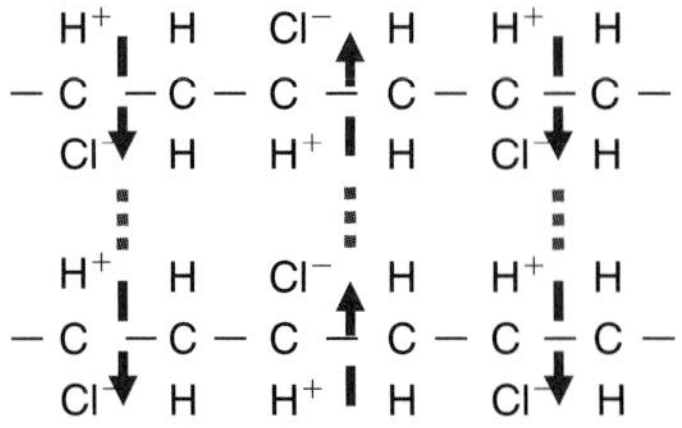

■ **FIGURE 9.6** Bonding between two PVC molecules. The ionic bond of chlorine forms a permanent dipole in the molecule. Attraction between the permanent dipoles forms a relatively strong secondary bond. Since a hydrogen atom is involved, this is also called a hydrogen bond.

primary bonds inside the polymer chain. Since this dipole involves a hydrogen atom, the resultant bond is also called a **hydrogen bond**.

Nylon also derives its strength from **hydrogen bonds** formed between neighboring molecules.

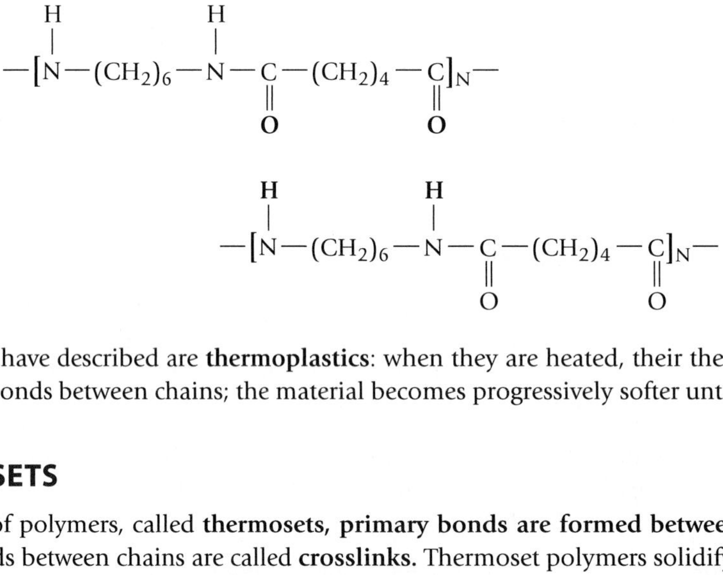

The polymers we have described are **thermoplastics**: when they are heated, their thermal energy overcomes the weak bonds between chains; the material becomes progressively softer until it liquefies.

9.4 THERMOSETS

In another class of polymers, called **thermosets, primary bonds are formed between the molecules**. The primary bonds between chains are called **crosslinks.** Thermoset polymers solidify (set) by a chemical reaction between two different substances. Usually, one of the substances consists of large hydrocarbon molecules and is called the **resin**; the other substance binds the resin molecules together and is called the **hardener**. With primary (strong) bonds between the chains, thermosets are stronger and stable to somewhat higher temperatures than the thermoplastics. These polymers are called thermosets because elevated temperatures accelerate the chemical reaction that solidifies the material. Thermosets do not soften at high temperatures but lose their hydrogen and transform into char. As a consequence, **thermosets are not recyclable** by simple heating. After use, they can be ground up and used as fillers in composites (Chapter 10). There is a push from the government to replace thermosets with thermoplastics as much as possible in order to permit recycling.

Thermosets are used in some applications where high-temperature stability is essential. They are also used as adhesives and as the matrix in fiber-reinforced composites (Chapter 10) because they can be solidified at room temperature, a property that is essential for the fabrication of large objects, such as boats, car bodies, or airplane parts. We now examine two important thermosets.

9.4.1 Epoxy

Epoxy glue comes in two tubes. When one mixes the content of the tubes, the glue hardens in a short time thanks to a chemical reaction between the two substances. The contents of the tubes are an epoxy resin and an ethylene diamine hardener.

The relevant chemical structure of epoxy is the **epoxide group** at its extremity.

A typical epoxy molecule has the form

(In this figure, Be stands for a benzene ring.) Note the degree of polymerization N.

When the epoxy is mixed with ethylene diamine, the oxygen bond is opened and the reaction binds two epoxy molecules by covalent bonds.

The reaction continues until a three-dimensional network is formed.

9.4.2 Unsaturated Polyester

This thermoset, which is less expensive than epoxy, is widely used in fiberglass composites. The resin is a linear polyester and the hardener is styrene. The C=C double bonds are opened in the reaction and create valences that bind the molecules. The open bonds react with another styrene molecule to continue the network.

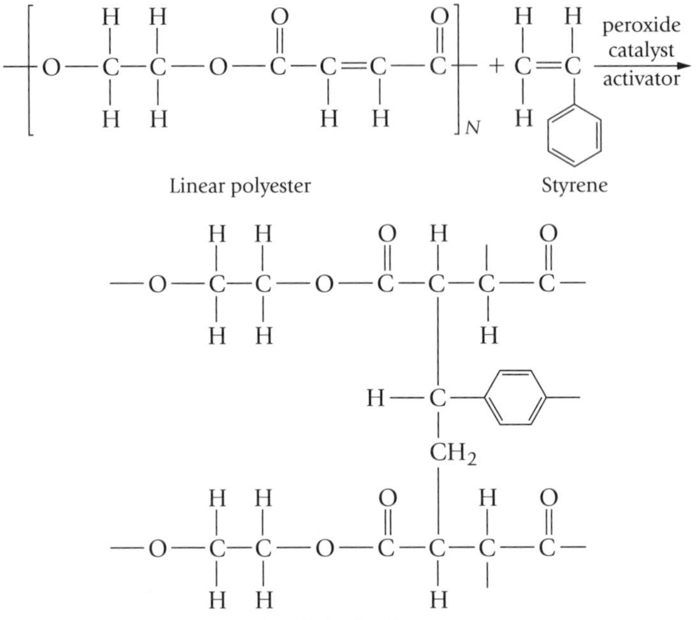

Linear polyester   Styrene

Crosslinked polyester

We note the relatively large spacing between the crosslinks. Therefore, thermosets, while harder than thermoplastics, remain much softer than metals or ceramics.

9.5 **RUBBER (ELASTOMER)**

Rubber is a natural or synthetic polymer with a molecular structure that allows it to stretch by large amounts. Natural rubber, which is *cis*-1,4 polyisoprene, has the following structure

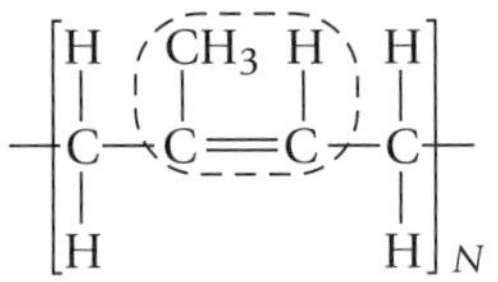

An isomer of this molecule, *trans*-1,4, polyisoprene has the structure

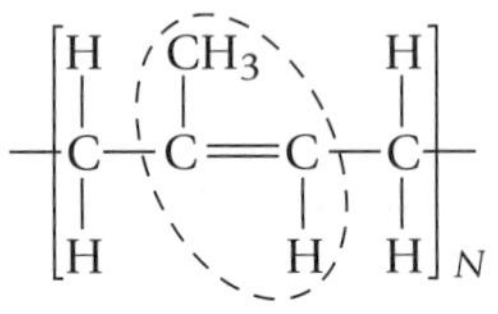

The latter is **gutta-percha** and is not a rubber. The synthetic rubbers polybutadiene and polychloroprene (also called neoprene) have the same *cis* geometry: the ligands lie on the same side of the double-bonded carbon atoms $C=C$.

Polybutadiene   Polychloroprene (neoprene)
structural unit

We note that the $C=C$ double bond involves sp^2 hybrid orbitals that lie in a plane: this gives the mer a flat geometry. The actual shapes of these two molecules are an arc for rubber and a straight molecule for gutta percha.

Rubber   Gutta-percha

The joining of arched mers of rubber permits the formation of a coiled polymer molecule that can deform in the same way as a coil spring under tension. This is not possible with the straight gutta-percha.

Since rubbers are elastic polymers, they are also called **elastomers**.

9.5.1 **Vulcanization**

Natural rubber is a thermoplastic polymer with secondary bonds between the chains. It is too soft for many applications. In 1844 Charles Goodyear obtained a patent for strengthening rubber by reacting it with sulfur. The process is called **vulcanization**. Sulfur reacts with the double bond in the rubber and establishes primary bonds between the chains, effectively transforming the rubber into a thermoset, as shown in Figure 9.7. Vulcanization can be performed to any desired degree, producing rubber of any required hardness and suitable for automotive tires.

9.6 **POLYMER STRUCTURE**

When cooled from the liquid state, thermoplastic polymers can solidify either into an amorphous or a partly crystalline structure. This structure has a large influence on the mechanical properties of the material.

9.6.1 **Amorphous Polymers**

An amorphous polymer is a random tangled ensemble of long chains. The solidification of an amorphous polymer is similar to that of a glass, which we have examined in Chapter 8. In the molten state,

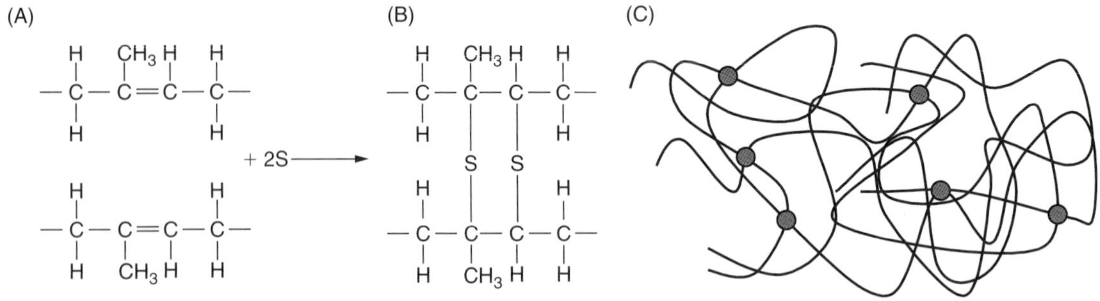

(A)

(B)

(C)

■ **FIGURE 9.7**   Vulcanization of rubber. (A) Fragments of two chains of rubber. (B) Crosslinking of the chains by sulfur. (C) Sketch of rubber chains with sulfur crosslinks. The crosslinking restrains the elastic deformation of the chains and hardens the rubber.

the molecules of the polymer are free to translate, rotate, extend, bend, twist, and so on. Through atacticity, crosslinking, and addition of random sidebranches, the formation of crystalline structure is suppressed. As the temperature is lowered, the movement of the polymer chains slows down, and the viscosity of the material increases in a manner similar to that of glasses. Volume contraction occurs because lower thermal agitation and mutual attraction allow molecules to pack more efficiently. The volume decreases and the viscosity increases until the glass transition temperature (TG) is reached. Below TG, the molecules are effectively immobilized, the viscosity is so large that the polymer solidifies as a rigid brittle solid similar to glass. The much smaller volume change with temperature reflects only the reduced amplitude of thermal vibrations. (See Figure 9.8.)

Polymers undergo large property changes in the vicinity of the glass temperature T_G. Above T_G the amorphous polymer is leathery and extends readily; below T_G it is rigid and brittle. This behavior has a profound influence on the mechanical properties we shall discuss in Section 9.8.

9.6.2 **Crystalline Polymers**

X-ray diffraction methods have demonstrated that many polymers, notably polyethylene and nylon, are crystalline. However, they possess none of the classic structural perfection of metal and ceramic crystals whose atoms maintain order over very long distances. The polymer molecules fold into parallel lengths of 5–50 nm to form crystalline platelets as shown in Figure 9.9. A fraction of the polymer molecule remains disordered so that crystallization is incomplete.

The shaggy crystalline platelets often organize themselves in spherulites by first attaching at some central crystalline nucleus (Figures 9.9 and 9.10). These plates extend radially outward, executing complex growth mechanisms and patterns. As the spherulites continue to grow, the plates splay or fan apart until eventually they butt against other spherulites. The resulting grain-like structure is exhibited by a number of thermoplastics.

In practice, polymers usually crystallize under melt-flow conditions rather than under the quiescent circumstances that lead to spherulitic growth. In such cases elongated crystals align along the flow direction

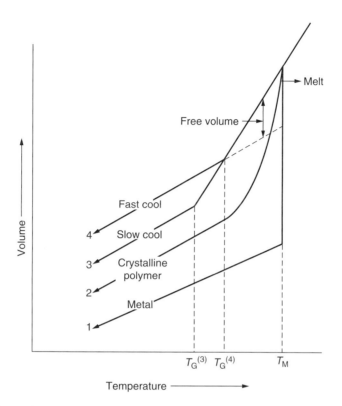

■ **FIGURE 9.8**   Volume changes of metals, crystalline polymers, and amorphous polymers as a function of temperature.

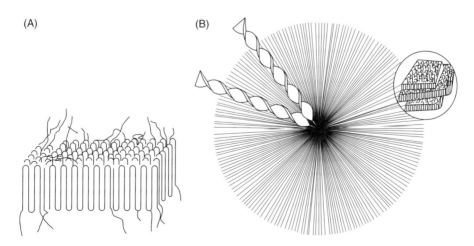

■ **FIGURE 9.9**   Crystallization of polymers. (A) The chains fold unto themselves to form platelets. (B) These platelets attach themselves to a central nucleus and fan out to form spherulites.

■ **FIGURE 9.10**   Spherulite structure of polyethylene. From L.C. Sawyer and D.T. Grubb, *Polymer Microscopy*, Chapman and Hall, London (1987), with kind permission of Springer Science and Business Media.

and a fibrous microstructure results (Figure 9.11). Orienting and extending chain crystals is desirable because both the modulus of elasticity and the strength of polymers are enhanced this way. Fiber-reinforced composites, tires, and industrial belts are some of the applications for high-modulus polymer fibers.

9.7 COPOLYMERS

When two or more kinds of mers are mixed within a single polymer chain, copolymers are produced. Synthesis of copolymers yields materials that combine the beneficial properties of both components.

Several important copolymers incorporate polystyrene

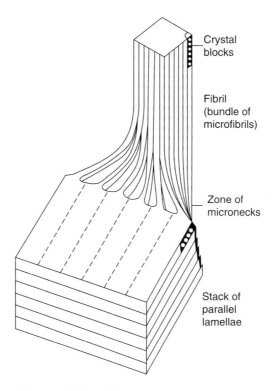

Crystal blocks

Fibril (bundle of microfibrils)

Zone of micronecks

Stack of parallel lamellae

■ **FIGURE 9.11** Fibrillar polymer crystallization under flow conditions.

as one of the ingredients. Polystyrene by itself is rather inflexible and brittle at room temperature, a condition largely caused by the steric hindrance of the benzene rings. But when it is copolymerized with some 3–10% of the stretchy rubber polybutadiene,

$$\left[\begin{array}{cccc} H & H & H & H \\ | & | & | & | \\ C & -C = C - C \\ | & & | \\ H & & H \end{array}\right]_N$$

one obtains a high-impact-strength polystyrene (HIPS). If butadiene additions can improve styrene, what about improving HIPS by further ternary alloying? This is what is done in the ABS family of thermoplastics. Acrylonitrile monomer is copolymerized with Styrene, and Polybutadiene is added to yield **ABS** polymer that possesses excellent properties. Acrylonitrile confers heat and chemical resistance, butadiene imparts the ability to withstand impact loading, and styrene provides rigidity and ease of processing or shaping. Drain and vent piping in buildings, and assorted auto and appliance parts, capitalize on the individual attributes of the components in this **terpolymer** (Figure 9.12).

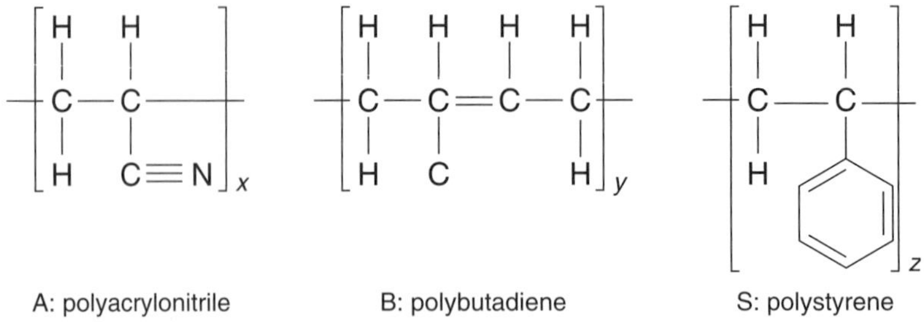

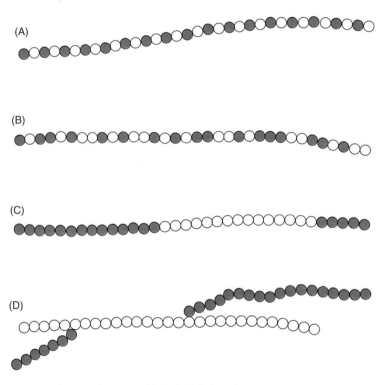

A: polyacrylonitrile   B: polybutadiene   S: polystyrene

■ **FIGURE 9.12**   ABS copolymer.

(A)

(B)

(C)

(D)

■ **FIGURE 9.13**   Copolymers. (A) Alternating. (B) Random. (C) Block. (D) Graft copolymer.

Copolymer structures are schematically depicted in Figure 9.13. At least three types—**alternating, random**, and **block**—copolymers can be distinguished. In addition, linear chains can interact by grafting to one another (Figure 9.13D).

The branching in graft copolymers (Figure 9.13D) can also occur in homopolymers (i.e., composed of one kind of mer only). Polymerization processes can be controlled to produce only linear chains (e.g., high density polyethylene, HDPE) or branched chains (e.g., low density polyethylene, LDPE). Branching strongly hinders alignment of neighboring chains, inhibiting crystalline ordering. Thus, LDPE is typically 50% crystalline while HDPE is ~80% crystalline.

9.8 MECHANICAL BEHAVIOR OF POLYMERS

9.8.1 The Strength of Plastics

Polymers respond to applied stresses in a unique fashion that reflects the nature of their chemical bond and their molecular structure. The glass transition temperature T_G plays a large role in the mechanical properties of polymers. Figure 9.14 shows the stress-strain curves for a typical amorphous polymer at various temperatures. The extremities of the curves represent rupture of the material.

We observe that, at 40°C, the material is brittle as glass: it breaks above a certain strain without any plastic deformation. With a modest increase of temperature (to 104°C), the material, still brittle, shows a marked decline in elastic modulus. The glass transition temperature T_G of the PMMA is 100°C. Below T_G, the molecules are practically immobile, and the material behaves mechanically like a glass. This is the **glassy regime**. Plastic drinking glasses and cases for compact disks break when deformed. They are made of polystyrene whose glass transition temperature is above room temperature.

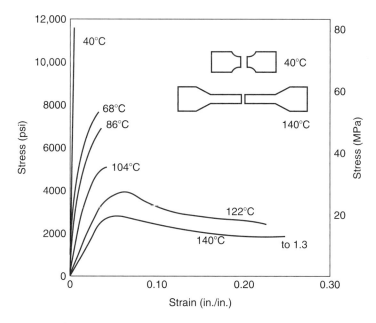

■ **FIGURE 9.14**   Stress-strain curves for polymethyl methacrylate (PMMA) at different test temperatures. After T. Alfrey, *Mechanical Behavior of Polymers*, Wiley-Interscience, New York (1967).

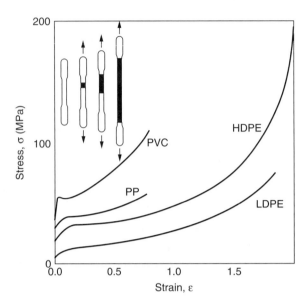

■ **FIGURE 9.15**   True stress-strain curves for several polymers above the glass transition temperature T_G. Insert shows uniform necking extension.

122°C and 140°C correspond to the **rubbery plateau**. After elastic deformation with still lower modulus, the material undergoes extensive plastic deformation (with a strain at rupture as large as 130% at 140°C). This deformation is viscous: the deformation **rate** $d\gamma/dt$, rather than the deformation γ, is proportional to the stress τ. The material feels **leathery** because of the combined elastic and viscous deformation. The plastic deformation occurs by straightening the polymer molecules; accordingly, the cross section of the material decreases with the elongation, as indicated in Figure 9.15. As the polymer chains are straightened, the strength of the material increases, as shown in the figure. The 1 gal milk or water containers, made of polyethylene, are examples of material in the leathery state ($T > T_G$).

At still higher temperatures, well above T_G, the polymer deforms like a viscous liquid. This is the **viscous regime** in which polymer objects are manufactured.

Figure 9.16 shows how the mechanical properties of polymers depend on temperature and how they are influenced by the volume fraction that is crystalline and by the amount of crosslinking. An increase in crystallinity raises the glass transition temperature. When the material is 100% crystalline, the viscoelastic modulus drops rapidly at the melting temperature. Increased crosslinking raises the elastic modulus and diminishes viscous flow.

Table 9.1 contains the mechanical properties of selected polymers.

9.8.2 The Viscoelasticity of Elastomers

We saw in Section 9.5 that elastomers derive their low elastic modulus and their capacity for large elastic deformation (to 700% elongation) from their molecular structure that gives the polymer chains the shape of coils. The elastic deformation occurs by straightening these coils. Elastomers operate above

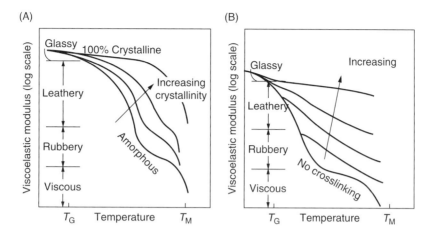

■ FIGURE 9.16 Typical temperature dependence of the viscoelastic modulus of long-chain polymers.(A) Effect of degree of crystallinity. (B) Effect of extent of crosslinking. Note that the vertical scale is logarithmic with a range of about 10 orders of magnitude. From S. Kalpakjian, *Manufacturing Processes for Engineering Materials*, 2nd ed., Addison-Wesley, Reading, MA (1991).

the glass transition temperature; therefore deformation also occurs by slow viscous flow. The deformation of elastomers is **viscoelastic**: the stress is not only proportional to the deformation but also to the deformation rate.

$$\sigma = \varepsilon E' + d\varepsilon / dt \, E'' \tag{9.1}$$

where E' is the elastic modulus and E'' the loss modulus because it expresses an absorption of energy. The response of a polymer to a varying stress is sluggish as shown in Figure 9.17.

The difference between elastic and viscoelastic deformation is easily observed: one can obtain a bell-like sound from a metal pan, a glass, or a ceramic pot, but not from a plastic pot. The vibrations of the latter are rapidly absorbed by the viscosity.

The viscoelasticity of polymers finds an important application in tires as illustrated in Figure 9.18. Roads are designed as a mixture of tar and protruding sharp stones. In rainy weather, a purely elastic tire would easily espouse the profile of the road and leave a continuous water film between them. This would lead to aquaplaning: the tire would never touch the road and slide on the water. The viscoelastic tire does not deform fast enough to espouse the sharp points of the stones; the latter penetrate the rubber and provide traction even in the presence of water. (Caution: At high speeds, the water film is thick enough to cause aquaplaning despite the viscoelastic rubber!)

Because of the viscous nature of deformation, it is not surprising that the moduli E' and E'' change with deformation rate and temperature. In a periodic deformation, as in a tire, this means that the moduli depend on the frequency of the deformation. The loss modulus E'' is largest at the glass transition temperature. The elastomers of tires are formulated to maximize the loss modulus at the expected speed and temperature of usage for maximum safety.

Table 9.1 Mechanical Properties of Selected Polymers.

Polymer	E (GPa)	ρ (g/cm³)	σ_o[a] (MPa)	T_G (°C)
Thermoplastics				
Polyethylene (PE)				
High-density	0.56	0.96	30	
Low-density	0.18	0.91	11	~20
Polyvinyl chloride (PVC)	2.0	1.2	25	80
Polypropylene (PP)	1.3	0.9	35	0
Styrene				
ABS[b]	2.5	1.2	50	80
Polystyrene (PS)	2.7	1.1	50	100
Polycarbonate (PC)	2.3	1.2–1.3	68	150
Polyethylene terephthalate (PET)	8	0.94	135	67
Polymethyl methacrylate (PMMA)	2.8	1.2	70	100
Polyesters	1.3–4.5	1.1–1.4	65	67
Polyamide	2.8	1.1–1.2	70	60
Polytetrafluoroethylene (PTFE)	0.4	2.3	25	120
Thermosets				
Epoxies	2.1–5.5	1.2–1.4	60	(107–200)
Phenolics	18	1.5	60	(200–300)
Polyesters (glass-filled)	14	1.1–1.5	70	200
Polyimides (glass-filled)	21	1.3	190	350
Silicones (glass-filled)	8	1.25	40	300
Ureas	7	1.3	60	80
Urethanes	7	1.2–1.4	70	100

[a]σ_o = yield strength, TG = glass transition temperature (maximum value)
[b]ABS, acrylonitrile = butadiene = styrene
After B. Derby, D.A. Hills and C. Ruiz, *Materials for Engineering*, Longman Scientific and Technical, London (1992)

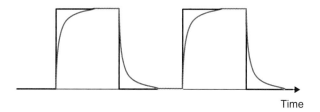

■ **FIGURE 9.17**   Viscoelastic deformation. Black: evolution of the stress. Color: evolution of the strain.

Elastic deformation   Viscoelastic deformation

■ **FIGURE 9.18**   Benefit of viscoelasticity in the safety of a tire running over a wet road. Left: unsafe deformation of a purely elastic body; the rubber espouses the contour of the road and permits a continuous water film. Right: The viscoelastic body does not deform fast enough to espouse the contour of the road; the sharp points penetrate the tire.

When an elastomer is cooled to $T < T_G$, it is hard and brittle.

9.8.3 **Fracture of Polymers**

The fracture of thermosetting polymers is much like that of other brittle solids. In the presence of surface flaws or sharp notches, critical levels of stress act to sever covalent bonds and cause fracture. Fracture of thermoplastic polymers is brittle below the glass transition temperature; at $T > T_G$ fracture is ductile and is preceded by **crazing**. In the large tensile stresses in front of a crack, plastic deformation occurs by straightening of the molecular chains; the local thinning of the material (see Figure 9.15) causes microvoids separated by fibrillar bridges. Figure 9.19 shows a craze in a polymer.

The energy necessary for the formation of the craze serves to increase the toughness of the material.

9.8.4 **Creep of Polymers**

Creep is a slow and steady increase in the deformation of a material subjected to a constant load. It takes place above the glass transition temperature. Creep in polymers can take place at room temperature and even below. Creep tests are conducted in the same manner as they are for metals and exhibit a similar behavior (see Section 2.10): one observes a rapid primary creep, a steady secondary creep, and an accelerated tertiary creep, followed by creep rupture (see Figures 2.27 and 2.28). Creep rates of polymers also increase with temperature and with applied stress. This deformation in polymers is called **viscoelastic creep** and is represented as a time-dependent **creep modulus** $E_c(t)$, which is the applied stress divided by the creep strain after a chosen time t. Usually the time t is 10 hr. A large creep modulus signifies small creep deformation and large creep resistance.

■ **FIGURE 9.19** Transmission electron micrograph of a craze in poly(phenylene oxide). Photo taken by R. P. Kambour, GE Corporate Research and Development.

9.9 APPLICATIONS OF POLYMERS

Polymers are so inexpensive, easy to manufacture, light, and generally corrosion resistant that they find application in practically all aspects of engineering, and their range of applications keeps increasing. Tables 9.2, 9.3, and 9.4 present some of the most important polymers and their applications.

Polyethylene, with a glass transition temperature below room temperature, is a flexible, easily processed material. It is widely used in thin film form as grocery and garbage bags and in packaging and lining. Most milk containers are polyethylene.

Polyvinyl chloride (PVC) is a strong, hard material used in buildings. When it is combined with phthalate ester, an oily substance serving as plasticizer, it is flexible and used in fabrics (mainly raincoats) and hoses.

Polystyrene is an inexpensive hard transparent material. When foamed with CO_2, it forms the well-known Styrofoam used in insulating containers (coffee cups). The styrene-butadiene copolymer is a tough material used, for instance, in computer casings.

Polytetrafluorethylene (PTFE), known under the trade name Teflon, is a crystalline high-temperature material. The strong C-F bond gives the material good temperature stability; it also makes it chemically inert. Consequently, PTFE does not adhere to other surfaces: it is used as a non-stick coating on pans and as a low-friction coating (particularly in the hinges of automobile doors). Since it does not adhere chemically, it is fastened to its substrate by mechanical interlocking (the surface of the substrate is rough).

Polyethylene terephthalate (PET) is used as a container for carbonated beverages because it is impervious to gases, a property that polyethylene lacks. It is also widely used for film packaging and as a fiber for fabrics, carpets, and ropes.

The hydrogen bonds between the chains of **nylon** provide it with great strength and a wide range of applications. Some typical uses of nylon are in machine parts such as unlubricated gears, in electrical connectors, and in hosiery.

Table 9.2 COMPOSITION AND USES OF THERMOPLASTICS

Thermoplastic	Composition of Repeating Unit	TG (K)	Uses
Polyethylene (PE) (Partly crystalline)		270	Tubing, film, sheet, bottles, packaging, electrical insulation
Polyvinylchloride (PVC) (Amorphous)		350	Window frames, plumbing piping, flooring, fabrics, hoses.
Polypropylene (PP) (Partly crystalline)		253	Same uses as PE, but lighter, stiffer, more resistant to sunlight.
Polystyrene (PS) (Amorphous) Foamed with CO_2 Toughened with butadiene		370	Inexpensive molded objects, insulating containers (Styrofoam), high-impact polystyrene (HIPS), packaging.
Polytetrafluroethylene Teflon (PTFE) (Amorphous)			Non-stick cookware, bearings, seals Container for corrosives
Polymethylmethacrylate Lucite (PMMA) (Amorphous)		378	Transparent sheet Aircraft windows windscreens.
Polyethylene Terepthalate (PET)			Bottles for carbonated beverages
Nylon (Partly crystalline when drawn)		340	Textiles, rope, gears, machine parts

Table 9.3 Composition and Uses of Thermosetting Polymers.

Thermoset	Composition of repeating unit	Uses
Phenolics Phenol-formaldehyde Bakelite (amorphous) See Eq. 4–7, Fig. 4–8	OH | —C$_6$H$_2$—CH$_2$— | —CH$_2$	Electrical insulation
Epoxy (amorphous)	—O— (ring) —C(CH$_3$)$_2$— (ring) —O—CH$_2$—CH(OH)—CH$_2$—	Fiberglass matrix, adhesives
Polyester (amorphous)	—C(=O)—(CH$_2$)$_m$—C(=O)—O—C(CH$_2$OH)$_2$—	Fiberglass composites, less expensive than epoxy
Melanin-formaldehyde	(triazine ring structure with N–H and N–C linkages)	Molded dinnerware adhesives and bonding resins for wood, flooring and furniture (usually cellulose-filled)

and similar units randomly
connected by a variety of links

The first widely used polymer was a thermoset, namely, **bakelite. Thermosets** find application when superior strength and utilization at relatively high temperatures are required. There is a trend to replace them with thermoplastics to allow recycling. The two most widely used thermosets are the epoxies and unsaturated polyester. These materials are used as adhesives and as matrix in composite materials, which we shall discuss in Chapter 10. In these applications, their great advantage is that they can be applied and processed at room temperature.

Table 9.4 Composition and Uses of Elastomers (Rubbers)

Elastomer	Composition of repeating unit	Uses
Polybutadiene	$$-\overset{\displaystyle H}{\underset{\displaystyle H}{C}}-\overset{\displaystyle }{\underset{\displaystyle H}{C}}=\overset{\displaystyle }{\underset{\displaystyle H}{C}}-\overset{\displaystyle H}{\underset{\displaystyle H}{C}}-$$	Tires, moldings; Amorphous except when stretched.
Polyisoprene (natural rubber)	$$-\overset{\displaystyle H}{\underset{\displaystyle H}{C}}-\overset{\displaystyle }{\underset{\displaystyle H}{C}}=\overset{\displaystyle }{\underset{\displaystyle CH_3}{C}}-\overset{\displaystyle H}{\underset{\displaystyle H}{C}}-$$	Tires, gaskets; amorphous except when stretched.
Neoprene	$$-\overset{\displaystyle H}{\underset{\displaystyle H}{C}}-\overset{\displaystyle }{\underset{\displaystyle H}{C}}=\overset{\displaystyle }{\underset{\displaystyle Cl}{C}}-\overset{\displaystyle H}{\underset{\displaystyle H}{C}}-$$	Oil-resistant rubber used for seals
Silicone rubber	$$-O-\overset{\displaystyle CH_3}{\underset{\displaystyle CH_3}{Si}}-O-\overset{\displaystyle CH_3}{\underset{\displaystyle CH_3}{Si}}-O-$$	Thermal and electrical insulation, components and coatings, foam rubber

9.10 MANUFACTURE OF POLYMERIC OBJECTS

The processing of thermoplastics exploits the relatively large viscosity the polymer acquires at moderate temperatures, and the fabrication of thermoset articles utilizes a chemical reaction at low temperatures that solidifies the material.

Thermoplastic melts are molded or blown at elevated temperatures where the viscosity is 10^3 to 10^5 Pa·s (these values are similar to those utilized in the processing of glass; see Figure 8.23). Pressure is generally required for polymers to fill molds. A variety of extrusion and molding operations are used to process polymers. In particular, the following processing methods now enable thermoplastic products of complex shape to be produced on a very large scale:

a. Extrusion.

b. Injection molding.

c. Blow molding.

d. Compression molding.

e. Calendering.

f. Thermoforming.

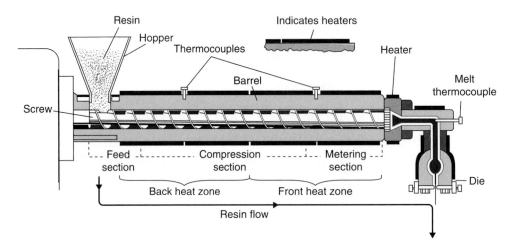

■ **FIGURE 9.20**   Schematic drawing of an extruder. The screw moves forward through the metering section to push the melt through the die. Modified from R.J. Baird, *Industrial Plastics*. Copyright (1976), Goodheart-Willcox — reproduced with permission.

Extrusion processing underlies both injection and blow molding of polymers and will, therefore, be discussed first.

9.10.1 **Extrusion**

Extruding polymers resembles extrusion of metals. Lengths of tubular, sheet, and rod products with simple or complex cross sections emerge from dies after pushing polymer feedstock through them. Thermoplastic materials (resins) in pellet form are fed in at one end of the extruder as shown in Figure 9.20. The polymer feed is conveyed down the extruder barrel by one or more long rotating screws where it is **compressed**. The polymer **melts** through contact with the heated walls and the mechanical action of the screw. Next, the screw is moved forward and forces the melt pressure through a tapered region and through a shaped die. The extrudate must have a sufficiently high viscosity when it leaves the die to prevent it from deforming mechanically or even collapsing in an uncontrolled way. Therefore, water or air sprays are used to cool the product.

9.10.2 **Injection Molding**

The process shown in Figure 9.21 starts the same way as extrusion. Instead of being forced through a die into the ambient, the melt is injected under pressure into a split die cavity. Injection molding is usually carried out at $\sim 1.5\ T_G$, where T_G is the glass transition temperature of the polymer. After the part cools below T_G (under pressure) the die opens and the part is ejected.

Thermosets are also processed by injection molding. In this case the chamber is essentially a chemical reactor that allows the mixed and heated ingredients to polymerize, blend, and crosslink during simultaneous shaping. Fibers and particles can also be introduced, enabling the production of composites.

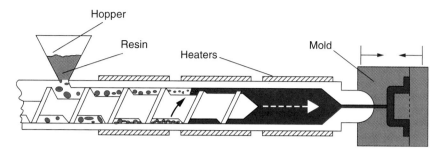

■ **FIGURE 9.21**   Injection molding. (A) Plastic granules are melted as they travel along the revolving screw. (B) The screw barrel is driven forward and injects the molten polymer in to the mold. (C) The screw barrel is retracted and the finished piece is ejected. Modified from R.J. Baird, *Industrial Plastics*. Copyright (1976), Goodheart-Willcox — reproduced with permission.

Injection molding is a high-production-rate process with typical cycle times ranging from seconds for thermoplastics to several minutes depending on the type of polymer and the part size. High injection pressures mean that good tolerances and surface finishes can be maintained. Among the parts produced by injection molding are containers, housings, plumbing fixtures, gears, telephone receivers, toys, and so on. Parts of comparable size to those produced in die-casting can be injection molded. This is why polymers have increasingly replaced metals in the above and other applications.

9.10.3 Blow Molding

The combination of extrusion and air pressure makes blow molding and the production of plastic bags and bottles possible. In the fabrication of bags and foils, a thin-walled tube of polymer is extruded vertically and expanded by blowing through the die until the desired film thickness is produced (Figure 9.22A). Wraps are made continuously by cutting open the cylinder; bags are made by periodically welding the tube with heat and perforating it.

In the fabrication of beverage bottles, the extruded tube, known as a **parison**, is pinched at one end, and clamped within a mold that is much larger than the tube diameter (Figure 9.22B). It is then blown outward until the polymer extends to the mold. A hot air blast at a pressure of 350–700 kPa (50–100 psi) is used for this purpose.

The plastic beverage bottle represents an interesting problem in materials design. For marketing purposes, the bottle must be highly transparent. In addition, it must be sufficiently strong (creep resistant) so that it does not lose its shape on the shelf. It must be impermeable to CO_2 so that gas loss does not make the contents go flat. To meet these needs the polymer PET (polyethylene terephthalate) is selected and blow molded. But, if cooled rapidly, PET is amorphous. It is quite clear, but its relatively loose molecular structure makes it permeable to CO_2 and susceptible to creep. At the other extreme of slow cooling, the polymer crystallizes. It is now strong and gas tight, but opaque. The answer is first to cool the parison rapidly below T_G (340°K) where it is amorphous. Then it is stretched by blow molding at 400°K, a temperature high enough for plastic flow but too low for appreciable crystallization. Stretching a polymer induces some crystallization and higher strength, and, if it is controlled, no loss in clarity occurs.

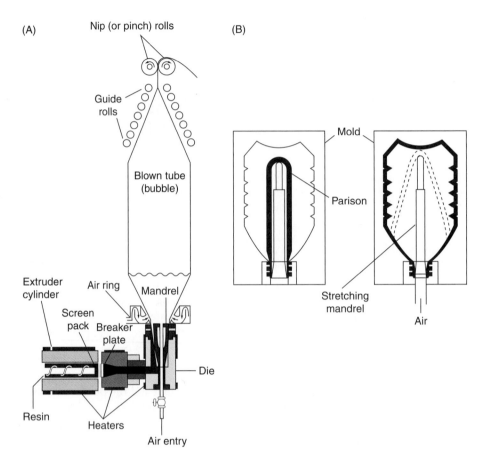

■ **FIGURE 9.22** Blow molding of polymers. (A) Forming a blown film or bag. (B) Forming a bottle: the parison is introduced into the mold and is expanded to fill the mold by compressed air. Modified from R.J. Baird, *Industrial Plastics*. Copyright (1976), Goodheart-Willcox — reproduced with permission.

9.10.4 **Compression Molding**

Where the expense of an extruder and injection molding dies is not warranted, compression molding of parts is practiced. In this process (Figure 9.23) pre-measured volumes of polymer powder or viscous mixtures of resin and filler are introduced into a heated multi-piece die and then compressed with an upper plug. Thermosetting polymers and elastomers are shaped by compression molding; products include electrical terminal strips, dishes, and washing machine agitators.

9.10.5 **Calendering**

Polymer sheets are often made by calendering. In this process, a warm plastic mass is fed through a series of heated rolls and stripped from them in the form of a sheet (see Figure 9.24).

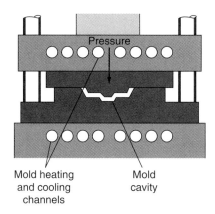

■ **FIGURE 9.23** Compression molding. Modified from R.J. Baird, *Industrial Plastics.* Copyright (1976), Goodheart-Willcox —reproduced with permission.

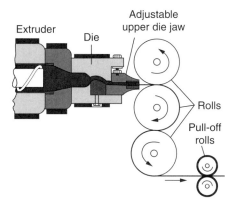

■ **FIGURE 9.24** Calendering. Modified from R.J. Baird, *Industrial Plastics.* Copyright (1976), Goodheart-Willcox—reproduced with permission.

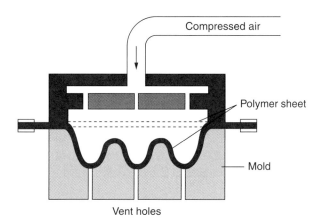

■ **FIGURE 9.25** Thermoforming. Modified from R.J. Baird, *Industrial Plastics.* Copyright (1976), Goodheart-Willcox —reproduced with permission.

9.10.6 **Thermoforming**

This process (Figure 9.25) parallels methods used to form sheet metals. Sheet polymer (e.g., formed by calendering) is preheated and then laid over a mold having the desired shape. Applied air pressure is normally sufficient to make the sheet flow plastically and cover the mold interior. Parts made this way include advertising signs, panels for shower stalls, appliance housings, and refrigerator linings.

■ SUMMARY

1. A polymer is a material that consists of macromolecules composed of repeating units called mers. A given polymer material contains molecules of different lengths. The degree of polymerization is the average number of mers contained in the chains. The molecular weight of a polymer is the product of the molecular weight of the mer and the degree of polymerization.

2. Atoms inside a polymer molecule are held together by primary bonds.

3. In thermoplastic polymers, the bonds between the macromolecules are secondary. They are van der Waals bonds in polyethylene and permanent dipole bonds in polyvinyl chloride (PVC). Van der Waals bonds are about 50 times weaker than primary bonds, and permanent dipole bonds are approximately 10 times weaker than primary bonds. These secondary bonds are responsible for the low melting temperatures and the softness of polymers. The term *thermoplastic* expresses the fact that these materials become plastic and finally melt as the temperature is increased. Thermoplastics can be recycled by simple heating.

4. In thermoset polymers, primary bonds are formed between the macromolecules. The term *thermoset* expresses the fact that heating accelerates setting, that is, the chemical reaction between resin and hardener that solidifies the material. Thermosets do not melt upon heating but decompose instead; they are not recyclable by reheating.

5. Elastomers are thermoplastic polymers that tolerate large elastic deformations. They obtain this property from the shape of the mer in which both ligands to the double-bonded carbons lie on the same size (i.e., *cis* isomer). Vulcanization is a process by which reaction with sulfur creates crosslinks between the elastomer chains. These crosslinks increase the stiffness of the rubber by impeding molecular deformation; they transform the rubber into a thermoset.

6. Polymers with interesting properties can be obtained by combining two or more kinds of mers. For instance, by combining the strong polystyrene with the elastomer polybutadiene, one obtains a strong and tough polymer. Copolymers exist in alternating, random, block, and graft forms.

7. The arrangement of the long polymer molecules is often disorderly: the polymer is amorphous. Amorphous materials do not have a distinct melting point at which they change from liquid to rigid solid; instead their viscosity increases as the temperature is lowered.

8. Amorphous materials are characterized by a glass transition temperature T_G. Above T_G, the polymer is flexible and leathery; below T_G, the material is brittle.

9. In crystalline polymers, the chains fold unto themselves to form a regular (crystalline) array of short parallel segments. Distinct crystalline regions in a polymer are spherulites. Polymers crystallize

only partly: they contain amorphous regions where some parts of their molecular chains remain disordered. Crystalline polymers are stronger and harder than amorphous polymers.

10. The mechanical properties of polymers depend on their glass transition temperature T_G. When T_G is below the utilization temperature, the polymers are soft and leathery; if T_G is above the utilization temperature, they are relatively hard and brittle.

11. Elastomers are viscoelastic: their deformation is simultaneously viscous and elastic. This property is utilized in the design of tires and roads: the sluggish deformation causes sharp points to penetrate the rubber and decreases hydroplaning on wet roads.

12. Fracture of polymers is accompanied by extensive straightening of polymer molecules (crazing). This causes energy absorption that toughens the polymers.

13. The relatively low glass transition temperature of polymers is responsible for creep even at room temperature.

14. In addition polymerization, a catalyst opens the double bond of the mer that becomes a radical capable of opening the double bond of anther mer to which it is attached. This reaction continues and forms the chain.

15. In condensation polymerization, an amine or alcohol reacts with an acid; the two are joined and reject a water molecule.

16. The processing of polymers utilizes the temperature-dependence of their viscosity. Polymers are typically formed by injection molding, blow molding, extrusion, and thermoforming. Thermosets utilize similar techniques, but one needs to give the material time to set.

■ KEY TERMS

A
ABS, 261
ABS, Acrylonitrile, Butadiene, Styrene Copolymer, 261
addition polymerization, 249, 250
alcohol, 250
alternating copolymer, 262
amine, 250
amorphous polymers, 257
applications of polymers, 268

B
bakelite, 270
benzoyl peroxide, 250
block copolymer, 262
blow molding, 273

C
calendering, 274

catalyst, 250
chain polymerization, 249
cis geometry, 257
compression molding, 274
condensation polymerization, 251
copolymers, 260
crazing, 267
creep modulus $E_c(t)$, 267
creep of polymers, 267
crosslinks, 254, 256
crystalline polymers, 258

D
degree of polymerization, 251

E
elastomers, 256
epoxide group, 255

epoxy, 255
extrusion, 272

F
fracture of polymers, 267

G
glass transition temperature (TG), 258
glassy regime, 263
grafting copolymer, 262
gutta-percha, 257

H
hardener, 254
hydrogen bonds, 254

I
injection molding, 272

■ REFERENCES FOR FURTHER READING

[1]   J.M.G. Cowie, V. Arrighi, *Polymers: Chemistry and Physics of Modern Materials*, 3rd ed., CRC Press, Boca Raton,
        FL (2007).

[2]   C. Hall, *Polymer Materials*, Macmillan Press, New York (1981).

[3]   S.L. Rosen, *Fundamental Principles of Polymeric Materials (Society of Plastics Engineers Monographs)*, Wiley
        Interscience, New York (1993).

■ QUESTIONS AND PROBLEMS

9.1.   The end-to-end distance of a stretched long-chain polymer molecule is much larger than the end-to-end dis-
        tance of the unrestrained molecule. Why?

9.2.   Write the formula for the condensation reaction between urea and formaldehyde to produce thermosetting
        urea-formaldehyde if urea has the following structure.

$$H-N-C-N-H$$
$$\;\;\;\;|\;\;\;\;||\;\;\;\;|$$
$$\;\;\;\;H\;\;\;O\;\;\;H$$

9.3.   What is the degree of polymerization of a polystyrene sample that has a molecular weight of 129,000?

9.4. A certain rubber is composed of equal weights of isoprene ($C_4H_5CH_3$) and butadiene (C_4H_6).

    a. What is the mole fraction of each of the rubber components?

    b. How many grams of sulfur must be added to 2 kg of this rubber to crosslink 1% of all of the mers? Note: One S atom crosslinks two mers.

9.5. If the degree of polymerization of PLVC is 729, what is its molecular weight?

9.6. A polypropylene polymer has equal numbers of macromolecules containing 450 mers, 500 mers, 550 mers, 600 mers, 650 mers, and 700 mers. What is the mass average molecular weight in amu?

9.7. A copolymer of polyvinyl chloride and vinyl acetate contains a ratio of 19 parts of the former to 1 part of the latter. If the molecular weight is 21,000 g/mol, then what is the degree of polymerization? Note: The acetate mer contains four C, six H, and two O atoms.

9.8. A kilogram of vinyl chloride polymerizes to polyvinyl chloride.

    a. What bonds are broken, and what bonds are formed?

    b. How much energy is released in the process? (C—C bond energy = 340 kJ/mole; C═C bond energy = 620 kJ/mole.)

9.9. For atactic polypropylene, the temperature (T in units of $°C$) dependence of the specific volume V_S (in units of mL/g) is given by:

$$V_S(L) = 1.137 + 1.4 \times 10^{-4}T \text{ (for amorphous solid at low temperature)}$$

$$V_S(H) = 1.145 + 8.0 \times 10^{-4}T \text{ (at high temperature)}$$

    a. What is the glass transition temperature of this polymer?

    b. What is the polymer density at 25°C?

    c. Suppose $V_S(L) = 1.15 + 1.4 \times 10^{-4}T$. Would this behavior signify a faster or slower cooling rate from the melt?

9.10. Plastic foams have many uses including padding, flotation devices, and insulation. They consist of large volumes of entrapped gas that can reside in either interconnected open cells or in isolated closed cells.

    a. Are open or closed cells more desirable from the standpoint of a water flotation device?

    b. A polymer has a specific gravity of 1.11 and is foamed to a density of 0.07 g/cm³. What is the percent expansion during foaming?

9.11. Which rubber, polyisoprene or polybutadiene, is more likely to be susceptible to atmospheric oxygen degradation?

9.12. Consider a soda can, a soda glass bottle, and a soda plastic bottle. Describe the fabrication of each and how this method is governed by the specific properties of each material. Comment on the similarities and differences between the methods.

9.13. What are the relative advantages and disadvantages of eyeglasses made of glass and of a polymer?

9.14. Plastic party glasses are made of polystyrene and soda bottles of PET. Compare their behavior as they are squeezed and explain it in terms of the relevant material property (Hint: this is a number).

9.15. Dip a piece of rubber into liquid nitrogen. It will crumble like glass. Explain why.

9.16. Why are polymers considered organic materials?

9.17. What does the word "polymer" mean?

9.18. You purchase polyethylene and specify a molecular weight of 200,000. Describe the structure of this material. What is its degree of polymerization?

9.19. What class of polymers are easily recycled and what class are not? Explain why.

9.20. Describe the chemical bonds in polyethylene.

9.21. Describe the chemical bonds in PVC.

9.22. Draw the mer of PTFE and describe the outstanding properties of this material. What is responsible for these properties?

9.23. Epoxy is sold in two tubes. What are the two materials? How does epoxy solidify?

9.24. Compare the chemical bonds in thermoplastics and thermosets.

9.25. Cite three examples of thermoplastics, draw their mers, and describe their applications.

9.26. Draw the mer of natural rubber $(CH_2—CCH_3==CH—CH_2)$ and explain what makes it an elastomer.

9.27. Describe vulcanization and its benefits.

9.28. Describe the copolymer polystyrene-polybutadiene. The latter is an elastomer. What benefit is derived from making it a copolymer?

9.29. Sketch an alternating, a random, a block, and a graft copolymer.

9.30. Sketch the mer of PVC, describe its special properties, and give an application. What provides PVC with these desirable properties?

9.31. Describe the effect of the glass transition temperature on the mechanical properties of polymers.

9.32. Plastic "glasses" often break in your hands. Polyethylene milk containers do not; they feel leathery. What can you say about their glass transition temperatures?

9.33. How could you make a plastic wrap crumble in your hands like a dry leaf? Explain why.

9.34. Describe the crystallization of nylon.

9.35. What are spherulites?

9.36. Compare the elasticity of a steel spring and a rubber band.

9.37. What property of elastomers contributes to the safety of tires?

9.38. Describe crazing.

9.39. Describe the polymerization of polyethylene.

9.40. Describe condensation polymerization. Compare it with the synthesis of NaCl.

9.41. How are plastic bottles made?

9.42. How are plastic garbage bags made?

9.43. Describe the fabrication of plastic toys.

9.44. Describe the fabrication of a plastic tube.

9.45. Compare the fabrication of thermoplastic and thermoset objects.

9.46. Compare what happens to a thermoplastic and a thermoset when you heat them to progressively higher temperatures.

Composites

A composite is a physical mixture of two or more materials. By combining the advantages of different material classes, a composite achieves properties that could not be obtained from any of its constituents alone. The best known composite is fiberglass, which consists of glass fibers embedded in a polymer matrix. Higher performance is obtained from carbon or graphite fibers in an epoxy matrix. Metal matrix composites contain ceramic particles distributed in a metal matrix. Ceramic matrix composites are reinforced with metal or ceramic fibers to increase fracture toughness. Reinforced and prestressed concrete are composites that combine the hardness of ceramics and the tensile strength of metal. Wood is a natural composite. The strength of these materials is dictated by the layout of the reinforcing phase.

LEARNING OBJECTIVES

After studying this chapter, the student will be able to:

1. Select the matrix and reinforcement for a specific application considering the requirements of performance and price.

2. Select the correct fabrication procedure for a given application.

3. Design the fiber configuration appropriate for an application.

4. Correctly place the rebars in reinforced concrete.

5. Prescribe the preparation of prestressed concrete structures.

6. Describe the structure and properties of "carbide" polymer tools.

7. Decide when plywood is preferable to bulk wood.

8. Correctly utilize wood for its strength.

10.1 **WHAT ARE COMPOSITES?**

"Graphite" tennis rackets, "fiberglass" bodies of cars and boats, hardmetal (tungsten carbide) tools, and reinforced concrete are all examples of composites. Graphite tennis rackets are made of graphite fibers embedded in a continuous epoxy matrix; fiberglass consists of glass fibers embedded in polyester; hardmetal is a mixture of tungsten carbide (ceramic) and cobalt (a metal); and reinforced concrete is a mixture of cement and gravel that is reinforced by steel bars.

A working definition for a composite is "a material that contains a physical mixture of two or more phases that are chemically different and separated by a distinct interface." A composite combines the attributes of the individual constituents in a synergistic way that achieves properties superior to those of any of its constituents.

Actually, the idea of using high strength fibers to strengthen a material is not new. The practice of embedding straw in mud bricks very likely predated the Hebrew slaves under Pharaoh.

The majority of composites are made up of a continuous phase, called a **matrix**, in which a stronger phase, consisting of **fibers** or **particles**, is embedded. Composites are usually designated by the matrix material. We shall examine here, in turn, the polymer matrix composites; metal matrix composites; ceramic matrix composites; concrete; and wood, which is a natural composite.

10.2 **POLYMER MATRIX COMPOSITES**

Composites, consisting of a continuous polymer matrix and reinforcing fibers, constitute by far the largest volume of composites in use today. They combine high strength and low weight, ease of manufacturing, and corrosion resistance. Different types of polymer matrix composites are in use, some because of their low cost, others because of their combination of high strength and low weight. Figure 10.1 shows the extensive use of such composites in modern aircraft design.

Fiberglass (Glass Fiber Reinforced Polymers GFRP)   Produced in the largest quantities of any polymer matrix composite, fiberglass consists of glass fibers in polyester or other polymer matrices. Glass fibers are readily drawn from the molten state; in addition to their high elastic modulus E they are chemically inert. Fiberglass applications include automotive and marine bodies, as well as walls and floorings in structures. Service temperatures do not extend much beyond 200°C, above which the matrix begins to soften.

Carbon and Other Fiber Composites; Carbon Fiber Reinforced Polymers (CFRP)   Carbon fiber has a higher specific modulus E/ρ than glass; it is also more resistant to elevated temperature and chemical exposure, but it is more costly. Therefore, CFRP composites are reserved for more demanding applications that can justify the added expense. In these applications, epoxy thermosets are used for the matrix. In aircraft these composites account for a weight savings of ~25% relative to metals. The widespread use of graphite-reinforced polymers in sporting equipment is well known. Tennis rackets, golf clubs, and sailboat masts are some of the applications.

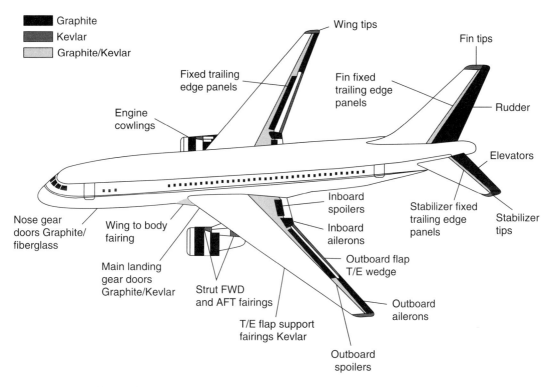

■ FIGURE 10.1   Use of high-performance polymer matrix composites on the Boeing 767 passenger jet. Courtesy of Boeing Corporation.

Polymers have also been impregnated with other fibers, notably Kevlar (KFRP), boron, and silicon carbide.

10.2.1 **Processing and Properties of Fibers**

The fibers used in composites have a diameter as small as 8–10 μm. Such thin fibers do not contain cracks or surface roughness to act as stress concentrators; therefore they resist high stresses before breaking and possess great strength as shown in Table 10.1.

Glass and polymers can be drawn into thin fibers because their viscosity increases rapidly as temperature decreases. Molten glass or polymer flowing through orifices in a container is pulled into fibers that are coated with a protective coating and wound into strands.

Carbon fibers are drawn by the same technique from a polymer precursor or from pitch. Heating to temperatures of 1,000–1,500 °C in an inert atmosphere eliminates hydrogen and leaves carbon fibers. Graphite fibers are obtained by further heating to about 1,800 °C where the carbon crystallizes into graphite. These fibers are stiffer and stronger than glass, but the treatments render them more expensive.

Table 10.1 Properties of Reinforcing Fibers, Polymer Matrices, and Composites.

Material	Density (g/cm3)	E (GPa)	σ_{UTS} (GPa)	Fiber radius (μm)	Fracture Toughness K_{1C} (MPa√m)
Fiber					
E-Glass	2.56	76	1.4–2.5	10	
Carbon (high-modulus)	1.75	390	2.2	8.0	
Carbon (high-strength)	1.95	250	2.7	8.0	
Kevlar	1.45	125	3.2	12	
Matrix					
Epoxy	1.2–1.4	2.1–5.5	0.063		
Polyester	1.1–1.4	1.3–4.5	0.060		
Composites					
CFRP (58%C in epoxy)	1.5	189	1.05		32–45
GFRP (50% glass in polyester)	2	48	1.24		42–60
KFRP (60% Kevlar in epoxy)	1.4	76	1.24		

After B. Derby, D.A. Hills and C. Ruiz, Materials for Engineering, *Longman Scientific and Technical, London (1992); and M.F. Ashby and D.R.H. Jones,* Engineering Properties 2- An Introduction to Microstructures, Processing and Design, *Pergamon Press, Oxford (1986).*

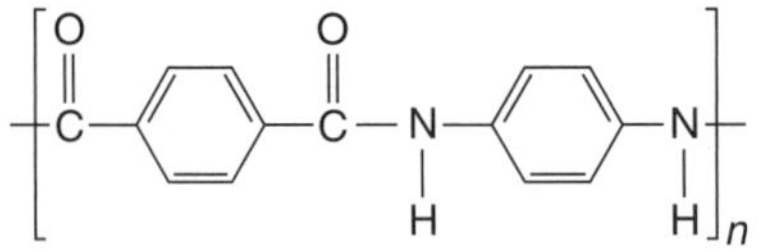

■ **FIGURE 10.2**   The mer of Kevlar.

Kevlar fibers are less stiff but significantly stronger than carbon fibers; they consist of polyamide, as shown in Figure 10.2. Since the polymer solidifies while being pulled, its molecules are straightened and aligned parallel to the fiber axis and crystallize (see Chapter 9). The amide (O=C-N-H) linkages in the molecule provide hydrogen bonding between the polymer chains in a manner similar to nylon, and the aromatic rings give the material its rigidity.

10.2.2 **Polymer Matrix Materials**

Most composites utilize thermoset polymer matrices that harden at room temperature. Epoxy resins are used when high performance is critical; the less expensive polyester is used for most other applications.

Because thermoset matrices do not permit recycling, there is presently a push to replace them by thermoplastics where convenient, namely, for small objects that can be processed at elevated temperature.

10.3 **FABRICATING POLYMER COMPOSITES**

Numerous methods have been developed to fabricate fiber-reinforced polymer matrix composites. The choice of method depends on the particular application.

10.3.1 **Hand Lay-up Process**

This method (Figure 10.3) is used in the fabrication of large structures in relatively small numbers, for instance the hulls of pleasure boats. One constructs an open mold that is first coated with a gel for subsequent easy removal of the composite piece. The fiberglass, in the form of a cloth or a mat, is laid

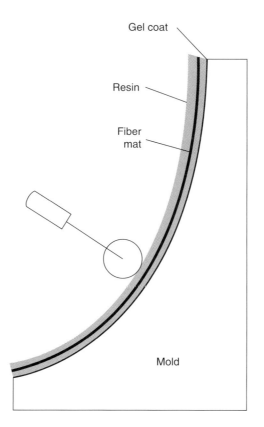

Gel coat

Resin

Fiber
mat

Mold

■ **FIGURE 10.3**   The hand lay-up process.

into the mold. The thermoset, consisting of resin and hardener, is applied by rolling to eliminate air bubbles. Repeated applications of glass cloth and resin achieve the desired thickness.

10.3.2 **The Spray-up Process**

This process (Figure 10.4) is utilized for the production of a large number of simple products such as bathtubs or seats. A spray gun is equipped with a device that chops the fibers into short filaments and sprays them into the mold together with the thermoset resin and hardener. It is recommended to densify the laminate with a roller in order to remove any trapped air.

10.3.3 **Pulltrusion**

This method is illustrated in Figure 10.5. It allows the fabrication of bars, beams, pipes, or tubing with composites. Continuous strands of fiber are pulled through a bath containing the polymer and hardener and through a heated die that gives the product the desired shape. Pulling the piece through the die ensures that the fibers are all parallel to the length of the piece.

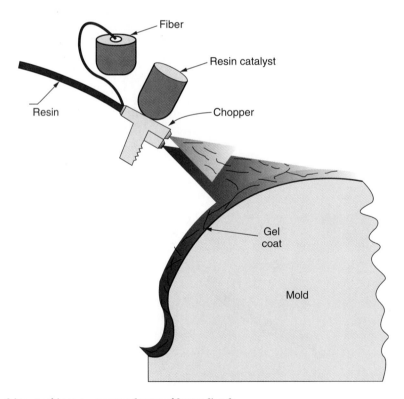

■ **FIGURE 10.4**   Schematic of the spray-up process. Courtesy of Corning Glass Co.

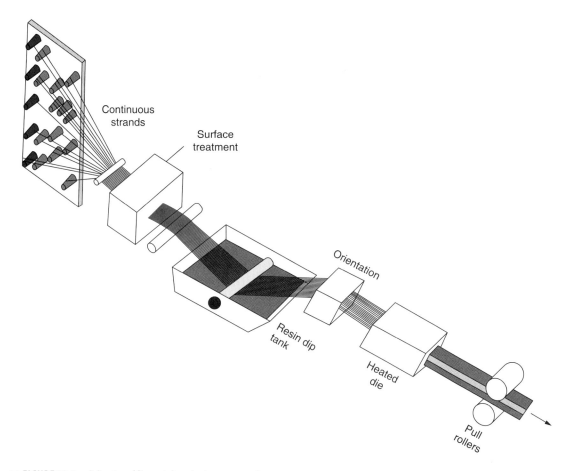

■ **FIGURE 10.5**   Pultrusion of fiber-reinforced polymer composites.

10.3.4 **The Filament Winding Process**

In the fabrication of strong hollow cylinders, such as the body of rockets or pressure tanks, the fibers are first pulled through a resin container and wound around a cylindrical mold until the required thickness is obtained. Once the polymer resin is cured, the composite is slipped off the mandrel. This is illustrated in Figure 10.6.

10.3.5 **Tape Prepregs**

Composite structures are often built with the help of tapes consisting of resin-impregnated fibers, called prepregs. To accomplish this, continuous fibers unwound from bobbins form strands that

■ **FIGURE 10.6**   Fabrication of a cylinder by the filament wind-up process.

are first coated with agents to promote good bonding to the matrix polymer. These strands are then continuously immersed in a bath of resin that will become the matrix that surrounds the fibers (Figure 10.7). Sheets can be compression molded or hand laid in molds to make laminates. Tape prepregs can also be wound over large shapes and cured to make large lightweight containers, in a manner similar to the filament winding technique.

10.3.6 Sheet Molding

Automobile body parts are built from prepreg sheets that are manufactured by an automated process shown in Figure 10.8. As in the spray-up process, fiber roving is chopped and sprayed together with the resin and hardener between two polyethylene films. This "sandwich" is then compacted by rolls and stored. The polymer is formulated in such a way that it does not cure completely at room temperature. The prepreg sheet obtains its final shape by pressing between heated molds that produce the final curing. The advantage of this method is that it can fabricate large numbers of pieces with excellent geometric uniformity.

10.3.7 Injection Molding

Shaped composite parts are commonly injection molded, as in Figure 9.21, by continuously adding chopped fiber (~1 cm long) to the polymer feedstock. This process is capable of producing composites with thermoplastic matrix. Controlling the fiber orientation is an important concern in this process.

10.4 METAL MATRIX COMPOSITES (MMCs)

10.4.1 Cermets

As their name implies, cermets are composites of ceramics and metals. They consist of a large proportion (≥90%) of ceramic particles embedded in a metal matrix. The most common is the tungsten

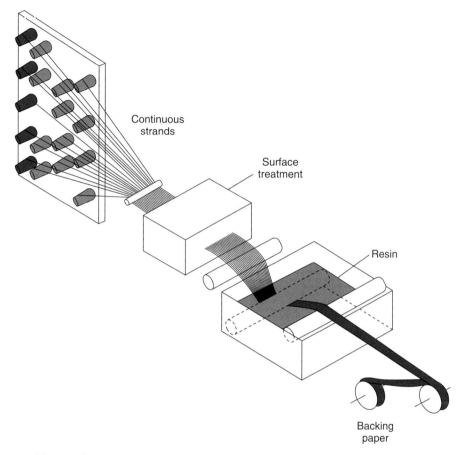

■ **FIGURE 10.7** Fabrication of a prepreg tape.

carbide-cobalt (WC-Co) system, but titanium carbide particles and a nickel matrix are also used. The combination provides these materials with the hardness of the ceramics and the fracture resistance of the metal phase. These materials are widely used as cutting tools for hard materials where they provide high cutting speeds and good surface finish. Figure 10.9 shows the microstructure of a WC-Co composite.

Cermets are produced by carburizing a tungsten-cobalt or a titanium-nickel alloy. The large affinity of carbon for tungsten or titanium leads to the production of the respective carbides and the segregation of cobalt or nickel.

10.4.2 Dispersion-Strengthened Alloys

We have seen earlier that precipitation hardening of aluminum with copper or silicon provides it with strength comparable to that of steel. This strengthening effect, however, is lost at elevated temperatures

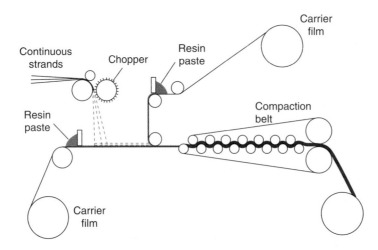

■ **FIGURE 10.8**   Sheet-molding process.

■ **FIGURE 10.9**   Microstructure of a WC-Co composite. The Co binder is dark in the figure.

because the precipitated particles grow at temperatures above 100°C. It is possible to produce aluminum that maintains superior strength at high temperature by dispersing in it very fine oxide particles that do not dissolve or grow as the temperature is reached. A significant improvement in resistance to thermal softening and creep can thus be achieved in **dispersion- strengthened** alloys. Aluminum-base MMC applications include rocket motor components (Figure 10.10), cylinder liners, control arms, connecting rods, wheels, bicycle frames, tennis rackets, and structural materials for space platforms.

■ FIGURE 10.10   This rocket motor component is the world's largest MMC forging. It consists of an aluminum alloy matrix containing 25 vol% SiC particulates, and was fabricated by powder metallurgy techniques. Courtesy of Advanced Composite Materials Corporation.

Dispersion strengthening is also common with copper and titanium. In all MMCs the basic challenge is to disperse particles uniformly throughout the metal. The typical processing route has involved the powder metallurgy techniques of pressing and sintering a blended mixture of metal and oxide powders, followed by extrusion. In this way, pure copper is strengthened with alumina to produce a high conductivity matrix that is resistant to softening up to ~900°C. Containing about 5 vol% Al_2O_3, this material is drawn into wire and used to support the very hot tungsten filament in incandescent lamps.

Attempts to extend the creep resistance of nickel-base superalloys to yet higher temperatures spawned an interesting new batch processing method known as **mechanical alloying**. Particle dispersion is accomplished by high-energy ball milling of a mixture of metal and ceramic powders. These dispersion-strengthened alloys show remarkable high-temperature strength after subsequent compaction, sintering, and extrusion.

10.4.3 **Fibrous Composites**

The idea of strengthening metals via the fibrous composite route has also gained momentum in recent years. Carbon, boron, silicon carbide, alumina as well as metal fibers have been incorporated in amounts of up to 50% by volume to stiffen Al, Mg, and Ti alloys. Fiber or whisker mats that are infiltrated with metals (by squeeze casting) can be oriented, providing strength and stiffness advantages

Table 10.2 Properties of Some Metal- and Ceramic-Matrix Composites and their Reinforcements

Material	Density (Mg/m³) (g/cm³)	E (GPa)	*σ_{UTS} (GPa)	Fibre radius (μm)	Fracture toughness K_{1C} (MPa$\sqrt{m}$)
Fiber					
Silicon carbide (Whisker)	3.2	480	3		
Silicon carbide (Fiber)	3.0	420	3.9	140	
Alumina	3.2	100	1.0	3	
Si_3N_4	3.2	380	2		
Boron	2.6	380	3.8	00–200	
Composite matrix (fiber)					
Si_3N_4 (10%SiC whisker)			0.45	—	~8
Al_2O_3 (10%SiC whisker)			0.45	—	~7.1
6061-Al (Continuous fiber-51%B)		231	1.42	—	
6061-Al (Discontinuous fiber-20% SiC)		115	0.48	—	
6061Al No reinforcement		69	0.31	—	

relative to particulate MMCs. The fiber surfaces must be specially treated to enhance metal wetting and interfacial adhesion and to limit undesirable reactions that degrade bonding.

Both types of MMC achieve 50% increases in tensile strength and Young's modulus and a similar decrease in density relative to the matrix metal alone. Actual property enhancements are indicated in Table 10.2; they depend on the particular composite and the volume fraction of particles or fibers. The utilization of fibrous composites is limited by the high price of their processing.

10.5 CERAMIC MATRIX COMPOSITES

Ceramics can be toughened by the use of composites. Very short, small-diameter fibers known as **whiskers** are incorporated into ceramic bodies up to a volume fraction of 25%. Because energy is expended in pulling whiskers out of the matrix and in bending or breaking them, crack propagation is hampered and the fracture toughness is effectively raised (Figure 10.11).

Cordierite glass-ceramic substrates, which are used as supports for computer chips, have been toughened with 1 μm diameter, 10 μm long Si_3N_4 whiskers. In another application, SiC fibers have toughened

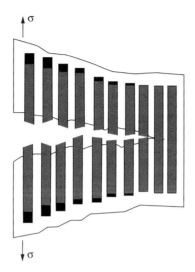

■ **FIGURE 10.11**   Cracking of a fiber-reinforced ceramic.

Al_2O_3 cutting tools used in machining superalloys. A number of reinforcements have been commercially available for some time including SiC, Al_2O_3, and boron fibers; typical fibers are 10–20 µm in diameter. Some of their properties and those of the ceramic matrices they strengthen are listed in Table 10.2.

10.6 MECHANICAL PROPERTIES OF COMPOSITES

In this section we examine the elastic behavior of composites and their strength. A simple analysis can be performed in the case of fiber-reinforced composites that consist of continuous, parallel fibers. The behavior of composites reinforced by randomly oriented fibers or by particles is more complex and will be considered later.

10.6.1 Young's Modulus

Longitudinal Loading   The case of loading parallel to the fibers is technically the most important; it is involved in the bending resistance of beams. When the composite is loaded along the fiber direction with force F_c (Figure 10.12A), the fibers and the matrix extend by the **same strain** ε. The fiber and matrix are assumed to behave elastically with respective moduli E_f and E_m. The force F_c carried by the composite is the sum of the forces carried by the two constituents, that is, $F_c = F_f + F_m$. These forces, in turn, are the product of the corresponding stresses, σ_f and σ_m, and the respective load bearing areas of fiber and matrix. For a composite of unit volume, these areas are equal to the respective volume fractions, V_f and V_m, where $V_f + V_m = 1$. Therefore, with the help of Hooke's law ($\sigma = \varepsilon E$)

$$\sigma_c = V_f\sigma_f + V_m\,\sigma_m = V_f E_f\varepsilon + V_m\,E_m\varepsilon \tag{10.1}$$

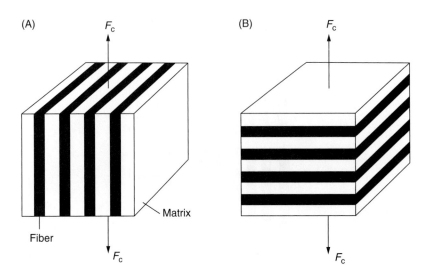

■ **FIGURE 10.12** (A) Model of a fibrous composite with fibers aligned parallel to the loading direction. (B) Model of a fibrous composite with fibers aligned perpendicular to the loading direction.

Young's modulus for the composite is $E_c = \sigma_c/\varepsilon$, with the result that

$$E_c = V_f E_f + V_m E_m = V_f E_f + (1 - V_f)E_m \qquad (10.2)$$

Transverse Loading When the fibers are oriented perpendicular to the loading axis, as in Figure 10.12B, both constituents support the same load or stress σ but strain by different amounts because their elastic moduli differ. The total composite strain ε_c is the sum of the individual strains ε_f and ε_m.

$$\varepsilon_c = \varepsilon_f + \varepsilon_m = \sigma V_f/E_f + \sigma V_m/E_m \qquad (10.3)$$

Again, Young's modulus for the composite is $E_c = \sigma/\varepsilon_c$, and therefore,

$$1/E_c = V_f/E_f + V_m/E_m \text{ or } E_c = E_m E_f/(V_m E_f + V_f E_m) \qquad (10.4)$$

The isostrain case, Equation (10.2), predicts upper bound (maximum) modulus values while lower bound or minimum values are described by the isostress approximation in Equation (10.4). Randomly oriented two-phase mixtures will exhibit a modulus that falls between these extreme values. Only in the case where the reinforcing phase is aligned along the loading axis is optimum elastic stiffness attained. In fiber reinforced polymers, the stiffness of the fibers and matrix differs by a factor of ≈ 50 so that the modulus of the composite perpendicular to the fibers is much lower than parallel to the fibers.

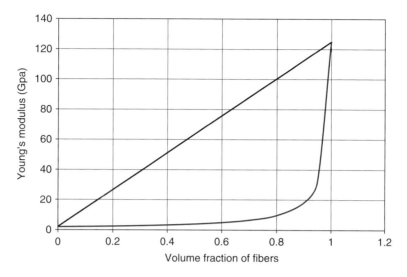

■ **FIGURE 10.13**   Young's modulus in a Kevlar-epoxy composite, parallel (black) and perpendicular (colored) to the fibers.

Figure 10.13 shows the variation of Young's modulus parallel and perpendicular to the fibers for a composite formed of Kevlar fibers ($E = 125\,\mathrm{GPa}$) and epoxy matrix ($E = 2.1\,\mathrm{GPa}$). When the fiber content is 50% per volume, for instance, Young's modulus is $63.5\,\mathrm{GPa}$ parallel to the fibers and $4\,\mathrm{GPa}$ perpendicular to the fibers.

When biaxial strength and stiffness are required, as in automobile bodies or boats and other surfaces, the fibers are arranged in both directions; they are either woven as cloth or mats, or in random orientation as in the spray-up process.

10.6.2 The Strength of Composites

Continuous Fibers, Longitudinal Loading   Beyond the elastic regime, the stress-strain behavior of the individual constituents and of the composite is schematically indicated in Figure 10.14 when the stress is parallel to the fibers. The overall response is complex. The fibers deform elastically until they fracture at the tensile stress $\sigma_f(\mathrm{TS})$ (upper line). Since fibers have a far higher tensile strength than the matrix yield strength $\sigma_m(\mathrm{YS})$, they continue to deform elastically while the matrix starts deforming plastically (dotted line). With further load, fibers begin to fracture one by one and the strength drops gradually. After all fibers break or are pulled out, only the matrix is left to carry the load until the composite fractures.

For design purposes, the peak or tensile stress of the composite $\sigma_c(\mathrm{TS})$ is of interest. An estimate of the composite strength is

$$\sigma_c(\mathrm{TS}) = \sigma_f(\mathrm{TS})V_f + \sigma_m(\mathrm{TS})V_m \tag{10.5}$$

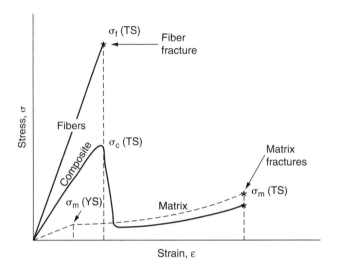

This expression reflects a simple weighted average of the fracture strength of the fibers and the yield strength of the matrix. Once all of the fibers break, the fracture strength of the matrix, $\sigma_m(TS)$, limits the strength of the composite. In fiber-reinforced polymer composites, the relative strengths of the fibers ($>1\,GPa$) and the matrix ($\sim 0.05\,GPa$) are such that the strength of the composite is essentially that of the fibers.

$$\sigma_c(TS) \approx \sigma_f(TS)V_f \tag{10.6}$$

Effect of Fiber Length It would be wrong to conclude from the previous discussion that only continuous fibers that span the entire structure will do. If this were the case, processing and forming difficulties would limit fibrous composites to very few applications. Fortunately, cut fibers that are convenient for molding and extrusion operations will impart substantial strengthening if they are longer than a critical length. To estimate this critical length, let us consider Figure 10.15, which depicts a chopped fiber embedded in a matrix.

Under a tensile load, the axial force is transmitted to the fiber via the matrix along the cylindrical fiber-matrix interface. On a segment of fiber dx long, the force transmitted is $dF = \tau_m(s)\,\pi d\,dx$, where $\tau_m(s)$ is the interfacial **shear** stress and d is the fiber diameter. At a distance x from the end the cumulative transferred force is

$$F = \int_0^x \pi d\tau_m(s)dx = \pi d\,\tau_m(s)x \tag{10.7}$$

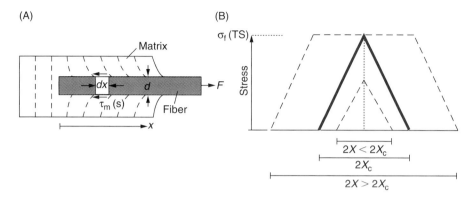

■ **FIGURE 10.15**   (A) Load transfer from the matrix to a discontinuous fiber. (B) Variation of the stress with position in cut fibers that are less than, equal to, or longer than the critical length.

The force that will break the fiber is

$$F_f = \sigma_f(TS)\, \pi d^2/4 \tag{10.8}$$

When these two expressions for force are equated, the critical distance x_c at which the fiber will break is

$$x_c = d\sigma_f(TS)/4\tau_m(s) \tag{10.9}$$

The critical fiber length is $l_c = 2x_c$. In the case of Kevlar-reinforced epoxy, where τ_m is half the σ_{UTS} of the matrix, $\tau_m = 0.032\,GPa$ and the strength of the fiber $\sigma_{UTS} = 3.2\,GPa$, the critical fiber length is 0.5 mm when the fiber diameter is 10 μm. If fibers are shorter than this length, they will not fail, and neither will they carry the maximum load they are capable of supporting.

Figure 10.15B shows that the average strength provided by a fiber of critical length $l_c = 2x_c$ is $1/2\sigma_f(TS)$; it is larger for longer fibers and less for shorter fibers. When $l >> l_c$, the strength of the composite is the same as if the fibers extended the whole length of the material. This is of great importance to the manufacture of fiber-reinforced composites, since the use of chopped fibers of sufficient length is much less expensive than incorporation of fibers extending the whole length of the piece. This also permits the fabrication of pieces with more complex geometries. Therefore, composites with fibers more than 15 times longer than l_c are considered **continuous**.

Transverse Loading   Theoretically, the strength of a composite with stress perpendicular to the fibers is the strength of the matrix. In reality, however, lower interface strength between matrix and fiber can lead to failure at lower stresses.

10.6.3 Toughness of Ceramic Matrix Composites

The purpose of reinforcing ceramics with fibers is to increase their resistance to fracture. Neither the ceramic matrix nor the fibers used to reinforce them are tough materials. It is not obvious why making

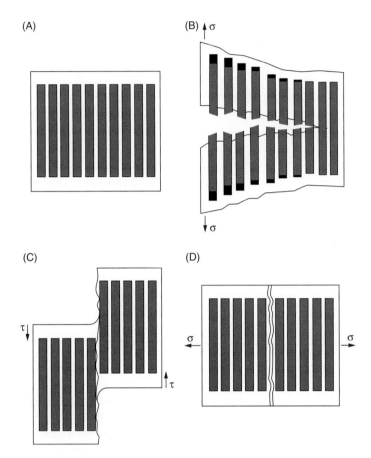

■ **FIGURE 10.16**   Fracture models in a fibrous composite. (A) Initial composite. (B) Under tensile pulling, fibers are pulled from the matrix and also fracture. (C) Interfacial fiber/matrix failure under applied shear stresses. (D) Matrix fracture under tensile stresses applied normal to the fibers.

a composite of the two should improve toughness; but this is nevertheless what happens. Attaining the right bond strength between the fiber and matrix is the key to the success of tougher ceramic composites. The bond must be strong enough to hold the ceramic, but it must be weak enough to allow the fibers to pull out instead of fracturing. The reason can be seen in Figure 10.16B where a crack has propagated transverse to the fiber axis. If the fibers are less than $2x_c$ in length, they do not break. Rather, they are pulled out of the matrix as the crack opens, a process that requires work and increases toughness. Longer fibers pull out as well and crack, and energy is likewise expended. Formulas that will not be derived here estimate the toughness as the work per unit area (W_c) of crack surface; they calculate the work required to pull fibers by shearing the fiber-matrix bonds.

For short fibers ($l < 2x_c$, which cannot be fractured)

$$W_c = \tau_m(s)\,V_f L^2/2d \tag{10.10}$$

On the other hand, if the fibers are long ($l > 15x_c$), they will break instead of pulling out. In this case,

$$W_c = [\sigma_f(TS)]^2 \, V_f d/8\tau_m(s) \tag{10.11}$$

Thus short fibers require high matrix shear strength to toughen the composite. Long fibers require high fiber strength and a low matrix shear strength.

Other toughening mechanisms include deflection of cracks as they impinge on fibers, and bridging of cracks by fibers that restrain them from opening and advancing further. Cracks travel more easily in a direction parallel to the fibers (Figure 10.15D); they can propagate through the matrix under tension, or along the fiber-matrix interface in response to shear stresses. Fibers that are parallel to the crack propagation direction (and normal to the stress) do not contribute to toughness.

10.7 **CONCRETE**

Concrete is certainly the composite that is used in the largest volumes. It contains cement and an **aggregate** that consists of sand (less than 2 mm in diameter) and larger particles of gravel and crushed rock. The smaller sand particles fit into the spaces between the larger gravel. Cement flows into the remaining spaces and forms a continuous matrix. A high proportion of sand in the aggregate is required for smooth finishes. The typical ratio of aggregate to cement is 5:1; the ratio of sand to gravel is about 3:2.

Concrete possesses the attractive properties of low cost, durability, fire resistance, resistance to the elements, and aesthetics. Its greatest advantage is that it is poured at the construction site and can take any desired shape. Concrete is a hard and brittle material; its fracture toughness is as low as $0.2 \, MPa\sqrt{m}$. Concrete possesses good compressive strength (around 35 MPa) but very low tensile strength.

The addition of steel bars (rebars) produces **reinforced concrete**, a remarkable material that combines the advantages of concrete with the ability to withstand tensile stresses. The rebars must be placed in the region where the load causes tensile stresses, as shown in Figure 10.17. The rebars usually have a square cross section that is twisted in order to provide mechanical interlocking with the concrete.

Prestressed concrete further increases the resistance to tensile stresses. This is achieved by putting the reinforcing steel into tension; the steel reinforcement then applies a compressive stress to the concrete that counteracts the tensile stresses resulting from an applied load. Prestressed concrete is fabricated either **pre-tensioned** or **post-tensioned**. Pre-tensioned concrete is manufactured by pouring the concrete into a bed that contains a steel tendon (usually a multi-stranded cable) that has been put in tension by a jack. When the concrete is hardened, the pressure is released and the cable, attempting to shorten, transfers a compressive stress to the concrete. This technique is used in prefabricated sections used in bridge construction as shown in Figure 10.18. In post-tensioned concrete, hollow conduits are placed into the form and concrete is poured. When the latter is hardened, a cable or a bar is introduced into the conduit and stressed by means of a jack. The two techniques are shown in Figure 10.19.

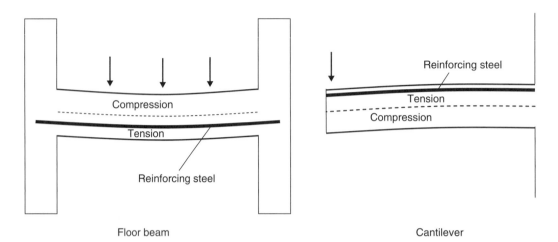

■ **FIGURE 10.17**   Stresses in a loaded beam and placement of the rebars in reinforced concrete. Left: Floor beam. Right: Cantilever.

■ **FIGURE 10.18**   Application of prestressed concrete in prefabricated bridge elements.

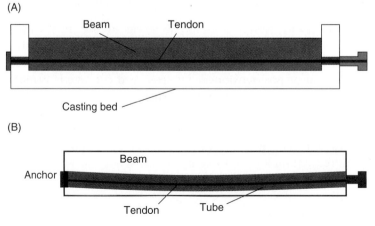

■ **FIGURE 10.19**   Fabrication of prestressed concrete. (A) Pre-tensioned concrete. (B) Post-tensioned concrete.

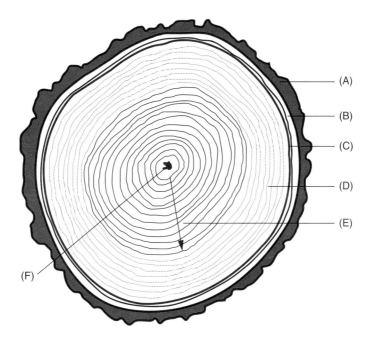

■ **FIGURE 10.20**   Cross section of a tree trunk. (A) Outer bark provides protection. (B) Inner bark carries food from the leaves to growing parts. (C) Thin cambium layer responsible for the growth of new rings. (D) Sapwood carries sap from the roots. (E) heartwood, dead material that is used. (F) Pith, a soft tissue in the center of the trunk.

10.8 **WOOD**

10.8.1 **The Nature of Wood**

Wood is described here because its mechanical properties are similar to those of a polymer composite reinforced with parallel fibers. It is the most widely used construction material in the United States. Wood is primarily composed of parallel, oblong hollow cellulose fibers that constitute the grain of wood and determine its mechanical properties.

Figure 10.20 shows the cross section of a tree trunk. Starting from the exterior, the trunk consists of the outer bark and inner bark, sapwood, heartwood, and pith. The outer bark is a dead layer that protects the tree. Its thickness and appearance are different for each species. The inner bark is soft and carries nutrients from the leaves to all growing parts of the tree. The cambium is a thin layer between the bark and the wood. All growth of the wood occurs within the cambium layer. Each year, the cambium layer produces a new ring of wood. The sapwood layer contains some living cells that carry sap from the roots to the leaves of the tree. The heartwood is no longer living; it constitutes the material that is used in construction. The pith is a soft tissue at the center of the tree.

Woods are divided into **soft woods** and **hard woods**. Softwood trees are the evergreen trees such as spruce, fir, pine, and cedar. Hardwood trees are deciduous; some examples are maple, oak, elm, birch, and fruit trees. As shown in Table 10.3, the hardwoods are generally stronger than the softwoods. Hardwood is preferred in furniture, and softwood is used predominately in housing construction.

Table 10.3 Mechanical Properties of American Woods after Drying.

Tree species	Average specific mass oven dry (g/cm³)	Static bending modulus of elasticity (E) (GPa)	Compress parallel to grain, max crushing strength (MPa)	Compress perpendicular to grain (MPa)	Shear parallel to grain, max shear strength (MPa)
Hardwoods					
Ash, Black	0.5	11.2	42	5.3	11
Beech	0.64	12	51	7	14
Cherry	0.50	10.4	50	4.8	11.9
Maple	0.57	11.3	46.8	7.1	12.7
Oak, White	0.68	12.5	52	7.5	14
Walnut	0.55	11.7	53	7.5	9.6
Softwoods					
Douglas-fir, Interior North	0.48	12.5	48	5.4	9.8
Fir	0.39	10.5	40.6	3.7	7.7
Pine	0.38	10.2	35.3	3.3	7.3
Redwood, Old-growth	0.40	9.4	43	4.9	6.55
Spruce, White	0.36	10	36.3	3.0	6.8

Source: U.S. Forest Products Laboratory.

10.8.2 **Mechanical Properties of Wood**

Hard and soft woods consist of fibers that are parallel to the axis of the trunk and the branches. Wood is 7 to 10 times stronger in the direction parallel to the fibers than perpendicular to them (Table 10.3 and Figure 10.21). This structure provides the trees with their bending strength. Applications of wood in beams and planks make judicial use of this property. In furniture construction, care must be exercised to avoid fracture of the wood when bending stresses perpendicular to the grain are applied.

The green wood of a living tree contains a large amount of water. Once the wood is cut, it dries and shrinks. Shrinkage causes distortion of the wood but also increases its mechanical strength. In order to maintain shape integrity, the wood must be dried either by lengthy storage or in an oven. Table 10.3 gives the mechanical properties of kiln-dried wood.

10.8.3 **Plywood**

Biaxial strength of wood is achieved in plywood. The latter is made by gluing together a number of thin sheets of wood, called plies, with the fibers aligned perpendicular to each other. For certain applications,

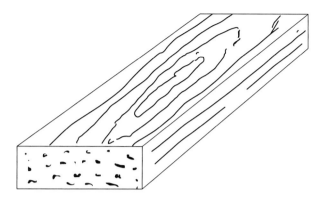

■ **FIGURE 10.21**   Typical orientation of a wooden plank. The long axis is parallel to the grain, the width is parallel to the growth rings, and the thickness is parallel to the radius of the trunk.

■ **FIGURE 10.22**   Examples of plywood.

the plies on the surface of the piece consist of an aesthetically pleasing wood, called veneer. The perpendicular alignment of the fibers also prevents warping when the water content changes because of variations in ambient humidity. Some samples of plywood are shown in Figure 10.22, and an application is illustrated in Figure 10.23.

■ **FIGURE 10.23**   Eames lounge chair (1956) built from plywood and leather. Reproduced with permission of Herman Miller Inc.

■ SUMMARY

1. A composite is a physical mixture of two or more materials.

2. Composites consist of a continuous matrix in which fiber or particle reinforcements are embedded.

3. Polymer matrix composites are made of a polymer (usually thermoset) matrix reinforced by glass, Kevlar, carbon, or graphite fibers.

4. Fibers are strong and fracture resistant because of their small diameter. They can be drawn to this shape because they are made of amorphous material whose viscosity increases as it cools.

5. Several fabrication methods exist for polymer matrix composites, depending on price and required performance. These are hand lay-up for large objects (boats); spray-up for mass production; filament

wind-up for rocket bodies; injection molding; sheet molding process used in car bodies; and pre-preg tapes that are later used in final fabrication.

6. Cermets are composites consisting of ceramic particles dispersed in a metal matrix. They combine the hardness of ceramics with the ductility of metals.

7. Dispersion-strengthened metals are used in high-temperature applications. They consist of fine ceramic dispersion in the metal matrix.

8. Ceramic matrix composites consist of fibers distributed in a ceramic matrix. Their aim is to increase the fracture resistance of the material.

9. A composite is strong in a direction parallel to its fibers and weak in the normal direction. When uniaxial strength is required (as in shafts and beams), the fibers are disposed along the shaft. When biaxial strength is required (as in sheets), the fibers are arranged in two directions.

10. The fibers need to have a minimum length in order to strengthen the composite. This length, however, is only a few millimeters.

11. Concrete is a composite of cement and aggregate (gravel); it has good compressive strength and low tensile strength.

12. Reinforced concrete obtains tensile strength by the incorporation of steel rebars; the latter must be placed where the slab is subjected to tensile stresses.

13. In prestressed concrete, the steel is put into tension; this generates a compressive stress in the concrete that counteracts the applied tensile stresses.

14. Wood is a natural composite that is strong along its fibers and weak across them.

15. Plywood consists of wood plies glued together with fibers in different directions. It possesses biaxial strength.

■ KEY TERMS

A
aggregate, 299

C
Carbon Fiber Composites
   (CFRP), 282
cement, 299
ceramic matrix composites, 292
cermets, 288
composites, 282
concrete, 299

D
dispersion-strengthened
   alloys, 289

E
effect of fiber length, 296

F
fiberglass (GFRP), 282
fibrous composites, 291
filament winding process, 287

H
hand lay-up process, 285
hard woods, 301

I
injection molding, 288

K
Kevlar, 284

L
longitudinal loading, 293

■ REFERENCES FOR FURTHER READING

[1]   B.D. Agarwal, L.J. Broutman and K. Chandrashekhara, *Analysis and Performance of Fiber Composites*, 3rd ed., Wiley, New York (2006).

[2]   B.T. Astrom, *Manufacturing of Polymer Composites*, CRC Press, Boca Raton, FL (1997).

[3]   K.K. Chawla, *Composite Materials Science and Engineering*, 3rd ed., Springer, New York (2001).

[4]   D. Hull, T.W. Clyne, *An Introduction to Composite Materials*, 2nd ed., Cambridge University Press, New York (1996).

[5]   A.B. Strong, *Fundamentals of Composites: Materials, Methods and Applications*, Society of Manufacturing Engineers, Dearborn, MI (1989).

■ PROBLEMS AND QUESTIONS

10.1.   A composite consists of 40% by volume of continuous E-glass fibers aligned in an epoxy matrix. If the elastic moduli of the glass and epoxy are 76 and 2.7 GPa, respectively,

   a. Calculate the composite modulus parallel to the fiber axis.

   b. Find the composite modulus perpendicular to the fiber axis.

10.2.   A composite consists of 35 vol% of continuous aligned Kevlar fiber in epoxy whose elastic modulus is 2.3 GPa.

   a. Calculate the composite modulus ($E_\|$) parallel to the fiber axis.

   b. What enhancement in ($E_\|$) occurs if the elastic modulus of the matrix is increased to 3.7 GPa?

   c. What increase in the percent fiber volume of the original composite will yield the same $E_\|$ enhancement as in part b?

10.3.   Consider a polymer with elastic modulus E_p that contains two different continuous fibers with moduli E_1 and E_2. Derive an expression for Young's modulus (E_c) that would be measured in the following composites if the corresponding fiber volume fractions are V_1 and V_2. (Assume V_1 and V_2 are relatively small in magnitude.)

   a. A composite where both fibers are aligned parallel to the axis of measurement.

   b. A composite where both fibers are aligned perpendicular to the axis of measurement.

   c. A composite where fiber 1 is aligned parallel and fiber 2 is aligned perpendicular to the composite axis.

10.4. It is desired that the fibers support 90% of the load in a composite containing aligned continuous high modulus carbon fibers embedded in an epoxy matrix. What volume fraction of fiber will be required? (For epoxy assume $\rho = 1.3\,g/cm^3$ and $E = 3\,GPa$.

10.5. In an aligned fiberglass composite, the tensile strength of the $10\,\mu m$ diameter fibers is $2.5\,GPa$, the matrix yield strength is $6\,MPa$, and the interfacial shear strength is $0.1\,GPa$.

 a. What is optimum cut fiber length?

 b. If 1 mm long chopped glass fiber is used, what volume fraction should be added to the matrix to produce a composite tensile strength of 1 GPa?

10.6. An optical fiber essentially consists of a 125 mm diameter glass filament (see Chapter 13) surrounded by a 250 mm diameter urethane-acrylate polymer.

 a. Young's modulus for the polymer is $10\,MPa$ and that for the glass fiber is $80\,GPa$. If an elastic load is applied to the composite, what fraction is carried by the fiber?

 b. To determine the polymer-glass interfacial bond strength, a 1 cm length of composite is embedded in a bracing medium and a force is applied to the glass. If a 2.5 kg force is required to extract the filament, what is the bond strength?

10.7. a. What is the value of E/ρ for a composite containing 35 vol% of continuous aligned Kevlar fibers ($E = 125\,GPa$, $\rho = 1.45\,g/cm^3$) in an epoxy matrix ($E = 3.2\,GPa$, $\rho = 1.3\,g/cm^3$)?

 b. High-strength carbon fibers are substituted for Kevlar in this composite. What volume fraction of continuous aligned carbon fibers will yield the same E/ρ value?

10.8. A single cylindrical fiber of diameter $2r$ is embedded into a matrix to a depth of d as suggested by Figure 10.15A. The ratio of the fiber tensile strength σ_f to the maximum interfacial shear stress σ_i between the fiber and matrix is 10. What fiber aspect ratio or value of d/r is required so that the fiber breaks before it is pulled out?

10.9. A sintered tungsten carbide cutting tool contains 82 wt% WC, 10 wt% TaC, and 8 wt% Co. What is the density of this composite? Assume component densities are additive in proportion to volume fractions. Note: $\rho_{WC} = 15.8\,g/cm^3$, $\rho_{TaC} = 14.5\,g/cm^3$, and $\rho_{Co} = 8.90\,g/cm^3$.

10.10. a. Why is composite formation a successful strategy to toughen polymers but not metals?

 b. Why is fiber incorporation more effective at strengthening metal and polymer matrices than ceramic matrices?

10.11. Could the Eames lounge chair of Figure 10.23 be built with solid wood instead of plywood with the same dimensions? Why or why not?

10.12. A circus artist breaks a wood plank with a chop of his bare hands. How does he cut the wood plank to make it possible?

10.13. How are the fibers aligned in the shaft of a golf club? How are they arranged in the hull of a sailboat? Sketch the arrangement of the fibers in a tennis racket. Explain your choices.

10.14. Compare the hand lay-up and the spray-up process (Figures 10.3 and 10.4). Which one would you use for the construction of a yacht and which one for the manufacture of bathtubs? Why not the opposite?

10.15. The hull of a sailboat is made of fiberglass, and its mast is fabricated from graphite-epoxy composite, which is 10 times more expensive. Justify the choice of both.

10.16. In polymer matrix composites, the adhesion of the matrix to the fiber must be as high as possible. In ceramic matrix composites, the adhesion of matrix to fibers must not be too strong. Explain the roles of the fiber in polymer matrix and ceramic matrix composites and justify the difference in fiber-matrix adhesion.

10.17. Compare wood, fiberglass, and a rolled steel. If a boat is made of wood planks 1 cm thick, how thick would fiberglass with 50% total fiber have to be to have the same elastic modulus? (Use the tables in Chapter 10.)

How thick would a rolled steel plate have to be to have the same elastic modulus against bending? Compare the weight of 1 m^2 of the three materials.

10.18. Look at a prestressed concrete bridge. How would a steel bridge of the same size look? Explain how the material makes the shape possible.

10.19. You wish to build a concrete building. Describe what material you choose and how the structure will be built.

10.20. What is the difference between cement and concrete?

10.21. What is a cermet? How is it made? What are the outstanding properties of cermets?

10.22. One says that fiber reinforcement of ceramics allows "graceful failure" of the material. Compare the fracture of a reinforced and a simple ceramic and justify the saying.

10.23. How would you make a fiberglass leaf spring?

10.24. Define a composite and its components.

10.25. Describe fiberglass: what it is made of, and what is its structure?

10.26. What is a graphite tennis racket made of? Describe its structure.

10.27. Compare the strength of a fiber and of the bulk of the same material.

10.28. How does one make a fiberglass sailboat?

10.29. How does one make a fiberglass bathtub?

10.30. How is the composite body of a rocket made?

10.31. Describe the fabrication of the fiberglass body of a car.

10.32. How do you arrange the fibers in a golf club?

10.33. How do you arrange the fibers in a fiberglass chair?

10.34. What fibers and matrix are used in aircraft parts?

10.35. Compute the elastic modulus of a composite consisting of 60% parallel fibers. The elastic modulus of the fibers is 200 GPa, and that of the matrix is 100 MPa.

10.36. A composite consists of 50% fibers with tensile strength 800 MPa. The matrix has a tensile strength of 20 MPa. What is its strength parallel to the fibers? What is the strength perpendicular to the fibers?

10.37. How long, approximately, do fibers have to be to provide adequate strengthening to a polymer?

10.38. What is a cermet? Name one widely used cermet and describe its structure and composition. What are the outstanding properties of cermets?

10.39. Describe a ceramic matrix composite.

10.40. What does concrete consist of?

10.41. Describe reinforced concrete.

10.42. Compare the mechanical properties of concrete and of reinforced concrete.

10.43. Describe prestressed concrete and its mechanical properties.

10.44. Where, in a structure, must you place the rebars in reinforced concrete?

10.45. Carpenters, when building a structure, are careful about the orientation of the wood fibers. Why?

10.46. A tree is a well-engineered composite. Explain in terms of applied stresses.

Part 3

Functional Materials

A large and increasing number of applications do not depend on the strength of the material but expect the material to perform various functions. Materials must conduct electricity with minimum losses, or offer enough electric resistance to serve as a heater; other materials must withstand large electric voltages as insulators, or provide capacitors with large enough electric storage capacity, or expand or contract under an electric signal. The semiconductors play a vital role in data processing and in communications; properly processed, they act as rectifiers, transistors, solar cells, and light emitters. Modern integrated circuits provide enormous computing power for a modest price. Magnetic materials function to improve the efficiency of transformers and motors, or they store and process data. Finally, optical materials must transmit light, act as filters or paints, serve in lenses, or emit light in large areas. Other materials are processed to store or produce electric power in batteries. These are the functional materials that will interest us in the remainder of this volume. Chapter 11 will be devoted to electric conductors, insulators, and semiconductors with their applications. In Chapter 12, we describe how integrated circuits and MEMS are processed. Chapter 13 is devoted to optical materials, Chapter 14 to magnetic materials, and Chapter 15 will describe the applications of materials in modern batteries.

Chapter 11

Conductors, Insulators, and Semiconductors

In this chapter, we examine what makes a material a good electric conductor, a useful resistance, an insulator, or a semiconductor. The partially filled valence band of a metal provides it with a large number of free electrons that can move under the action of an electric field. The electric resistance of the metal is provided by lattice imperfections that scatter the electrons, mostly impurity atoms and thermal vibrations. To make a metallic resistor, one uses an alloy. Solids with a filled valence band, ceramics and polymers, are insulators. An electric field polarizes an insulator by a slight shift of its charges. The resultant dielectric constant is used in capacitors. In certain materials this shift of ionic charges results in a change in size; this is piezoelectricity. Semiconductors are insulators with a small energy-band gap. They can be doped to conduct electricity by mobile electrons (n-type) or holes (p-type). The p-n junction is a versatile device; it can be an electric rectifier, a light detector, a solar cell, a light-emitting diode (LED), or a laser. Transistors are p-n-p or n-p-n double junctions. Certain organic materials and polymers have a small energy gap between the occupied π orbitals and a higher band. They can be doped and used as semiconductors. Their performance is not as good as that of silicon and III-V compounds, but they hold much promise for large-area applications.

LEARNING OBJECTIVES

After studying this chapter, the student will be able to:

1. Distinguish the contributions of charge carrier density and electron scattering in the electrical conductivity of solids.

2. Select and process a metal for either low or high resistance and evaluate its resistance at any temperature.

3. Describe the principles of a superconductor.

4. Select a material intended for electric insulation. Calculate the thickness of an insulator that must hold a certain voltage.

5. Distinguish dielectric strength and the dielectric constant.

6. Design a capacitor and select the proper dielectric.

7. Utilize piezoelectric materials in design as actuators or measuring devices.

8. Select materials used as semiconductors in electronic and optoelectronic applications.

9. Specify the doping of semiconductors that makes them n-type or p-type conductors and evaluate the amount of doping necessary to achieve a desired conductivity.

10. Design a solid-state rectifier.

11. Describe the various electronic and optoelectronic applications of p-n junctions.

12. Describe the bipolar transistor and the MOSFET and their operation.

13. Select either a bipolar transistor or a MOSFET for a specific application.

14. Select a material that is an organic semiconductor, describe its operation and how it is doped n- or p-type.

11.1 INTRODUCTION

Electric conductors, insulators, and the electronic devices based on semiconductors play a dominant role in our technology. We need excellent **conductors** to minimize losses in power lines and electric machinery. But we need metals with sufficient **electric resistance** to produce heaters and incandescent light bulbs. We need a technology to achieve the desired resistance in metals.

Electrical appliances and machinery need **insulators** for safety and to avoid electrical shorting. In this case the question is what materials make good insulators and how these should be processed and designed so that a high voltage can be applied before breakdown occurs.

Perhaps the most spectacular range of applications is that of **semiconductors**. We can process these materials to make a rectifier that conducts current in one direction only. On the basis of this technology, we can make rectifiers, transistors, light detectors, solar cells, light-emitting diodes (LED), and solid-state lasers.

In this chapter we will first examine the nature of the electric current and the material properties that determine whether a material is a conductor, an insulator, or a semiconductor. We will then examine, in turn, metallic resistors and their applications, insulators, their relevant properties and how these are achieved, and the nature of semiconductors and how they are processed. Finally, we describe the p-n junction and its many applications.

11.2 BASIC CONCEPTS OF ELECTRIC CONDUCTION

11.2.1 Ohm's Law

When an electrical potential difference (a voltage) V is applied to a solid, an electric current I flows through it. The applied voltage V required for a given current I is determined by the **electric resistance** R of the solid.

$$V = IR \tag{11.1}$$

The electric resistance of a body depends on its length L, its cross section A, and the **resistivity** ρ of the material (Figure 11.1),

$$R = \rho \frac{L}{A} \tag{11.2}$$

We will find it useful to define the electrical **conductivity**, $\sigma = 1/\rho$, which is the inverse of the resistivity. In this chapter, we are interested in ρ and σ; they measure the ability of a material to conduct electricity.

We can rewrite Ohm's law in terms that are independent of the conductor's size and that express what is going on in the material: The applied voltage V sets up an **electric field** $E = V/L$ in the solid; the field generates the **current density** $j = I/A$, so that Equation (11.2) becomes

$$E = \rho j \text{ or } j = \sigma E \tag{11.3}$$

The resistance is measured in units of Ohms (Ω), the resistivity ρ is expressed in Ωm, and the conductivity σ is expressed in $(\Omega\text{m})^{-1}$, also called Mho's.

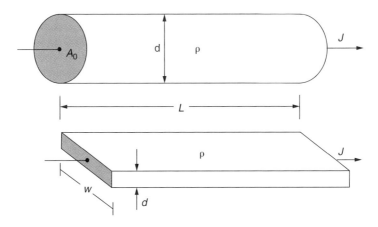

■ **FIGURE 11.1** The electrical resistance of a conductor with resistivity ρ, length L, and uniform cross section A. Top: $A = \pi d^2/4$. Bottom: $A = wd$.

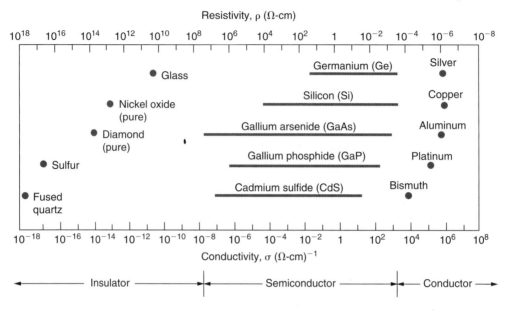

■ **FIGURE 11.2**   The resistivity of selected materials.

Figure 11.2 shows the resistivity at room temperature of a number of metals, ceramics, and polymers. Note that the resistivities are given here in Ωcm rather than in Ωm ($1\,\Omega$cm $= 10^{-2}\,\Omega$m). These units are commonly used with electronic equipment. We see that the range of resistivities is enormous. Note that this range is 25 orders of magnitude. (By comparison, the distance from earth to the sun is "only" 10^{21} times larger than the distance between two atoms in a solid.) We distinguish three groups of materials in the figure. The metals are very good conductors of electric current: their resistivity is around 10^{-8} to $10^{-7}\,\Omega$m; their resistivity is commonly expressed in nano Ohm meters (nΩm) or micro Ohm centimeters ($\mu\Omega$cm). They are the obvious choice for electric power transmission. At the other extreme, the electric resistivity of insulators ranges from 10^{10} to $10^{16}\,\Omega$m. The resistivity of semiconductors, which is intermediate, can be varied by large amounts (a factor of 10^8) by doping with selected elements.

11.2.2 **The Electric Current**

An electric current is produced by the flow of particles, each carrying the electric charge q. In the materials of interest here, these charged particles are electrons. (In other cases, such as the electrolytes of batteries and in some gas sensors, the mobile charge carriers are ions.) When no current flows, for instance in a metal without applied voltage, the electrons still move rapidly, but for every electron moving in one direction, there is another moving at the same velocity in the opposite direction, so that their average velocity is zero (Figure 11.3A).

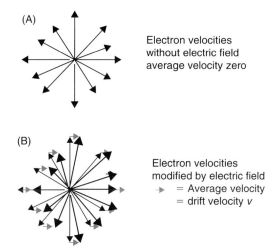

(A)

Electron velocities
without electric field
average velocity zero

(B)

Electron velocities
modified by electric field
→ = Average velocity
= drift velocity *v*

■ **FIGURE 11.3** (A) Velocities of electrons in a solid without electric field: the average velocity is zero. (B) The thin lines are the same as above; they represent the velocities without an electric field. The small colored arrows pointing to the right represent the drift velocity, Equation (11.3). The thick lines represent the velocity of the electrons in the electric field.

In an electric field E, the electrons experience a force $F = Eq$ and an acceleration $a = F/m = Eq/m$. After a time τ, the velocity of all electrons is **modified** by the amount

$$v = \frac{q\tau}{m} E \qquad (11.4)$$

This is shown in Figure 11.3B.

Electrons moving through a solid are **scattered** by various obstacles that we shall examine later. They are scattered into random directions so that the average velocity of the electrons after scattering is again zero. If we take τ to represent the time of free flight between scattering events, Equation (11.4) represents the average velocity change of the electrons, which is also called the **drift velocity**.

If n electrons per unit volume, each carrying the charge q, move with the average velocity v, the **current density** is

$$j = nqv = n\frac{q^2}{m}\tau E \qquad (11.5)$$

Comparing Equations (11.3) and (11.5) we find that the electric conductivity σ and the resistivity ρ are

$$\sigma = n\frac{q^2}{m}\tau \quad \text{and} \quad \rho = \frac{m}{nq^2} f \qquad (11.6)$$

where $f = 1/\tau$ is the electron scattering frequency.

Equations (11.4) and (11.6) allow us to define the **mobility** μ of electrons.

$$v = \mu E; \quad \mu = \frac{q^2}{m}\tau \qquad (11.7)$$

The conductivity σ is the product of the density n of free electrons, their charge and their mobility

$$\sigma = nq\mu \qquad (11.8)$$

We have obtained the important insight that **the electrical resistivity is determined by the density n and the scattering frequency f of mobile electrons**.

11.3 THE DENSITY OF MOBILE ELECTRONS AND THE PAULI EXCLUSION PRINCIPLE

We saw in Chapter 1 that in each energy band, there are exactly as many orbitals (i.e., electron trajectories) as there are atoms in the solid. Because of the Pauli exclusion principle an energy band can contain at most two electrons per atom.

When the band of valence electrons is not fully occupied (Figure 11.4A,B), we are in the presence of a **metal**. Electrons can be accelerated into empty energy levels; they acquire a drift velocity v according to Equation (11.4). The number n of mobile electrons is very large, on the order of one electron per atom (i.e., 10^{22} to 10^{23} electrons/cm^3). Consequently, the conductivity σ is large, and the electrical resistivity ρ in Equation (11.6) is very small.

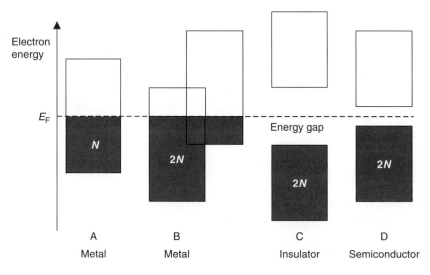

■ **FIGURE 11.4**   The electron energy band structure of metals, insulators, and semiconductors.

When all energy bands are full (Figure 11.4C), all possible trajectories, velocities, and energies of the electrons are occupied and cannot be modified because of the Pauli exclusion principle. The change in the velocity of an electron by the electric field, expressed by Equation (11.3), is not possible. The number n of **mobile** electrons in Equation (11.6) **whose velocity can be changed** by the electric field is **zero**, and so is the conductivity. We are in the presence of an **insulator**. This is the case of most ceramics and polymers. (If you prefer, one can also state that the drift velocity the electrons can acquire is zero.)

In semiconductors, as in insulators, the valence band is fully occupied by electrons, but the energy gap separating it from the nearest empty band is relatively small as shown in Figure 11.4D. By the introduction of judicious impurities, we can introduce electrons in the otherwise empty band, which is now called the **conduction band** or remove some electrons from the **valence band**. Both processes create mobile carriers and electric conductivity. Thus, electrical conductivity in semiconductors is manipulated by processing. We now discuss these three classes of materials in more detail.

11.4 **ELECTRON SCATTERING AND THE ELECTRIC RESISTANCE OF METALS**

Because of their partially filled electron band, metals have a large fixed density n of mobile electrons (n has values of 10^{22} to 10^{23} cm^{-3}). The electrical resistance of metals results from the **scattering** of the electrons. Electrons have the dual nature of particles and waves; quantum mechanics teaches that a perfect crystal does not scatter the electronic waves; **disorder in the crystal structure scatters electrons**. The most important causes of disorder are foreign atoms (impurities or alloying elements) and the slight disorder introduced by the thermal vibrations of the atoms; these scatter the electrons and contribute to the electrical resistivity of the metal. The different types of crystal defects act independently and their effects are additive, so that

$$\rho = \frac{m}{nq^2}\left(f_{\text{imp}} + f_{\text{th}}\right) \quad \text{or} \quad \rho = \rho_{\text{imp}} + \rho_{\text{th}} \tag{11.9}$$

This is known as Matthiesens' rule. (The subscripts stand for "impurities" and "thermal vibrations"). Structural defects such as dislocations and grain boundaries also contribute to scattering, but they are not of technical interest.

11.4.1 **Resistance Increase Due to Impurities and Alloying**

The effects of chemical purity or alloying are relevant in the design of power transmission lines and heating elements, for example. Figure 11.5 shows that even small amounts of a foreign element can cause quite large increases in resistivity. If a metal with low resistance is wanted, for instance for the transmission of power, the engineer will choose a pure metal. Silver has the lowest resistance of all pure metals, as shown in Table 11.1, but its cost is prohibitive. The second best conductor is pure copper. Copper refiners make great efforts to remove elements like phosphorus and iron, which cause large increases in electrical resistance (Figure 11.5). For most electrical applications, one utilizes oxygen-free high conductivity (OFHC) copper. It has a purity of 99.995% Cu. If alloying is necessary for mechanical strength, as in overland power transmission lines, one selects a metal that causes small

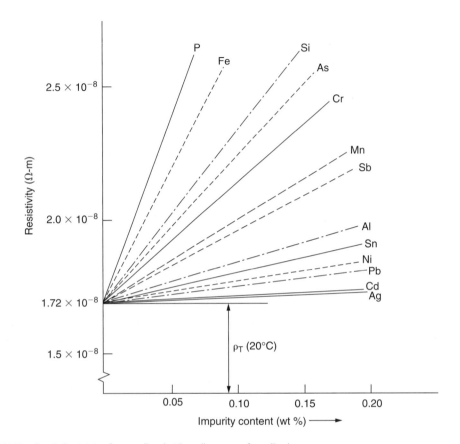

■ **FIGURE 11.5**   Electrical resistivity of copper alloyed with small amounts of metallic elements.

increases in resistivity. Solid solution strengthening with small addition of silver and precipitation hardening with zirconium are two ways in which copper is strengthened.

If a metal with high resistivity is required, as in electrical furnaces, one utilizes an alloy. The addition of 50% nickel to copper increases the resistivity at 20°C from 16 to 470 nΩm. The resistivity of the alloy decreases again to 68 nΩm as the nickel content increases to 100%. Obviously, the additional requirement of oxidation resistance must be considered when using an alloy in furnaces. Nichrome (80%Ni-20%Cr), with a resistivity of 1,100 nΩm, is the most widely used metal for heating wires.

11.4.2 Temperature-Dependence of the Electric Resistance of Metals

The higher the temperature, the more pronounced the disorder introduced by thermal vibrations of atoms and the higher the scattering frequency f and the electrical resistance. The resistivity of metals is approximately linear in temperature above about 200°C and can be represented by the equation

Table 11.1 Electrical Resistivity of Metals.

Metal	Resistivity at 20°C ρ (Ωm) 10^{-9} $(n\Omega m)$	Temperature coefficient of resistivity α $(°C^{-1})$
Aluminum (annealed)	28.3	0.0039
Copper (annealed standard)	17.2	0.0039
Brass (65% Cu 35% Zn)	70	0.002
Gold	24.4	0.0034
Silver	15.9	0.0041
Contact alloy (85% Ag, 15% Cd)	50	0.004
Platinum (99.99%)	106	0.0039
Iron (99.99 + %)	97.1	0.0065
Steel (wire)	107–200	0.006–0.0036
Silicon iron (4% Si)	59	0.002
Stainless steel (18% Cr, 8% Ni)	73	0.00094
Nickel (99.95% + Co)	68.4	0.0069
Nichrome (80% Ni, 20% Cr)	1,000	0.0004
Invar (65% Fe, 35% Ni)	81	0.0014
Constantan (55% Cu, 45% Ni)	490	0.0001
Tungsten	55	0.0045

The resistivity in this table is indicated in $n\Omega m$. Another popular unit for metals is in $\mu\Omega cm$ ($10^{-6}\Omega cm$); the resistivity of copper is $1.72\,\mu\Omega cm$.

$$\rho = \rho_0(1 + \alpha\Delta T) \tag{11.10}$$

where ρ_0 is the resistivity at room temperature and $\Delta T = T - 20°C$.

Table 11.1 shows the temperature coefficients α for a number of metals. We note that the temperature coefficient of pure elements can be approximated roughly as

$$\alpha \approx 0.004C^{-1} = \frac{1}{250°C}$$

The resistivity in Table 11.1 is indicated in units of $n\Omega m$. Another popular unit for metals is $\mu\Omega cm$ ($10^{-6}\Omega cm$); the resistivity of copper is $1.72\,\mu\Omega cm$.

The temperature-dependence of the resistance must, of course, be considered in the design of furnaces and incandescent lightbulbs.

In Table 11.1 we notice that the temperature-coefficient of resistivity α is much smaller for alloys that have a larger value of ρ at room temperature. This is easily understood with Matthiessen's rule, Equation (11.7), as illustrated in Figure 11.6. According to Equation (11.10), we define the coefficient α as

$$\alpha = \frac{1}{\rho_o} \frac{d\rho}{dT} \tag{11.11}$$

The increase in resistivity due to the temperature is the same as for the pure metal, Equation (11.9), but the value of ρ_o is much larger because of alloying. Therefore the relative temperature-dependence α is smaller. The alloy constantan (45% Cu and 55% Ni) with a resistivity at room temperature $\rho = 490\,\text{n}\Omega\text{m}$ derives its name from a very small coefficient α. This property makes it the material of choice for strain gauges.

11.4.3 Superconductors

In 1911, the Dutch physicist Kammerling-Onnes discovered that mercury becomes a **superconductor** at 4°K; its electric resistance drops abruptly to zero. It is really zero, not very small: by induction, a current was generated in the laboratory at the beginning of the twentieth century; it is still flowing

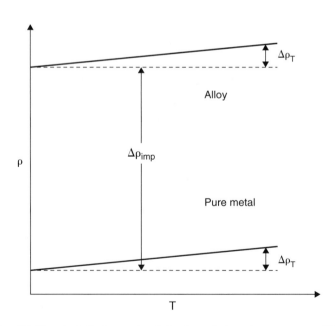

■ **FIGURE 11.6**   Illustration of Mathiessen's rule. The increases in resistivity due to alloying and temperature are additive.

undiminished. Later, Bardeen, Cooper, and Schrieffer explained that the resistivity of superconductors disappears because **their electrons cannot be scattered** at these low temperatures. A number of other superconductors have been discovered since, and the critical temperature for some of them is as high as 120°K, well above the temperature of liquid nitrogen, which is not expensive. The most important "high-temperature" superconductors are cuprates such as $La_{1.85}Ba_{0.15}CuO_4$ and YBCO (Yttrium-Barium-Copper-Oxide). They are used in powerful electromagnets, in particular those used in MRI machines. The latter use metallic superconductors that can be used as wires and therefore operate at liquid helium temperature or below. The most important materials for these applications are NbTi and Nb_3Sn because they withstand large magnetic fields before losing their superconductivity. The expected applications in superconductive magnetic levitation trains and electric motors, however, have met difficulties in execution because the high-temperature superconductors are ceramics, and therefore brittle and difficult to produce in wire form.

11.5 **INSULATORS**

Insulators are used for their ability to withstand high voltages without conducting any appreciable current. Porcelain insulators are used in overland power transmission lines; silicon oxide and hafnium oxide are the insulator materials of choice in integrated electronic circuits. Polymer insulators are used in electric cables and machinery.

Insulators have a completely filled valence band that is separated from an empty energy band by a gap large enough to prevent the excitation of electrons either by thermal energy or the absorption of visible light. Consequently, they have no mobile charge carriers: in Equation (11.5), $n = 0$. As Table 11.2 shows, the electric resistivity of insulators ranges from 10^{12} to 10^{18} Ωm.

Electric current can also be conducted by ions. An important case is the electric conduction by all but the purest water. Any wet insulator can conduct electricity through the water film. For this reason power-line insulators have the characteristic ribbed shape that forms "umbrellas" to keep their surface dry (Figure 11.7). For this same reason, electric appliances are dangerous in bathrooms.

Let us examine three important properties of insulators: their dielectric constant, their dielectric strength, and piezoelectricity.

11.5.1 **Dielectric Strength**

When designing an electrical system with high voltage, the engineer must be concerned with the possibility of electric breakdown. Just as an electric arc (or a lightning bolt) can proceed through air, a similar discharge can move through an insulator and usually destroys it. **The dielectric strength** of an insulator is the maximum electric field it can withstand without destruction. Dielectric strengths of technically important insulators are given in Table 11.2.

When designing for dielectric strength, keep in mind that the surface of the metallic conductor must be smooth: any asperity creates an enhanced electric field. This "tip effect" was used to produce the breakdown illustrated in Figure 11.8.

Table 11.2 Dielectric Properties of Selected Materials (at 300°K).

Material	Dielectric constant (at 10^6 cycles)	Dielectric strength (kV/mm)	Resistivity (Ωm)
Air	1	3.0	
Alumina	8.8	1.6–6.3	$>10^{13}$
Glass (soda-lime)	7.0–7.6	1.2–5.9	10^{10}–10^{11}
Mica	4.5–7.5	2.0–8.7	
Nylon	3.0–3.5	18.5	10^{12}–10^{13}
Polyester resin	3.1	17.1	10^{13}
Polyethylene	2.3	18.1	10^{15}–5.10^{16}
Polystyrene	2.4–2.6	19.7–27.6	$>10^{14}$
Porcelain (high voltage)	6–7	9.8–15.7	
Pyrex glass		500	10^{13}
Rubber (butyl)	2.3	23.6–35.4	
Silica thin films (gate oxide)	3.8	700	10^{16}
Silicone molding compound (glass-filled)	3.7	7.3	
Titanates (Ba, Sr, Mg, Pb)	15–12,000	2.0–11.8	
Titanium dioxide	14–110	3.9–8.3	

Data from various sources.

■ **FIGURE 11.7** Suspension insulators used in high-voltage power lines.

■ **FIGURE 11.8** Dielectric breakdown in glass. Courtesy of Corning Inc.

11.5.2 **Dielectric Constant**

While an electric field cannot accelerate electrons and ions in an insulator, it can displace these charges by small distances. Since positive and negative charges are displaced in opposite directions, the electric field generates an electric polarization that creates an internal electric field opposite to the one that is applied. The total electric field inside the body is then decreased by the factor κ, called the dielectric constant. This polarization can be due to the distortion of electronic orbitals, to the displacement of ions, or to the reorientation of existing dipoles in the solid. This is illustrated in Figure 11.9.

When an insulator with dielectric constant κ separates the two metallic surfaces of a capacitor, the charge Q stored is multiplied by κ (Figure 11.10). A high value of κ is desirable in the manufacture of electric capacitors and in the insulating layer in the gate of a MOSFET. The latter will be explained in Section 11.6.6. A high value of κ is undesirable for materials used to insulate the electric leads in an integrated circuit because it causes prohibitive capacitive signal losses at the very high frequencies of modern computers. Therefore the production of "high k" and "low k" insulators is an active research field at this time.

11.5.3 **Piezoelectricity**

Certain ceramics expand or contract when an electric potential difference is applied to them. These same materials generate a voltage when they are compressed or extended. These materials are known as piezoelectrics.

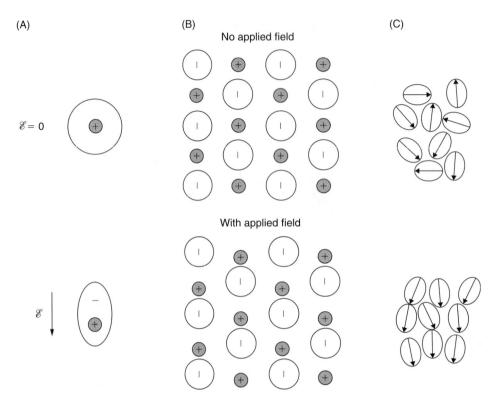

■ **FIGURE 11.9** Polarization of a dielectric by an applied field. ε is the applied field, ε_d is the field due to polarization. $\varepsilon - \varepsilon d = \varepsilon/\kappa$ is the internal field. (A) Electronic polarization. (B) Ionic polarization. (C) Orientational polarization.

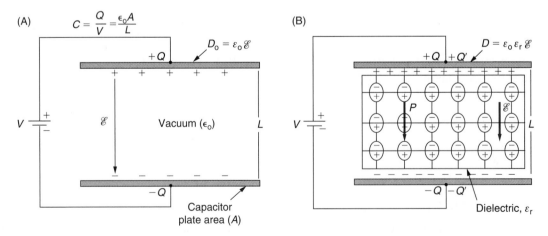

■ **FIGURE 11.10** (A) Capacitor plates with an intervening vacuum space. (B) Capacitor filled with a dielectric. In this case more charge is stored on the plates for the same voltage. (Note: the dielectric constant is often labeled as ε rather than κ.)

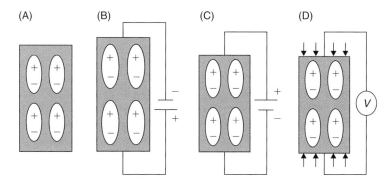

■ **FIGURE 11.11**   Schematic of electrostriction and piezoelectric effect. (A) No applied field. (B) With applied electric field, specimen elongates. (C) With reverse electric field, specimen shortens. (D) Applied stress induces polarization and external voltage.

Quartz watches are the most familiar application of piezoelectricity. Piezoelectrics are also used as frequency stabilizers in broadcasting equipment, as generators of ultrasound, for the control of extremely small movement in atomic force microscopes, and in sonar equipment of submarines. So what is piezoelectricity?

We saw in Figure 11.9 that a voltage applied across an insulator pushes the positive and negative charges in different directions. In certain ionic materials, this can lead to an elongation or a compression of the molecules and therefore of the entire solid: the piezoelectric expands or contracts under the effect of the applied voltage, as illustrated in Figure 11.11B,C. (This is called electrostriction.) Conversely, a mechanical deformation, compressive or tensile, generates a voltage on the electrodes applied to the piezoelectric. This is illustrated in Figure 11.11D. Thus piezoelectrics can be used as mechanical actuators or for the measurement of forces or pressures.

Quartz is an important piezoelectric used in watches and electronic circuits as a frequency stabilizer. Quartz vibrates with very low mechanical losses: its vibration resonance peak is extremely narrow. A quartz crystal inserted in an electronic oscillator provides a very precise and stable resonance frequency. (It is a high Q resonator.)

11.6 SEMICONDUCTORS

An extraordinary scientific and technological revolution occurred after World War II. Physicists discovered the role of certain impurities in electric conduction in semiconductors; they used this knowledge to fabricate solid-state diodes and soon invented the transistor. This development opened the door to smaller and much more powerful electronic circuits. Interplay of new materials processing techniques, solid-state science, and electrical engineering led to the development of the integrated circuits that are found in all electronic equipment and made the modern computer possible. The results of this effort have been extraordinary. Integrated circuits are produced with more than a billion transistors, each of which is no more than 250 nm in size. Progress is still so rapid that this information is bound to be out of date when you read this. It is advisable to obtain the latest information from trade magazines or the Internet. Nevertheless, the principles outlined below remain valid.

Semiconductors, like insulators, have a filled energy band, called the **valence band**, that is separated by a band gap from an empty energy band, which is called the **conduction band**. In semiconductors, the band gap is relatively small (see Table 11.3). The most useful property of semiconductors is that one is able to place electrons into the conduction band (making it an n-type semiconductor) or remove electrons from the valence band (to produce a p-type semiconductor) by the addition of impurities, thereby producing mobile charge carriers capable of electric conduction. By juxtaposing n-type and

Table 11.3 Semiconductor Properties

Material	Lattice parameter (nm)	Melting point (K)	Energy Gap ~(eV at 25°C)	Wavelength of emitted light (μm)
Diamond	0.3560	4300	5.4	
Si	0.5431	1685	1.12 I	
Ge	0.5657	1231	0.68 I	
AlAs	0.5661	1870	2.16 I	
AlSb	0.6136	1330	1.60 I	
GaP	0.5451	1750	2.26 I	
GaAs	0.5653	1510	1.43 D	0.867
GaSb	0.6095	980	0.67 D	1.85
InP	0.5869	1338	1.27 D	0.976
InAs	0.6068	1215	0.36 D	3.44
InSb	0.6479	796	0.165 D	7.515
ZnS	0.5409	3200	3.54 D	0.35
ZnSe	0.5669	1790	2.58 D	0.48
ZnTe	0.6101	1568	2.26 D	0.549
CdTe	0.6477	1365	1.44 D	0.86
HgTe	0.6460	943	0.15	
CdS	0.5832	1750	2.42 D	0.512
SiC	HCP or FCC	2730	3.23 (3.03)	

From S.M. Sze, Semiconductor Devices; Physics and Technology, J. Wiley & Sons, NY (1985), and B. R. Pamplin in Handbook of Chemistry and Physics, R.C. Weast, Ed., CRC Press, Boca Raton, FL (1980). I = Indirect gap, D = Direct gap. Direct gap semiconductors are necessary to form light-emitting diodes.

p-type semiconductor regions, one creates p-n junctions, which can act as electric rectifiers, transistors, light detectors, light-emitting diodes, and solid-state lasers.

Silicon is the material that forms the basis of all common integrated circuits. Other semiconductors of technical interest are III-V compounds, such as GaAs, GaP, and II-VI compounds such as ZnS or CdS. The semiconductors are listed in Table 11.3.

For simplicity, we shall use silicon to discuss semiconductor materials and devices. Silicon is an element in column IV of the periodic table. Each silicon atom has four valence electrons that it shares with four nearest neighbors in covalent bonds (Figure 11.12A). Its full valence band is separated from the nearest empty band, the conduction band, by a gap of 1.12 eV. The electrical resistivity of pure, **intrinsic** silicon is too low to make it a practical insulator and too high to make it a usable conductor. It becomes extraordinarily useful when **doped** by the addition of very small amounts of elements such as phosphorus or boron, becoming, respectively, **n-type** and **p-type** silicon. Figure 11.12 illustrates intrinsic and doped silicon. Figure 11.13 shows the electron energies in the same materials.

11.6.1 **n-type Semiconductors**

Figure 11.12B is a two-dimensional representation of silicon containing a phosphorous impurity. The phosphorous atom is incorporated into the crystal, replacing a silicon atom (a substitutional impurity). Phosphorus has five valence electrons. Four of these form covalent bonds with its four silicon neighbors. The fifth electron is so loosely bound to the phosphorus that thermal energy at room temperature is enough for it to break loose and travel freely through the material. In doing so, it leaves an immobile P^+ ion behind. The silicon now is an electric conductor in which the density n of mobile electrons is equal to the density of the positive phosphorous ions, and this is approximately equal to the density N_D of phosphorous atoms we introduced. The electric conductivity of n-type semiconductors is

$$\sigma = n \frac{q^2}{m} \tau \tag{11.12}$$

The amount of phosphorus introduced into silicon is very small, between 10^{14} and 10^{17} atoms per cm^3 (one in 500 million to one in 500,000 Si atoms!); therefore, the introduction of phosphorus is called **doping** rather than alloying. Note the extraordinary demands on the purity of the material: the concentration of electrically active impurities must be smaller than that of the doping atoms; modern silicon technology utilizes a purity of 99.999999999% silicon.

The mobile electrons introduced by doping reside in the conduction band. Since the electrons carry a negative electric charge, the phosphorous-doped silicon is an **n-type semiconductor**, and phosphorus is an electron **donor**. While phosphorus is the most used donor, any element of column V, namely, phosphorus, arsenic, or antimony, can be used to dope n-type silicon.

Equation (11.12) points to a fundamental difference between semiconductors and metals. In semiconductors, we increase the conductivity σ by introducing free carriers through doping; in metals, we increase the resistivity $\rho = 1/\sigma$ by scattering of electrons through alloying. In semiconductors, we manipulate the density of conductors n; in metals we manipulate the scattering frequency $f = 1/\tau$.

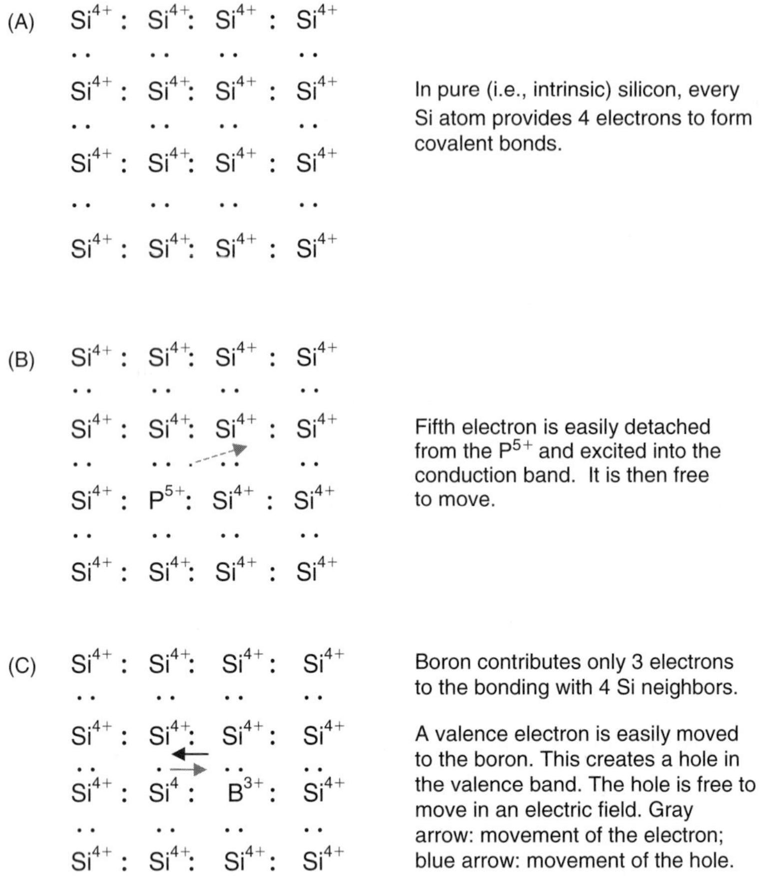

(A)

Si^{4+} : Si^{4+}: Si^{4+} : Si^{4+}

Si^{4+} : Si^{4+}: Si^{4+} : Si^{4+}

Si^{4+} : Si^{4+}: Si^{4+} : Si^{4+}

Si^{4+} : Si^{4+}: Si^{4+} : Si^{4+}

In pure (i.e., intrinsic) silicon, every Si atom provides 4 electrons to form covalent bonds.

(B)

Si^{4+} : Si^{4+}: Si^{4+} : Si^{4+}

Si^{4+} : Si^{4+}: Si^{4+} : Si^{4+}

Si^{4+} : P^{5+}: Si^{4+} : Si^{4+}

Si^{4+} : Si^{4+}: Si^{4+} : Si^{4+}

Fifth electron is easily detached from the P^{5+} and excited into the conduction band. It is then free to move.

(C)

Si^{4+} : Si^{4+}: Si^{4+}: Si^{4+}

Si^{4+} : Si^{4+}: Si^{4+}: Si^{4+}

Si^{4+} : Si^{4} : B^{3+}: Si^{4+}

Si^{4+} : Si^{4+}: Si^{4+}: Si^{4+}

Boron contributes only 3 electrons to the bonding with 4 Si neighbors.

A valence electron is easily moved to the boron. This creates a hole in the valence band. The hole is free to move in an electric field. Gray arrow: movement of the electron; blue arrow: movement of the hole.

■ **FIGURE 11.12** (A) Schematic of the covalent bonding of silicon. (B) Creation of a free electron by a substitutional phosphorous atom from column V. (C) Creation of a free hole by a substitutional boron atom from column III of the periodic table.

11.6.2 **p-type Semiconductors**

Let us now dope silicon with boron, as shown in Figure 11.12C. Boron is in column III of the periodic table and has three valence electrons. Boron substitutes for a silicon atom and forms covalent bonds with its four neighbors, but one electron is missing. Very little energy is needed for a valence electron from a neighboring silicon atom to be moved into the B-Si covalent bond (black arrow). This creates a stationary B^- ion and removes an electron from the valence band. The absence of a valence electron is called a **hole**. The hole carries a positive charge, consisting of a Si^+ ion surrounded by only three electrons. Holes are mobile because valence electrons move easily from one Si atom to another. As the electrons move, it is easy to see that the hole travels in the opposite direction (color arrow in Figure 11.12C). Under an electric field, the holes move and transport a positive charge, and produce an

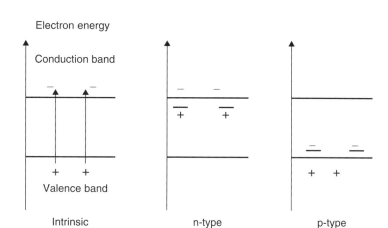

■ FIGURE 11.13 Energy diagrams of intrinsic, n-type and p-type semiconductors. The diagram shows the electron energies. The colored symbols represent free electrons and holes. The black symbols represent the fixed charges on donors ($+$) and on acceptors ($-$). Intrinsic semiconductor: free electrons in the conduction band and holes in the valence band are created by direct excitation of valence electrons into the conduction band, either by thermal energy or by absorption of light; their number is exceedingly small at room temperature. n-type semiconductor: free electrons in the conduction band have been excited out of the donor levels (short horizontal lines). p-type semiconductor: free holes in the valence band created by excitation of valence electrons into the fixed acceptor levels.

electric current. This creates a **p-type semiconductor**. Boron is called an **acceptor** since it receives an electron from the valence band. The density of holes is denoted as p; it is approximately equal to the density of acceptors:

$$p \approx N_A.$$

The electrical conductivity of a p-type semiconductor is

$$\sigma = p \frac{q^2}{m} \tau \qquad (11.13)$$

The density of acceptors introduced by doping is in the range of 10^{14} to 10^{17} cm^{-3} and the corresponding resistivity ranges from 0.01 to 1 Ωm. Figure 11.14 shows the electrical resistivity of silicon and gallium arsenide as a function of the concentration of donors and acceptors. (In the semiconductor industry, the resistivity is usually measured in Ωcm; 1 Ωm = 100 Ωcm.)

11.6.3 **Intrinsic Semiconductor**

At elevated temperatures, in silicon above 100°C, thermal energy is sufficient to excite electrons from the valence band into the conduction band. This generates an equal number of free electrons and holes that are intrinsic to the material, independent of doping. These free carriers move under an electric field and provide the silicon with **intrinsic conductivity**. The latter increases rapidly with rising

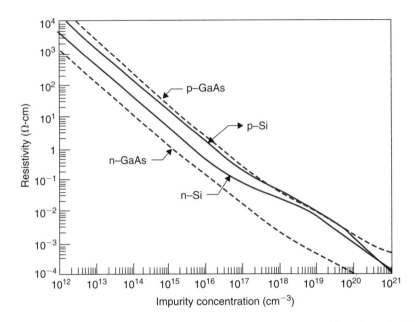

■ **FIGURE 11.14**   Electrical resistivity of Si and GaAs semiconductors as a function of the concentration of donors (n–type) and acceptors (p–type) at 300°K. From S.M. Sze. *Semiconductor Devices: Physics and Technology,* Wiley New York Copyright © 1985 by AT&T; Reprinted with permission of John Wiley & Sons, Inc.

temperature (Figure 11.15). There is no technical application of intrinsic semiconductors. When the density of intrinsic carriers is larger than that introduced by doping, intrinsic conductivity prevents the operation of the electronic devices. It limits the useful temperature range of semiconductor devices.

11.6.4 The p-n Junction

The real usefulness of semiconductors resides in the p-n junction, namely, the juxtaposition of a p-type region with an n-type region of the semiconductor. Depending on its physical configuration, the p-n junction acts as an electric rectifier, a light detector, a photocell, a light-emitting diode (LED), or a solid-state laser. It also forms the basis for the transistor. It is illustrated in Figure 11.16.

The p-n junction is a diode, which expresses the fact that it is composed of two electrodes, namely, the p-type and the n-type region. Figure 11.16A shows an n-type and a p-type semiconductor before they are joined. The n-type material contains free electrons and fixed, positively charged phosphorous atoms. The p-type side contains mobile holes in the valence band and fixed boron ions. Each material is electrically neutral because it contains an equal number of free and fixed charges. When the two sides are joined, as shown in Figure 11.16B, the mobile electrons and holes **recombine** near the junction. In other words, the free electrons fill the holes in the valence band. This sets up a region at the junction where only fixed charges remain. This region is electrically charged by fixed positive charges in the

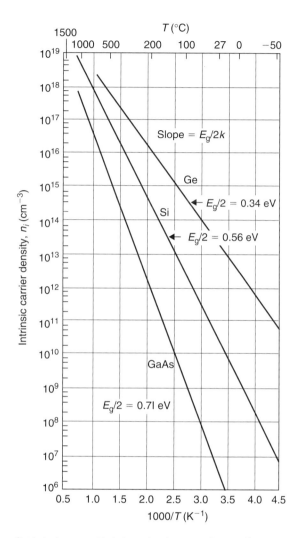

■ **FIGURE 11.15**   Concentrations of intrinsic electrons and holes in semiconductors as a function of temperature.

n-type material and fixed negative charges in the p-type. This space charge region forms an electric dipole and creates an electric field that keeps the electrons in the n-type side and the holes in the p-type side.

Let us apply a **forward bias** to the p-n junction: We apply a voltage to the junction with the negative terminal on the n-type side and the positive terminal at the p-type region (Figure 11.17B). Under the action of the electric field, the electrons move over to the p-type region where they recombine with the holes and disappear. The holes move to the n-type region where they also disappear by recombination. In order to maintain electrical neutrality, the electrons and the holes are continually replaced: electrons enter the n-type side from the applied wire, and electrons move from the p-type side into the

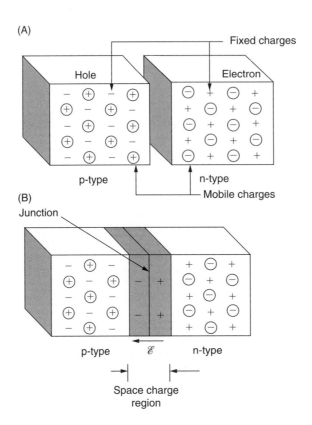

■ **FIGURE 11.16**   The formation of a p-n junction.

attached wire and leave holes behind. The result is an uninterrupted flow of electric current through the diode.

When we apply a **reverse bias** to the p-n junction by making the n-type positive and the p-type electrode negative (Figure 11.17), both types of majority carriers are pulled away from the junction into the wire. They cannot be replaced and no current can flow.

In the **n-type** region, the **electrons** are called **majority carriers**, and the **holes** are **minority carriers**. In the **p-type** semiconductor, the holes are the majority carriers and the electrons are the minority carriers.

11.6.5 **Applications of the p-n Junction**

Numerous electronic devices are based on the semiconductor p-n junction. These are commonly called **solid-state devices**. A number of these applications involve a single p-n junction; these are the **diodes**, namely the rectifier, the photodiode, the light-emitting diode (LED), the solar cell, and the semiconductor laser.

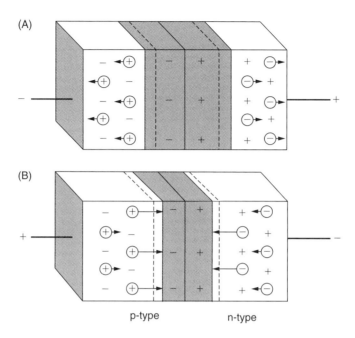

■ **FIGURE 11.17**   Electric conduction in a p-n junction. (A) Reverse bias, no current flows. (B) Forward bias, electric current flows.

Rectifier   We have just shown that the p-n junction is an electric rectifier: it allows current to flow in one direction and not in the other. It provides a DC current when an AC voltage is applied as illustrated in Figure 11.18. Solid-state rectifiers find numerous applications in electrical engineering. They provide a DC circuit powered by the 60-cycle AC power grid. They also permit the transmission of low-frequency signals on high-frequency carriers.

Electro-optical Devices   When light with photon energy larger than the band gap is absorbed by a semiconductor, it creates electron-hole pairs by exciting electrons from the valence band to the conduction band. Conversely, electron-hole recombination in certain semiconductors (not silicon or germanium) can lead to the emission of light with photon energy equal to the band gap, $h\nu = E_g$. These are illustrated in Figure 11.19. Combining these phenomena with the properties of the p-n junction, we obtain electro-optical devices such as the photodiode, the light-emitting diode (LED), the solid-state laser and the solar cell. These are briefly described here and will be treated in more detail in Chapter 13 on Optical Materials.

The Photodiode   The photodiode is a reverse-biased p-n junction. Refer to Figure 11.17. No current flows when the diode is in the dark. When the junction is illuminated by photons with an energy $h\nu$ larger than the band gap, the photons are absorbed and create electron-hole pairs (Figure 11.19) that generate an electric current. Photodiodes have numerous applications, transforming light into electric signals, including the soundtrack in movies, optical fiber telephone transmission, optical scanners, automatic door openers, and many more.

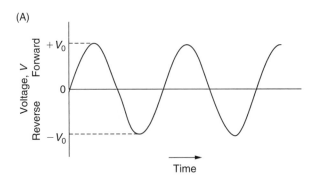

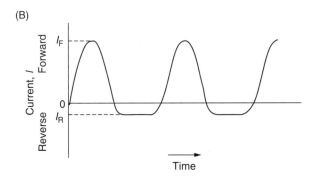

■ **FIGURE 11.18**   Rectification of electric current. (A) applied voltage varying with time. (B) Current flowing through the device. This figure is schematic; the reverse current I_R is much smaller than shown.

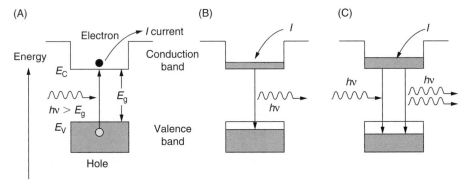

■ **FIGURE 11.19**   Interaction of light with semiconductors. (A) A photon is absorbed and excites an electron from the valence band to the conduction band, creating an electron-hole pair. This is used in photodiodes and solar cells. (B) Electrons recombine with holes and get rid of the energy by emitting a photon. This is used in light-emitting diodes. (C) The recombination of electrons with holes is stimulated by incoming light. The light emitted by this process is in phase with the incoming light. This phenomenon is the basis of the laser.

The Solar Cell   The solar cell is a large-area p-n junction to which no external voltage is applied. Illumination of the junction generates electron-hole pairs. The built-in electric field set up by the space charge at the p-n interface sweeps the electrons into the n-type region and the holes into the p-type region. The excess carriers produce a negative electric charge in the n-type electrode and a positive charge on the p-type side. These charges set up a potential difference V between the electrodes and a flow of current through the external circuit, generating the power $W = IV$.

The Light-Emitting Diode (LED) and Solid-State Lasers The light-emitting diode is a forward biased p-n junction. Electrons moving into the p-type region and holes moving into the n-type region recombine with the majority carriers. In III-V compounds (Table 11.3), the recombination energy is dissipated by emission of light of energy

$$hv = E_g$$

11.6.6 **Transistors**

Transistors are triodes consisting of two p-n junctions. They exist as n-p-n or p-n-p structures (Figures 11.20 and 11.21). A constant potential is applied between the extreme electrodes, and a small variable potential applied to the middle electrode controls the amount of current flowing through the transistor. Bipolar transistors are used as power amplifiers, and field effect transistors serve as on-off switches in digital circuits.

The Bipolar Transistor The bipolar transistor, illustrated in Figure 11.20, is a p-n-p or n-p-n triode. It consists of three electrically contacted regions: the emitter, the base, and the collector. The transistor forms two p-n junctions: one is the junction between the emitter and the base, and the other is the junction between the base and the collector. Let us choose an n-p-n transistor to examine how it functions. The junction between the base and the collector is always reverse biased, with a constant, rather large, voltage V_C (usually $V_C = 15\,V$); no current flows from the base to the collector. When the junction between the emitter and the base is forward biased (a small negative voltage is applied to the emitter), electrons flow from the emitter into the base. These electrons drift through the base and are attracted into the collector. Thus, an electron current I flows through the transistor. When the emitter-base junction is reverse-biased, no current flows.

Now why is this interesting? The transistor is a power amplifier. The input voltage on the emitter V_E is small, usually a fraction of a volt, and input power consumed is IV_E. The output voltage V_C on the

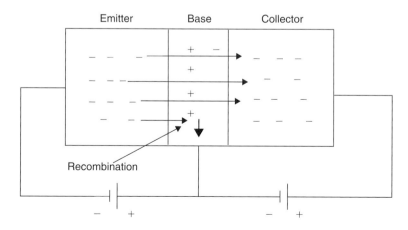

■ **FIGURE 11.20** Operation of the bipolar junction transistor.

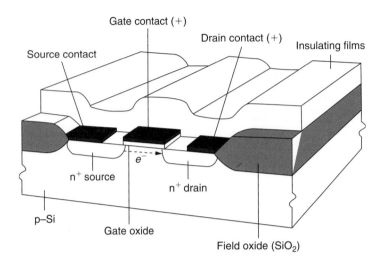

■ FIGURE 11.21 The MOSFET (Metal Oxide Field Effect Transistor). In the absence of a bias applied to the gate, electrons cannot travel from the source to the drain because of the p-type region separating them. When a positive bias is applied to the gate, electrons are attracted toward the gate and can proceed to the drain as shown. From W.E. Beadle, J.C.C. Tsai, and R. D. Plummer, *Quick Reference Manual for Silicon Intergrated Circuit Technology*, Published by John Wiley and Sons, Copyright © 1985 by AT&T; Reprinted with permission of John Wiley & Sons, Inc.

collector is 15 V, and the power output is IV_C. The current through both is the same, so that the power amplification is the ratio of the output to the input voltage.

The drawback of the bipolar transistor is that a fraction of the electrons drifting through the p-type base is lost by recombination. This causes a loss of current and a low input impedance. In order to minimize recombination, one reduces the number of acceptors in the base: the latter is made thin and lightly doped. Mechanical defects, particularly grain boundaries and dislocations, are active centers for electron-hole recombination. For this reason, the silicon is prepared in single crystal form with extreme care so that it contains a very small density of dislocations.

The MOS Field Effect Transistor   This transistor is used in digital electronics. It gets its name from the Metal Oxide Semiconductor structure of the device, and from the electric field that is placed transverse to it. A typical device structure is shown in Figure 11.21; it consists of three electrically contacted regions: the **source**, the **drain**, and the **gate**. Source and drain regions are doped alike (either n or p) and separated by the gate region that is oppositely doped (p-type when source and drain are n-type and vice versa). The gate region sits directly under a very thin gate insulator covered with a gate electrode.

The drain is made positive with respect to the source, but no current flows horizontally because of the reverse-biased junction formed by the gate and source. When a positive voltage is applied to the gate contact, electrons are drawn to the interface, and the holes of the p-type material are repelled. At a critical **threshold voltage** the electrons drawn to the interface form a thin **n-type channel** that joins the source and drain and allows electrons to flow. The device is turned on, going from 0 to 1 or from nonconducting to conducting state. The current can be switched off by applying a negative potential to the gate. These transistors are used to switch signals in computers.

The MOSFET has the advantage of large input impedance. Since an insulator separates the source from the gate, the loss currents are avoided.

In Section 11.5.2 we have mentioned that a high dielectric constant κ is desirable for the gate insulator of the MOSFET. The gate electrode and the underlying semiconductor of the MOSFET can be considered a parallel plate capacitor. The carriers drawn to the thin n-type channel by the gate voltage are, in fact, the charge of this capacitor. A large capacity resulting from a large κ reduces the critical threshold voltage and the possibility of electric breakdown as the transistors become smaller and the gate insulators thinner.

11.6.7 **Organic Semiconductors**

The recent discovery of organic semiconductors presents an interesting development because these materials can be processed with the techniques of printing and polymer chemistry, which are much less expensive and energy-intensive than photolithography.

Organic semiconductors include oligomers (i.e., large molecules) and polymers. Oligomers include anthracene, pentacene, and rubrene (Figure 11.22). Semiconducting polymers include polyacetylene, poly(p-phenylene vinylene) (see Figure 11.23), and others.

Examining the two figures, we note that all these molecules possess double bonds that involve π orbitals. The latter overlap and provide mobility for the electrons. It is also known that these substances are black; that is, the gap between the filled valence band of π electrons and the band of lowest unoccupied orbitals (LUMO) is less than the photon energy of visible light, that is, less than 2 eV. The pure materials are insulators since their valence band is full. Heeger, MacDiarmid, and Shirakawa, who were awarded the 2000 Nobel Prize for this discovery, were able to dope polyacetylene. They obtained a p-type material by oxidizing the polyacetylene in iodine vapor and an n-type material by reducing it with sodium. Recall that oxidation is defined as the removal of electrons (in this case the creation of holes) and reduction is the addition of electrons.

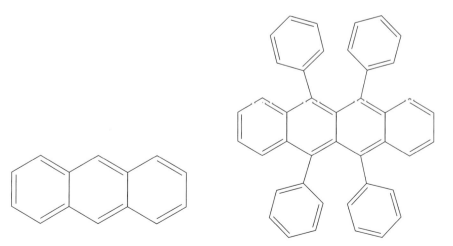

■ **FIGURE 11.22**   *The molecules anthracene (left) and rubrene (right).*

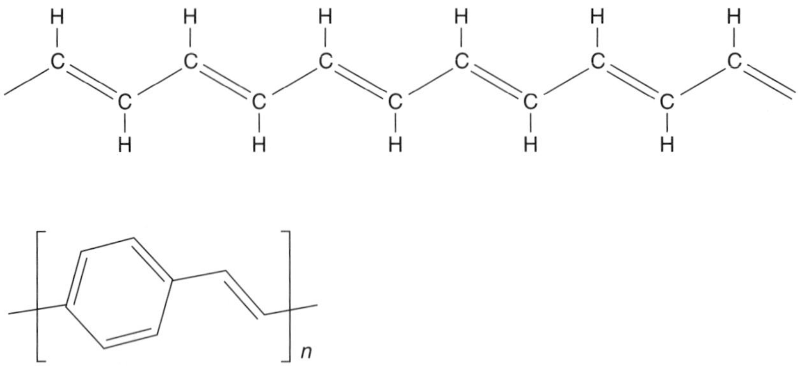

■ **FIGURE 11.23**   Molecular structure of polyacetylene (top) and poly(p-phenylene vinylene).

With this treatment, the electric conductivity of the material is increased by a factor of 10 million and the color turns from black to silvery. The density of charge carriers is so high that it gives the material the electric and optical properties of a metal.

The technologically important fact is that these materials can be doped to be n- or p-type semi-conductors and can be used in many of the applications described above. We will see in Chapter 13 that organic semiconductors find important applications in opto-electronic devices, especially Organic Light-Emitting Devices (OLED) and electroluminescent light sources. They are also the object of much research for the purpose of making large-area solar cells.

■ SUMMARY

1. An electric current consists of the flow of electrons that are accelerated by an applied electric field. The magnitude of the current depends on the density of electrons and their average velocity, called drift velocity.

2. The drift velocity of electrons is limited by the scattering of the electrons.

3. Defects include impurities (alloying), thermal vibrations (temperature), and defects due to plastic deformation scatter electrons. A perfect crystal does not scatter electrons.

4. The valence band of a metal is only partially filled. Its electrons can change their velocity and energy. A metal has a large, fixed number of free electrons. Its electric resistance is determined by the scattering of the electrons.

5. Pure metals have very low resistivity, 14–100 nΩm. Metals with increased resistivity are obtained by alloying.

6. The electrical resistivity of metals increases with temperature.

7. Solids with covalent-ionic bonding have a filled valence band separated from an empty band by an energy gap. According to the Pauli principle, electrons in a filled band cannot modify their energy or their velocity. Materials with a large band gap (5–10 eV) are insulators. Their electrical resistance is very large (up to 10^{16} Ωm).

8. The dielectric constant of an insulator increases the electric capacity of a capacitor.

9. The dielectric strength is the electric field an insulator can withstand without breaking down.

10. Piezoelectrics modify their size under an applied voltage. They generate a voltage when strained. They are used as force transducers, actuators, and force sensors. Piezoelectric quartz crystals are used as frequency standards.

11. Semiconductors have a filled valence band separated by a small energy gap (0.2–3 eV) from an empty conduction band.

12. Doping of a semiconductor with donors (Group V elements for silicon) creates free electrons in the conduction band, making it an n-type conductor.

13. Doping of a semiconductor with acceptors (Group III elements for silicon) creates mobile holes in the valence band, making it a p-type conductor.

14. A p-n junction conducts electricity in one direction only. It is used as an electric rectifier.

15. A photodiode is a reverse-biased p-n junction that conducts electricity only when illuminated. The absorbed light excites electrons from the valence band to the conduction band, creating free electrons and holes.

16. The light-emitting diode (LED) is a forward biased p-n junction: electrons recombining with holes lose their energy by emitting light. LEDs are made of III-V and II-VI semiconductors.

17. A bipolar transistor consists of two p-n junctions: the base-collector junction is reverse biased; the emitter-base junction has variable bias driving the transistor current. The electron (or hole) current originates from the emitter, drifts through the base, and reaches the collector. The transistor is a power amplifier.

18. The MOSFET is an n-p-n (or p-n-p) double junction that conducts a current when an appropriate bias is applied to the gate.

19. Organic semiconductors consist of oligomers or polymers that contain double bonds involving π orbitals that are delocalized. They can be doped by oxidation or reduction.

■ KEY TERMS

■ REFERENCES FOR FURTHER READING

[1]   C. Kittel, *Introduction to Solid State Physics*, 7th ed., Wiley, New York (1995).

[2]   L. Solymar, and D. Walsh, *Electrical Properties of Materials*, 7th ed., Oxford University Press, Oxford, UK (2004).

[3]   S.M. Sze, *Semiconductor Devices: Physics and Technology*, Wiley, New York (1985).

[4]   S.O. Kasap, *Principles of Electronic Materials and Devices*, McGraw-Hill (2006).

■ PROBLEMS AND QUESTIONS

11.1   Design the filament of a 60 W lightbulb for 110 V. The lightbulb operates at 1,500°C.

   a.   Briefly describe a commercial lightbulb in order to recognize the design constraints.

   b.   Select the material, considering the high temperature needed.

   c.   Calculate the resistance needed for 60 W; then calculate the dimensions of the wire. You will need to make a decision concerning the length or the thickness of the wire. In doing so, be realistic. It will be necessary to visualize these dimensions in order to obtain a reasonable solution.

   d.   In commercial lightbulbs, the filament has the shape of a double winding. Propose the dimensions of such an arrangement.

11.2   Design the heating element of an electric furnace of 1 kW at 110 V. The wire operates at 900°C. Select the material and propose a reasonable thickness and length of the wire.

11.3   Explain why incandescent lightbulbs burn out usually as they are switched on. Hint: the filament burns out because a constriction becomes very hot on switching the bulb on and evaporates. First make a sketch of the time-evolution of the voltage, that of the temperature, that of the temperature-dependent resistivity, and finally that of the current flowing. Then examine what happens in a segment of reduced diameter. (When a failed lightbulb has a dark area on the glass, it failed the way you described, the dark area is evaporated metal; when the lightbulb has a white deposit, it failed because air leaked into the bulb and oxidized the tungsten.)

11.4 Find out what material the heating elements of an electric stove are made of.

11.5 Make a graph showing the electrical resistivity of Cu-Ni alloys between pure Cu and pure Ni. At small Ni concentrations, use Figure 11.5. Explain the shape of the curve. Compare this curve to that of Problem 4.3 of Chapter 4 and remark on the physical phenomena underlying both graphs.

11.6 Using the equations at the beginning of this chapter, design a strain gauge. A strain gauge measures the elastic elongation of an object by means of an electric signal.

11.7 Design a remote electrical thermometer that uses a metal. It is called a positive thermistor.

11.8 By using Ohm's law, it is possible to measure the stress on structural element and the temperature. How would you do it? How can you minimize "cross-talk" between stress and temperature?

11.9 Examine Mathiesen's rule for constantan. Compare the effect of thermal scattering on the resistivity of constantan and of copper.

11.10 What is the electric resistance, at room temperature, of a 5 m long copper power cord of 1 mm diameter?

11.11 When a current of 10 A flows in the cord of Problem 11.10, what is the drift velocity of the electrons?

11.12 A cable consists of three strands, each 4 m long and 250 μm in diameter. What is the room temperature resistance of the cable?

11.13 Approximately how many collisions do electrons make each second with lattice scattering sites in pure silver at 25°C?

11.14 A milligram of arsenic is added to 10 kg of silicon.

   a. How many unit cells of Si are associated with an As atom?

   b. Estimate the electrical resistivity of this semiconductor.

11.15 Silicon is doped to a level of 10^{18} boron atoms/cm^3. If all of the acceptors are ionized at 350°K, determine

   a. The carrier concentrations.

   b. What fraction of the total current is carried by holes in this semiconductor?

11.16 Roughly design an optical-fiber telephone system: I give you a microphone (that transforms sound waves into electric signal) and a loudspeaker; select the solid-state devices in between (you will also need to amplify the signal) and for each device, describe what it does and how it does it.

11.17 An electron microscope operates at 300 kV. Design the electric cable that provides this potential: select an insulator and determine how thick it must be to avoid electric breakdown. Use a safety factor of two: make it break down at 600 kV. Hint: you will need to calculate the electric field at the surface of a cylinder. Design the diameter of the cable and of the insulator. What is more important, the diameter of the conductor or the thickness of the insulator?

11.18 Design a 100 pF capacitor. A capacitor consists of two metallic surfaces separated by a thin insulator. If A is the area of the metallic surfaces, t the thickness of the insulator, and κ its dielectric cons tant, the capacity is

$$C = \kappa \varepsilon_0 \frac{A}{d}$$

where $\varepsilon_0 = 8.854 \times 10^{-12}$ F/m. Select a dielectric for this capacitor from Table 11.2. If I want the capacitor to operate safely at 1,000 V, how thick does the insulator have to be? How large does the area of the capacitor have to be?

11.19 Using Ohm's law, explain why overland power lines operate at high electric potentials to minimize electric losses. If the potential were 110 V, what diameter of copper cable would be necessary to transmit 1,000 kW over 50 km so that not more than 1% of the power would be lost to ohmic resistance? What would the diameter be at 300 kV? (This explains why Tesla, with his AC power plant at Niagara Falls, won over Edison who preferred DC power.)

11.20 Calculate the mobility of electrons and holes in silicon from Figure 11.14. Do so at doping concentrations of 10^{15} and 10^{19} cm^{-3}. Can you give a reason for the different values at high doping concentrations?

11.21 Consider a Si-based n-type field-effect transistor with a 2 μm channel length. If 5 V is applied between source and drain, how long will it take electrons to traverse the channel? What is a rough upper limit to the frequency response of a circuit employing transistors of this type?

11.22 Using Figure 11.15, estimate the maximum temperature at which a silicon-based transistor can still function if the critical electrode (the base) is doped with 10^{15} dopant atoms/cm^3.

11.23 Movie films have a soundtrack next to the pictures. It consists of band with varying density (opacity). Sketch how it works.

11.24 Describe in your own words how a solar cell functions. Design (sketch) a solar cell.

11.25 It is hoped that, in the future, solar cells can be painted on roofs. Describe what materials would be used and how they could be made.

11.26 What determines the electric resistance of a metal?

11.27 What scatters electrons in a metal?

11.28 What is the drift velocity of electrons?

11.29 The electric resistivity of copper is 16 nΩm, and that of nickel is 70 nΩm. What, approximately, is the resistivity of a 50% Cu-Ni alloy?

11.30 Compare the electron band structure of a metal, an insulator, and a semiconductor.

11.31 What can be done to increase the resistance of a metal?

11.32 Define the dielectric constant and the dielectric strength of an insulator.

11.33 In what application is the dielectric constant important? In what application is the dielectric strength important?

11.34 Describe a piezoelectric material and name a use for it.

11.35 How does one make an n-type semiconductor?

11.36 How does one make a p-type semiconductor?

11.37 Describe a p-n junction and name one use for it.

11.38 How does the electric rectifier (diode) work?

11.39 Describe a photodiode and how it functions.

11.40 Describe a bipolar transistor and how it functions.

11.41 Describe a MOSFET and how it functions.

Fabrication of Integrated Circuits and Micro Electro-Mechanical Systems (MEMS)

A computer chip contains about one billion transistors. Each transistor consists of microscopic pockets of n- or p-type doped silicon, insulating layers, and metallic interconnects on the surface of a silicon chip. In a modern plant (called a foundry), more than a hundred chips are manufactured simultaneously on the surface of a silicon wafer by photolithography. The wafer is cut from a single crystal boule of highly pure and defect-free silicon. Photolithography transfers the designs of the transistor elements by shining light onto a photoresist that becomes soluble and exposes the underlying material for further processing. Doping is done by ion implantation, and interconnects are made by metallic vapor deposition. Several surface modification and material deposition techniques are used in the processing of integrated circuits; these are described in some detail. Photolithography is also used for the fabrication of MEMS (Micro Electro-Mechanical Systems) that are used as microscopic sensors and actuators such as the accelerometers in airbags.

LEARNING OBJECTIVES

After studying this chapter, the student will be able to:

1. Prescribe the successive steps in the fabrication of an integrated circuit.

2. Describe the fabrication of a silicon wafer.

3. Describe the steps in photolithography.

4. Describe the methods used for doping.

5. Describe the various surface technologies used in semiconductor and MEMS processing and select the appropriate technique.

6. Define a micro electro-mechanical system and describe its fabrication.

12.1 **A CHIP AND ITS MILLIONS OF TRANSISTORS**

Modern electronics mostly consists of **integrated circuits,** which means that all the elements of an electronic circuit are placed and manufactured on a silicon chip. The microprocessor of a modern computer, for instance, is an integrated circuit containing about 1.5 billion transistors; it is fabricated on a silicon chip approximately 11 × 11 mm (about 1/2 in square) in size as shown in Figure 12.1. Moore's law is often cited in the development of computers. In 1965, Gordon Moore, a co-founder of Intel, predicted that the density of transistors on integrated circuits would double every 18 months. Figure 12.2 shows that his prediction was quite accurate. Since 1970 the number of transistors on an Intel chip has grown from about 2,000 to one and a half billion, which corresponds to a doubling every 21 months. With such a high density, the transistors must be exceedingly small; each one is less than 0.2 μm on a side. It is expected that the size of transistors will still decrease and their density on the chips will further increase in the future. How is it possible to manufacture 1.5 billion microscopic transistors on a chip, and this for an affordable price? This is the subject of this chapter.

The transistors are made on the surface of the silicon chip as shown in Figure 12.3. Each element of a transistor is a pocket where the silicon has been doped with boron or arsenic to make it p-type or n-type. The gate of the MOSFET of Figure 12.3B has a length of 35 nm and its oxide insulator is 1.2 nm thick.

Integrated circuits are fabricated on the surface of a single crystal silicon wafer that has a diameter of 300 or 450 mm (12 or 18 in) and a thickness less than one millimeter (see Figure 12.4). This wafer contains hundreds of chips, which are all processed simultaneously by photolithography.

■ **FIGURE 12.1** Chip on a computer motherboard.

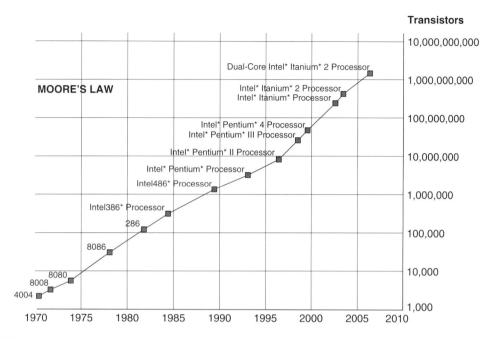

Transistors

■ **FIGURE 12.2**   Moore's Law: the evolution of the capacity of integrated circuits. From Intel web site.

12.2 **GROWTH OF SILICON SINGLE CRYSTALS**

The material requirements for the wafer on which the integrated circuits are fabricated are quite extraordinary. The silicon must be of extreme purity, containing less than 1 impurity per billion silicon atoms. The material must be a single crystal with very few dislocations; any departure from a perfect crystal will deteriorate the performance of the circuit because defects facilitate electron-hole recombination.

Silicon is obtained by reducing silicon oxide, usually sand, in an electric furnace. The reaction of SiO_2 with wood or charcoal at $T > 1,900\,°C$ yields molten silicon according to the reaction

$$SiO_2 + C \rightarrow Si + CO_2 \tag{12.1}$$

The **metallurgical grade silicon** so obtained is polycrystalline and about 98% pure. This material is purified by converting it to a gas, usually trichlorosilane $HSiCl_3$, that can be purified by multiple fractional distillation. At high temperature, this gas decomposes and deposits high-purity silicon with the reaction

$$2HSiCl_3 \rightarrow Si + 2HCl + SiCl_4 \tag{12.2}$$

Polycrystalline silicon obtained by this process has impurity levels of 1 part per billion or less.

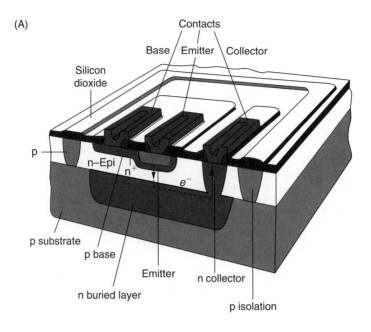

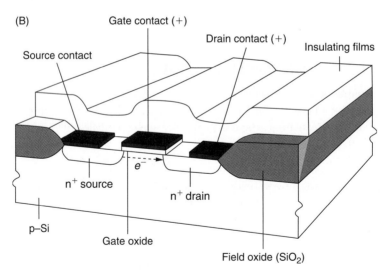

■ **FIGURE 12.3**   Schematic of a bipolar (A) and a MOSFET transistor (B) on the surface of a chip. In a modern transistor, the distance between the centers of the source and drain is 0.25 μm and the length of the gate is 35 nm. From W.E. Beadle, J.C. C. Tsai and R.D. Plummer, *Quick Reference Manual for Silicon Integrated Circuit Technology* (1985), Published by John Wiley and Sons, Copyright © 1985 by AT&T; reprinted with permission.

■ **FIGURE 12.4**   300 mm silicon wafer with integrated circuits.

Bulk single crystals are grown by the **Czochralski** method, which was described in Chapter 7 and Figure 7.12. A charge of pure silicon is melted in a silica crucible. A small cooled single crystal seed is lowered into it. Silicon solidifies on the seed and continues the seed's crystal structure. The cylindrical crystal that solidifies is simultaneously pulled up and rotated in such a way that it extends its length and maintains a constant diameter. The heat released in the solidification of the melt is dissipated by conduction along the seed and by radiation from the crystal.

The resulting crystal is called a **boule**. It has a diameter exceeding 450 millimeters (18 in.) and a length approaching 1 m. The boule is ground to the desired constant diameter and cut into thin wafers. The wafers are polished and etched to remove the impurities and mechanical defects introduced by the dicing operation. Each wafer is now ready for further processing where the designs of the circuits of hundreds of chips are transferred to the wafer by photolithography, the n- and p-type elements of the transistors are doped by ion implantation, the oxide layers for processing and for electric insulation are grown or deposited by various techniques, and the interconnects (i.e., the wiring) are deposited by vapor deposition; each step treats all the elements of all the transistors of all the chips on the wafer simultaneously.

12.3 **PHOTOLITHOGRAPHY**

This Greek term translates into English as "drawing on stone with light." After the layout of the chip and all drawings necessary for the processing have been completed by computer-aided design (CAD), they are optically reduced to the proper scale. Fabrication of the chip uses a light-sensitive film called photoresist. The photoresist is soluble where illuminated; it permits the definition of the submicroscopic areas where the wafer is to be processed, simply by shining the microscopic designs of the transistor elements on it. Let us first look at the steps involved in this technology. We will describe each step in more detail later. These steps are illustrated in Figure 12.5.

1. Make a single crystal of silicon with 99.999999999% purity (one part per billion); cut it into thin wafers and polish them to remove all surface defects.

2. Grow a protective oxide layer on the surface of the wafer (by oxidation in a furnace).

3. Spread a photoresist on the surface of the wafer. A photoresist is a substance (usually a polymer) whose solubility in an organic solvent is modified by illumination.

4. Apply a mask that has been designed to be transparent where one wants to illuminate the photoresist. The mask simultaneously defines the area of a given element (source, gate, or drain) for all the transistors of one chip.

5. Illuminate the photoresist through the mask. One uses ultraviolet light ($\lambda = 139$ nm) or soft X-rays. (The wavelength of visible light, 400–600 nm, does not provide sufficient resolution.) This illumination is repeated for each chip on the wafer.

6. Develop the photoresist (by placing the wafer in a solvent). This removes the photoresist where it was illuminated and exposes the oxide.

7. Etch away the oxide. The oxide is only dissolved where the photoresist has been removed. Now we have areas of exposed silicon to be doped. Steps 3 to 7 are illustrated in Figure 12.5.

8. Dope by driving the dopant (B for p-type, P or As for n type) into the surface by ion implantation. (This is done in a machine that accelerates the ions with high voltage, 30–100 kV, to drive them into the surface.) The ions will only penetrate where the silicon is exposed and will be stopped where the oxide remains.

9. Heat to anneal the wafer (repair the damage done by implantation and drive the dopant in deeper by diffusion).

Now we have made one element (e.g., source) of all the transistors on the wafer at once. The above sequence is repeated for each other doped pocket (device element) on the wafer.

How are the interconnects (the "wires" that connect the transistors) made on the wafer? This is again done by photolithography. By use of suitable masks, the photoresist is illuminated and removed where the interconnect will be deposited. A suitable metal is then deposited on the wafer by evaporation. Removing the remaining photoresist lifts off the unneeded metal. In other processes, the metal is deposited on the whole wafer and a photoresist defines the areas where the metal is to be removed by etching. A modern microprocessor possesses up to eight layers of interconnects separated by oxide insulation.

The processing of a wafer often requires 10–15 cycles of photolithography and processing until all the electrodes and connections are made. Let us examine, for example, the fabrication of a MOSFET transistor.

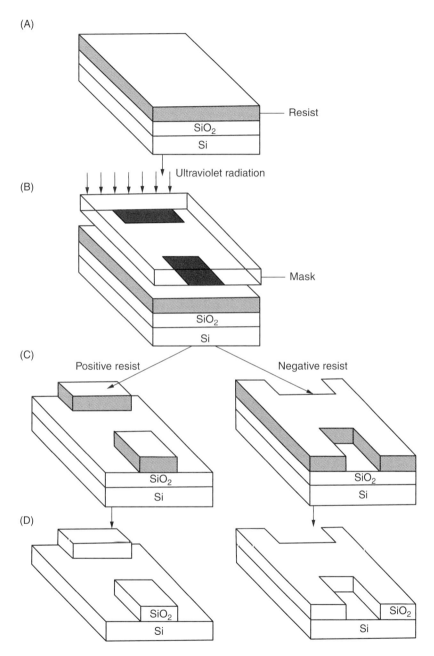

(A)

Resist

SiO$_2$

Si

Ultraviolet radiation

(B)

Mask

SiO$_2$

Si

(C)

Positive resist

Negative resist

SiO$_2$

Si

SiO$_2$

Si

(D)

SiO$_2$

Si

SiO$_2$

Si

■ **FIGURE 12.5** Schematic of the lithography process to define patterned oxide regions on the wafer surface. (A) A photoresist layer is first applied to an oxidized Si wafer. (B) Ultraviolet light passing through a mask exposes the protoresist to define the undesired pattern geometry. (C) In a positive resist, the illuminated area of the photoresist and SiO$_2$ layer are removed. (D) In a negative resist, the opaque region of the mask defines the removal of the photoresist and the SiO$_2$ layer.

12.3.1 **Fabrication of an NMOS Field Effect Transistor**

The different steps in the fabrication of a NMOS transistor are illustrated in Figure 12.6 and described below.

1. A lightly doped p-type (100) Si wafer is the starting point. The wafer is first thermally oxidized (~50 nm thick), and then covered with a deposited Si_3N_4 layer (100 nm thick). Photoresist is then applied and patterned (i.e., illuminated through a mask and developed).

2. The photoresist serves as the mask that shields the eventual transistor region against the implantation of boron into the surrounding area. Boron is implanted; it serves to limit the channel length and isolate the transistor electrically from neighboring transistors.

3. The Si_3N_4 layer not covered by photoresist is etched away, and the resist is also removed. A layer of SiO_2 is thermally grown everywhere except in the Si_3N_4-protected areas. This 0.5–1 μm thick **field**

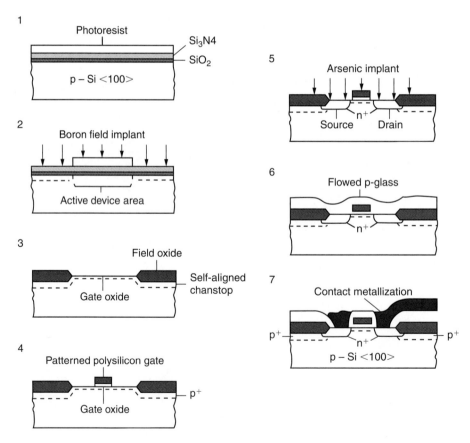

■ **FIGURE 12.6**   The fabrication of a NMOS transistor in an integrated circuit. Adapted from S.M. Sze, *Semiconductor Devices Physics and Technology*, Wiley, New York (1985), Copyright © 1985 by AT&T, reprinted by permission of John Wiley and Sons.

oxide film insulates the individual transistors from one another. The remaining Si_3N_4 and underlying SiO_2 films are both stripped.

4. The bare Si surface is oxidized to produce a ~20 nm **gate oxide** across the channel region. Higher performance can be achieved by depositing a layer of hafnium oxide instead of the silicon oxide. The hafnium oxide possesses a higher dielectric constant κ; it can be deposited as a thicker layer than SiO_2 for the same capacity to avoid electric breakdown of the gate. In this case, the transistor is designated as a Metal Insulator Semiconductor Transistor (MISFET).

5. A heavily doped polycrystalline Si layer is deposited by chemical vapor deposition. The **polysilicon** is lithographically patterned and removed everywhere but in a small central region where it will become the gate electrode.

6. This gate serves as the mask for the ion implantation of an n-type dopant (usually As) that creates the source and drain regions simultaneously. In this way the source and drain are symmetrically located and **self-aligned** with respect to the gate.

7. A phosphorous-doped oxide glass is then deposited and heated to make it flow viscously to **planarize** or smooth the overall topography of the structure. This glass provides additional insulation. Contact windows are then defined lithographically and etched to expose the semiconductor surface.

8. The last step is metallization to form the electrical connections (wires) to the transistor. Metal is evaporated over the entire structure. A final lithography step is required where metal is to be removed in order to delineate the conductor stripe network.

Modern devices require several layers of metallic interconnects, separated by insulating oxide layers. Metallic interconnects also provide an electric link to contact pads deposited on the sides of the chip.

12.4 **PACKAGING**

Once all the transistors and interconnects have been fabricated by photolithography, the chips must be separated and prepared for application. This is called packaging (Figure 12.7). The wafer is cut to separate the individual chips. Each chip is mounted on a support that is equipped with electrical leads. The leads are gold-plated to provide good electrical contact. Small wires are welded to the leads of the support and to the lead pads on the chip. Finally, the whole structure is covered with a polymer molding compound to protect it from mechanical contact and corrosion.

We now examine in more details the individual processing methods employed in the fabrication of a chip. Our starting point is the silicon wafer prepared according to Section 12.2.

12.5 **OXIDE LAYERS**

Oxide layers are grown on the surface of the wafer for two very different reasons. They can be grown as electric insulators, for instance, as the field oxide and the gate in a MOSFET (steps 3 and 4 in Figure 12.6).

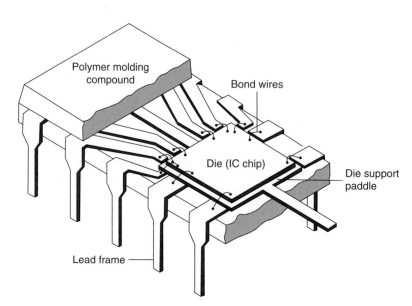

■ **FIGURE 12.7** Packaging of a chip. Such a chip, without its polymer coating, is also shown in Figure 12.1.

More frequently, they are grown as a tool for processing: the oxide layers serve as a barrier for the penetration of dopant atoms where they are not wanted. Oxidation is performed by heating the wafer at 1,100 to 1,200°C in humid air inside a fused-quartz tube as shown in Figure 12.8.

12.6 **PHOTORESIST**

The photoresist permits the fabrication of extremely small items (i.e., sources, gates, drains, and metallic interconnects) by optical reduction of much larger designs. Like the emulsion on a photographic film, the photoresist undergoes a photochemical reaction under the influence of light. **Positive photoresists undergo photolysis**: a large molecule (usually a polymer) is dissociated into smaller fragments when illuminated. The smaller molecules are soluble in an organic solvent. **A negative photoresist** consists of small, soluble molecules that are **crosslinked** under the action of light and become large and insoluble. The photoresist is **developed** by dissolving the soluble material. Photoresist is spread over the wafer by spin coating: the wafer is rotated at 2,000–6,000 rpm, and the resist, dissolved in a liquid solvent, is applied to the center and spreads over the wafer by centrifugal force.

12.7 **THE MASK**

The masks define the area to be illuminated on the photoresist. Masks consist of highly polished silica plates on which a chrome layer has been deposited with the required pattern. Often photolithography

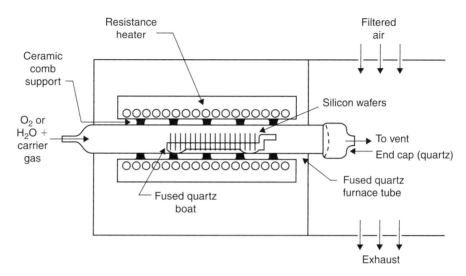

■ FIGURE 12.8   Schematic of a reactor used to oxidize silicon wafers.

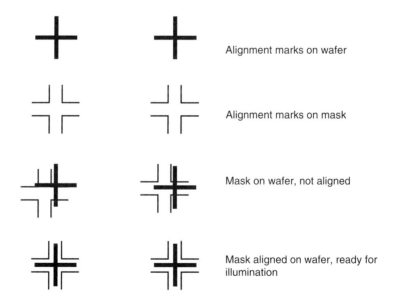

Alignment marks on wafer

Alignment marks on mask

Mask on wafer, not aligned

Mask aligned on wafer, ready for illumination

■ FIGURE 12.9   Alignment of a mask on a wafer by means of marks on the wafer (top line) and on the mask. Third line: mask on wafer, not aligned. Bottom line: The mask and wafer are aligned, ready for illumination.

is used to pattern the masks as well. Since the fabrication of an integrated circuit requires up to 20 repeated applications of photolithography, it is important to align the masks on the wafer with great precision. Precise alignment is achieved by placing marks on the wafer and on the mask; the marks are aligned in a microscope as illustrated in Figure 12.9.

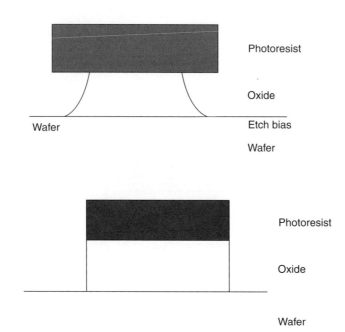

■ **FIGURE 12.10**   Top: Undercutting resulting from wet etching of the oxide layer. Bottom: Clean oxide removal by dry etching.

12.8 **ETCHING**

Access to the bare silicon surface is obtained by etching the oxide layer. **Wet etching** in HF has the disadvantage of undercutting the photoresist in ways that are not acceptable in the fabrication of the very small modern devices. (See Figure 12.10.)

For this reason, the oxide layer is usually patterned by **dry etching**. In this technique, reactive gases are excited in a plasma and are driven at relatively high energies onto the surface. In the geometry of the two electrodes, the electric field and the velocity of the ions are perpendicular to the wafer. This is illustrated in Figure 12.11.

12.9 **DOPING BY ION IMPLANTATION**

Doping is performed by ion implantation. In this technique, the appropriate ions (B for p-type, As or P for n-type) are accelerated to high energies, 30–100 kV, depending on the depth of the desired doping profile and driven into the surface. This technique avoids the high temperatures that would be necessary for diffusion of the ions and provides good control of the penetration depth. Figure 12.12 shows the penetration of some ions as a function of their kinetic energy; Figure 12.13 is a sketch of an

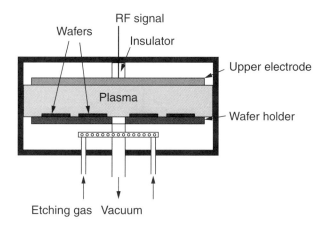

■ FIGURE 12.11   Dry etching apparatus.

(A)   (B)

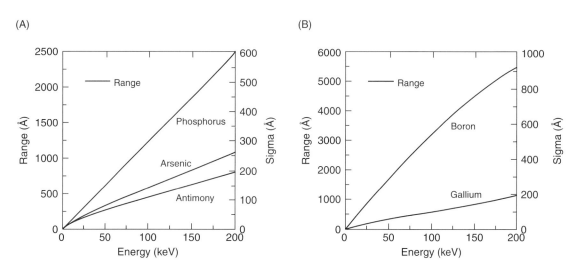

■ FIGURE 12.12   Depth of penetration of various ions into silicon as a function of implantation energy. Adapted from Stephen A. Campbell, *The Science and Engineering of Microelectronic Fabrication* Oxford University Press, Oxford, 2000.

ion implanter. The ion source is an arc chamber in which a gas of BF_3, AsH_3, or PH_3 is decomposed and ionized. The ions are accelerated by a potential of 30–100 kV. The ion beam is bent in a magnetic field at angles depending on the mass of the ions. This serves to separate the desirable ions from impurities. Finally, horizontal and vertical deflection plates guide the beam to the proper place on the wafer. Modern implantation machines can treat several wafers mounted on a spinning disk.

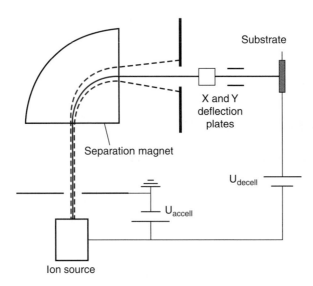

■ **FIGURE 12.13** Schematic of an ion implanter. The separation magnet eliminates undesirable elements.

The high impact energy of the ions has the undesired effect of causing lattice damage. For this reason, the wafer is **annealed** after each implantation. This annealing also has the beneficial effect of moving the implanted ions into substitutional sites and rendering them electrically active as donors and acceptors.

12.9.1 **Compensation**

Examine Figure 12.3. In the MOSFET, highly doped n-type sources and drains must be made where a p-type layer had previously been made. This is possible because of the phenomenon of **compensation**. The p-type layer is made first by the implantation of, say, N_A boron atoms. This introduces $p \approx N_A$ holes. Subsequent implantation of N_D phosphorous atoms generates $n \approx N_D$ electrons. If the number of donors is larger than the number of acceptors, the holes recombine with electrons, leaving

$$n \approx N_D - N_A \tag{12.3}$$

electrons in the conduction band: the source is n-type.

12.10 **DEPOSITION OF INTERCONNECTS AND INSULATING FILMS**

There is a need to **contact** and **interconnect** all of the doped semiconductor electrodes (e.g., emitter, base, and collector in bipolar transistors, or source, drain, and gate in field effect transistors) to other devices on the chip.

One first deposits **contacts** or **plugs** into holes made for this purpose in the oxide layers. Many materials have been tried; the ones used to make the contacts with the silicon include Al and Al alloys, W, PtSi, Pd_2Si, $TiSi_2$, and so on. The **interconnection lines** between contacts are deposited in a separate step; they are invariably made of Al or Al-Cu alloys.

The materials are chosen because they possess the following desirable properties:

1. High electrical conductivity.

2. Low contact resistance to Si.

3. No tendency to react chemically with, or diffuse into, Si.

4. Resistance to corrosion or environmental degradation.

5. Ease of deposition.

6. Compatibility with other materials and steps (e.g., lithography, etching) in the fabrication process.

12.10.1 **Physical Vapor Deposition**

The metallic interconnects are deposited by the physical vapor deposition (PVD) techniques of **evaporation** and **sputtering**. Both techniques have been discussed in Chapter 6. Evaporation is the simpler of these techniques and involves heating the source metal until it evaporates at appreciable rates. Electron-beam evaporation is utilized generally in semiconductor processing because it avoids the deposition of material from the heating elements. A great advantage of sputtering is the maintenance of stoichiometry in deposited alloy films.

Sputtering and evaporation are line-of-sight methods: the atoms or molecules travel in straight lines toward the material to be coated. When the latter has a complex geometry, with holes or overhangs, part of the surface may not be reached by the straight-moving atoms. For such geometries, one employs chemical vapor deposition.

12.10.2 **Chemical Vapor Deposition**

In microelectronic processing, insulating films are deposited by chemical vapor deposition. In this process, the coating is deposited by a chemical reaction of precursor gases. Typical reactions with precursor gases and products include:

$$\text{for } SiO_2: Si\,(OC_2H_5)_4 \rightarrow SiO_2 + \text{by-products @ 700°C}$$
$$\text{for silicon nitride: } 4NH_3 + 3Si_2Cl_2H_2 \rightarrow Si_3N_4 + 6HCl + 6H_2 \text{ @ } \sim 750°C$$
$$SiH_4 + NH_3 \rightarrow SiNH + 3H_2 \text{ @ 300°C}$$
$$\text{for phosphosilicate glass:}$$
$$SiH_4 + 4PH_3 + 6O_2 \rightarrow SiO_2 \cdot 2P_2O_5 + 8H_2 \text{ @ 450°C}$$
$$\text{for tungsten plugs in contacts: } WF_6 + 3H_2 \rightarrow W + 6HF$$

The wafer must be heated for the reaction to proceed. The formation of a quality film requires that the reaction producing the solid coating occur only on the surface to be coated and not in the gas phase in order to avoid the formation of grains of solid that would fall onto the wafer. This is achieved either by a low gas pressure or by diluting the precursor gases in an inert gas. Reactors employed are similar to those used to oxidize Si (Figure 12.8). There are several variants of CVD: hot wall reactor in which the reactor tube is heated together with the substrates; cold wall reactors where only the wafers are heated; and plasma-enhanced CVD where a plasma stimulates the reaction.

12.10.3 Epitaxy

When atoms are deposited onto a solid surface and reach thermal equilibrium, they continue the crystal structure of the substrate. This is called **epitaxy** (Figure 12.14). The deposition of silicon on a silicon substrate is called **homoepitaxy**; **heteroepitaxy** is the deposition of crystalline layers of dissimilar materials that possess similar crystal structures. The film and substrate must have very closely matched lattice constants; otherwise the interface will contain dislocation defects that impair charge transport or seriously degrade light emission and absorption. For high quality devices, the lattice constants of the two materials must differ by less than 0.2%.

Epitaxial films are deposited by chemical vapor deposition (Figure 12.15) or by molecular beam epitaxy (MBE, Figure 12.16). Many semiconductors, especially the III-V compounds used in LEDs or solid state lasers, are grown by **metalorganic vapor phase epitaxy**. Deposition of the layer occurs in a CVD reactor with the use of metalorganic gases such as phopshine (PH_3), arsine (AsH_3), trimethylindium

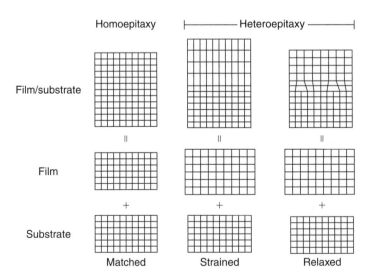

■ FIGURE 12.14   Atomic arrangements in the film and substrate during homoepitaxy, strained film heteroepitaxy (no interfacial defects), and heteroepitaxy. Large lattice misfits give rise to interfacial dislocations that may propagate into the film.

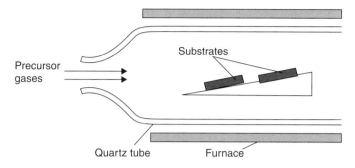

■ **FIGURE 12.15**   Schematic eptiaxial growth by chemical vapor deposition. The gases react on the surface of the substrates and deposit the desired material.

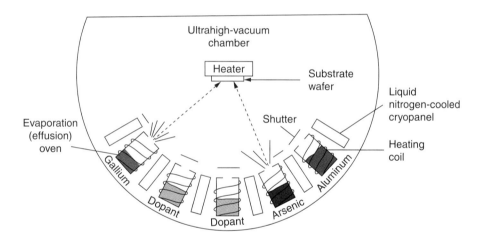

■ **FIGURE 12.16**   Schematic of molecular beam epitaxy.

$(In(CH_3)_3,$ trimethyl gallium, or trimethyl aluminum. The **molecular beam epitaxy (MBE)** method depends on the highly controlled simultaneous thermal evaporation of the atoms involved. These emanate as beams from different heated sources under extraordinarily clean high vacuum conditions. By MBE methods a 1 μm thick film is typically grown in an hour; CVD methods take about a minute to deposit the same film thickness.

A combination of a low film nucleation rate and a high film growth rate is critical to ensuring single crystal formation. This is achieved by low deposition rates and high substrate temperatures. A low gas supersaturation and deposition rate will create few nuclei. High substrate temperatures enhance diffusion rates and facilitate atomic incorporation into lattice sites. This promotes the early lateral extension of the single crystal film and growth perfection as it thickens.

12.11 MEMS (MICRO ELECTRO-MECHANICAL SYSTEMS)

The surface processing techniques that have been developed for the integrated circuits are now being used for the fabrication of other microscopic devices such as the accelerometers that control the safety bags in automobiles, the microscopic mirrors in some television displays, microscopic pumps, and valves. Since this technology is very recent, a rapid development into other applications can be expected.

The fabrication of these devices is based on the photolithography of silicon. This feature permits the integration of the MEMS and the associated electronic circuit on the same silicon chip (Figure 12.17). Some differences exist between the processing of electronic circuits and that of MEMS. One such difference is the need to remove large amounts of material in shaping MEMS. Therefore, chemical etching is used more prominently than in integrated circuits. The more important modification is the need to fabricate hollow structures in a pressure sensor (Figure 12.18) or moving parts in an accelerometer (Figure 12.19). The processing of these elements makes use of a sacrificial layer.

■ **FIGURE 12.17** Microchip integrating a MEMS pressure sensor (center) with the associated electronic circuits. Gordon Bitko and Andrew McNeil, Motorola, Randy Frank, ON Semiconductor.

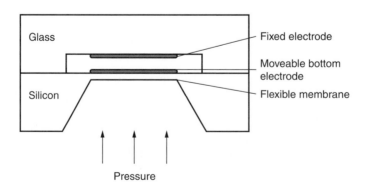

■ **FIGURE 12.18** Schematic presentation of a pressure sensor based on changes in capacitance.

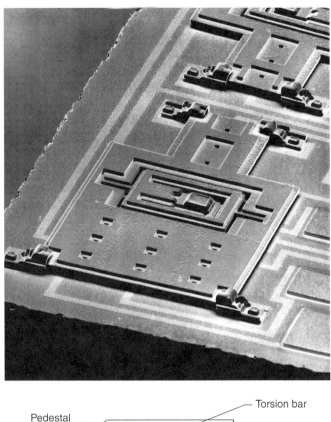

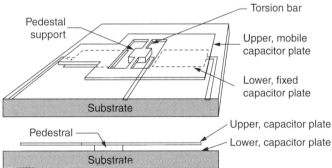

■ **FIGURE 12.19**   Photograph and diagram of a MEMS accelerometer. Courtesy of Silicon Designs. Inc.

12.11.1 **Sacrificial Layers**

In order to create a void in a device or a free space underneath a cantilever, one deposits a sacrificial layer consisting of a material that is easily removed by dissolution. Materials used for sacrificial layers include photoresist, aluminum, or SiO_2. Silicon oxide has the advantage of resisting the high temperatures required in other processing steps. It is generally deposited by low-pressure CVD and doped with phosphorus to create a phospho-silicate glass. Once the sacrificial layer has been deposited and patterned by photolithography, one continues the processing of the device by depositing polycrystalline silicon on top of it. The sacrificial layer is then removed by wet etching. Figure 12.20 shows the steps in the fabrication of a cantilever beam. In the fabrication of a structure with a void, one provides holes in one of the layers in order to permit the etchant to reach the sacrificial material.

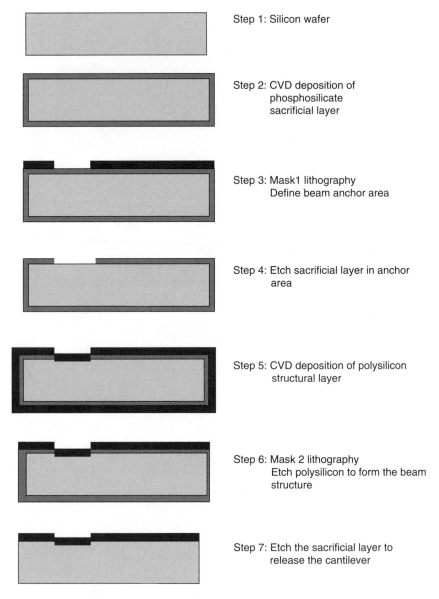

Step 1: Silicon wafer

Step 2: CVD deposition of
phosphosilicate
sacrificial layer

Step 3: Mask1 lithography
Define beam anchor area

Step 4: Etch sacrificial layer in anchor
area

Step 5: CVD deposition of polysilicon
structural layer

Step 6: Mask 2 lithography
Etch polysilicon to form the beam
structure

Step 7: Etch the sacrificial layer to
release the cantilever

■ **FIGURE 12.20**   The fabrication of a cantilever beam with help of a sacrificial layer.

Most liquids wet the surface of silicon. As the etching liquid recedes from the void, the surface tension pulls the two surfaces together. If the two surfaces touch each other, the surface tension of the remaining liquid prevents the two surfaces from separating and the device is useless. This is known as **stiction** and can greatly reduce the yield of the MEMS fabrication.

■ SUMMARY

1. Integrated circuits are fabricated on the surface of a very pure single crystal silicon wafer.

2. The microscopic transistors are fabricated by photolithography where the design is transferred to the chip by light.

3. A photolithographic sequence is as follows: grow an oxide layer, deposit a photoresist, apply a mask, illuminate, develop the photoresist, etch the oxide, dope, and anneal. As many as 15 such sequences may be required in the processing of a chip.

4. Very pure silicon is obtained by transforming it to a gas that is purified by fractional distillation.

5. Single crystal boules are made by the Czochralski method where the single crystal grows around a seed that is immersed in the melt.

6. Oxide layers, used for the processing of the transistors, are formed by heating in humid air.

7. Masks define the area to be illuminated. They must be carefully aligned on the chip.

8. The oxide layer is removed by dry etching in an active gas plasma.

9. Doping is achieved by ion implantation where ions are driven into the surface after being accelerated to high energies.

10. The doping of silicon can be changed by compensation where the dopant with the higher concentration determines the conduction type.

11. Metallic interconnects are deposited by physical vapor deposition, either by evaporation or by sputtering of the metal.

12. Insulating layers are deposited by chemical vapor deposition where a reactive gas decomposes on the surface and deposits the desired material.

13. Epitaxy is the deposition of a layer that continues the crystal structure of the substrate.

14. MEMS are microscopic mechanical devices; they are fabricated by the same techniques as integrated circuits.

■ KEY TERMS

A
Al, 357
Al–Cu alloys, 357
annealing, 348, 356

B
boule, 347

C
chemical vapor deposition, 351, 357

chip, 344
compensation, 356
contact, 356
crosslink, 352

■ REFERENCES FOR FURTHER READING

[1]   S.A. Campbell, *The Science and Engineering of Microelectronic Fabrication*, 2nd ed., Oxford University Press, New York (2001).

[2]   C.R.M. Grovenor, *Microelectronic Mateirals*, Adam Hilger, Bristol, UK (1989).

[3]   R.C. Jaeger, *Introduction to Microelectronic Fabrication*, Vol. 5, Addison-Wesley, Reading, MA (1988).

[4]   M. Madou, *Fundamentals of Microfabrication: The Science of Miniaturization*, 2nd ed., CRC Press, Boca Raton, FL (2002).

[5]   N. Maluf, K. Williams, *An Introduction to Microelectromechanical Systems Engineering*, 2nd ed., Artech House Publishers, Norwood, MA (2004).

[6]   J.W. Mayer, S.S. Lau, *Electronic Materials Science: For Integrated Circuits in Si and GaAs*, Macmillan, New York (1990).

[7] M. Ohring, *The Materials Science of Thin Films*, 2nd ed., Academic Press, San Diego, CA (2001).

[8] W.S. Ruska, *Microelectronic Processing*, McGraw-Hill, New York (1987).

[9] D.L. Smith, *Thin-Film Deposition: Principles and Practice*, McGraw-Hill Professional, New York (1995).

■ PROBLEMS AND QUESTIONS

12.1. When a chip is a square of 15 mm length, what length does a single transistor have if the chip contains 1.5 billion transistors?

12.2. Describe how to etch your signature onto a piece of metal by photolithography. Fabricate the mask and select the photoresist (positive or negative): (A) so that your signature is engraved into the metal and (B) so that your signature stands proud of the metal. (C) How would you proceed to make your signature not larger than 1 mm?

12.3. Explain why Intel is developing an extreme ultraviolet (EUV) photolithography technique in order to continue the trend of Moore's law.

12.4. Explain why wet etching is out of the question to produce modern integrated circuits. Sketch how the etching proceeds with time in wet and in dry etching.

12.5. One Si wafer has a concentration of 2×10^{18} antimony atoms/cm^3, while a second Si wafer contains 1.5×10^{18} gallium atoms/cm^3.
   a. What is the majority carrier in each wafer?
   b. What is the majority carrier in a third wafer where the same area is doped with both elements at the above concentrations?
   c. Estimate the resistivity of the three wafers (use Figure 11.13).

12.6. Explain what would happen to the contour of a single crystal if, during Czochralski growth, the melt temperature were to suddenly rise due to a faulty temperature controller. What would happen if the melt temperature were to drop momentarily?

12.7. Calculate the value of the misfit and describe the state of stress in the following single crystal film-substrate combinations:
   a. An AlAs film layer on a GaAs substrate.
   b. A CdS film layer on an InP substrate.
   c. A ZnSe film on a GaAs substrate.

12.8. Explain how ion implantation can be used to make a p-n junction.

12.9. Semiconductor technology has helped to spin off new high-tech materials industries, processes, and equipment. Give one example of each.

12.10. In the deposition of contacts and connecting lines, Section 12.10, which of the criteria 1 to 6 apply to the contacts and which apply to the interconnecting lines? Explain the different choices of materials on the basis of these criteria.

12.11. Explain why the same mask can be used in the deposition of interconnects for either limiting metal vapor deposition to interconnects or for depositing the metal everywhere and etching away the unwanted material.

12.12. Why does one deposit the contact metal (the plug) and the interconnects in two different photolithography steps?

12.13. At first the doping of semiconductor devices was done by diffusion of the dopants from the surface. Why is one forced to use ion implantation in modern integrated circuits?

12.14. Sketch the penetration of high-energy dopant ions into a silicon lattice and justify the need to anneal the material after implantation to repair the damage done to the crystal and to ensure that the dopant atoms sit in electrically active substitutional sites.

12.15. Compare sputter deposition and chemical vapor deposition and cite the advantages and disadvantages of each. (Include the cost of equipment and process.)

12.16. Describe the sequence of processing steps in the fabrication of the pressure sensor (Figure 12.18).

12.17. Find and describe the size of transistors used today **and** the radiation used in photolithography.

12.18. Find and describe a present example of MEMS.

12.19. Find and describe a present example of NEMS (Nano Electro-mechancial System).

12.20. How is it possible to fabricate integrated circuits containing more than 1 billion transistors and do it for a reasonable price?

12.21. What are the dimensions of a wafer on which integrated circuits are made? What materials is it made of? What is its purity?

12.22. Why must one make the silicon with as few structure defects as possible?

12.23. How is silicon for integrated circuits purified?

12.24. What is the name of the process used in the fabrication of integrated circuits?

12.25. What is a photoresist? What is it used for and how does it function?

12.26. What is a mask, how is it used, and how does it function?

12.27. How is the design of a transistor element reduced from a macroscopic design to its nanometer dimensions on the wafer?

12.28. Describe the sequence of steps in the photolithographic fabrication of, say, the source of a transistor?

12.29. How many transistors are made simultaneously on a wafer?

12.30. With the speed of a modern computer, how far can light travel between two impulses?

12.31. What is the role of the oxide layer in the processing of integrated circuits? How is this oxide layer deposited?

12.32. How is doping achieved in modern fabrication?

12.33. Why must one anneal the wafer after ion implantation?

12.34. How and why is it possible to make a p-type element in a material that has been doped n-type?

12.35. Describe chemical vapor deposition.

12.36. Describe epitaxy.

12.37. What is a MEMS, and by what processes is it fabricated?

12.38. What is a sacrificial layer in MEMS processing?

Chapter 13

Optical Materials

Optical materials form a large and important class of functional materials. Metals are used as mirrors. Insulating materials, when transparent, refract the light and are used in lenses and reflectors; insulators absorb a fraction of the visible spectrum and serve as paint or light filters. When they are excited with electrons, insulators are capable of emitting light; these are the phosphors. Semiconductors, mainly through the use of p-n junctions, form the class of electro-optical devices. They serve as light detectors and solar cells; they are also capable of acting as cold light sources in the form of light-emitting diodes (LEDs), solid-state lasers, and electroluminescent light sources. Liquid crystals are used as light polarizers in optical displays, including computer monitors and flat TV screens. Organic semiconductors are being developed for use in large-area solar cells and light-emitting panels. Research is underway to develop nanostructured composites to produce solar cells with increased efficiency.

LEARNING OBJECTIVES

After studying this chapter, the student will be able to:

1. Explain how any color we perceive can be created with three light sources or three pigments and how this fact is related to the physiology of the human eye.

2. Relate the electronic band structure of a solid to its optical properties.

3. Relate electron excitation to absorption of light and the energy-loss of electrons to light emission.

4. Relate light refraction, reflection, and total internal reflection to the polarization of the solid by the electromagnetic field of light.

5. Select a proper material for efficient light reflection.

6. Select a material as a polarizer of light.

7. Describe the operation of a phosphor and its uses.

8. Describe the fabrication and operation of a light detector and a solar cell.

9. Describe the fabrication and operation of LEDs and solid-state lasers.

10. Select materials and describe the fabrication of a large-area OLED.

11. Describe the operation of the screen of a television set or a computer monitor.

12. Describe the properties and manufacture of optical fibers.

13.1 USES OF OPTICAL MATERIALS

Let us examine the **flat screen monitor** of your computer. Touch it, it is not hot: its light is provided by an electroluminescent material that emits light without heat. The screen is composed of a large number of "pixels" or units of display. Each pixel consists of three cells equipped one with a red, the other with a green, and the third with a blue filter. Figure 13.1 shows the operation of one of the cells. The light from the electroluminescent panel is linearly polarized by a polarizing film P and passes through a thin layer of liquid crystal that rotates its plane of polarization by 90°; then it passes through a second polarizer (A) oriented at 90° to the first. The light can thus be transmitted. The liquid crystal is placed between two electrodes. When an electric field E is applied, the liquid crystal molecules are oriented as shown in Figure 13.1B and are unable to rotate the polarization of the light. Because the polarizers are crossed, the light is not transmitted and the screen is dark. In order to provide the color of the screen, the light passes ultimately through colored filters. (We have selected a blue filter in Figure 13.1.) When all three cells in a pixel transmit, the eye perceives the added colors as white. Different intensities of the light transmitted through the three filters provide the vast array of colors we see. For the flat screen monitor, we thus need four different optical materials: an electroluminescent light source, polarizers, liquid crystal, and filters.

Telephone conversations are transmitted by an **optical fiber** system (Figure 13.2). In this technology, the electrical signals carrying the information are transformed into optical signals by a solid state laser; the optical signal is transmitted by an optical fiber. Every 45 km, a repeater, consisting of an optically pumped laser, amplifies the light. Finally, a photodiode transforms the light signal again into an electric signal. This system requires optical fibers that are extremely transparent and configured in such a way that the light is reflected internally instead of being absorbed. It also requires a laser that emits the light at a wavelength corresponding to the highest transparency of the glass fiber; it also requires a photodiode and a light amplifier consisting of an erbium-doped glass laser that is optically pumped.

These are just two applications of optical materials. Other applications include solar cells, mirrors, lenses, and eyeglasses, heat-reflecting glass panels, anti-reflex coatings on camera lenses and, finally, paint.

In the following, we first review the nature of light and color vision, then we examine the interactions of light with materials, and finally we describe the major applications of these phenomena.

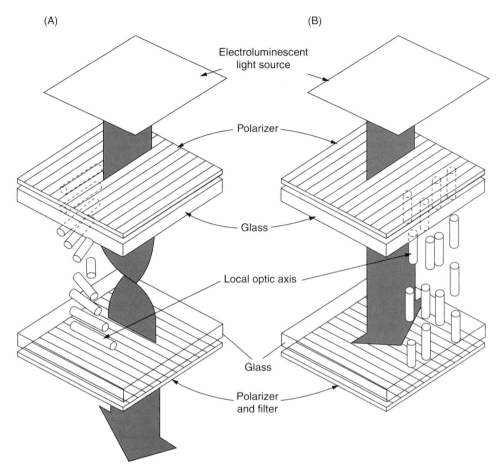

■ FIGURE 13.1 Liquid crystal display. The display unit consists of an electroluminescent plane, two crossed polarizers, two glass plates that orient a nematic liquid crystal (in such a way that, at rest, their orientation is twisted by 90°). Left: no field applied, the liquid crystal rotates the polarization of the light so that the unit is transparent. A filter produces the light of selected color. Right: an electric field E orients the nematic molecules so that they do not interact with the light. The polarization is not rotated, and the light is blocked by the crossed polarizers; the unit is dark.

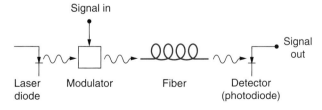

■ FIGURE 13.2 Sketch of an optical communication system.

13.2 **LIGHT AND VISION**

Light has a dual nature: it is a high-frequency electromagnetic wave with electric and magnetic field vectors perpendicular to its propagation; it consists of individual photons that each carry the energy

$$E_{ph} = h\nu = \frac{hc}{\lambda} \tag{13.1}$$

where E is the photon energy, ν the frequency, λ the wavelength, and c the speed of the light; h is Planck's constant. When λ is expressed in micrometers, the photon energy in electron volts is

$$E_{ph}(eV) = 1.24/\lambda \ (\mu m) \tag{13.2}$$

The electromagnetic spectrum ranges from radio waves to the infrared, visible, ultraviolet, X-ray, and gamma ray regions. The wavelength of visible light ranges from about 0.4 to 0.6 μm and its photon energy from about 2 to 3 eV.

We perceive the color of light by three cones in the eye that respond to different regions of the spectrum as shown in Figure 13.3. The sensitivity of the red cones peaks at 564 nanometers, that of the green cones at 533 nm, and that of the blue cones at 437 nm. The more sensitive night-vision rods respond to light at 498 nm. All colors that we perceive are generated in our brain by the addition of the signals from these three cones. Adding red to green signals, for instance, produces our perception of yellow. **This mode of vision permits the production of any color with three light sources of wavelengths roughly corresponding to the peak sensitivity of the cones.**

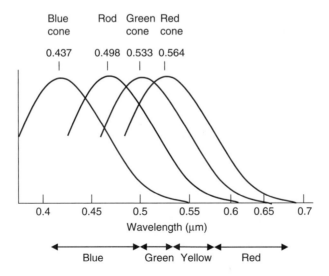

■ **Figure 13.3** The spectral sensitivities of the three cone types and the rods in the primate retina.

13.3 **INTERACTION OF LIGHT WITH ELECTRONS IN SOLIDS**

All classes of engineering materials—metals, ceramics, semiconductors, and polymers—have representatives with useful optical properties. These are all based on the interaction of electromagnetic fields of the light with the electrons of the material. On this basis, optical properties are determined by the electronic structure of the material.

13.3.1 **Absorption of Light**

When light acts on an electron that has energy E, a photon can be absorbed if there are empty orbitals at energy $E + h\nu$ into which the electron can be excited. If no such energy level is available to the electron, the photon is not absorbed (see Figure 13.4).

When light is absorbed, its intensity (commonly in units of power density, e.g., W/m^2) decreases as it travels a distance x through an absorbing medium

$$I = I_o \exp - \alpha x \tag{13.3}$$

where α is the absorption coefficient (in units of m^{-1} or cm^{-1}). The absorption coefficient is proportional to the density of electrons capable of absorbing the photon; therefore it changes with the wavelength of the light. It is useful to define the dimensionless **index of absorption** (k) that is related to α and the wavelength of light λ by

$$k = \alpha\lambda/4\pi \text{ or } \alpha = 4\,\pi k/\lambda \tag{13.4}$$

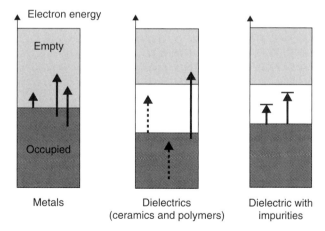

■ **FIGURE 13.4**   Light absorption in solids. Left: Metals are characterized by a partially filled band; their electrons are free to acquire any higher energy; they absorb all light. Middle: Dielectrics possess a filled band separated by an empty band by a relatively large energy gap; a pure dielectric can absorb light only with photon energy larger than the band gap: $h\nu > E_G$. Right: Impurities introduce local energy levels in the band gap to or from which electrons can be excited by absorbing light. These levels absorb light of specific wavelengths and are responsible for the color of transparent materials.

In **metals**, a great number of electrons are available to interact with electromagnetic radiation. They occupy a partially filled band that contains a continuum of empty states into which the electrons can be excited. Therefore **metals absorb all visible and infrared light**; α and k are large at all wavelengths.

Ceramics and polymers have a full valence band separated from a higher, empty band by an energy gap E_G. Light is strongly absorbed only when its photon energy is larger than the band gap

$$hv > E_G$$

Many ceramics and polymers have a band gap that exceeds the energy of visible light; these are transparent to visible light and absorb only in the ultraviolet. Figure 13.5 shows the spectral regions of high transparency in important optical materials.

When the band gap E_G corresponds to a photon energy in the visible region, the short-wavelength light is absorbed and the rest is transmitted: the material is colored. Cadmium sulfide, for example,

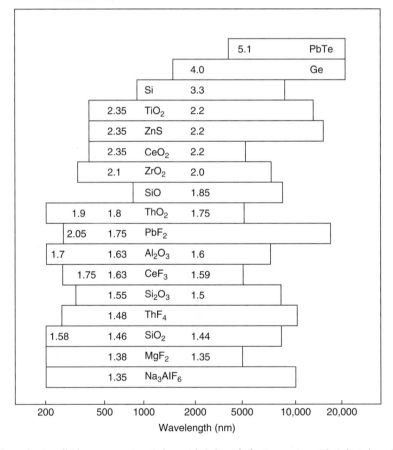

■ **FIGURE 13.5**   Spectral region of high transparency in optical materials. Indices of refraction are given at the indicated wavelength. Reprinted with permission from G. Hass and E. Ritter, *Journal of Vacuum Science and Technology*, Vol. 4, p.71 (1967) Copyright [1967], American Vacuum Society.

has a band gap $E_G = 2.43\,\text{eV}$ that corresponds to a wavelength of 510 nm. Cadmium sulfide absorbs light of shorter wavelength and is transparent to red, yellow, and green light. This excites the red and green cones in the human eye: transmitted light gives the material a yellow color. Other solids, with smaller band gaps, transmit red light only.

Impurities introduce local orbitals with discrete energy levels within the band gap. These absorb light with a specific photon energy (Figure 13.4, right). This absorption decreases the transparency of the material but can also produce a color as in pigments and gemstones.

13.3.2 **Color**

Silica glasses are colored by incorporating transition metal ions during melting in amounts ranging from 0.1 % to ~5%. The ions substitute for matrix atoms and create new levels in the band gap of the glass. Specific wavelengths are now preferentially absorbed at these impurity levels, giving rise to optical transitions at specific photon energies. Gold and copper form nanometer-sized particles that absorb light at frequencies that depend on their size (see Chapter 18). Specific wavelengths are now preferentially absorbed at these impurity levels giving rise to optical transitions at specific photon energies. Thus V^{4+} as well as colloidal Au and Cu color glass red; Fe^{2+}, Co^{2+} color glass blue; Cu^{2+} colors glass blue-green; Fe^{3+} colors glass yellow green; Mn^{3+} colors glass purple, and so on.

In the production of color, we distinguish between pigments, which are not soluble in the support, and dyes, which are soluble. Pigments are ceramics (minerals) containing impurities. Dyes are large organic molecules, either natural (such as chlorophyll) or synthetic. As a rule, the organic dyes are subject to bleaching: the absorbed light induces photochemical reactions that transform the molecules and remove the absorbing energy levels. Ceramic pigments are not susceptible to photochemical reactions and do not bleach. This accounts for the remarkably vivid colors in medieval paintings which were painted with mineral pigments.

13.3.3 **Refraction**

When the light cannot be absorbed, it nevertheless interacts with the electrons in the solid. We saw in Chapter 11 that an electric field causes a polarization in insulators. This polarization decreases the electric field inside the solid by the dielectric constant κ. The electric field of the light produces a similar polarization, with frequency ν, that decreases the propagation velocity of the light by the factor $n = \sqrt{\varepsilon}$. The change in light velocity on entering the solid causes it to change direction; **light is refracted**. Refraction is expressed by Snell's law

$$n(1)/n(2) = \sin\phi_2/\sin\phi_1 \qquad (13.5)$$

where $n(1)$ and $n(2)$ are the refractive indices in adjacent media 1 and 2; ϕ_1 is the angle of incidence on the solid and ϕ_2 is the angle of propagation in the solid (see Figure 13.6). Note that n also varies with the wavelength of the light, a phenomenon known as **dispersion**. Blue light has a higher index of

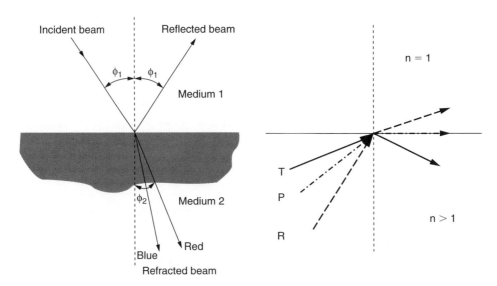

■ FIGURE 13.6 Optical effects at the surface of a transparent medium with index of refraction n. Left: The incident beam is reflected at the surface with intensity expressed by Equation (13.6). The penetrating beam is bent (refracted) according to Snell's law, Equation (13.5). The index of refraction for red light is smaller than for blue light, causing the dispersion of the refracted light. Right: Light transmitted from medium 2 with $n > 1$ into air ($n = 1$). R: refracted beam, P: light exits parallel to the surface; T: total internal reflection, beam cannot escape and is totally reflected back into the glass.

refraction n than red light: it travels more slowly and bends more as shown in the figure. (This wavelength-dependence of refraction is the cause of rainbows and dispersion by prisms). Table 13.1 presents the refractive indices of the main optical materials.

The largest application area of refraction is in lenses for eyewear and optical equipment. High-performance lenses are usually made of glass. Eyeglasses are made of polymers because of their lower weight. Polyethylene is the least expensive material. Polycarbonates (Figure 13.7) are more expensive but hard and impact resistant.

Photochromic lenses become dark under illumination by ultraviolet light. This effect is obtained by incorporating silver halide crystals in the glass.

13.3.4 **Reflection of Light**

Wave theory shows that an abrupt change in the propagation of light at an interface, by refraction or by absorption, causes a certain fraction R of the wave to be reflected. (This is as true for sound, radar, and water waves as for light.)

When light impinges from air ($n = 1$) onto the surface of a solid with index of refraction n and index of absorption k, the reflection coefficient is

$$R = [(n - 1)^2 + k^2]/[(n + 1)^2 + k^2] \qquad (13.6)$$

Table 13.1 Refractive Indices of Various Materials

Glasses		Polymers		Ceramics	
Material	**Refractive index**	**Material**	**Refractive index**	**Material**	**Refractive index**
Silica glass	1.46	Polyethylene	1.51	CaF_2	1.434
Soda lime glass	1.51	Polypropylene	1.49	MgO	1.74
Borosilicate glass	1.47	Polystyrene	1.60	Al_2O_3	1.76
Flint optical glass	1.6–1.7	Polymethyl methacrylate	1.49	Quartz (SiO_2) Rutile (TiO_2)	1.55 2.71
				Litharge (PbO)	2.61
				Calcite ($CaCO_3$)	1.65
				$LiNbO_3$	2.31
				PbS	3.91
				Diamond	2.42

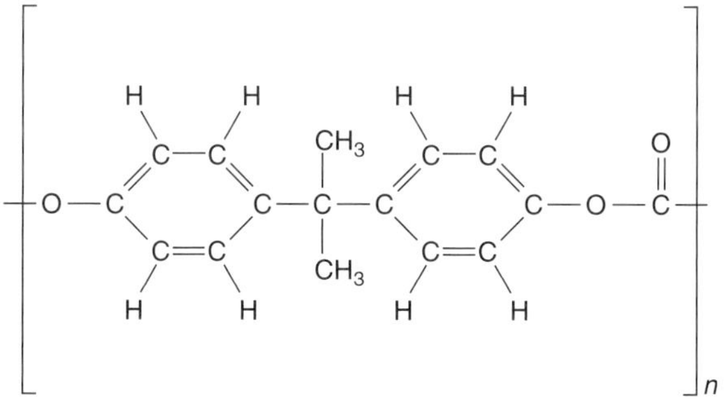

■ **FIGURE 13.7** The chemical formula of Polycarbonate A (Lexan ® by General Electric).

In transparent materials, where $k = 0$, the equation simplifies to

$$R = (n - 1)^2/(n + 1)^2 \qquad (13.7)$$

At the interface between two transparent materials with refraction indices n_1 and n_2, the index of reflection is

$$R = (n_1 - n_2)^2/(n_1 + n_2)^2 \qquad (13.8)$$

For glass with $n = 1.5$, $R = 0.5^2/2.5^2 = 0.04$. Thus 4% of the light is reflected from the surface. Even this small amount is an intolerable loss in many optical applications; as we shall see later, it is reduced through the use of antireflection coatings.

Metallic Mirrors   Metals have a high index of absorption, and their reflection coefficient can approach 99% according to Equation (13.6). Modern mirrors consist of thin metal films or coatings (typically less than 1 μm thick) deposited on the back of glass substrates. High reflectivity in the visible spectrum is the chief property required of metals for mirror applications, and Figure 13.8 provides such information for the most widely used metals. The absorption coefficients α and k are proportional to the density of electrons capable of acquiring the photon energy $h\nu$. Depending on the details of the electron band structure, the absorption and reflection can vary with the photon energy. Thus, gold and copper (and alloys of these metals) are significantly colored. Aluminum and chromium have a faint bluish, and nickel a weak yellowish, coloration. Most other metals reflect all portions of the visible spectrum and thus appear white.

In front-surface mirrors, resistance to mechanical scratching (of Au, Ag, Cu, Al), oxidation (of Al), and tarnishing (of Ag) are additional concerns. Despite its relatively low reflectivity, rhodium, a metal that is more expensive than gold, has found application in telescope mirrors, optical reflectivity standards, and in mirrors for medical purposes. The reason is due to the high hardness and environmental stability of this metal.

EXAMPLE 13.1   *(A) The optical constants of silver at a wavelength of 0.55 μm (550 nm) are n = 0.055 and k = 3.32. What is the reflectance (i.e., fraction of light reflected) of silver at this wavelength? (B) How thin does silver have to be to transmit 50% of the light that is not reflected?*
ANSWER   (A) Silver is an absorbing material, so Equation (13.6) applies. Substitution yields

$$R = [(0.055 - 1)^2 + 3.32^2]/[(0.055 + 1)^2 + 3.32^2] = 0.982$$

The high reflectivity of silver earmarks its use for mirrors.

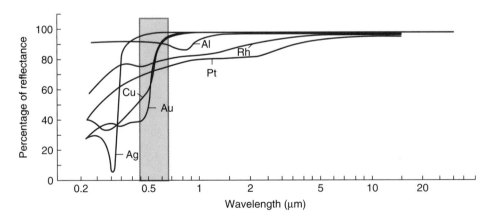

■ **FIGURE 13.8**   Reflection coefficient of some metals. (Visible range is shaded.)

(B) From Equation (13.3), $x = -\alpha^{-1}\ln I/I_o$ or $x = -(4\pi k/\lambda)^{-1}\ln(I/I_o)$. Substituting,

$$x = -(4\pi \times 3.32 / 0.55)^{-1} \ln (0.5) = 9.14 \times 10^{-3} \text{ } \mu\text{m or } 9.14 \text{ nm}$$

A very thin layer of silver absorbs much of the light. Silvered mirrors that transmit some light in assorted optical applications are composed of a thin film deposited on a glass substrate. Such films are usually prepared by evaporation.

13.3.5 **Total Internal Reflection**

When the light moves from a medium with **larger** index of refraction to one with **small** index, for instance when it exits a transparent solid, the refracted light makes a larger angle with the surface normal, as shown by the rays marked R in Figure 13.6 at the right. There is an incident angle, marked P, at which the refracted light is parallel to the surface. Equation 13.5, with $\phi_2 = 90°$ and $n = 1$ in air, gives $\sin\phi_P = 1/n$ in the solid. When the incident light makes a larger angle with the surface normal (marked T), the light cannot escape the solid and is totally reflected. This phenomenon is used in the fabrication of mirrors in optical devices (cameras), jewelry, in cut glass, and in optical fibers.

An interesting modern variation is **frustrated total internal reflection**. When the surface of the glass is in contact with an absorbing material, the light is not reflected by the surface, but absorbed. This is used in modern fingerprinting machines where one presses the finger against the surface of a prism and takes a photograph of the reflected image that is dark where the finger touches the glass. (Figure 13.9).

Jewels and Cut Glass   Diamond has a relatively large index of refraction $n = 2.42$. Light impinging the surface from within with an angle larger than 25° from normal is totally reflected and captured inside the diamond. The diamond is cut so that it can take advantage of this fact: light enters the diamond through all facets but can escape through few facets, throwing the well-known lights of diamond. Zircon ($ZrSiO_4$) is a less expensive transparent material with an index of refraction $n = 2$. It is therefore used to produce the same effect as diamond for much lower cost.

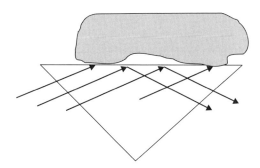

■ **FIGURE 13.9**   Taking fingerprints by frustrated total internal reflection. Light falling on the body is absorbed and not reflected.

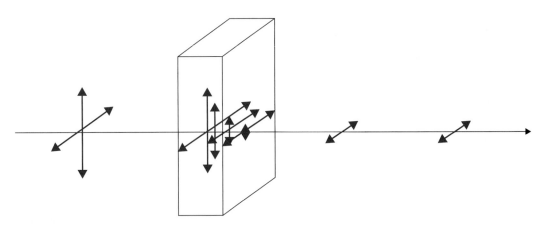

■ **Figure 13.10**   Linear polarization of light by absorption. The vertical component of the electric field is absorbed by the polarizer and the horizontal component is transmitted.

Lead oxide has a large index of refraction $n = 2.61$. Glass of high refractive index is produced by dissolving PbO. This is the "crystal glass" used in cut glass, which is cut in a way to produce the same scintillating effect as diamond. (Note that "crystal glass" is not crystalline, but amorphous.)

13.3.6 **Polarization**

In natural light, the electric field is oriented in all directions perpendicular to its propagation (see Figure 13.10). When a material is anisotropic its response to the electromagnetic field of light can be anisotropic as well. Either refraction or absorption can vary with the orientation of the light's electric field with respect to the material. These effects can be utilized to produce polarized light as shown in the figure.

Polaroid is the best-known example of such a material. It consists of aligned molecules of poly(vinyl alcohol) doped with iodine. The molecules are aligned by heating and stretching a sheet of the material. The sheet is subsequently dipped in iodine. The latter attaches to the polymer molecules and provides free electrons; these can move along the molecules, but not perpendicularly to them. These free electrons absorb light with electric vector along the molecule but not light with electric field perpendicular to it.

Liquid crystals (see, e.g., Figure 13.11) consist of long, rigid, polar organic molecules that are free to move as in a liquid but are aligned parallel to a vector called the **director**. The long hydrocarbon molecules absorb light whose electric vector is parallel to their axis but does not absorb light with the electric field perpendicular to it. A nematic liquid crystal, therefore, is a polarizer.

When the liquid crystal is **twisted**, it rotates the polarization of linearly polarized light. This is shown in Figure 13.1. In **liquid crystal displays**, a liquid crystal is contained between two glass plates. One plate has parallel grooves that align the molecules in contact with it. The second plate contains grooves oriented at 90° with respect to the ones in the first plate. The molecules in the liquid crystal are slightly

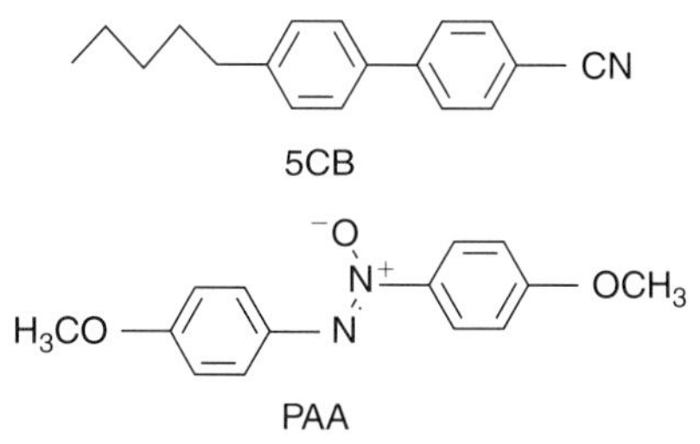

■ Figure 13.11   Structures of two typical liquid crystal molecules.

twisted with respect to each other until the director has been rotated by $90°$. When linearly polarized light traverses the liquid crystal, its plane of polarization is rotated by $90°$.

The display unit is equipped with crossed polarizers as shown in Figure 13.1. Light enters the first polarizer, has its polarization plane rotated by the liquid crystal, and exits the second polarizer. The system is transparent.

When an electric field is applied across the display unit, it orients the molecules parallel to itself, as shown on the right side of Figure 13.1. With this orientation, the molecules no longer rotate the polarization of the light; the latter is blocked by the second polarizer and the unit is dark.

In monochromatic LCDs, as in calculators or watches, properly shaped electrodes provide the light and dark fields of the display. In color displays, used in flat screen computer monitors and flat screen television sets, each pixel consists of three liquid crystal units as in Figure 13.1, one equipped with a red, one with a green, and one with a blue filter.

13.4 DIELECTRIC OPTICAL COATINGS

13.4.1 Antireflection (AR) Coatings

To appreciate the importance of reducing the reflectivity at optical surfaces consider the following example.

EXAMPLE 13.2   *An optical system consists of 20 air-glass interfaces at lenses, prisms, beam splitters, and so on. The glass employed has an index of refraction of $n = 1.50$. (A) Neglecting absorption, what is the transmission of the system? (B) If the reflectivity at each interface is reduced to 0.01, what is the transmission?*
ANSWER   (A) At interface 1, the reflectivity is R_1 and the transmission is $(1 - R_1)$. At interface 2 the transmission is $(1 - R_1)(1 - R_2)$. Therefore, for 20 interfaces the transmission is $T = (1 - R_1)(1 - R_2)(1 - R_3) \ldots (1 - R_{20})$. Because the same glass is used $T = (1 - R)^{20}$. With $n = 1.5$, $R = 0.04$, from Equation (13.6). Substituting $T = (1 - 0.04)^{20} = 0.442$.

Thus, less than half of the light incident is transmitted through the system.

(B) For $R = 0.01$, $T = (1 - 0.01)^{20} = 0.818$. The transmission is almost doubled, representing a significant increase in optical performance.

Antireflection (AR) optical coatings are the practical method for reducing reflectivity at glass surfaces. A common example is the plum-purple colored AR coating on camera lenses. They are also used in solar cells and on eyeglass, microscope, telescope, and binocular lenses. The **wave interference** effect shown in Figure 13.12 is responsible for antireflection properties. Imagine that light impinges normally on a planar surface containing a transparent AR coating of thickness d. Rays that bounce off the top and bottom (interfacial) coating surfaces will be **out of phase** because of the difference in light path difference $2d$. When the latter is equal to **half** a wavelength, that is, $2d = \lambda/2$, the two waves interfere destructively and the incident light is not reflected. An AR coating has thickness $\lambda/4$. For light of $0.6\,\mu m$, a $0.15\,\mu m$ thick coating is required. Films this thin are usually deposited by thermal evaporation.

Theory shows that an AR coating with an index of refraction equal to the square root of that of the glass yields the lowest reflectivity. Thus for a glass lens with $n = 1.5$, an AR film with an index of refraction equal to $(1.5)^{1/2} = 1.22$ should be chosen, other things being equal. The fact that n for MgF_2 is reasonably close to the optimum value is one reason for its widespread use.

13.4.2 **Dielectric Reflectors**

It is possible to create dielectric coatings that are highly reflecting and behave as mirrors. They act as mirrors when their index of refraction n is **larger** than that of the glass substrate. Reflectivity is enhanced by **constructive** interference between the two rays leaving the coating surfaces. This occurs

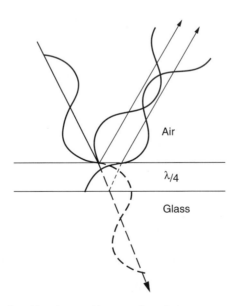

■ **FIGURE 13.12** Interference of light reflected from the top and bottom surfaces of a quarter-wave antireflex coating. Antireflection coatings are designed for normal incidence of the light. Oblique incidence is chosen in this figure for clarity.

when the path difference ($2d$) is equal to an integral number of **full** wavelengths. Therefore $\lambda/2$ coatings reflect the light. Practical mirrors are usually coated with several layers of $\lambda/2$ thickness. Dielectric mirrors are employed in high-performance optical systems such as lasers. Unlike metal mirrors that absorb some light, dielectric mirrors depend on interference effects, so that there is very little absorption.

13.4.3 **Filters**

Light filters that transmit in some regions of the spectrum but not in others have been employed in slide and movie projectors, energy saving devices, and assorted photography applications. Filters can be produced by utilizing the optical properties of the material or interference effects. Figure 13.13 shows some of the responses that are achieved by modern coatings. Filtering infrared radiation out of visible light is important in projectors in order not to damage photographic emulsions with heat. Thin optical films and coatings also help to conserve energy. For solar water heaters, high thermal absorption coatings enable efficient capture of the sun's heat with little loss due to re-radiation. In another application, optically transparent windowpanes coated with a so-called **heat mirror** reflect heat back into the home in winter.

13.4.4 **Phosphors**

A material that emits light when excited by high-energy electrons, electric fields, X-rays, or ultraviolet light is called a **phosphor**, a word derived from the element phosphorus which glows in the dark as it oxidizes. One objective for display applications is to get them to emit colored light in desired regions of the visible spectrum. To see what is required to do this, let us consider the band diagram of a phosphor shown in Figure 13.14.

Phosphors are basically large band gap semiconductors. Excitation with absorbed light or electrons of energy E_g or greater causes the usual transition of an electron from the valence band to the conduction band.

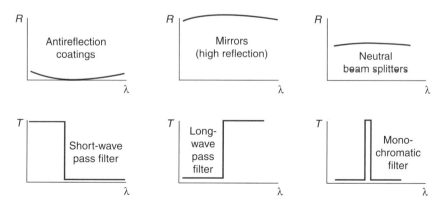

■ **FIGURE 13.13** Optical response as a function of wavelength for assorted applications.

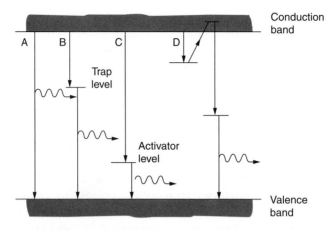

■ **FIGURE 13.14** Energy band diagram illustrating the photon emission in a luminescent material like ZnS. (A) Photon emission with full band gap energy. (B,C) Photon emission with less than full band gap energy because of either trap or activator level transitions. (D) Complex photon emission process (an excited electron is trapped (D), later released by thermal energy, falls into an activator level from where it recombines with a hole and emits a photon.

The excited electron can recombine with a hole by emitting light in several ways shown in Figure 13.14. When the material is pure, direct recombination will occur with emission of a photon with energy equal to the band gap ($h\nu = E_G$). (Case A in the figure). For pure ZnS, a popular phosphor material, E_g is 3.54 eV, corresponding to a wavelength of 0.350 μm in the UV. Clearly, to make ZnS emit light in the visible spectrum, energy level spacings less than 3 eV have to be created. There are two potential strategies to accomplish this. The first involves alloying ZnS with a semiconductor having a smaller energy gap. CdS ($E_g = 2.42$ eV) has been successfully alloyed to ZnS to reduce the width of the energy gap to emit visible light. By these means emission anywhere from red to blue has been obtained in ZnS-CdS alloys. The second approach is to dope ZnS with donor and acceptor atoms that create levels throughout the energy gap (Figure 13.14B and C). In the case of ZnS, additions of ~0.01 at% Ag produces blue light emission (~0.48 mm) while Cu gives a green light (~0.54 μm). If 0.005 at% Ag is added to a 50-50 ZnS-CdS alloy, yellow light emission is obtained.

Electrons initially excited to the conduction band can execute a variety of odysseys during de-excitation. They can alternately populate conduction band and trap levels and then descend to recombine with vacant activator states and emit a photon (Figure 13.14D). The greater the number of electrons participating in the process, the greater is the light emission intensity. There are many variations on the theme of recombination, and the longer it takes to occur, the greater is the **persistence** of the phosphor.

13.5 **ELECTRO-OPTICAL DEVICES**

Electro-optical (also called opto-electronic) phenomena rely on the active coupling between electrical and optical effects. Semiconductors are the primary materials in which these effects are large enough for practical devices. The three basic radiative transitions are illustrated in Figure 13.15.

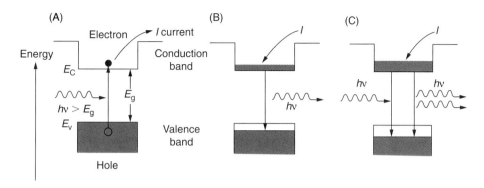

■ Figure 13.15 Three basic electron transition processes between valence and conduction bands of semiconductors. (A) Absorption of light. (B) Spontaneous emission of light. (C) Stimulated light emission.

1. **Absorption** of an incident photon excites an electron from the valence to the conduction band, generating mobile carriers. These carriers can provide an electric signal in a **photodiode detector**, or electric power in a **solar cell**.

2. Photon **emission** occurs during de-excitation of electrons from the conduction to valence band (i.e., electron-hole recombination). In **light-emitting diodes** (LEDs), for example, the emission is triggered by an electric current that passes through the junction, causing electron-hole recombination and a corresponding emission of monochromatic light. The light can now be viewed directly in a display, made to pass through lenses of an optical system, or enter an optical fiber.

3. The last case in Figure 13.15 is **stimulated** light emission, which was predicted by Einstein in 1907. When a photon with energy $h\nu = E_C - E_V$ impinges on an electron in the conduction band, the latter can be **stimulated** to recombine with a hole and emit a photon. This photon has the **same energy and phase** as the first one. This phenomenon is utilized in lasers.

These phenomena are utilized in p-n junctions to produce the photodiode, the solar cell, the light-emitting diode (LED), and the solid-state laser.

13.5.1 **The Photodiode**

The photodiode transforms light into an electric signal. It is a reverse biased p-n junction. No current flows when the diode is in the dark. When the junction is illuminated by photons with an energy $h\nu$ larger than the band gap, the photons are absorbed and create electron-hole pairs that generate an electric current as shown in Figure 13.16. Photodiodes have numerous applications including the soundtrack in movies, optical fiber telephone transmission, optical scanners, and automatic door openers.

13.5.2 **The Solar Cell**

The solar cell is a large-area p-n junction to which no external voltage is applied. Illumination of the junction generates electron-hole pairs as in the photodiode (Figure 13.17). The built-in electric field

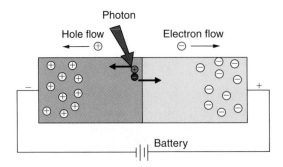

■ **Figure 13.16** Operation of a photodiode: absorption of a photon creates electron–hole pairs which produce an electric current in the reverse biased diode.

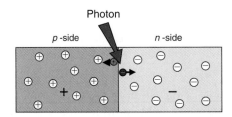

■ **Figure 13.17** Operation of a solar cell. Light creates electron–hole pairs which are separated by an internal electric field and create electric charges in n and p regions.

at the p-n interface sweeps the electrons into the n-type region and the holes into the p-type region. These excess carriers produce a negative electric charge in the n-type electrode and a positive charge the p-type side. These charges set up a potential difference V between the electrodes and a flow of current through the external circuit, generating the power $W = IV$.

Solar cells have been fabricated from various semiconductor materials, but only silicon cells have been commercialized on a large scale. Polycrystalline and even amorphous Si cells can be purchased. The efficiency of energy conversion is now about 12% for crystalline cells, meaning that of the 1 kW/m^2 of solar power density falling on earth, 120 W of electric power would be generated from 1 m^2 of cell area. Silicon solar cells are expensive and energy-intensive in their construction. They must operate about four years before they produce as much energy as was used in their manufacture. These devices are suitable for special applications, but the energy they produce is still more expensive than conventionally produced electricity. Solar cells can be made with polycrystalline or even amorphous silicon. These are less expensive to produce but have lower efficiency.

Much research activity now exists with the purpose of producing solar cells that can generate economically competitive electric power. One direction endeavors to increase the energy conversion efficiency by using multiple junctions of III-V compounds.

Efforts are underway to fabricate solar cells with large areas using inexpensive techniques such as printing. Solar cells utilizing **organic semiconductors** are one such approach but, at this writing, their efficiency is at most 4%. In addition, polymer semiconductors have limited longevity as they are susceptible to chemical degradation by light. An approach to solar cells that use **carbon nanotubes** or **nanocrystalline semiconductors** is being investigated by several consortia of university laboratories and companies. These materials can be dissolved in solvents; consequently, the solar cells could be manufactured by painting or printing methods. Theoretically, larger efficiencies could be expected in the future from nanostructured solar cells, in part because they would utilize a larger fraction of the solar spectrum. Plastic and nanostructured solar cells are already available in the market, but their application for large-scale energy production is still some time off. In this climate of finite hydrocarbon resources and need to reduce greenhouse gases, research is intense.

13.5.3 **The Light-Emitting Diode (LED)**

The light-emitting diode is a forward-biased p-n junction. Electrons and holes move into the recombination zone where they recombine and release their energy by emitting a photon (Figures 13.15B and 13.18).

For quantum mechanical reasons that are beyond the scope of this text, crystalline silicon and germanium are unsuitable for light-emitting diodes. Electron-hole recombination does not generate light in these materials. They are termed **indirect band gap semiconductors**. For light-emitting diodes, one relies on compound semiconductors that possess a **direct band gap**. In these materials, electron-hole recombination occurs through the emission of a photon with energy equal to the band gap: $h\nu = E_G$. The band gaps and wavelengths of emitted light of semiconductors are shown in Table 11.3.

Variable Energy Band Gaps   Electro-optical devices operate at a fixed wavelength that corresponds to the magnitude of the energy gap of the semiconductor. Applications of LEDs, however, require specific wavelengths of the emitted light. Optical data transmission, for instance, requires a wavelength of 1.55 μm, which corresponds to the maximum transparency of the fibers. Other applications require LEDs generating blue, green, and red light whose combination produces white light or any desired color.

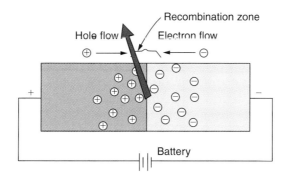

■ **FIGURE 13.18**   Operation of a light-emitting diode (LED): energy released by electron-hole recombination is emitted as a photon.

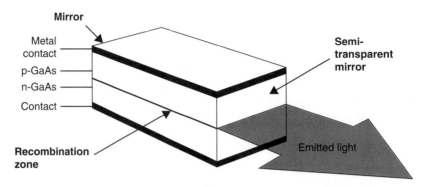

It is possible to synthesize semiconductor alloys that have the energy gap value for a requisite wavelength of the emitted light. This is achieved by alloying III-V compounds to form ternary or quaternary compounds. The light source for optical communications at $1.55\,\mu$m wavelength is made of a quaternary InGaAsP material grown on an InP substrate. White light for space illumination is achieved by a combination of blue (GaN), red, and yellow LEDs or by an ultraviolet GaN LED combined with a phosphor.

13.5.4 **The Solid-State Laser**

The solid-state laser is an LED that makes use of the phenomenon of stimulated emission. It differs from an ordinary light-emitting diode in two respects: (1) it emits much more intense light, and (2) the emitted light is coherent, which means that the waves of all the emitted photons are in phase with each other. The solid-state laser (Figure 13.19) is a highly doped p-n junction that is forward biased. The heavy doping provides a high density of electrons in the n-type region and holes in the p-type region; these provide a large number of sites for the recombination with minority carriers and the emission of intense light. The laser is equipped with two mirrors that reflect the light back into the device.

The junction region where electron-hole recombination occurs is the horizontal plane in the middle of the laser. At first, light is emitted in all directions (Figure 13.20A). The light is reflected back and forth inside the laser where it stimulates further electron-hole recombination (Figure 13.20B). The latter generates light emission in the same direction as the reflected light and in phase with it. A very high intensity of light builds up in the laser; this in turn stimulates more recombination and emission of coherent light. One of the mirrors is semi-transparent (Figure 13.20C) and allows the escape of some light, which is shown in Figure 13.19 as the large blue arrow. The coherent, stimulated, emission of light is set up in an extremely short time.

Actual solid-state lasers have a more complex geometry. They are made of different semiconductor materials and have a small active region for applications where a small beam is required, such as in

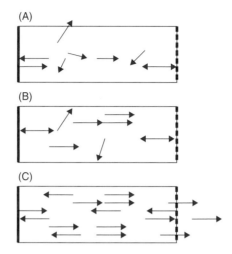

■ **Figure 13.20** Schematic of laser action. The plane represents the recombination zone of the p-n junction (see Figure 13.22). (A) Initial light emission, incoherent. (B) Light emitted toward mirrors is reflected and builds up in the laser, creating stimulated light emission. (C) Laser action: intense light in the laser causes stimulated emission of coherent parallel light.

optical recording (CDs). Figure 13.21 shows the configuration of a GaAlAs laser. Note the shape of the active region that forms a narrow beam of light.

13.5.5 **Electroluminescent Light Sources**

Electroluminescence is "the property of emitting light on activation by an alternating current" (American Collegiate Dictionary). Like LEDs, electroluminescent light sources are cold sources in that they emit light at ambient temperature. The sources consist of two conducting electrodes separated by a phosphor as shown in Figure 13.22. One electrode, the back electrode, is a low-work function metal (such as aluminum or calcium). The front electrode is transparent; it consists of a glass plate or transparent polymer coated with indium-tin oxide (ITO). ITO consists of 90% In_2O_3 and 10% SnO_2; it is transparent as a thin film and is an electric conductor. Under the applied potential, the back electrode emits electrons into the phosphor. The ITO injects holes into the phosphor. The energetic electrons emitted by the back electrode combine with the holes in the phosphor and emit light. Since the material between the conducting electrodes is an electric insulator, an AC voltage is required to discharge the device to avoid a decreasing luminosity because of charging. Typical phosphors (also called luminophores or lumophores) are zinc sulfide (ZnS) doped with copper or silver; or III-V semiconductors such as InP, GaAs, or GaN. Phosphors are chosen for the color of the light.

Electroluminescent lamps have the advantage of being area light sources. They emit light uniformly over an area as large as 18×24 in (40×60 cm). By combining panels, one can obtain large uniform source of light. The back light sources of LCD displays (Figure 13.1) are often electroluminescent.

(A)

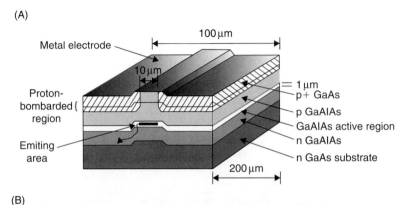

(B)

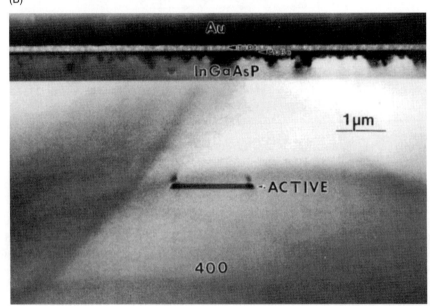

■ **FIGURE 13.21**   (A) Configuration of a GaAlAs solid-state laser. (B) Cross section of an indium phosphide semiconductor diode laser structure taken by electron microscope. A. From "*The Science and Engineering of Microelectronic Fabrication*," 2nd ed., Stephen A. Campbell, Oxford University Press, Oxford, UK, 2000. B. Courtesy of S. Nakahara, AT&T Bell Laboratories.

13.5.6 Organic Light-Emitting Diode (OLED)

OLED is a planar light-emitting diode based on organic semiconductors. The principle is that of an electroluminescent source with the difference that the use of organic semiconductors permits the operation with a DC power source (Figure 13.23). Electrons are supplied by the metallic back electrode

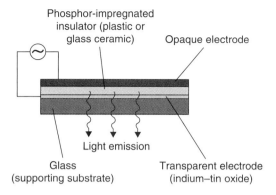

■ **FIGURE 13.22**   Schematic representation of a luminescent light source. The glass substrate is coated with a transparent conductive coating of indium-tin oxide. A phosphor-impregnated insulator separates the plate from an opaque electrode. Electrons emitted by the opaque electrode and holes emitted by the ITO recombine in the phosphor molecules to emit light.

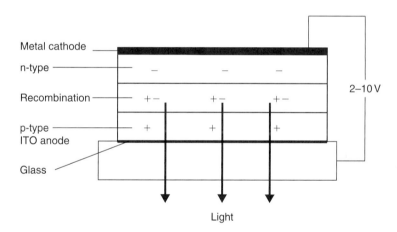

■ **FIGURE 13.23**   Schematic representation of an organic light-emitting diode (OLED). The glass substrate is coated with a transparent conductive coating of indium-tin oxide (ITO). Organic semiconductor layers transport electrons from the metal cathode and holes from the ITO into the recombination layer

and are transported in an n-type organic semiconductor. Holes are supplied by the ITO anode that is covered with a p-type organic semiconductor. Electron-hole recombination occurs in the phosphors that are selected to emit light of the desired color. When deposited on a transparent polymer sheet instead of a glass plate, an OLED can form a large-area flexible light source.

Active Matrix OLEDs (AMOLED) are used in flat screen television or computer monitors. In these devices, the OLED is combined with an array of thin film transistors that provide the voltage applied to every light-emitting cell. This structure allows the formation of active color emitters such as television screens or computer monitors. Each pixel contains three cells equipped with a red, a blue, and a green emitting phosphor.

Organic devices possess a number of advantages. When the glass plate is replaced by a transparent polymer, the device becomes a flexible OLED, which can be produced by simple inkjet printing methods.

The OLED sources have the advantage that they emit the light directly and do not require a backlight as do the liquid crystal display devices (LCD). They are now used in the monitors of digital cameras.

13.6 **OPTICAL RECORDING**

Compact Discs (CDs) and Digital Video Discs (DVDs) utilize lasers to burn (i.e., write) and to read digital signals. The recording medium is a 1.2 mm thick disc of polycarbonate (Figure 13.24) coated with a thin film of aluminum to increase optical reflection. In Read Only Memories (CD-ROMs)

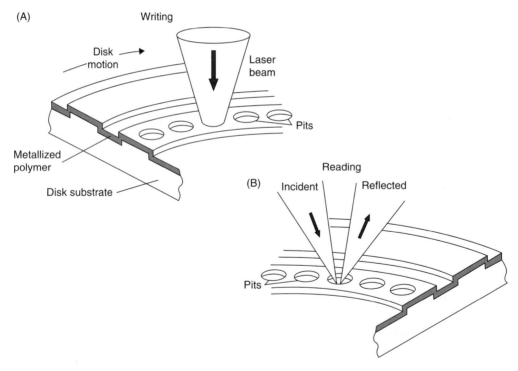

■ **Figure 13.24**   Schematic of the optical recording and playback in a CD-ROM. (A) During recording, information is stored in the form of pits that are burned into the polymer disc by a laser beam. (B) In playback, a weak laser reflected from the disk surface is detected. Reflection from the pit is weak because of destructive interference.

data are recorded by a laser that burns little indentations ("pits") into the polymer as shown in Figure 13.24A. A weaker laser light reflected from the disk provides the optical signal that is converted back into an electric signal (Figure 13.24 B). In mass-produced CDs for the reproduction of music, the pits are stamped into the disc from a master. The depth of the pits is approximately 1/4 of the wavelength of the reading light, which causes destructive interference as described in Section 13.4.1.

Read-write or erasable discs (CD-RW) are coated with the alloy AgInSbTe. When heated to 400°C by a laser pulse and cooled rapidly, this alloy becomes amorphous. It is crystallized by annealing at a lower temperature. The two phases have a different optical reflectance so that the disc can be read with the same reflecting laser beam as the CD-ROM. On these discs the recorded data can be erased by re-crystallizing the area.

13.7 **OPTICAL COMMUNICATIONS**

The very high frequency of light permits the transmission of many communications simultaneously over the same fiber. Optical communications have the following additional advantages over wire transmission: capacity, small size, low weight, quality transmission (minimal crosstalk), low cost, and generally good security. In the development and improvement of optical communications technology, materials have played a critical role, none more perhaps, than the glass fiber to which we now turn our attention.

13.7.1 **Optical Fibers**

Light Transmission   Optical communications require two important attributes of the fiber, namely that light rays be confined to it, and that long distance transmission occur with minimal loss in light intensity. The optical waveguide (Figure 13.25A) consists of an active inner cylindrical fiber core

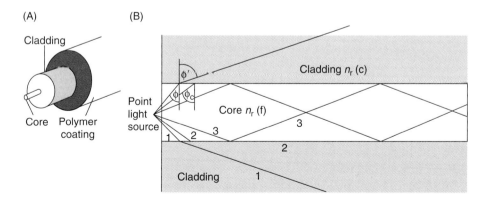

■ **FIGURE 13.25**   (A) Structure of a single-mode optical fiber. The diameter of the core, cladding, and polymer coating are 8.5, 125, and 250 μm, respectively. (B) Refraction and propagation of light in an optical fiber.

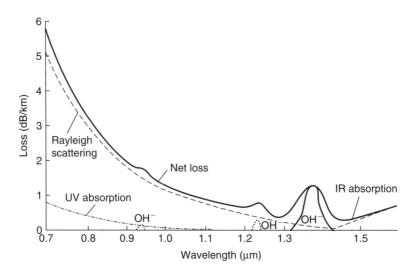

■ **FIGURE 13.26**   Absorption of light in glass as a function of wavelength.

(8.5 μm diameter) and a surrounding cladding (~125 μm outer diameter), whose index of refraction $n(c)$ is lower than that $n(f)$ of the core. The cladding is silica and the core is silica doped with GeO_2 (germania), which raises $n(f)$ about a percent above $n(c)$. To confine the light to the core, it is launched at an angle of incidence that is less than the critical angle ϕ_C (Equation (13.4)) so that it is confined to the core fiber by total internal reflection (e.g., rays 2 and 3 in Figure 13.25B) and propagates in zigzag fashion down the fiber. Signal loss by frustrated internal reflection (see Section 13.3.5) is avoided by the transparent cladding.

Loss Factor   Light intensity in fibers declines exponentially with distance, according to Equation (13.3). It is customary to express this loss in terms of decibels (dB), where the attenuation in dB/km is defined as

$$\text{Loss(dB/km)} = \frac{-10}{L} \cdot \log \frac{I}{I_o} \tag{13.9}$$

In this definition I and I_o are the light intensity at the detector and source, respectively, while L is the distance between them in kilometers. Since the introduction of optical communications, the transparency of glass has made enormous progress.

Absorption   In optical fibers, the absorption coefficient is the result of **bulk** and **intrinsic losses** that vary with the wavelength as shown in Figure 13.26. Bulk absorption has been attributed to light-absorbing impurities (e.g., Fe^{2+} and OH^- ions). Scattering of light by imperfections (e.g., air

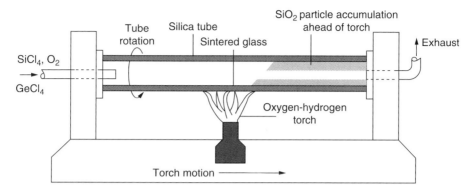

■ Figure 13.27   Modified chemical vapor deposition process for producing optical fibers.

bubbles, scratches, etc.) and density and compositional fluctuations are sources of intrinsic loss. The latter contribution, known as Rayleigh scattering, varies as λ^{-4}. Because this term is large at small wavelengths, UV and visible light are not used in optical communications. The loss is at its minimum at $\lambda \sim 1.3$–$1.5\,\mu m$, and this infrared wavelength is selected for light transmission in optical fibers.

13.7.2 **Fiber Fabrication**

The first step in producing fiber is to make a preform. A modified chemical vapor deposition (MCVD) technique, schematically shown in Figure 13.27, is widely used for this purpose. Very pure precursor gases of $SiCl_4$ and $GeCl_4$ are mixed with oxygen, transported down a rotating silica tube and reacted at temperatures of $\sim 1,600\,°C$. As a result of oxidation, small solid "soot" particles of SiO_2 doped with GeO_2 are produced. The cotton candy-like soot deposits on the tube walls, and when enough has collected it is densified by heating to temperatures of $\sim 2,000\,°C$. At this very high temperature, surface tension and external pressure promote the sintering that causes the tube to collapse radially until the center hole is eliminated. The dopant profile is carefully controlled during deposition to ensure that the subsequently drawn fiber will have a larger index of refraction at the core than in the cladding. Phosphorus and fluorine are also introduced to facilitate processing and assist in achieving the required index of refraction profile across the preform.

Next, the $\sim 2\,cm$ diameter, $\sim 1\,m$ long preforms are drawn to fiber in a process that remotely resembles wire drawing of metals (Figure 13.28). High draw towers produce miles of fiber from a single preform. The latter are heated again, this time to temperatures greater than $2,000\,°C$. Fiber of uniform diameter is then drawn in a way that faithfully reproduces the radial profile of refractive index on a much smaller scale. Immediately after the fiber cools sufficiently, its pristine surface is coated with a protective polymer coating. The intent is to prevent mechanical reliability problems posed by defects and cracks.

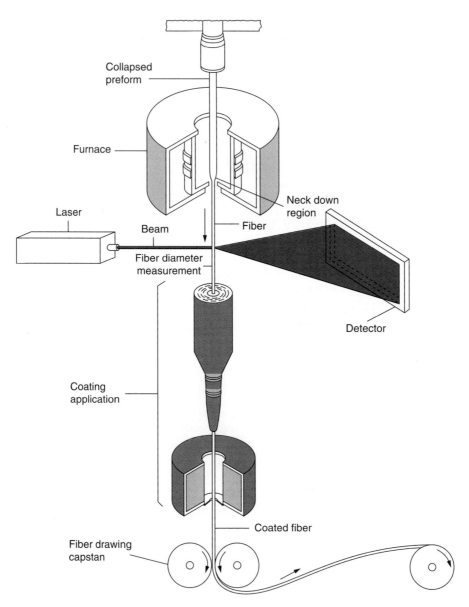

■ **FIGURE 13.28** Schematic of the fiber drawing process. From D.H. Smithgall and D.L. Myers, *The Western Electric Engineer,* (Winter 1980)

■ SUMMARY

1. Light is a high-frequency electromagnetic radiation.

2. The energy of light is quantized; light consists of photons that carry the energy $E = h\nu$.

3. The photon energy of visible light lies between 2 and 3 eV, corresponding to wavelengths between 0.6 and 0.4 μm.

4. The human eye possesses three types of cones that are sensitive to blue, green, and red light. Low-intensity light is perceived by one type of rod. The perception of all colors results from the different excitations of the three cones.

5. Light is absorbed by electrons capable of acquiring the energy $E = h\nu$; it is emitted by electrons losing the energy $E = h\nu$. Absorption and emission of light is therefore related to the electronic structure of the solid.

6. In transparent solids, light induces an electric polarization that reduces its velocity and refracts it. Refraction of light is used in lenses.

7. Optical fibers are highly transparent glass fibers in which the light is trapped by total internal reflection.

8. When the propagation of light is modified in a solid, by refraction or absorption, the light is reflected at the interface. Metals absorb light and therefore reflect it. They are used as mirrors.

9. Optical polarizers consist of long parallel molecules that absorb the light polarized in one direction. Polaroid and liquid crystals are important polarizers; liquid crystals can be oriented by an electric field.

10. Antireflex coatings have a thickness equal to 1/4 the wavelength of light. They suppress reflection by interference.

11. Phosphors are insulators that emit light when excited by high-energy electrons or photons. They possess optically active energy levels in their band gap that are caused by selected impurities.

12. Semiconductor p-n junctions are opto-electric materials.

13. A forward-biased junction constitutes a light-emitting diode (LED) or, when properly configured, a solid-state laser. LEDs are made of III-V compounds and, increasingly, of organic semiconductors (OLEDs).

14. A reverse biased p-n junction is a photodetector that transforms light into an electric current.

15. The solar cell is an unbiased p-n junction. Light absorbed in it produces electron-hole pairs that are separated by the electric field of the junction. Electrons migrate to the n-type and holes to the p-type electrode. This charge separation creates an electric potential.

16. Present solar cells are made of silicon. Organic semiconductors and nanostructured composites are being developed as large-area cells.

17. In CD-ROM discs, data bits are small cavities in a polymer. These are "burned" with a laser that evaporates the material locally or embossed mechanically in mass production.

18. In CD-RW discs, data bits consist of amorphous material produced by a laser pulse. The data can be erased by re-crystallizing the material with a laser beam with lower power.

■ KEY TERMS

A
absorption, 371, 383, 392, 395
absorption coefficients, 371, 376
Active Matrix OLED, 389
AgInSbTe, 391
aluminum, 376
AMOLED, 389
antireflection, 380
antireflexion, 379

B
band gap, 381

C
cadmium sulfide, 372
carbon nanotubes, 385
CD, 387, 390
CD-ROM, 390
CD-RW, 391
cell, 368
ceramics, 372
chromium, 376, 398
Color, 373
compact discs, 390
cones, 370
copper, 373
crystal glass, 378
cut glass, 377

D
diamond, 377
dielectric optical coatings, 379
dielectric reflectors, 380
digital video discs, 390
direct band gap, 385
director, 378
dispersion, 373

DVD, 390
dyes, 373

E
efficiency, 384, 385, 399
electroluminescent, 368, 387, 388
electromagnetic spectrum, 370
electro-optical devices, 382
eyeglasses, 374

F
fiber fabrication, 393
filters, 368, 381
flat screen monitor, 368
front-surface mirrors, 376

G
GaAlAs laser, 387
gamma ray, 370
$GeCl_4$, 393
GeO_2, 392
gold, 373, 398

H
heat mirror, 381

I
index of absorption, 371
indirect band gap, 385
indium-tin oxide, 387
infrared, 370, 372
interference, 380, 381, 391, 395
ITO, 387, 389

J
jewels, 377

L
LCD, 379, 387, 390
lead oxide, 378
LED, 368, 383, 385, 386, 387, 395, 400
lenses, 368, 374, 379, 380, 383, 395, 400
light, 370
light-emitting diode, 385
liquid crystal, 368, 378, 379, 390
liquid crystal displays, 378
loss factor, 392

M
metallic mirrors, 376
metals, 372
MgF_2, 380

N
nanocrystalline semiconductors, 385
nematic, 379

O
OLED, 368, 388, 389, 390, 398, 400
optical communications, 391
optical fibers, 368, 383, 391
optical recording, 390
organic semiconductors, 385

P
PbO, 378
phase, 383
phosphor, 367, 381, 382, 386, 387, 390, 398, 400
photochromic lenses, 374
photodiode, 383

■ REFERENCES FOR FURTHER READING

[1] P. Bhattacharya, *Semiconductor Optoelectronic Devices*, Prentice-Hall, Englewood Cliffs, NJ (1994).

[2] M. Fox, *Optical Properties of Solids (Paperback)*, Oxford University Press, New York (2001).

[3] O.S. Heavens, *Optical Properties of Thin Solid Films (Paperback)*, 2nd ed, Dover Publications, Mineola, NY (2007).

[4] J. Nelson, *The Physics of Solar Cells (Properties of Semiconductor Materials)*, Imperial College Press, London (2003).

[5] J.I. Pankove, *Optical Processes in Semiconductors (Paperback)*, 2nd ed, Dover Publications, Mineola, NY (1975).

[6] J.P. Powers, *An Introduction to Fiber Optics Systems*, Irwin, Homewood, IL (1993).

[7] J. Simmons, K.S. Potter, *Optical Materials*, Academic Press, San Diego, CA (2000).

[8] A. Yariv, *Introduction to Optical Electronics*, 2nd ed., Holt, Rinehart and Winston, New York (1976).

■ PROBLEMS AND QUESTIONS

13.1. By means of approximate curves of transmission versus wavelength curves, distinguish the following materials: (A) a transparent colorless solid (B) a transparent blue solid (C) an opaque solid (D) a translucent colorless solid.

13.2. Consider a glass slide with an index of refraction of $n = 1.48$ and light incident on the glass-air interface.

   a. What is the speed of light in the glass?

   b. At what angle of incidence (measured from the normal) will light undergo total internal reflection in the glass?

   c. Can light undergo total internal reflection when traveling from air to into the glass?

13.3. Six percent of the light normally incident on a nonabsorbing solid is reflected back. What is the index of refraction of the material?

13.4. Light is incident on the surface of a parallel plate of material whose absorption coefficient is 10^3 cm^{-1} and index of refraction is 2.0. If 0.01 of the incident intensity is transmitted, how thick is the material?

13.5. Sapphire, which is essentially pure Al_2O_3, has no absorption bands over the visible spectral range and is therefore colorless. When about 1% of Cr^{3+} ion is added in the form of Cr_2O_3 there are two broad absorption bands, one near 0.4 mm and the other around 0.6 mm. Roughly sketch the absorption spectrum of chromium-doped alumina in the visible range. What is the color of this material?

13.6. The optical constants of the colored metals copper and gold are tabulated at the indicated wavelengths:

	0.50 μm		0.95 μm	
Wavelength	N	k	N	k
Copper	0.88	2.42	0.13	6.22
Gold	0.84	1.84	0.19	6.10

For both materials calculate the fraction of incident light reflected from the surface, or the reflectance at both wavelengths.

Compare your values of R with those plotted in Figure 13.8.

13.7. What are the optical and metallurgical advantages and disadvantages of using gold, silver, aluminum, and rhodium metals for mirror applications?

13.8. Which of the optical coating materials displayed in Figure 13.5 could be used to make a one-layer AR coating that is to operate at 0.50 μm on a glass lens having an index of refraction of 1.52?

13.9. A germanium crystal is illuminated with monochromatic light of 1 eV energy. What is the value of the reflectance at the incident surface?

13.10. What types of oxides added to silica melts will tend to increase the index of refraction of the resulting glass? What scientific principles underscore the influence of specific metal ions?

13.11. Specifications for the design of an optical system containing 17 glass-air interfaces call for a total minimum transmittance T of 0.72 at $\lambda = 0.45$ μm, $T = 0.91$ at $\lambda = 0.60$ μm, and $T = 0.75$ at $\lambda = 0.77$ μm. The same antireflection coating will be used on each optical glass substrate of refractive index 1.52. What is the reflectance of this AR coating at each wavelength?

13.12. A certain material absorbs light strongly at a wavelength of 160 nm and nowhere else, being transparent up to 5 μm. Is this material a metal or nonmetal? Why?

13.13. The relaxation time for a zinc-based phosphor bombarded with an electron beam is 3×10^{-3} s.

a. How long will it take the light to decay to 20% of the initial light intensity?

b. If 15% of the intensity is to persist between successive beam scans on a television screen, how many frames per second must be displayed?

13.14. Phosphorescent materials have long persistence times. Ultraviolet light incident on such a phosphorescent material results in re-emitted light after the UV is turned off. If the re-emitted light decays to half the original intensity in 5 min, how long will it take for the intensity to decline to 1/10 of its original value?

13.15. Solar cells are made from silicon or GaAs. Cells fabricated from GaAs can be much thinner than Si cells, thus conserving gallium and arsenic resources. Why are thinner GaAs cells possible?

13.16. a. At a wavelength of $0.4\,\mu m$, what is the ratio of the depths that 98% of the incident light will penetrate silicon relative to germanium?

b. What is the corresponding ratio of depths at a wavelength of $0.8\,\mu m$?

c. Such light causes Si and Ge to become electrically more conductive. Why?

13.17. By consulting Table 12.1, determine the wavelength of light associated with the energy band gap of Si, GaP, and ZnSe.

13.18. Compound semiconductors containing binary, ternary, and quaternary combinations of Ga, In, As, and P are, by far, the easiest to grow and fabricate into high quality electro-optical devices.

a. On this basis explain why infrared devices commonly employ these elements.

b. Explain why blue light-emitting devices are not common.

c. Similarly, explain why ultraviolet light-emitting semiconductor devices are rare.

13.19. There has been a great deal of interest in developing a blue-light semiconductor laser operating in the range of $2.6\,eV$. By consulting Table 12.1, select a possible candidate material and suggest a substrate that is closely lattice-matched to it.

13.20. The junction depth of a silicon solar cell lies $200\,nm$ below the surface. What fraction of $550\,nm$ wavelength light incident on the device will penetrate to the junction? (Assume no light reflection at the surface.)

13.21. How would the answer to the previous problem change if the solar cell were fabricated from GaAs?

13.22. A measure of the electric power delivered by a solar cell is the fourth quadrant area enclosed by the I-V characteristics. Specifically, the power is reported as the area of the largest rectangle that can be inscribed in the third quadrant. The cell "fill factor," a quantity related to efficiency, is the ratio of this maximum power rectangle to the product of Voc and Isc. Suppose the I -V characteristic is a straight line joining Voc and Isc. What are the values of the maximum cell power and fill factor?

13.23. Comment on the purity or monochromatic character of the light emitted during an electron transition in an atom compared to that in an extended solid source of many atoms.

13.24. a. Consider a $100\,W$ lightbulb viewed at a distance of $1\,m$ through a $1\,mm$ diameter aperture. What radiant power falls on the eye?

b. A $5\,mW$ He-Ne laser emits a $1\,mm$ diameter beam that does not spread appreciably. What laser light power falls on the eye when viewed at $1\,m$?

(From the answer it is clear why we should never look at a laser beam directly.)

13.25. For an optical recording application, it is desired to vaporize some 100,000 hemispherical pits per second in a polymer having a density of $1.5\,g/cm^3$. About $2\,kJ/g$ is required to vaporize this polymer. If each pit is assumed to have a diameter of $5\,\mu m$, what laser power is required for this application?

13.26. One way to make a short planar waveguide is to dip a glass slide (containing ~12% Na) into a $AgNO_3$ melt for a period of time. The waveguide produced (at each surface) consists of the unaltered glass slide and air regions sandwiching the altered slide surface (core) layer in between. Explain what optical property the core layer must have. Suggest how exposure to $AgNO_3$ produces the required property.

13.27. A planar optical waveguide is modeled by a stack of three glass slides in perfect contact. The two outer (cladding) slides of refractive index $n = 1.46$ sandwich a (core) slide of refractive index $n = 1.48$ in between. Surrounding the waveguide is air. For total internal reflection to occur in the core, determine the angle of incidence ϕ_c (measured from the normal) for light.

a. Incident on the core-cladding interface originating from the core side.

b. Incident on the air-cladding interface originating from the air side.

    c. Sketch, by means of a ray diagram, what happens when the angle of incidence is slightly greater or slightly smaller than the correct answer to part b.

13.28. a. Silica optical fibers for communications purposes require 1.3–1.5 μm light for most efficient operation. Why?

    b. Optical communications applications have largely driven the development of opto-electronic devices operating in the infrared region of the spectrum. Explain why.

13.29. If the core of a glass fiber has an index of refraction 1% larger than that of the cladding, by what angle can light entering the end of the fiber deviate from the fiber axis direction and still be confined to the fiber? (See Figure 13.25.)

13.30. Repeater stations for underwater cables containing optical fibers are spaced 45 km apart. The fiber loss is known to be –0.2dB/km. What is the ratio of the light intensity leaving a repeater to that entering it?

13.31. What is the range of photon energies and wavelengths of visible light?

13.32. Describe how we perceive colors and explain why this allows us to reproduce all colors with three dyes in print and three phosphors on screens.

13.33. Describe the mechanism by which light is absorbed or emitted in solids.

13.34. Describe how the electron energy structure of solids determines their optical properties.

13.35. When light passes through a transparent solid, how does that solid modify its propagation?

13.36. How does light modify a transparent solid?

13.37. What optical properties of materials are used in lenses?

13.38. How do optical fibers transmit light even when they are bent?

13.39. How does one measure fingerprints without ink?

13.40. What causes the reflection of light?

13.41. Why do metals make good mirrors?

13.42. How do Polaroid and liquid crystals polarize light?

13.43. Describe how antireflex coatings work. What thickness do they have?

13.44. What is a phosphor? How does it emit light under the impact of a high-energy electron?

13.45. Describe the structure and the functioning of a light-emitting diode (LED). What type of material is it made of?

13.46. Describe the structure and functioning of a solid-state laser.

13.47. What is an OLED?

13.48. Describe the structure of a photodetector and how it functions.

13.49. Describe the structure of a solar cell. How does it transform light into electric energy?

14

Magnetic Materials

Magnetic materials are used in electric motors, transformers, loudspeakers, cranes, data processing, and in households. Hard magnets must retain their magnetization even in stray magnetic fields, and soft magnets must change their magnetization with the lowest possible resistance. In this chapter we explore the different magnetic materials and the processing that endows them with the desired properties. In order to do that, we first review the concepts of magnetism and examine how a material becomes magnetic.

LEARNING OBJECTIVES

After studying this chapter, the student will be able to:

1. Define whether a soft or hard magnet is required for important classes of applications.

2. Draw a *B-H* curve and identify the coercive field and the remnant magnetization.

3. Draw the *B-H* curves for a hard and a soft magnetic material.

4. Describe magnetic energy losses in terms of hysteresis and eddy currents.

5. Describe the origin of ferromagnetism in terms of electron spins and self-alignment of magnetic moments.

6. Relate changes in magnetic domain structure to different regions of the *B-H* curve.

7. Describe the salient properties and the processing of soft magnets.

8. Describe the salient properties and the processing of hard magnets.

9. Name the most important magnetic materials.

10. Select an appropriate material for a given application.

14.1 **USES OF MAGNETS AND THE REQUIRED MATERIAL PROPERTIES**

We are all familiar with magnets: they are used in the home to attach notes to the refrigerator door, as magnetic catches holding the refrigerator door closed, or to pick up small metallic objects. With these magnets, we require that the magnetization be strong enough and that it be permanent; we do not wish the magnetization to be weakened or modified by contact with other magnets or by stray magnetic fields. Such devices use **hard ferromagnetic** materials. Advances in the magnetic strength of these materials allow for greater efficiency and miniaturization in motors, loudspeakers, and tool holders. Judging by Figure 14.1, progress in the last century has been impressive.

Another class of materials is employed on a large scale in generators, transformers, and inductors for the generation, transmission, storage, or conversion of electric power. These important materials perform the critical function of shaping and concentrating magnetic flux in alternating current equipment, a task that must be accomplished with minimum energy loss. This requires that the magnetization be changed and reversed with a minimum expenditure of energy. These are the **soft ferromagnetic materials**. An almost two orders of magnitude decline in energy loss in magnets since the beginnings of

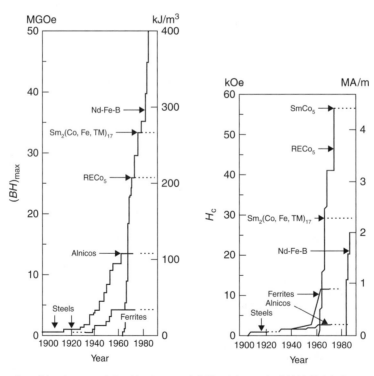

■ **FIGURE 14.1** Chronology of the advances made in raising the strength (left) and the coercive field H_C (right) of permanent magnets. From National Materials Advisory Board Report NMAB-426 on Magnetic Materials 1985.

the electric power industry has been responsible for huge energy savings. Still, the annual 60 Hz core energy losses in the U.S. are currently estimated to be about 10^{14} Whr—a magnitude equal to the output of several large utility power generating plants.

What magnetic materials are available? What are their properties and how are these obtained? How does one select the proper material for a given application and how does one use it correctly? These are the questions we shall address in this chapter. We will concentrate our interest on the ferromagnetic and ferrimagnetic materials, which are of engineering interest.

14.2 MAGNETIC FIELDS, INDUCTION, AND MAGNETIZATION

The ability to attract steel and redistribute iron filings in a characteristic pattern is familiar evidence that a magnetic field (H) is established in space by a bar magnet (Figure 14.2A). A magnetic field is distributed geometrically the same way by a solenoid as shown in Figure 14.2A and B.

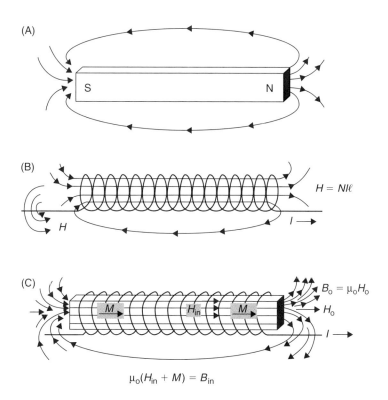

■ **FIGURE 14.2**   (A) Magnetic field surrounding a bar magnet. (B) Magnetic field surrounding an air-filled solenoid. (C) Magnetic field surrounding a solenoid with an iron core.

14.2.1 **Magnetic Fields**

We start with the hollow solenoid of Figure 14.2B, which has N turns over a length ℓ. When a current I flows through the wire, a **magnetic field** H is produced having a magnitude

$$H = IN / \ell \tag{14.1}$$

inside the solenoid. The magnetic field is proportional to the current and the density N/ℓ of turns per meter in the solenoid. H has the dimensions of ampere turns per meter, or simply (A/m). It is interesting to note that the magnetic field does not depend on the length or the diameter of the solenoid, but only on the density of coils.

14.2.2 **Induction**

If we now wind a second solenoid with N_2 turns over the first one and **vary** the magnetic field intensity (for instance by varying the current), a voltage is **induced** in the second coil and has the magnitude

$$V = -N_2 \ A \ \mu_o \ dH/dt \tag{14.2}$$

where A is the area inside the coils ($A = \pi r^2$ if r is the radius of the coils) in the second solenoid. Note that the induced voltage is proportional to the number of turns, not their density. This is illustrated in Figure 14.3.

It is useful to express the magnetic field in different units by multiplying H with μ_o:

$$B = \mu_o H \tag{14.3}$$

where $\mu_o = 4\pi 10^{-7}$ Vs/Am is the permeability of free space, also called the induction constant. B is called the magnetic **induction** and has the dimension of volt seconds per square meter (Vs/m^2) or **Tesla** (T). Thus a field of 1 A/m corresponds to $1.257.10^{-6}$ T. With this definition, the induced voltage is

$$V = -NA \frac{dB}{dt} \tag{14.4}$$

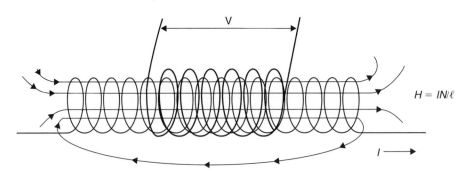

■ **FIGURE 14.3** Induced voltage in a varying magnetic field.

The integral $\int BdH$ is the energy density of the magnetic field; it is measured in J/m^3. In free space, it is $\frac{1}{2}\mu_0 H^2$ or $\frac{1}{2}BH$. (We suggest that you quickly verify that the product BH indeed has the dimensions of J/m^3.)

These units are in the new International Unit System. The old cgs units are still widely used in magnetism; it is useful to present them here.

In cgs, the magnetic field H is measured in oersteds, $1\,\text{A/m} = 4\pi 10^{-3}$ oersted; the induction is measured in gauss, 1 tesla $= 10^4$ gauss.

14.2.3 Magnetization

When an iron bar fills the air space of the solenoid core (Figure 14.2C), the iron is magnetized by the applied field and the external magnetic field strength is dramatically increased but exhibits the same spatial character as for the empty solenoid. The **magnetization** M of the iron adds itself to the magnetic field and creates a much larger induction

$$B = \mu_o(H + M) \tag{14.5}$$

The magnetization has the same dimensions (A/m) as the magnetic field.

For ferromagnetic materials the magnetization is much larger than the applied magnetic field. In a commercial iron, for instance, a magnetic field of 100 A/m causes a magnetization $M = 1,600,000\,\text{A/m}$ and an induction $B = 2$ Tesla; in the empty solenoid it would produce only $B = 100 \times 4\pi \times 10^{-7} = 1.25 \times 10^{-4}$ Tesla, which is 16,000 times smaller.

When the iron that fills the solenoid has the shape of a closed loop as in Figure 14.4A, the magnetic field does not flare out into space, but remains concentrated inside the iron. Stray magnetic fields are

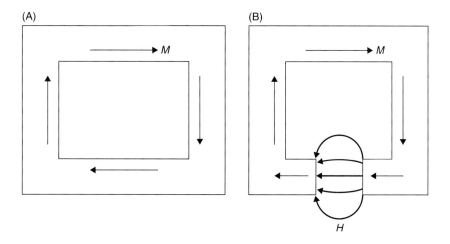

■ **FIGURE 14.4**   (A) Magnetic field is concentrated inside closed loop of iron. (B) A gap in the iron loop concentrates the magnetic field.

suppressed, and so is the spatial magnetostatic energy $\frac{1}{2}BH$ integrated over space. This is the geometry used in transformers. If a gap is introduced in the loop, as in Figure 14.4B, the magnetic field (and the induction) is concentrated in the gap. This feature is used, for instance, in magnetic recording.

14.2.4 Hysteresis Curves

The magnetization M varies according to the hysteresis curve shown in Figure 14.5. It is customary to plot the induction B (in Vs/m^2), rather than the magnetization M (in A/m). B is the quantity of engineering interest. Let us start with a material that is not magnetized. As the field H is increased from zero, the induction increases along the segment shown as a dotted line. The induction increases rapidly at first, then reaches **saturation** B_s. When the magnetic field is then reduced to zero, the induction does not disappear but decreases only slightly to the value B_r, known as the **remanent induction**. When we reverse the magnetic field B decreases, reaching the value $B = 0$ when $H = -H_c$. H_c is known as the **coercive field**. Increasing the intensity of the reversed field H increases the induction in the opposite direction until saturation is reached. When we decrease H, in the direction of the arrow, B decreases to the **remanent induction** B_r when H is zero. Increasing the magnetic field in the positive direction causes $B = 0$ again at the coercive field H_c. Further increase of H increases B until saturation B_s. In an alternating magnetic field H, the induction B follows the hysteresis curve.

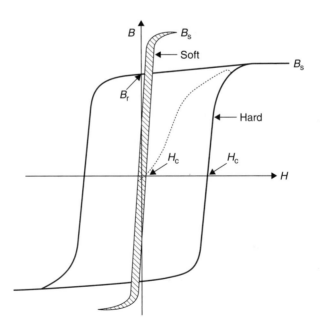

■ **FIGURE 14.5**   Schematic hysteresis curve of a soft and a hard magnet. In reality, the coercive field of a hard magnet is thousands of times larger than that of a soft magnet. See Tables 14.1 and 14.2.

14.2.5 **Energy Losses in an Alternating Magnetic Field**

When the ferromagnet is subjected to an alternating magnetic field, such as in a transformer, energy is lost at each cycle by two mechanisms: hysteresis losses and eddy currents.

Hysteresis Losses   At each cycle of the alternating field, an energy corresponding to the area inside the hysteresis curve is lost and transformed into heat. If, for instance, the coercive field is $H_C = 500\,A/m$ and the saturation induction is $B_S = 2\,T$, the energy loss is $4{,}000\,J/m^3$ at each cycle. With a frequency of 60 Hz, this represents $240\,kW/m^3$.

Eddy Currents   The magnetic core can be considered to be made of many loops in which the time variation dB/dt induces a voltage according to Equation (14.4). If the magnet is electrically conductive, this causes an electric current to flow. This current is proportional to the frequency. This current heats the material and constitutes an important loss of energy, especially at high frequencies. (Such energy absorption is used in heating by inductive coils and in microwave ovens.)

14.2.6 **Soft Magnets**

When the magnetization is produced by an AC current, as for instance in electric transformers, one uses magnetic materials with a **small coercive field** to reduce hysteresis losses. These are the **soft magnets**. Table 14.1 presents the properties of a number of commercial soft magnets.

Table 14.1 Properties of Soft Magnets.

Materials and composition (wt%)	Saturation induction B_S Tesla (gauss)	Coercive field A/m (oersted)	Hysteresis loss per cycle J/m³	Resistivity ($\mu\Omega m$)
Commercial iron ingot (99.95Fe)	2.2 (22,000)	80 (1)	270	0.1
Silicon-iron, oriented (97Fe, 3Si)	2.01 (20,100)	12 (0.15)	40	0.47
45 Permalloy (45Ni, 55Fe)	1.6 (16,000)	4 (0.05)	120	0.45
Supermalloy (79Ni, 15Fe, 5Mo)	0.8 (8,000)	0.16 (2.10^{-3})	–	0.6
Ferroxcube A (MnZn ferrite)	0.25 (2,500)	0.8 (0.01)	~40	2.10^6
Amorphous Fe-B-Si	1.6 (0.02)	~0	~0	10^3

14.2.7 **Hard Magnets**

When it is desired that a magnet conserve its magnetization in the presence of external magnetic fields, one selects a material with a **high coercive field H_c**. These are the **hard magnets**, which are represented in Table 14.2. Note that the coercive field in the hard magnets is indicated in kA/m and in A/m for soft magnets.

*The BH **Product of Hard Magnets*** Good hard magnetic materials possess a high saturation induction and a high coercivity. These properties are combined in a single quantity, the maximum BH product or $(BH)_{max}$, which is used as a convenient figure of merit for permanent magnet materials. $(BH)_{max}$ is derived from the demagnetization portion, or second quadrant of the B-H curve shown in Figure 14.6. It is the largest rectangle $(B \times H)$ that can be inscribed in the hysteresis curve. (The right side of Figure 14.6B is a plot of BH versus B from which $(BH)_{max}$ is readily obtained.) This is the way the values reported in Table 14.2, and noted in Figure 14.1, were obtained. Since it is derived from the hysteresis curve, the BH product is also known as the energy (density) product with units of J/m^3 cycle. In the cgs system, the strength of magnets is measured in mega-gauss-oersted (MGOE). It is often also indicated as N followed by the number of MGOE's. (A magnet N32 has a strength of 32 MGOE.)

Table 14.2 Properties of Hard Magnetic Materials.

Material and composition (wt%)	Remanent magnetization Tesla (gauss)	Coercivity H_C kA/m (oersted)	BH_{max} kJ/m³ (MGOe)	Curie temperature (°C)
Carbon steel (0.9C, 1 Mn)	0.95 (9,500)	4 (50)	1.6 (0.2)	~768
Cunife (20Fe, 20Ni, 60Cu)	0.54 (5,400)	44 (550)	12 (1.5)	410
Alnico V (50Fe, 14Ni, 25Co, 8Al, 3Cu)	0.76 (7,600)	123 (1,550)	36 (4.5)	887
Samarium cobalt (SmCo₃)	0.9 (9,000)	600 (7,540)	140 (18)	727
Nd₂Fe₁₄B	1.1 (11,000)	900 (11,300)	220–330 (28–42)	347
Barium ferrite (Ba-6Fe₂O₃)	0.4 (4,000)	264 (3,300)	28 (3.5)	447

Note the change in scale from Table 14.1. Coercivity is here measured in k-A/m. Data from various sources.

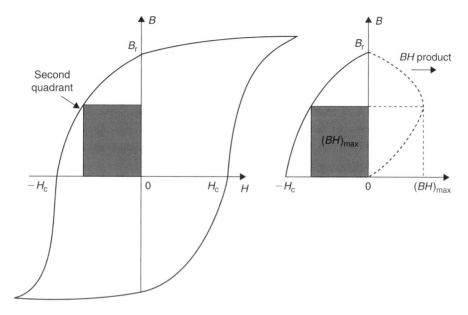

■ **FIGURE 14.6**   Method for obtaining the strength $(BH)_{max}$ of a hard magnet from the hysteresis curve. $(BH)_{max}$ is the product of the *B* and *H* lengths defining the largest rectangular area that can be inscribed within the second quadrant of the *B-H* loop.

14.3 **FERROMAGNETIC MATERIALS**

How is it possible to fabricate hard and soft magnetic materials whose coercive fields differ by a factor of 10,000? The processing of magnets is best understood if we first examine what causes the magnetization of these materials and how this magnetization responds to an applied magnetic field.

The magnetization *M* of a material has its origin in the properties of its electrons.

14.3.1 **Magnetic Moments**

Magnetic Moment of an Electron   Every electron has a spin: just like the earth, it spins around its axis and carries a magnetic moment. The magnetic moment of an electron is the **Bohr magneton** μ_B ($\mu_B = 9.27 \times 10^{-24}\,A.m^2$). In addition, the electrons turn around the nucleus according to the quantum number ℓ. This angular orbit acts as a one-loop magnet and creates an **orbital magnetic moment**. These moments are illustrated in Figure 14.7.

Magnetic Moment of an Atom   In a single atom the total magnetic moment is the vector sum of the orbital and spin magnetic moments of all its electrons. These moments have different directions and

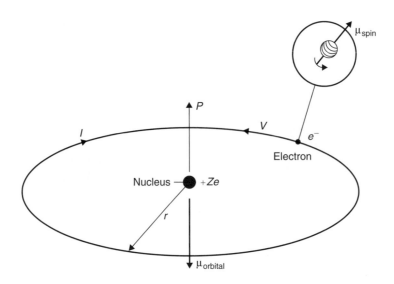

■ FIGURE 14.7   Origin of the magnetic moments in atoms due to an electron orbiting the nucleus ($\mu_{orbital}$) and to the electron spin about its axis ($\mu_{spin} = \mu_B$).

Ion	Number of electrons	Electronic structure 3d shell					Magnetic moment (Bohr magnetons)
Cr^{2+}, Mn^{3+}	22	↑	↑	↑	↑	☐	4
Fe^{3+}, Mn^{2+}	23	↑	↑	↑	↑	↑	5
Fe^{2+}	24	↑↓	↑	↑	↑	↑	4
Co^{2+}	25	↑↓	↑↓	↑	↑	↑	3
Ni^{2+}	26	↑↓	↑↓	↑↓	↑	↑	2
Cu^{2+}	27	↑↓	↑↓	↑↓	↑↓	↑	1
Zn^{2+}, Cu^+	28	↑↓	↑↓	↑↓	↑↓	↑↓	0

Atom							4s	
Fe	26	↑↓	↑	↑	↑	↑	↑↓	4
Co	27	↑↓	↑↓	↑	↑	↑	↑↓	3
Ni	28	↑↓	↑↓	↑↓	↑	↑	↑↓	2

■ FIGURE 14.8   Electronic structure and magnetic moments of ions and atoms in the transition metal series. The arrows show the spins of the electrons, and the boxes identify the electron orbitals.

may cancel each other. As shown in Figure 14.8, iron has six 3d electrons, of which two have opposite spins, which results in a spin magnetic moment of four Bohr magnetons. Similarly, cobalt has three unpaired spins and nickel has two. Because of the addition of the orbital and spin moments, the net magnetic moments of these atoms are

$$\mu_A = 2.22 \; \mu_B \text{ for Fe}$$

$$\mu_A = 1.72 \; \mu_B \text{ for Co}$$

$$\mu_A = 0.60 \; \mu_B \text{ for Ni.}$$

14.3.2 **Ferromagnetism**

In most solids, the magnetic moments of the atoms are randomly oriented, so that the total magnetization of the material is zero in the absence of a magnetic field and very small when a magnetic field is applied. In iron, cobalt, and nickel, the magnetic coupling aligns all the atomic moments μ_A so that they are parallel to each other; this results in a spontaneous magnetization

$$M_S = N_A \mu_A$$

where N_A is the density of atoms in the metal (m^{-3}) and μ_A is the magnetic moment of an atom (Am^2).

Iron has a density of $7.8 \times 10^3 \, kg/m^3$, and an atomic weight of $55.86 \times 10^{-3} \, kg/mole$; Avogadro's number is 6.023×10^{26} atoms/mole so that $N_A = 8.38 \times 10^{28}$ atoms/m^3. We calculate a saturation magnetization $M_S = 8.38 \times 10^{28} \times 2.22 \times 9.27 \times 10^{-24} = 1.72 \times 10^6 \, A/m$. This corresponds to $B_S = 2.17 \, T$, which compares well to the measured value $2.2 \, T$ given in Table 14.1. This is the magnetization of iron when all atomic moments are aligned; it is the saturation magnetization in the hysteresis curve.

14.3.3 **Ferrimagnetism**

Ferrimagnets are ceramics, generally oxides. They have a smaller saturation magnetization than the metallic ferromagnets; these materials are electric insulators and thereby avoid the losses by eddy currents.

Magnetite, Fe_3O_4, is a magnetic oxide, or **ferrite**. It is composed of one Fe^{2+} and two Fe^{3+} ions. It could be rewritten as Fe_2O_3FeO, or $Fe^{2+}Fe^{3+}{}_2O^{2-}{}_4$. Magnetite has the inverted spinel structure shown in Figure 14.9. It can be considered an FCC lattice of O^{2-} ions in which one Fe^{3+} ion, with spin down, is placed in a tetrahedral interstice, the other Fe^{3+} with spin up is in an octahedral interstice, and the Fe^{2+} ion occupies another octahedral site. The magnetic moments of the two Fe^{3+} cancel each other, so that the Fe^{2+}, carrying four Bohr magnetons, provides the magnetic moment of the whole unit cell. The saturation magnetization of magnetite can be estimated as

$$M_S = 4 \times 9.27 \times 10^{-24} / (0.419 \times 10^{-9})^3 = 5.04 \times 10^5 \, A/m$$

The measured value is $M_S = 5.3 \times 10^5 \, A/m$, corresponding to a saturation induction $B_S = 0.67 \, T$.

Other magnetic ferrites have inverted spinel structures in which the divalent cation Fe^{2+} is replaced by Ni^{+2}, Co^{+2}, or Zn^{+2} which provide the magnetic moment. For example, in $Co^{2+}Fe^{3+}{}_2O^4$ a mole of CoO is alloyed with one of Fe_2O_3. The subunit cell magnetization depends only on the Co^{2+} ion moment of $3 \, \mu_B$. Because of moment cancellation of the two Fe^{3+} ions, the saturation magnetization of ferrites is generally less than that of metal magnets.

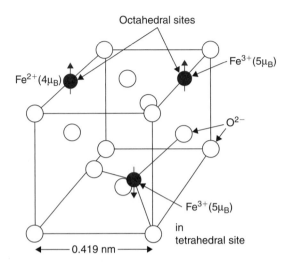

Octahedral sites

$Fe^{2+} (4\mu_B)$

$Fe^{3+} (5\mu_B)$

O^{2-}

$Fe^{3+} (5\mu_B)$
in
tetrahedral site

0.419 nm

■ **FIGURE 14.9** Crystal structure of the inverted spinel Fe_3O_4.

Hexagonal ferrites are an important class of magnetic ferrites that have a complex hexagonal structure. These ferrites have the formula $MO.6Fe_2O_3$ or $MFe_{12}O_{19}$, where M is Ba, Sr, or Pb. In each formula unit, the 12 Fe^{3+} ions are arranged on several different sites: four Fe^{3+} ions have spin up and eight have spin down. There are thus four uncompensated Fe^{3+} ion moments each of magnitude $5\mu_B$. Therefore, the net moment is $4 \times 5\mu_B = 20\mu_B$ per formula. The hexagonal ferrites are used to make all kinds of permanent magnets and will be addressed again below.

14.3.4 Temperature-Dependence of Magnetism

Ferromagnetic materials are spontaneously magnetized, but this does not mean that every last moment is oriented in the same direction. Certainly the overwhelming majority are, and at $0°K$ all moments are perfectly aligned. In this case the magnetization saturates or reaches its maximum value M_s $(T = 0°K)$. Above $0°K$ the alignment of the spins is imperfect, a trend that increases as the temperature rises higher and higher. Finally at the **Curie temperature** (T_c) the magnetization vanishes because thermal agitation energy overcomes the magnetic interaction energy and the orientation of the magnetic moments is random.

14.3.5 Magnetic Domains

If magnetic interaction aligns all the atomic magnetic moments of the atoms, how is it possible to demagnetize a ferromagnetic material? Look at Figure 14.10. When all the atomic moments are aligned, as in Figure 14.10A, the magnetization has the saturation value M_S we just calculated. Outside of the magnet, the magnetic field extends into space and carries the spatial magnetization energy $1/2 \int BH.dV$. This spatial energy can be reduced if the orientation of the magnetic moments organizes itself into **domains**. With the domain structures B to D, the net magnetization of the bar is zero; the

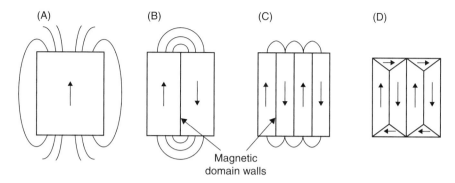

■ FIGURE 14.10   Ferromagnetic domain structure accompanying the reduction in magnetostatic energy. (A) Single domain (this occurs in very small magnetic particles). In domain structures B–D, the material is demagnetized. (B) Two domains of opposite magnetization. (C) Four domains alternating in magnetization. (D) Closure domains minimize magnetostatic energy.

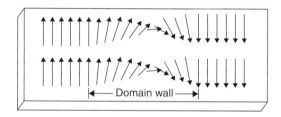

■ FIGURE 14.11   Formation of a domain wall by rotation of individual atomic moments.

material is demagnetized. In the domain structure D, the magnetic space energy vanishes as well. Remember the difference: in a demagnetized ferromagnet, neighboring atomic moments remain parallel, but they are arranged in a domain structure such that the total magnetization vanishes; above the Curie temperature, the magnetization vanishes because individual magnetic moments are oriented at random.

Domain Walls   The energy that aligns the atomic moments is too large to permit an abrupt change in direction at the border between two domains. In this border, neighboring atomic moments rotate by small amounts until the new direction is achieved. This is illustrated in Figure 14.11. The region in which the rotation occurs is a **domain wall**. The thickness of the domain wall represents the compromise with the smallest energy.

14.3.6 Interaction with a Magnetic Field and Hysteresis Curve

A magnetic field H applied to a demagnetized ferromagnet tends to align the atomic moments in the direction of H. This occurs by the movement of the domain walls as shown in Figure 14.12. Figure 14.12A depicts a demagnetized ferromagnet. When a magnetic field is applied, the domain walls move, the region that is magnetized parallel to H increases, and so does the total magnetization M.

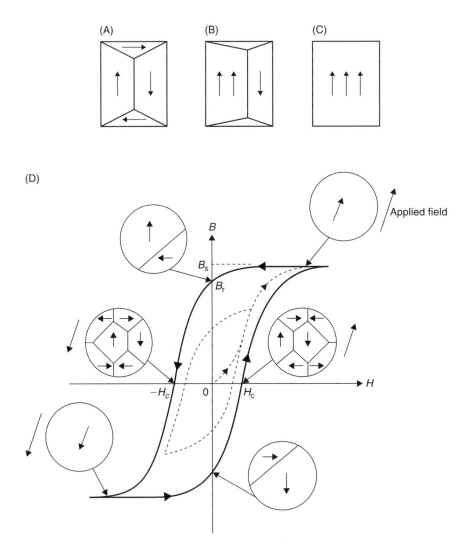

■ **FIGURE 14.12** The movement of magnetic domain walls increases the magnetization in response to an applied magnetic field. (A) Demagnetized material: the total magnetization is zero. (B) A magnetic field H is applied (vertical in the figure); the domain wall moves to the right so that more atomic moments are parallel to the magnetic field. (C) When the applied field H is large enough, the domain wall has moved to the border of the particle which now contains a single domain; all magnetic moments are parallel and saturation magnetization has been reached, $M = M_S$. (D) Schematic structure of the domains as a function of magnetization.

Figure 14.12C shows the saturation magnetization where all atomic moments are parallel. When this is reached, saturation magnetization M_s is achieved and the magnet is a single domain.

Domain walls move easily in a material that is perfectly homogeneous. Grain boundaries and other crystalline defects are obstacles to the movement of domain walls. With grain boundaries and defects this movement now absorbs energy. When the magnetic field H is reduced to zero, complete demagnetization (Figure 14.12A) would correspond to the lowest energy, but it cannot be reached by the spatial magnetostatic energy alone. A certain magnetization (or induction $B_{R)}$ remains at $H = 0$; this is the **remanent induction** B_r. A magnetic field in the opposite direction is needed to demagnetize the material; this is the **coercive field** H_C. Figure 14.12D shows schematically the configuration of domains as the magnetization is cycled by an AC magnetic field. The total energy necessary to move the domain walls through an entire cycle of the hysteresis curve is the **hysteresis loss per cycle**; it corresponds to the area circumscribed by the hysteresis curve as we have seen in Section 14.2.5.

14.4 PROPERTIES AND PROCESSING OF MAGNETIC MATERIALS

We now understand that a soft magnet is one in which the domain walls move with little resistance, and a hard magnet is one that requires a large energy to move the domain walls. In this section we examine the processing methods that allow us to obtain soft or hard magnets.

14.4.1 Soft Magnets

Domain walls move easily in a perfectly homogeneous material. Crystal defects such as grain boundaries, dislocations, impurities, or precipitates are obstacles to the motion of domain walls. Therefore, soft magnets are processed to contain as few defects as possible. This is achieved in three types of magnets: iron silicon steels, soft ferrites, and metallic glasses.

Iron-Silicon Electrical grade steel alloys for the generation and distribution of electrical energy (i.e., transformers) are the magnetic metals produced in the largest tonnage. In 1900 the benefit of silicon additions to iron was discovered. It was found that 3 wt% Si in Fe not only increases the permeability, but simultaneously reduces the coercive force relative to unalloyed iron. By using laminations rather than bulk metal cores, eddy current paths were interrupted, further lowering core losses. Later progress in lowering hysteresis losses was made by reducing the carbon content through hydrogen annealing, adjusting the grain size, rolling the iron into sheets, and by introducing tensile stresses in the sheet. Through incremental improvement over a century, core losses dropped from 8 W/kg of core metal to 0.4 W/kg.

Ferrites The development of soft ferrites dates to the mid 1930s when important advances in magnetic materials began to unfold. Over the years important compositions were developed; they included $(MnZn)Fe_2O_4$, $(MnCu)Fe_2O_4$, and $(NiZn)Fe_2O_4$.

Ferrite products are produced by powder pressing and sintering, and one of the common shapes produced is the toroid. When wound with wire, toroids make efficient transformer cores, and because they are electrically insulating, there is very little eddy-current loss even when they are operated at microwave frequencies. This, coupled with high saturation magnetization, has enabled ferrites to find applications in television and radio components such as line transformers, deflection coils, tuners, rod antennas, as well as in small to medium power supplies.

Metallic Glasses These relatively recent metal alloys typically contain one or more transition metals (e.g., Fe, Ni, Co) and one or more metalloids (e.g., B, Si, C). Melts quenched at extremely high rates are continuously cast to yield ribbons that are approximately as thick as a sheet of paper (i.e., $75\,\mu m$). Several alloys (e.g., 80Fe-20B) possess all the properties desired of soft magnetic materials: high saturation induction, low coercivity, high permeability, and low core loss from DC to several MHz. Their low anisotropy and lack of grain boundaries means there is little to prevent the movement of domain walls. Although they are potentially attractive as a replacement for electrical steels, their low ductility, high cost, and potential instability during ageing are issues of concern.

14.4.2 **Hard Magnets**

Grain boundaries and the defects introduced by heavily deformed or strained materials do impede the motion of domain walls and increase the coercive field. By far the most effective way to achieve a high coercive field is the use of shape anisotropy. Consider the magnetic needle of a compass, with the magnetization along the long axis or perpendicular to it, Figures 14.13A and B. When the magnetization is perpendicular to the long axis (Figure 14.13B), the magnetic field (or induction) outside the material covers a much larger space and carries a much larger energy $E = 1/2\int\mu_0 H^2 dV$ than it does when the magnetization is parallel to the long axis (Figure 14.13A). Therefore, the magnetization will spontaneously orient itself along the needle axis, and a large energy is required to rotate it: the coercive field of such a magnet is large. This method is particularly effective when the magnetic needle is so small that it contains a single magnetic domain. The strategy to produce high-coercive-field magnets is to consolidate needle-shaped single-domain particles. There are several methods that achieve this aim.

Alnico The name of one of the most widely produced permanent magnet materials is derived from the **al**uminum, **ni**ckel, and **co**balt additions to the iron base metal. The first step involves solution treatment at $1,300\,°C$. Upon cooling to $800\,°C$, a two-phase decomposition into α and α' occurs. Rich

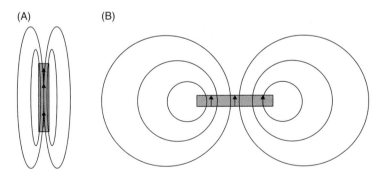

■ **FIGURE 14.13** Magnetization of a needle: (A) Magnetization along the long axis produces induction in a small volume and small magnetostatic energy. (B) (impossible): Magnetization perpendicular to the long axis would produce an induction field *B* over a large volume: magnetostatic energy would be large; the magnetization spontaneously rotates to configuration (A). Rotation of the magnetization downward would require going through (B), which requires a large amount of energy.

in Fe and Co, α' has a high magnetization. When the material is cooled in a magnetic field, the α' phase precipitates in the shape of needles and provides a high coercive field H_C (Figure 14.14).

Rare Earth Magnets   Magnetism in these alloys is derived microscopically from unpaired $4f$ electrons of Sm; on a macroscopic basis the coercivity arises because of domain pinning at grain boundaries and surfaces. The most important rare earth magnet is $SmCo_5$. In single phase form, this alloy has an energy product four to six times that of the Alnicos and coercivities that are 10 times higher. This allows miniaturization of magnets used in motors, surgically implanted pumps and valves, and electronic wristwatch movements. Sm-Co alloys are fabricated by powder metallurgy methods where the magnetism is aligned during pressing in a magnetic field. Sintering is done at low temperatures to prevent grain growth. There are other similar Sm-Co alloys containing copper or iron that precipitate magnetic phases and can be heat treated to yield extraordinarily high $(BH)_{max}$ products.

The most recent significant advance in permanent magnet materials has been the development of neodymium-iron-boron ($Nd_2Fe_{14}B$) in 1984. These materials have $(BH)_{max}$ products of $\sim300\,kJ/m^3$, the highest yet obtained. Produced by both rapid solidification and powder metallurgy processing, these magnetic materials are expected to find applications requiring low weight and volume. High cost is a disadvantage of rare earth magnets.

Ferrites   Hard ferrites account for slightly over 60% of the total dollar value and over 97% of the total weight (250,000 tons) of permanent magnets produced. These magnets are largely based on barium and strontium ferrite with compositions of $BaO.6Fe_2O_3$ and $SrO.6Fe_2O_3$. They are made by

■ **FIGURE 14.14**   Microstructure of Alnico. The white areas represent the magnetic phase. TEM image taken by Y. Iwama and M. Inagaki. Courtesy of JEOL USA Inc.

pressing and sintering powders whose grain sizes are large compared with the domain size. Thus the magnetization is believed to occur by domain nucleation and motion, rather than by the rotation of single domain particles. Hard ferrites have low $(BH)_{max}$ products compared to other permanent magnets. Nevertheless, their low cost, moderately high coercivity, and low density are responsible for widespread use in motors, generators, loudspeakers, relays, door latches, and toys.

14.5 ILLUSTRATION: MAGNETIC RECORDING

In magnetic recording systems, information is digitally stored as 1 and 0 states corresponding to positive and negative magnetization levels. To better appreciate the material requirements necessary to store and read large densities of information at high speeds cheaply and reliably, it is helpful to understand the recording process.

Time-dependent electrical input signals are converted into spatial magnetic patterns (**writing**) when the **storage medium** translates relative to a recording head as schematically shown in Figure 14.15. The medium is either a magnetic tape or a flat disc. It must be a hard magnet that possesses high coercivity. The latter prevents the magnetization from being erased or altered when exposed to external magnetic fields. The head is a gapped soft ferrite with windings around the core portion. Input electrical signals are converted into magnetic flux that is concentrated in the head. Fringe flux flaring out into the narrow gap intercepts the nearby magnetic medium, magnetizing it over small regions; the smaller the region, the higher the storage density.

During playback (**reading**) the magnetized medium is run past the head. The moving magnetic fields, originating from the medium, readily induce voltage signals, as in Equation (14.4), in the stationary pick-up coils of the head (Figure 14.15A). More recent read heads utilize the giant magnetoresistance (GMR) effect described in the next section (Figure 14.15B). Electrical amplification regenerates the sound, picture, or information recorded or stored in the first place.

From the foregoing it is apparent that magnetic recording systems require opposite but complementary magnetic properties, that is, soft magnetic materials for the recording and playback heads and hard magnetic materials for the storage media. The write and read heads of modern devices are extremely small and are produced by techniques similar to those described in Chapter 12.

Traditionally, high permeability soft MnZn ferrites and alloys like permalloy have been used for heads. Tape and disc media have been dominated by the use of very small ($<0.5\,\mu m \times 0.06\,\mu m$) elongated particles of γFe_2O_3, $Co\gamma Fe_2O_3$ or CrO_2 that are essentially single domains. The high coercivity in these high aspect ratio domains is due to **shape anisotropy**. They are deployed in a passive binder on flexible polymer tapes, or on more rigid metal discs.

The increasing use of thin magnetic film media has been driven by the desire to reduce the size of personal computers and portable video recording and playback systems. Thin films enable higher recording densities to be achieved. The reason is due to the combined efficiency of 100% packing of magnetic material in films—compared to 20–40% in particulate media—and the generally higher magnetization possible with cobalt alloy films.

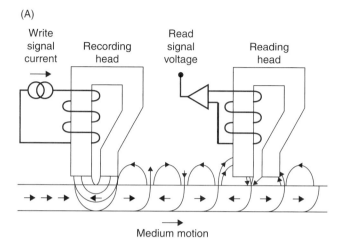

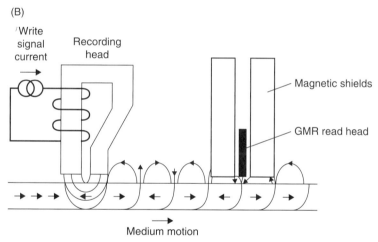

■ **FIGURE 14.15** Magnetic recording on tapes of computer discs. A signal in the recording head orients the magnetization in a recording bit. (A) The movement of the medium induces a voltage in the reading head loop every time the magnetization changes. (B) In modern computers, the reading head uses giant magnetoresistance (GMR) head instead of electromagnetic induction.

14.5.1 **Giant Magnetoresistance (GMR)**

Modern read-heads of computers use a phenomenon called giant magnetoresistance. The key structure in GMR materials is a spacer layer of a nonmagnetic metal, a few nanometers thick, between two magnetic metals (Figure 14.16). The magnetization of one layer is fixed, and that of the other layer is free to move (it is a soft magnet). When a magnetized bit of information on the medium approaches the free layer, the magnetization of the latter rotates, either parallel or antiparallel to the fixed magnetic layer. The chief source of GMR is "spin-dependent" scattering of electrons. Electrical resistance is due to scattering of electrons within a material. A single-domain magnetic material scatters electrons with

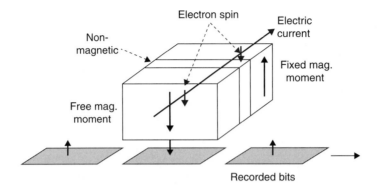

■ **FIGURE 14.16** Structure of a GMR read head. The structure is formed of very thin magnetic layers and a nonmagnetic (copper) spacer. The magnetization of one layer is pinned (by apposing it to a strong antiferromagnetic material). The magnetization of the "free layer" is rotated by the magnetic bit on the recording medium.

spin parallel or antiparallel to its magnetization differently. When the magnetic layers in GMR structures are aligned antiparallel, the resistance is high because electrons that are weakly scattered in one layer are strongly scattered in the other. When the magnetic fields of the layers are parallel to each other, the electrons that are weakly scattered in one layer are weakly scattered in the other layer as well; as a consequence the electric resistance is lower than in antiparallel magnetization. The change in electrical resistance, as the recorded magnetic bit passes below the structure, produces an electric signal that is transmitted to the computer.

The extraordinarily small GMR structures are fabricated by depositing very thin layers of different materials onto a substrate. Their structure and fabrication will be described in more detail in Chapter 18 on Nanomaterials.

The name "giant magnetoresistance" is understood as follows. All conductors exhibit a small magnetoresistance. It is well known in physics that a magnetic field imposes a curvature on the trajectories of electrons. This curvature reduces their drift velocity and increases the resistance. The effect of the structure used in read-heads is very much larger than in ordinary conductors, hence the name.

■ SUMMARY

1. A magnetic field H is generated by a solenoid in which a current flows. The magnetic field has the units ampere turns per meter or ampere per meter (A/m). The same field is also expressed as induction B in volt second/square meter (Vs/m^2), which is also called Tesla (T); it is obtained by multiplying H with the induction constant $\mu_o = 4\pi \times 10^{-7}$ Vs/Am. Thus $1\,\text{A/m} = 4\pi \times 10^{-7}$ T $= 1.2566 \times 10^{-6}$ T. The integral $\int B dH$ is the energy density in J/m^3 of the magnetic field. In free space, it is $\frac{1}{2}\mu_o H^2$ or $\frac{1}{2}BH$. The old cgs units are still widely used in magnetism. The magnetic field H is measured in oersteds, $1\,\text{A/m} = 4\pi \times 10^{-3}$ oersted; (1 oersted $\approx 80\,\text{A/m}$) and the induction is measured in gauss, 1 tesla $= 10^4$ gauss. The magnetic energy is customarily measured in megagauss-oersted MGOE.

2. A magnetic material acquires a magnetization M when subjected to a magnetic field. This magnetization is measured in the same units of H (A/m) or B (T). The total induction then becomes $B = \mu_o(H + M)$. In ferromagnetic materials M is vastly larger than H.

3. Every atom carries a magnetic moment μ_A that is the sum of the magnetic moments of the electronic spins and electronic orbitals. In ferromagnetic materials, these atomic moments are aligned parallel to each other.

4. A ferromagnet is divided into magnetic domains. A domain is a region in which all atomic moments are parallel. Its magnetic moment is the sum of all the atomic moments. The magnetic moments of different domains are oriented in different directions. The total magnetization M of a magnet is the vector sum of the magnetic moments of the domains. Saturation magnetization M_S or B_S is achieved when all magnetic moments are aligned, that is, when the magnet consists of a single domain.

5. The magnetization M changes under the influence of an external magnetic field. It does so by the movement of domain walls.

6. The motion of domain walls is inhibited by crystalline defects and by spatial magnetic anisotropy. The magnetization M (or induction B) does not follow the applied magnetic field H exactly.

7. The relationship between B and H is described by the hysteresis curve. A saturated magnet exposed to a diminishing field H retains the remanent magnetization M_R or remanent induction B_r when $H = 0$. A coercive magnetic field H_C in the opposite direction of its moment is required to demagnetize a ferromagnet.

8. In a complete cycle an amount of energy $\int BdH$ corresponding to the area inside the hysteresis curve is lost.

9. A soft magnet has a small coercive field H_C of a few A/m; its domain walls move easily. It is used in transformers, read-write storage heads, and other AC applications.

10. A hard magnet has a large coercive field H_C of up to 900,000 A/m. It is used in permanent magnets.

11. The strength of a permanent magnet is its BH product, measured in kJ/m^3 or MGOE. The commonly used indication N32 signifies a strength of 32 MGOE.

12. Soft magnets are processed with as few crystalline defects as possible in order to minimize the energy loss $\int BdH$ under the hysteresis curve in AC service.

13. Hard magnets are processed in the form of needle-shaped particles that are aligned in a magnetic field during processing.

14. The strongest ferromagnets are alloys of rare earths, particularly neodymium-iron-boron.

15. Eddy currents cause an additional energy loss in AC service of a magnet. They are proportional to the AC frequency. Eddy currents are minimized by fabricating power transformers in the form of iron sheets separated by insulating films and, at high frequency, by the use of ceramic ferrimagnets.

16. Ferrimagnets are electrically insulating oxides that contain ions with large magnetic moments. Some of these moments are antiparallel, so that the total magnetic moment is smaller than in metallic ferromagnets. Their advantages are very low eddy current losses and low price.

17. Giant magnetoresistance (GMR) allows the measurement of magnetization on a recording medium. The electrical resistance of the material is caused by the scattering of the electrons. The latter depends on the orientation of the electron spin with respect to the magnetization of the conductor. When spin and magnetization are parallel, there is less scattering than when they are antiparallel.

■ KEY TERMS

γFe_2O_3, $Co\gamma Fe_2O_3$, 418
μ_B, 409

A
Alnico, 416
atomic moment, 411

B
$BaO.6Fe_2O_3$, 417
BH product, 408
$(BH)_{max}$, 408
Bohr magneton, 409

C
coercive field, 406, 407, 408, 415, 416, 417, 421, 425, 426
CrO_2, 418
Curie temperature, 412, 413, 423, 424

D
domain walls, 413

E
eddy currents, 407, 421
energy density, 405, 420

F
Fe_3O_4, 411
ferrimagnetism, 411
ferrites, 411, 415, 417
ferromagnetism, 411

G
gauss, 405
giant magnetoresistance, 418, 419

GMR, 418, 419

H
hard ferromagnetic, 402
hard magnets, 408, 416
hexagonal ferrites, 412
hysteresis, 406
hysteresis curve, 413
hysteresis losses, 407, 415

I
induction, 403, 404, 405, 406, 407, 408, 411, 415, 416, 420, 421, 423, 426
induction constant, 404
iron, 403, 405, 410, 415, 416, 417, 421, 424, 426
Iron-Silicon, 415

M
magnetic domains, 412
magnetic fields, 403, 404
magnetic moment, 409, 410, 411, 421
magnetic recording, 418
magnetite, 411
magnetization, 403, 405
magnetoresistance, 418
magnetostatic energy, 415
metallic glasses, 416
$MFe_{12}O_{19}$, 412
MGOE, 408
$(MnCu)Fe_2O_4$, 415
$(MnZn)Fe_2O_4$, 415

N
$Nd_2Fe_{14}B$, 417
$(NiZn)Fe_2O_4$, 415

O
oersted, 405
orbital magnetic moment, 409

P
permeability of free space, 404

R
rare earth magnets, 417
reading, 418
remanent induction, 406, 415

S
saturation, 406, 407, 408, 411, 412, 414, 415, 416, 424, 425
shape anisotropy, 418
$SmCo_5$, 417
soft ferromagnetic, 402
soft magnets, 407, 415
solenoid, 403, 404, 405, 420, 423, 425
$SrO.6Fe_2O_3$, 417
storage medium, 418

T
temperature-dependence, 412
Tesla, 404

W
writing, 418

■ REFERENCES FOR FURTHER READING

[1] R.M. Bozorth, *Ferromagnetism*, Wiley-IEEE, New York (1993).

[2] P. Campbell, *Permanent Magnet Materials and Their Application (Paperback)*, Cambridge University Press, Cambridge, UK (1996).

[3] R.C. O'Handley, *Modern Magnetic Materials: Principles and Applications*, Wiley Interscience, New York (1999).

[4] N.A. Spaldin, *Magnetic Materials: Fundamentals and Device Applications*, Cambridge University Press, Cambridge, UK (2003).

■ PROBLEMS AND QUESTIONS

14.1. It is required to design a solenoid that will develop a magnetic field of 10 kA/m when powered with 1 A in vacuum. The solenoid dimensions are 0.3 m in length and 2 cm in diameter.

    a. How many turns of wire are required?

    b. If the solenoid is wound with 0.5 mm diameter copper wire, what DC voltage is required to power it?

14.2. Explain the physical difference between magnetization M and induction B.

14.3. Schematically sketch the hysteresis loop for a ferromagnet at a temperature close to $0°K$, just below the Curie temperature, and just above the Curie temperature.

14.4. Provide reasons for, or examples that illustrate, the following statements:

    a. All magnets are ferro- or ferrimagnetic materials but not all ferromagnetic materials are magnets.

    b. Not all elements with incomplete 3d bands are ferromagnetic.

    c. Ferromagnetic domain boundaries do not coincide with grain boundaries.

    d. Mechanical hardness in a magnetic metal promotes magnetic hardness; similarly mechanical softness promotes magnetic softness.

14.5. The *B-H* curve for a hard ferromagnet can be described by Figure 14.17A

    a. What is the total hysteresis energy loss per cycle?

    b. What is the $(BH)_{max}$ product?

14.6. The *B-H* curve for a hard ferromagnet outlines the parallelogram of Figure 14.17B.

    a. What is the total hysteresis energy loss per cycle?

    b. What is the $(BH)_{max}$ product?

14.7. Mention two ways in which soft and hard magnets differ.

14.8. Magnetic shielding materials protect electrical instruments and systems from interference by external magnetic fields, for example, the earth's magnetic field.

    a. Are soft ferromagnetic or hard ferromagnetic materials suitable for magnetic shielding applications?

    b. What specific property is important for magnetic shielding?

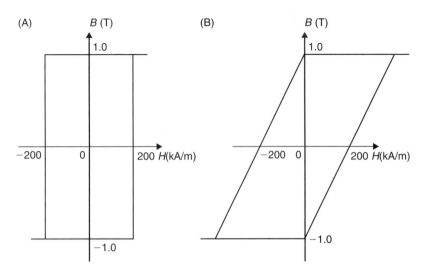

■ **FIGURE 14.17** Hysteresis curves for use in Problems 14.5 and 14.6.

14.9. For a permanent magnet material, the hysteresis loop second quadrant BH product varies parabolically with B as

$$BH \ (J/m^3) = -2 \times 10^5 B^2 + 2 \times 10^5 B \ \text{(units of } B \text{ are T).}$$

a. Plot BH vs. B.

b. What is the maximum value of BH or (BH)max?

c. What is the value of H at $(BH)_{max}$?

14.10. When is the electrical resistance of a ferromagnet an issue of concern in applications? What materials are available to address required electrical resistance needs?

14.11. A soft ferrite core weighing 20 g has rectangular hysteresis characteristics with $B_S = 0.4$ T and $H_c = 0.7$ A/m. The ferrite density is 5 g/cm^3 and the heat capacity is 0.85 J/g C.

a. Estimate the temperature rise after one magnetizing cycle if the process is carried out adiabatically, that is, with no loss of heat to the outside.

b. At 60 Hz operation, how long would it take for the magnet temperature to rise from 25°C to 420°C, the Curie temperature?

c. How will the temperature vary above the Curie temperature?

14.12. Predict the value of the saturation magnetization of body-centered cubic iron metal at room temperature?

14.13. The ferrite $NiFe_2O_4$ is a derivative of magnetite with Ni^{2+} substituting for Fe^{2+}.

a. If the lattice constant is unchanged, what is M_S for this ferrite?

b. Suppose that Zn^{2+} substituted for half of the Ni^{2+} so that the ferrite formula is now $Ni_{0.5}Zn_{0.5}Fe_2O_4$. What is M_S for this ferrite?

14.14. Computer memories require stable binary states (1 or 0 , or $+M$ and $-M$) for operation. Electrical methods must be provided to write in a 1 or 0, and read out a 1 or 0. Explain how magnetic bars plus wires (solenoids, etc.) can be configured to work as memory elements. What material hysteresis loop would be ideal for this application?

14.15. Distinguish between hysteresis losses and eddy current losses in soft ferromagnets.

    a. When do the former dominate the latter?

    b. When do the latter dominate the former?

    c. Mention an application that makes use of eddy current losses.

14.16. An Alnico magnet bar whose coercive field is $120\,kA/m$, is contained within a $0.2\,m$ long, 2,000 turn, air solenoid, far from either end. The axis of the bar is oriented parallel to the solenoid axis. Approximately what current will cause the Alnico magnet to demagnetize and essentially go on to reverse its polarity?

14.17. a. What magnetic properties are required of computer discs?

    b. What magnetic properties are required of reading and writing heads?

    c. Suggest ways to increase magnetic storage capacity.

14.18. How would you make a flexible magnet that would conformally hug a sheet steel surface?

14.19. For the following applications state whether high or low values of B_S and H_c are required.

    a. Strong electromagnet.

    b. Strong permanent magnet.

    c. GMR read-head.

    d. Loudspeaker.

    e. Transformer core.

    f. Compass needle.

    g. Write-head of hard-drive.

    h. Computer hard-drive disc.

14.20. From the tables in the chapter, select a material for the following, considering magnetic properties, manufacturability, and price. Justify your choice in terms of magnetic properties.

    a. The write-head of a computer.

    b. A VCR tape.

    c. A large power transformer.

    d. A high-frequency transformer.

    e. A magnet for pinning pictures on a refrigerator. State requirements for the refrigerator door.

    f. The hard disc of a computer.

14.21. What produces the magnetization in a ferromagnet?

14.22. Draw the hysteresis curves of a hard and a soft ferromagnet. Identify the coercivity and the saturation magnetization. Is the comparison of the two magnets in your figure to scale? Explain.

14.23. Why does one require a low coercive field in a transformer?

14.24. Compare a ferromagnet and a ferrimagnet. What is the relative advantage of each?

14.25. How does the hysteresis curve measure the hysteresis loss?

14.26. What are the units of a magnetic field?

14.27. What are the units of induction?

14.28. How is the core of a large transformer made? It is cast out of a block of iron. Why is it made this way?

14.29. How does the magnetization of a ferromagnet change?

14.30. What is the principle used to fabricate a soft magnet? What, approximately, is the coercive field of a modern soft magnet?

14.31. What is the principle used to fabricate a hard magnet?

14.32. What, approximately, is the coercive field of a modern hard magnet?

14.33. What material are the soft magnets of large transformers made of?

14.34. What material are the soft magnets of small high-frequency transformers made of?

14.35. What materials are the strongest permanent magnets?

14.36. How does one measure the strength of a magnet? What is the strength of an N32 magnet?

14.37. What is an eddy current? Where must it be avoided? Where does one make use of it?

14.38. Describe a ferrimagnet. Give an example.

14.39. Describe giant magnetoresistance and what causes it.

Chapter 15

Batteries

Electrical batteries are assemblies of electrochemical cells in which the free energy released in a chemical reaction is transformed directly into electrical energy. They consist of electrochemical cells in which two electrodes react with an electrolyte. In primary batteries, the electrode materials are consumed in an exothermic reaction. These batteries cannot be recharged. The chemical reaction occurs in two half reactions: The anode is dissolved in the electrolyte as positive ions, leaving the electron charge in the anode. The positive ions from the electrolyte are neutralized on the cathode, which acquires a positive charge. The result is a voltage between the electrodes that stops the chemical reaction when the circuit is open. When a current flows through an external load, the reaction continues until the battery is discharged because one of the electrodes is consumed by the reaction. Secondary batteries can be recharged because the chemical reaction is reversible. In primary batteries the anode is zinc or lithium and the cathode is MnO_2. Rechargeable batteries include the lead acid, nickel metal hydride, and lithium ion batteries.

Novel developments in ultracapacitors promise to store as much energy as batteries. If these become economically viable, they present the advantages of very high charging rates that are desirable for regenerative breaking in electric vehicles.

LEARNING OBJECTIVES

After studying this chapter, the student will be able to:

1. Describe the components of an electrochemical cell.

2. Explain how the free energy of a reaction is transformed into electrical energy.

3. Describe the properties and role of the electrolyte.

4. Use the Table of Standard Electrode Potentials to compute the open circuit voltage of a cell.

5. Define the theoretical and practical specific charge and specific energy of a battery.

6. Define the standard electrode.

7. Describe the structure and operation of primary batteries.

8. Describe the structure and operation of rechargeable batteries.

9. Describe the structure and operation of ultracapacitors.

15.1 **BATTERIES**

Electrical batteries are devices that transform the energy released by exothermic chemical reactions directly into electric power. They have at least twice the energy efficiency of gas turbines or internal combustion engines because they do not involve the Carnot cycle. In addition, batteries do not need fuel, do not emit gases or heat, are easily transportable, and can be made so small that they can be implanted in the human body (for pacemakers). From small button batteries that power hearing aids, low-weight rechargeable batteries for portable telephones, computers, and power tools, and the lead-acid batteries used in automobiles to the large batteries that power electric or hybrid automobiles, these devices will play an ever increasing role in modern technology. With such high demand, much technical progress can be expected in the near future.

Let us start by examining the lead-acid battery that is used to power the starter, lighting, and ignition (LSI) in practically all automobiles. The **12 volt battery** contains six galvanic cells connected in series; each cell generates 2 V. Each cell consists of a lead plate and a lead oxide plate immersed in sulfuric acid. There is an electric potential difference of 2 V between the plates. When the battery is connected to the starter or the lamps, an electric current flows while the lead on one plate and the lead oxide on the other are transformed into lead sulfate $PbSO_4$ and the sulfuric acid is changed to water. The overall reaction is

$$Pb + PbO_2 + 2H_2SO_4 \rightarrow 2PbSO_4 + 2H_2O \qquad (15.1)$$

This reaction is exothermic; it releases 405 kJ/mole of lead. This energy is released in the form of electric power. Reaction (15.1) is reversible: by applying a voltage somewhat larger than 12 V to the battery (or 2 V per cell), we can recharge the battery. This forces the current to flow in the reverse direction and the lead sulfate to revert to lead on one plate and to lead oxide on the other so that the battery is ready to provide power again. All this is made possible by the phenomenon of **electrochemistry**, which we examine next.

15.2 **PRINCIPLES OF ELECTROCHEMISTRY**

Electrochemical reactions occur in electrolytic cells. Each electrolytic cell consists of two electrodes and an electrolyte. One electrode (the lead plate in the battery) is the **anode, which is oxidized** in the spontaneous reaction; the other electrode is the **cathode** which is **reduced**. A **separator** placed between the two electrodes, consisting of a porous membrane, prevents liquid flow but permits ionic current from one side of the cell to the other. This is shown in Figures 15.1 to 15.3.

The electrolyte is a material, usually a liquid, which conducts electricity by mobile ions but not by electrons. In most cases, the electrolyte consists of water in which an acid, a base, or a salt is dissolved and dissociated into positive and negative ions. In the case of the lead acid battery, sulfuric acid is dissociated into H^+ and SO_4^{--} ions by the water. This is illustrated in Figure 15.4. Dissociation occurs because the water molecule, consisting of H^+ and O^{--} ions, forms a strong electric dipole. These dipoles give liquid water a large dielectric constant, $\kappa \approx 80$. As a consequence, the coulombic field binding the H^+ and SO_4^{--} ions together is reduced by a factor of 80. Thermal energy easily overcomes the now weakened

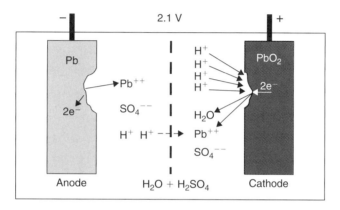

■ **FIGURE 15.1**   The lead-acid cell in open circuit. Lead is oxidized to Pb^{++} at the anode which is negatively charged, and lead oxide is reduced at the cathode which is positively charged. In the electrolyte, H_2SO_4 is dissociated into H^+ and SO_4^{--} ions.

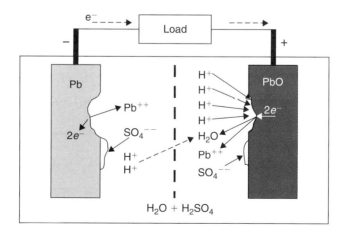

■ **FIGURE 15.2**   Discharging the lead-acid battery cell.

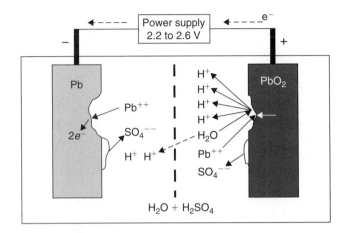

■ **FIGURE 15.3** Charging the lead–acid battery cell.

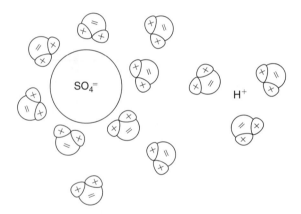

■ **FIGURE 15.4** Dissociation of H_2SO_4 in water. The water molecules constitute strong electric dipoles that decrease the electric field emanating from the charges in H^+ and SO_4^{--}. This decreases the ionic bond sufficiently to dissociate H_2SO_4.

ionic bond and the H^+ and SO^{--} separate, each surrounded by a shell of water dipoles that screen the electric field emanating from them. The same phenomenon is responsible for the dissolution of salt and other ionic materials.

Let us examine what happens in an electrolytic cell, and use the example of the lead-acid battery, Figure 15.1.

Water is also capable of dissolving lead from the anode; the lead goes into solution as Pb^{++} ions; each ion leaves two electrons in the solid.

$$Pb \rightarrow Pb^{++} + 2e^- \qquad (15.2)$$

The transformation of Pb into Pb^{++} is an **oxidation** reaction; it is exothermic and proceeds spontaneously. As a consequence, the anode acquires a negative charge and the electrolyte a positive charge. The separation of charges establishes an electric potential between the anode and the electrolyte. This increases as more lead is dissolved. The potential creates a current of ions back to the plate. When the potential is large enough, it recalls as many ions as are being dissolved, equilibrium is reached, and the dissolution **stops. This equilibrium potential E^o_A is equal to the lowering of free energy (measured in eV per ion) resulting from the dissolution.** Remember that we measure such energies in joules per mole or in eV per atom. The energy in joules per mole is $\Delta G = NeV$ where N is Avogadro's number.

A similar reaction takes place at the cathode where PbO_2 goes into solution as Pb^{2+} and $2O^{2-}$. In the process, lead that is Pb^{4+} in the oxide is reduced to Pb^{2+} in the electrolyte. H^+ ions in solution react with the O^{2-} ions to form water.

$$PbO_2 + 2e^- \rightarrow Pb^{++} + 2O^{--}$$
$$2O^{--} + 4H^+ \rightarrow 2H_2O$$

with the overall reaction

$$PbO_2 + 4H^+ + 2e^- \rightarrow Pb^{++} + 2H_2O \tag{15.3}$$

This reaction removes two electrons from the solid, which acquires a positive charge. Losing four H^+ ions and obtaining one Pb^{++} ion, the solution acquires a negative charge. The charges again establish an electric potential between the solid and the electrolyte. The reduction proceeds until the potential reaches an equilibrium value E^o_C.

> *Note that the symbol E^o is used in electrochemistry to denote a potential difference (not an electric field). In this chapter, we will use the symbol E for the electrode potential between electrode and electrolyte and V for the measured voltage at the battery terminals. Both E and V are expressed in volts.*

15.2.1 Open Circuit Voltage

In the complete cell we now have a negatively charged lead plate with a positively charge electrolyte in its vicinity and a positively charged lead oxide plate surrounded by a negatively charged electrolyte. The separator (a porous membrane) prevents the flow of liquid electrolyte but permits the H^+ ions to flow from the anode region to the cathode side so there is no electric potential difference inside the electrolyte.

The two potential differences between electrolyte and the electrodes add up to

$$V^o = E^o_C - E^o_A \tag{15.4}$$

between the electrodes; this is 2.1 V in the case of the lead-acid battery cell. This is the **open circuit voltage** of the cell. The free energy of the exothermic reaction has been transformed into a potential electric energy instead of heat.

15.2.2 **Battery Discharge**

When the electrodes are connected to an exterior load (Figure 15.2) an electron current I flows from the anode to the cathode and the voltage V between the electrodes decreases below its equilibrium value V^o. At the same time, H^+ ion current flows from the anode region to the cathode region of the electrolyte and the dissolution of Pb^{++} ions continues at both electrodes. For the ion current to flow, the electrolyte must have sufficient electrical conductivity σ_{ion}; the latter is proportional to the concentration of dissolved ions. Providing conductivity is the main task of the sulfuric acid. The electric resistivity $\rho = 1/\sigma_{ion}$ of the electrolyte is responsible for the **internal resistance** of the battery. The potential difference necessary to drive the H^+ flow is responsible for the drop of V from V^o.

The discharge voltage in the lead-acid battery varies between 1.8 and 2 V, depending on the current flowing and the state of charge.

The battery provides the power $W = I \cdot V$.

When the concentration of Pb^{++} is larger than can be dissolved in the electrolyte, the lead ions react with the SO_4^{--} ions to form solid $PbSO_4$ (Figure 15.2).

$$Pb^{++} + SO_4^{--} \rightarrow PbSO_4 \tag{15.5}$$

The lead sulfate precipitates on the electrode. The overall reaction in the cell is

$$Pb + PbO_2 + 2H_2SO_4 \rightarrow 2PbSO_4 + 2H_2O \tag{15.1}$$

The lead is oxidized and the lead oxide is reduced, both to $PbSO_4$; sulfuric acid is consumed and water is produced. The battery is discharged by the flow of current. It is exhausted when all the lead or lead oxide is transformed into lead sulfate or all the sulfuric acid is consumed.

15.2.3 **Charging the Battery**

The chemical reactions in the lead-acid battery are reversible. A voltage higher than the equilibrium potential E_o of 2.1 V reverses the current (Figure 15.3); it supplies electrons to the negative electrode and removes electrons from the positive electrode so that Pb^{++} is drawn back as Pb to the negative electrode. At the cathode, likewise, the Pb^{++} reacts with the water to create PbO_2 and four H^+ ions plus two electrons. The PbO_2 and the two electrons return to the anode and the four H^+ ions to the electrolyte. As the Pb^{++} concentration in the electrolyte decreases, it is replenished by the dissociation of $PbSO_4$ to Pb^{++} and SO_4^{--}.

The charging potential of the lead-acid battery is 2.2–2.6 V per cell.

Overcharging Once the battery is fully recharged and the potential is still applied to the electrodes, it will cause electrolysis of the water

$$H_2O \rightarrow \tfrac{1}{2}O_2 + 2H^+ + 2e^- \text{ at the cathode} \tag{15.6}$$

and

$$H^+ + 2e^- \rightarrow H_2 \text{ at the anode} \tag{15.7}$$

An explosive mix of gases is released by the battery. Hence the warning on car batteries to charge them outdoors or in a well-ventilated place.

Electrochemical reactions differ from ordinary chemical reactions in several ways:

1. The free energy of the reaction is released in the form of electrical energy instead of heat.

2. The reaction proceeds in two half reactions that occur on separate electrodes which are separated by an electrolyte.

3. The half reactions involve an exchange of electrons between the electrolyte and the electrodes.

15.2.4 Stored Charge and Stored Power (Faraday's Law)

How much energy can be released by the battery? The power released is the voltage (2 V per cell) multiplied by the current. The total energy stored in the cell is the voltage multiplied by the total charge released. The latter is easily computed:

Every lead atom releases two electrons. One mole of Pb releases two times Avogadro's number of electrons.

$$Q = neN = nF \text{ (coulomb)} \tag{15.8}$$

Here, n is the number of electrons released per atom ($n = 2$ with lead), $e = -1.6 \times 10^{-16}$. As is the charge of an electron, $N = 6.023 \times 10^{23}$ is Avogadro's number; $F = eN = 96,500$ coulombs or 26.8 Ampere hours (Ah) is a constant known as the **Faraday** (often also called **Faraday's constant**). The charge released by the lead-acid battery is $2F = 53.6$ Ah/mole; it requires the reaction of one mole lead (207.2 g) and one mole lead oxide (239.2 g), namely a mass equal to 446.4 g. The theoretical charge capacity of the lead-acid battery is

$$Q_S = nF/(M_{\text{mcathode}} + M_{\text{manode}}) \tag{15.9}$$

or $53.6/0.4464 = 120$ Ah/kg.

The **specific energy** is the energy released per unit mass of the battery. The **theoretical specific energy** is the energy released divided by the mass consumed in the reaction.

$$E_S = Q_S V_o = \frac{nF V_o}{(M_{\text{mcathode}} + M_{\text{manode}})} \tag{15.10}$$

Here V_o is the electric potential of the cell (2.1 V). The theoretical specific energy of the lead-acid battery is $120 \times 2.1 = 252$ Wh/kg.

The **practical** specific energy is the energy released divided by the **total mass** of the battery. So we count not only the electrode material that participates in the reaction but add the mass of the electrolyte, container, electric leads, and so on. The practical specific energy of the lead-acid battery is around 35 Wh/kg.

Modern applications in hand-held power tools and in hybrid vehicles demand higher amounts of stored energy and a lower weight of the batteries. According to Equation (15.10), the way to obtain this is in materials with a higher electrochemical potential E_o and a lower molecular mass. So we must explore the electrochemical potentials of other materials.

15.2.5 **Standard Electrode Potentials**

From Equation 15.4 we know that the open circuit voltage of the battery is the difference of the equilibrium electrode potentials

$$V^o = E^o{}_C - E^o{}_A$$

The open cell voltage V^o is easily measured with a voltmeter, but can we measure $E^o{}_C$ and $E^o{}_A$? Not directly.

If we wish to measure the electrochemical potential between a solid and an electrolyte, we must insert a second metallic electrode into the liquid. But the second metal will establish its own electrochemical potential, and a voltmeter will read the difference of the two electrochemical potentials: $V = E_{M1} - E_{M2}$. The measured voltage will depend on the metal chosen for the second electrode. It will also depend on how many ions of the electrode material are already dissolved in the electrolyte. If their concentration is high, the potential needed to call them back is lower than when their concentration is low. This is Nernst's law, which we will examine in Chapter 16. Clearly, we cannot measure the electrode potentials directly. We need to choose a standard reference electrode in well-defined electrolyte concentrations against which we can determine **standard electrode potentials**.

To determine the standard electrode potentials of different materials, one constructs an electrolytic cell depicted in Figure 15.5. The **standard reference electrode** has been chosen to be chemically inert platinum immersed in a **1 molar solution of HCl**. (A 1 molar solution is defined as 1 mole dissolved in 1 L of water.) The HCl dissociates in the water into H^+ and Cl^- ions and constitutes the electrolyte.

At the platinum electrode, the reaction

$$2H^+ + 2e^- \leftrightarrow H_2 \tag{15.11}$$

takes place. In the forward reaction, H^+ ions are neutralized on the surface of the platinum, which donates one electron to each ion and acquires a positive charge. The hydrogen atoms combine and bubble away as H_2. The missing H^+ ions produce a corresponding negative charge in the liquid. In the reverse reaction, H_2 bubbled into the half cell decomposes on the platinum surface where the H atoms are ionized and dissolved as H^+, donating an electron to the platinum. This standard is called a **hydrogen electrode**.

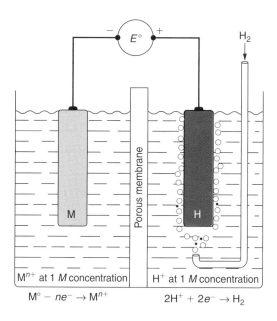

■ **FIGURE 15.5** Standard electrolytic cell. The standard hydrogen electrode on the right is saturated with H_2. On the left, metal electrode M is immersed in a 1 M solution of M^{n+} ions. E^o is the standard electrode potential of the metal.

In the left half cell, the metal M is immersed in an electrolyte consisting of a **1 M solution of its own ions** (for instance, 1 mole of $FeCl_2$ dissolved in 1 L of water in the case of iron). Each $FeCl_2$ molecule in the solution dissociates to one Fe^{++} and two Cl^- ions.

An electrometer measures the voltage between the metal M and the platinum when no current flows between the two electrodes. This voltage represents the **standard electrode potential** of the metal. The potential at the hydrogen electrode is **defined** as zero; it is used as the reference potential.

Table 15.1 shows the values that are obtained in this fashion; it represents the **standard electromotive force series**. In the standard electrolytic cell, the electrode assuming a positive charge is the **cathode** and the negatively charged electrode is the **anode**.

A number of items are worth noting about the information in this table.

1. All values refer to 1 molar solutions of the metal at 1 atm pressure and 25°C.

2. The electrode potential values (E^o) are measured when no current flows.

3. Metals that have a positive electrode potential behave as cathodes relative to hydrogen. Ions of such metals in solution tend to be reduced to metal and be deposited on the electrode. The higher the metals are situated in the table, the more pronounced is their tendency to remain cathodes and to be reduced. These are the **noble metals**.

Table 15.1 Standard Electrode Potentials at 25°C.

Electrode reaction			Electrode potential, E^o (V), relative to standard hydrogen
Au	$\rightarrow$	$Au^{3+} + 3e^-$	+1.498
$PbSO_4 + 2H_2O$	$\rightarrow$	$PbO_2 + SO_4^{2-} + 4H^+ + 2e^-$	+1.69
$Pb^{2+} + 2H_2O$	$\rightarrow$	$PbO_2 + 4H^+ + 2e^-$	+1.46
$2H_2O$	$\rightarrow$	$O_2 + 4H^+ + 4e^-$	+1.229
$Mn^{2+} + 2H_2O$	$\rightarrow$	$MnO_2 + 4H^+ + 2e^-$	+1.23
Pt	$\rightarrow$	$Pt^{2+} + 2e^-$	+1.200
Ag	$\rightarrow$	$Ag^+ + e^-$	+0.799
2Hg	$\rightarrow$	$Hg_2^{2+} + 2e^-$	+0.788
Fe^{2+}	$\rightarrow$	$Fe^{3+} + e^-$	+0.771
$Li_{0.5}CoO_2 + 0.5Li^+$	$\rightarrow$	$LiCoO_2 + 0.5e^-$	+0.7
$Ni(OH)_2 + OH^-$	$\rightarrow$	$NiOOH + H_2O + e^-$	+0.52
$4(OH)^-$	$\rightarrow$	$O_2 + 2H_2O + 4e^-$	+0.401
Cu	$\rightarrow$	$Cu^{2+} + 2e^-$	+0.337
Sn^{2+}	$\rightarrow$	$Sn^{4+} + 2e^-$	+0.150
H_2	$\rightarrow$	$2H^+ + 2e^-$	0.000
Pb	$\rightarrow$	$Pb^{2+} + 2e^-$	−0.126
Sn	$\rightarrow$	$Sn^{2+} + 2e^-$	−0.136
Ni	$\rightarrow$	$Ni^{2+} + 2e^-$	−0.250
Co	$\rightarrow$	$Co^{2+} + 2e^-$	−0.277
$Pb + SO_4^{2-}$	$\rightarrow$	$PbSO_4 + 2e^-$	−0.36
Cd	$\rightarrow$	$Cd^{2+} + 2e^-$	−0.403
Fe	$\rightarrow$	$Fe^{2+} + 2e^-$	−0.440
Cr	$\rightarrow$	$Cr^{3+} + 3e^-$	−0.744
Zn	$\rightarrow$	$Zn^{2+} + 2e^-$	−0.763
MH	$\rightarrow$	$H^+ + M + e^-$	−0.83
Al	$\rightarrow$	$Al^{3+} + 3e^-$	−1.662
Mg	$\rightarrow$	$Mg^{2+} + 2e^-$	−2.363
Na	$\rightarrow$	$Na^+ + e^-$	−2.714
LiC_6	$\rightarrow$	$Li^+ + 6C + e^-$	−2.8
Li	$\rightarrow$	$Li^+ + e^-$	−3.05

4. Metals that have a negative electrode potential behave as anodes relative to hydrogen and would tend to oxidize by dissolving in solution. The lower the metals are situated in the table, the more pronounced is their tendency to become anodic and to dissolve.

Practical cases represent cells consisting of two materials; these can both lie above or below hydrogen, or one can lie above and the other below. As an example of the latter consider the combination of half cells composed of copper in a 1 mole solution of Cu^{++} and iron in a 1 mole solution of Fe^{++}.

$$Fe^o = Fe^{2+}(1\,M) + 2e^- \qquad E^o(Fe) = -0.440\,V \qquad (15.12a)$$

$$Cu^o = Cu^{2+}(1\,M) + 2e^- \qquad E^o(Cu) = +0.337\,V \qquad (15.12b)$$

The overall reaction is obtained by subtracting the half reactions to eliminate the electronic charge.

$$Fe^o = Fe^{2+}(1\,M) \qquad (15.13)$$

$$- Cu^o = Cu^{2+}(1\,M)$$
$$E^o_{cell} = -0.440 - (+0.337) = -0.777\,V.$$

As we have seen with the lead-acid battery, when we let a current pass in the wire connecting the two electrodes the potential difference will be lower than 0.777 V, equilibrium is lost, an electric current flows, iron is dissolved in the liquid, and copper ions are drawn to the copper bar which increases its mass. When we **impose** a voltage larger than 0.777 V (making the iron, say, $-1\,V$ with respect to the copper), the current is reversed; copper is dissolved and iron is deposited on the iron electrode.

From the above, we draw the following conclusions:

1. When an electric current is allowed to flow in a two-metal system in contact with an electrolyte, the less noble metal is dissolved. Dissolution can be stopped by preventing a current from flowing, for instance by insulating the two metals from one another.

2. Electrical batteries are such electrolytic cells; the electric energy released by the battery is equal to the change in free energy in the total reaction. The power released is $W = I \cdot V$ where I is the current flowing and V the electrochemical potential.

3. Dissolution of iron (or other metals) in electrochemical systems is a mechanism of **corrosion**. The electrolyte is water that contains enough dissolved salts to be electrically conductive. It can be salt-water, or practically all but highly purified water. Corrosion is driven by the cell voltage but can be controlled if one can impose an appropriate voltage.

4. The electrolytic system can also be used for **electroplating** a metal onto a metallic surface. In the above system, copper is plated onto the copper bar when the iron and copper are connected. If we impose a larger voltage than $-0.777\,V$, (say, $-1\,V$), the current flows in the reverse direction and iron metal is plated onto the iron bar. If we replace the Fe^{++} electrolyte by a dissolved gold salt and apply the appropriate potential, gold will be plated onto the metal.

Corrosion and electroplating will be discussed in Chapter 16.

15.3 **PRIMARY BATTERIES**

Desirable properties of batteries are: high specific energy (low weight), safe operation, high enough current density, and long shelf life. Electrochemical reaction rates are usually quite low so that large surfaces are required to obtain sufficient currents. This requires porous electrode materials.

Not all electrochemical reactions are reversible in practice; either the charging reaction is not possible because gas is evolved during discharge and it is impractical to supply the gases for charging, or the charging reaction is dangerous. Thus we distinguish primary batteries that are not rechargeable and are discarded when completely discharged, and secondary batteries that are rechargeable. (Spent batteries must not be discarded in the garbage but returned to collection sites: they contain toxic chemicals as we shall see below).

Most commercial primary batteries are based on zinc or on lithium anodes. Other materials, for instance sodium or aluminum, have been tried but have not resulted in usable batteries. Batteries based on mercury or cadmium provide excellent performance, but they are being phased out because of the toxicity of these materials.

15.3.1 **The Leclanché Battery**

The first primary battery sold was invented in 1866 by Georges Lionel Leclanché, a French signal engineer. With some improvements, it has been the standard battery until recently. Its anode consists of zinc, its cathode is a natural MnO_2 powder, and its electrolyte is a mixture of NH_4Cl and $ZnCl_2$. More modern versions utilize zinc chloride electrolytes only. Its open circuit voltage is 1.6 V; its practical energy density is 85 Wh/kg, and its operating voltage 1.5 V. The chemical reactions in this battery are complex and still not understood in detail. The zinc anode is oxidized to ZnO and the cathode is reduced, first to MnOOH, and later to Mn_2O_3.

$$Zn + 2MnO_2 \rightarrow ZnO + Mn_2O_3 \tag{15.14}$$

These reactions will be described in more detail in the next section.

Household batteries consist of one cell in which a Zn can is the anode. The construction of the battery is shown in Figure 15.6. The can is filled with the cathode material, which is a fine mixture of MnO_2 and carbon powders; the carbon pin in the center of the can serves as the positive current collector. The electrolyte is absorbed in the separator and cathode, leaving practically no free liquid. The separator is a starch paste and paper. The zinc can is the anode.

It is interesting that Leclanché already introduced the most versatile cathode material that is used in most modern batteries in a more refined form.

15.3.2 **The Alkali Battery**

The alkaline battery has replaced the Leclanché battery because it has a higher specific energy (145 Wh/kg), a lower internal resistance, and a longer shelf life. Its anode is a 99.99% pure zinc, its cathode is

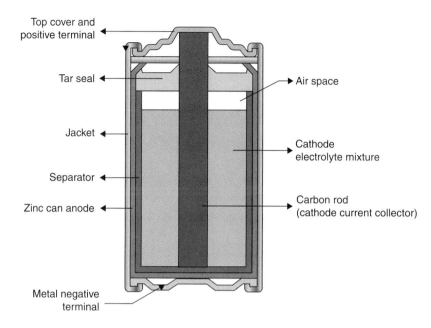

Top cover and
positive terminal

Tar seal

Air space

Jacket

Cathode
electrolyte mixture

Separator

Zinc can anode

Carbon rod
(cathode current collector)

Metal negative
terminal

■ **FIGURE 15.6** Construction of the Leclanché battery.

an electrolytic MnO_2, and its electrolyte is an alkaline solution, namely KOH in water. The discharge of the battery occurs in two stages.

Up to one electron per MnO_2 molecule, the reaction at the cathode is

$$MnO_2 + H_2O + e^- \rightarrow MnOOH + OH^- \qquad (15.15)$$

The cathode acquires a positive charge because it contributes an electron to the reaction.

At the anode, the reaction is

$$Zn + 2OH^- \rightarrow Zn(OH)_2 + 2e^- \qquad (15.16)$$

The Zn^{2+} ions dissolve from the anode and react with the OH^- to form $Zn(OH)_2$; the anode acquires $2e^-$. The overall reaction is

$$2MnO_2 + Zn + 2H_2O \rightarrow 2MnOOH + Zn(OH)_2 \qquad (15.17)$$

As drainage continues to a total of 1.33 electrons per molecule of MnO_2, the cathode reaction continues to

$$3MnOOH + e^- \rightarrow Mn_3O_4 + OH^- + H_2O \qquad (15.18)$$

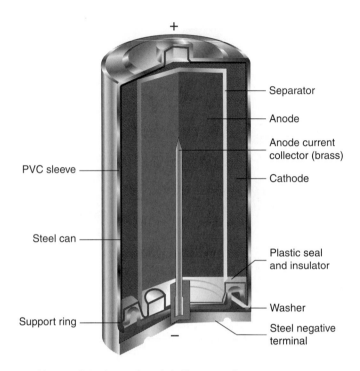

■ **FIGURE 15.7** Construction of the 1.5 V alkaline battery. This is the bobbin-type configuration.

Finally the anode dehydrates to ZnO

$$Zn(OH)_2 \rightarrow ZnO + H_2O \tag{15.19}$$

The total discharge of the battery occurs as

$$3MnO_2 + 2Zn \rightarrow Mn_3O_4 + 2ZnO \tag{15.20}$$

The open circuit potential of this cell is 1.5–1.65 V and the average operating voltage is 1.2 V.

The cell construction is shown in Figure 15.7. The cathode consists of electrolytic manganese dioxide and carbon. This purified material is more expensive than the MnO_2 ore used in the Leclanché battery, but it provides a higher capacity because the cathode can be packed very densely. The anode consists of pure zinc powder with a brass current collector. The separator is normally made of a combination of vinyl polymers, polyolefins, or fibrous regenerated cellulose. The electrolyte is absorbed in the anode, the cathode, and the separator.

Alkaline batteries also exist in button configuration as shown in Figure 15.8.

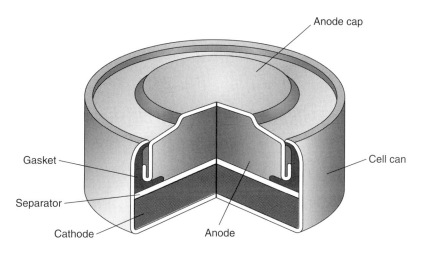

Anode cap

Gasket

Cell can

Separator

Cathode

Anode

■ **FIGURE 15.8** Miniature button configuration of a battery.

15.3.3 **Lithium Batteries**

Table 15.1 shows that lithium is an obvious choice as an anode material for batteries. It has a very high electrode potential (3.1 V against the hydrogen electrode) and is the lightest solid element. Lithium, however, is very reactive toward water: it requires a nonaqueous electrolyte and care in its handling. Physical damage to the battery exposes the lithium to air and can cause fire or explosion. Attempts to recharge a primary lithium battery, likewise, present the risk of explosion. The reason is that lithium, plated back onto the anode, acquires a large surface area and becomes dangerously reactive.

Lithium primary batteries utilize a lithium foil as anode. Batteries exist with soluble or solid cathodes. Solid cathodes consist of a mixture of MnO_2 and carbon powder mounted in a conductive grid. MnO_2 has an open crystal structure that can be intercalated by lithium. (Intercalation is described in Section 15.4.2, on the rechargeable lithium ion battery.) The electrolytes consist of LiBr, $LiClO_4$ or $LiCl_2SO_2$ salt dissolved in an organic solvent such as acetonitrile ($H_3CC{\equiv}N$).

The chemical reaction is

$$Li \rightarrow Li^+ + e^- \tag{15.21}$$

at the anode and

$$Mn^{IV}O_2 + Li^+ + e^- \rightarrow LiMn^{III}O_2$$

at the cathode.

The lithium reduces the Mn^{IV} to Mn^{III}.

The open circuit voltage of the cell is about 3.3 V.

The attractive properties of this battery are as follows:

1. The cell has a high voltage.

2. It has a very high specific energy, above 230 Wh/kg.

3. It has a long shelf life of at least 10 years. (It does not discharge.) This results from a chemical reaction between the lithium and the electrolyte that deposits an insulating layer on the anode.

Lithium batteries are built as miniature buttons similar to those in Figure 15.8, in the bobbin configuration as illustrated in Figure 15.7, and as a spirally wound cylindrical cell (the "jelly roll configuration") shown in Figure 15.9. In this configuration the cathode, anode, and separator are made of long sheets that are wound in a spiral. The large surface area and small space between the electrodes permits much larger currents than the bobbin. The latter, however, has a higher energy density.

Soluble Cathode Lithium Batteries   Lithium batteries are also manufactured with a soluble cathode. These are used in space and military applications. In these batteries, the electrolyte serves also as the active cathode material. The positive electrode is a porous carbon that is saturated with SO_2 or $SOCl_2$. The absence of a solid cathode makes for a low total mass and the highest energy density of any battery. In these batteries the anode is a lithium foil.

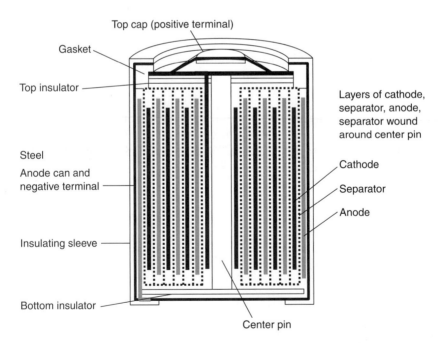

■ **FIGURE 15.9**   Spirally wound battery. A four-ply layer consisting of cathode, separator, anode, and separator is wound around a central pin; contact tabs connect cathode to metal cap and anode to metal can.

In the **Li/SO$_2$** cell, the chemical reaction is

$$2Li + 2SO_2 \rightarrow Li_2S_2O_4 \tag{15.22}$$

The reaction product is a solid that precipitates. Since SO_2 is a gas, the pressure in the cell is initially $3-4 \times 10^5$ Pa (i.e., 3–4 atm) at room temperature. The cells must be built strong enough to contain this pressure; they are equipped with a vent to release the gas in case the pressure becomes excessive at elevated temperature. The open circuit voltage of the Li/SO$_2$ battery is 2.95 V. During discharge, the battery maintains a remarkably constant voltage that depends on the discharge current. The specific energy of the Li/SO$_2$ battery is 260 Wh/kg.

The **Li/SOCl$_2$** battery has a very high open circuit voltage of 3.6 V. Specific energy ranges up to 590 Wh/kg. The anode is a lithium foil. A solution of $SOCl_2$ and $LiAlCl_4$ in an inorganic solvent serves as the electrolyte and active cathode. The latter is absorbed in a porous carbon electrode that collects the current. The chemical reaction is

$$4Li + 2SOCl_2 \rightarrow 4LiCl + S + SO_2 \tag{15.23}$$

The solid LiCl precipitates during discharge. Precipitation of sulfur in the carbon limits the cell's service.

The soluble cathode batteries are manufactured in the bobbin configuration (Figure 15.7) for maximum energy density or in the spiral wound configuration (Figure 15.9) for high discharge currents. The Li/SOCl$_2$ battery also exists in a flat disk configuration that is similar to the button shown in Figure 15.8.

15.4 SECONDARY OR RECHARGEABLE BATTERIES

15.4.1 Lead-Acid Battery

The lead-acid battery was invented and first produced in 1860 by R.F. Planté. He already recognized that high enough currents are not possible with the reaction of solid lead plates. His design is still used today, with improvements of detail (Figure 15.10A). Both electrodes are made of lead grids that are filled with a porous reactive paste. The paste consists of small particles of red lead oxide Pb_3O_4 and a solution of sulfuric acid. Once the paste-filled plates are dry, they are stacked in the cell with porous separators between them. Modern separators are sintered polyvinyl chloride (PVC) or microporous polyethylene.

Once the battery is assembled, it is filled with sulfuric acid and charged. The sulfuric acid has a typical concentration of 37% acid by weight (4.4 molar). The electric charge reduces the anode to lead and oxidizes the cathode to PbO_2. Its operation was described in Sections 15.1 and 15.2.

The metallic grids are produced either by casting or by mechanical formation. A modern grid consists of crimped lead foil that is expanded as shown in Figure 15.10B. Mechanical strength of the lead grids is accomplished by alloying with 0.1–0.3% calcium and tin.

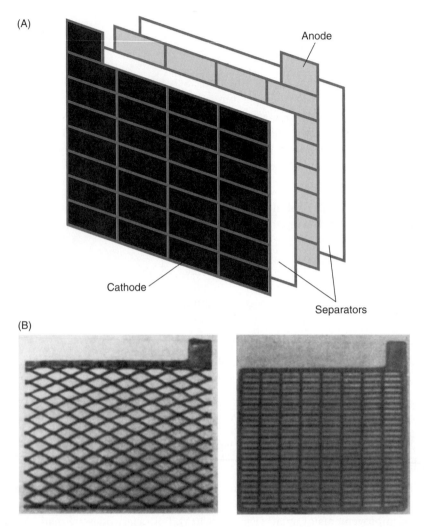

■ **FIGURE 15.10** (A) Prismatic construction of the lead-acid battery. (B) Grids for lead-acid battery. Left: Expanded wrought grid. Right: Cast grid.

The lead-acid battery has a specific energy up to 40 Wh/kg and a high electrical turnaround efficiency of 75–80%.

15.4.2 The Nickel-Metal Hydride Battery

This relatively new secondary battery was developed in the 1980s and has a practical energy density of 75 Wh/kg. Its positive electrode consists of nickel oxyhydroxide NiOOH in the charged state, and its anode is a metal hydride. The electrolyte is an aqueous solution of potassium hydroxide KOH.

During discharge, the NiOOH reacts with water and is reduced to the divalent $Ni(OH)_2$. Its standard electrode potential is 0.52 V.

$$NiOOH + H_2O + e^- \rightarrow Ni(OH)_2 + OH^- \qquad E = 0.52 \text{ V} \qquad (15.24)$$

The anode of the battery is a metal hydride that is oxidized to the metal state in discharge

$$MH + OH^- \rightarrow M + H_2O + e^- \qquad E = 0.83 \text{ V} \qquad (15.25)$$

The overall reaction is

$$MH + NiOOH \rightarrow M + Ni(OH)_2 \qquad E = 1.35 \text{ V} \qquad (15.26)$$

with an open-circuit voltage of 1.35 V.

The metal used in the cathode is an alloy, known as AB_5, where B is nickel and A is Mischmetal, a naturally occurring mixture of rare earths Ce, Nd, Pr, Gd, and Y. When hydrogen is formed during the charging reaction at the anode

$$H_2O + e^- \rightarrow OH^- + \tfrac{1}{2} H_2 \qquad (15.27)$$

it is not released as a gas but is absorbed by the Mischmetal to form a hydride

$$M + \tfrac{1}{2} H_2 \rightarrow MH \qquad (15.28)$$

Electrolysis of water to form oxygen and hydrogen must be avoided. To this end, the battery contains an excess of metal hydride. The oxygen diffuses from the positive to the negative electrode and reacts with the hydride to form water and avoid the production of gaseous hydrogen.

The positive electrode is a highly porous nickel plate into which the hydroxide is pasted and converted into the active material by electrochemical reaction. The negative electrode, likewise, is a nickel grid onto which the hydrogen storage alloy is bonded by plastic. The separator is a porous synthetic material. The battery contains a minimum of liquid electrolyte which is absorbed into the electrodes and the separator. The lack of liquid facilitates the diffusion of the oxygen toward the hydride during overcharging; it also permits operating the battery in any position

Rechargeable nickel metal hydride batteries exist in many forms. Most people are familiar with the small cylindrical batteries. Large prismatic batteries are used in hybrid automobiles, notably the Toyota Prius and the Ford Escape.

15.4.3 The Rechargeable Lithium Ion Battery

There are no rechargeable batteries with solid lithium metal anodes. During the charge cycle, metallic lithium would be deposited in a porous form. The high surface area of his material renders it chemically very active and prone to explosion.

The **lithium ion** battery does not utilize a proper chemical reaction, but depends on the transfer of lithium ions from one electrode to the other.

The anode, or negative electrode, consists of graphite bonded to copper plates. The graphite possesses a layered structure. It is an intercalation compound. Li atoms can "intercalate" the graphite, which means that they insert themselves between the graphite layers.

The cathode (positive electrode) of the battery consists of $LiCoO_2$ or

$LiNi_{(1-x)}Co_xO_2$ bonded to an aluminum foil. These compounds also possess a layered structure, shown in Figure 15.11. They, likewise, act as intercalation compounds in the sense that the Li atoms freely leave and reenter the structure.

In equilibrium (i.e., the lowest energy state), the cathode is fully intercalated; it exists as $LiCoO_2$ and the anode is graphite.

The charge and discharge reactions are illustrated in Figure 15.11. When, during charging, a negative potential is applied to the anode and a positive to the cathode, Li^+ ions are transferred from the $LiCoO_2$ to the graphite.

Cathode

$$LiCoO_2 \rightarrow Li_{1-x}CoO_2 + xLi^+ + xe^- \tag{15.29}$$

Anode

$$C + xLi^+ + xe^- \rightarrow + CLi_x \tag{15.30}$$

Overall

$$LiCoO_2 + C \rightarrow Li_{1-x}CoO_2 + CLi_x \tag{15.31}$$

In the discharge of the battery, the Li atoms leave the graphite, are oxidized to Li^+, and return to the cathode.

The charging voltage is 4–4.5 V and the discharge is 3.7 V. The specific capacity is 155 Ah/kg for $LiCoO_2$. When the cobalt is partially replaced by nickel, the specific capacity increases; it is as high as 220 Ah/kg for $LiNi_{0.9}Co_{0.1}O_2$.

The electrolyte is composed of a $LiPF_6$ salt dissolved in organic solvents. The latter must consist of polar molecules that result in a high dielectric constant so that large amounts of an ionic salt are dissolved and dissociated into Li^+ and PF_6^- ions. Practical solvents consist of a mixture of ethyl carbonate (EC), propyl carbonate (PC), dimethyl carbonate (DMC), methyl ethyl carbonate (MEC), and diethyl carbonate (DEC). Their molecular structures are shown in Figure 15.12.

The separator consists of a microporous polyethylene or polypropylene. This material has the advantage of providing protection against thermal runaway: when the temperature reaches 135°C for polyethylene or 165°C for the polypropylene, the material flows, loses its porosity and prevents further current.

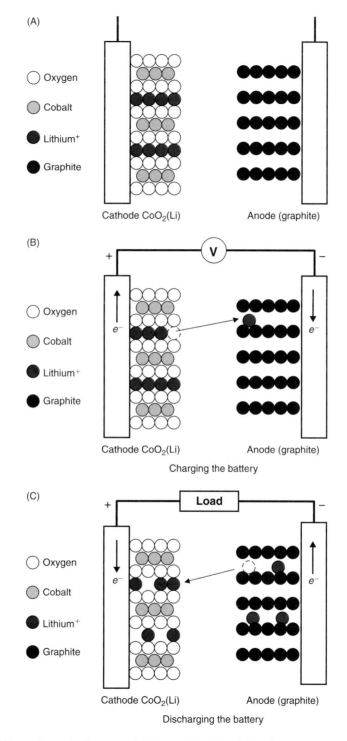

■ **FIGURE 15.11**   The lithium ion battery. (A) Construction of the battery. Left: LiCoO$_2$ cathode, right: graphite anode. (B) Charging the battery: Li$^+$ ions are transferred to and intercalated in the graphite. (C) Discharging the battery: Li$^+$ ions transfer back to LiCoO$_2$ while electron current flows through the load.

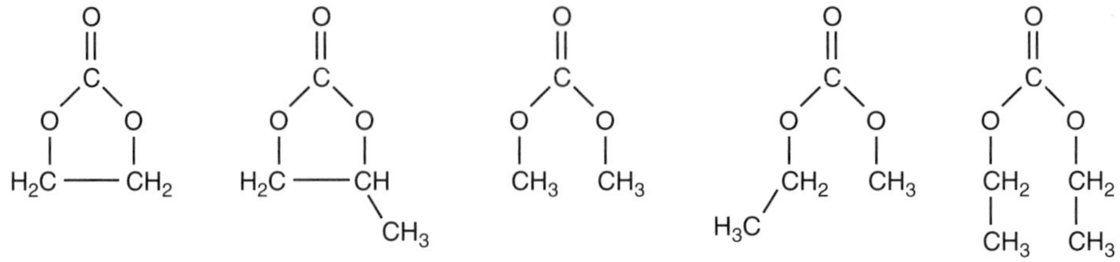

	EC	PC	DMC	EMC	DEC
Dielectric constant	89.6	64.4	3.12	2.9	2.82

■ **FIGURE 15.12** Solvents for the electrolyte in lithium ion batteries.

The electrolyte is mostly absorbed in the separator and the electrolytes so that very little liquid is present.

A modern variant of the lithium ion battery is the polymer cell in which the electrode materials are bonded to expanded metal grids and the electrolyte is absorbed in a polymer. Grid materials are Al for the cathode and Cu for the anode. The whole battery is then encapsulated in a polymer. This design is very robust and immune to physical abuse.

Innovation in lithium ion batteries is active. Cathodes of layered compounds where cobalt is replaced by nickel or manganese are being developed and are already in some commercial use. These materials are less expensive than cobalt, but their synthesis is more difficult and they are subject to discharge. The Lithium Iron Phosphate ($LiFePO_4$) battery offers low cost, non-toxicity, excellent thermal stability and high specific capacity of $170\,A.h/kg$. The initial problem of low electric conductivity of iron phosphate is overcome by doping with aluminum, niobium, and zirconium. There is also considerable work to replace the graphite anode with lithium titanate nanoparticles. This material permits much larger charging and discharging currents.

15.5 FUEL CELLS

A fuel cell is an electrochemical device that operates on the same principle as the battery with the crucial difference that an external fuel is consumed in the production of electric power instead of the electrode material (Figure 15.13). The hydrogen electrode described in Section 15.2.5. serves as the anode in a hydrogen fuel cell with the reaction

$$H_2 \rightarrow 2H^+ + 2e^- \tag{15.32}$$

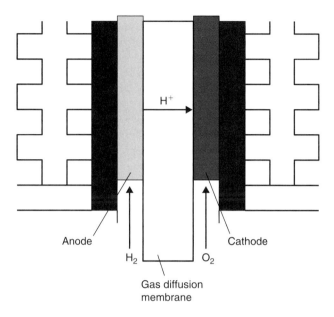

■ **FIGURE 15.13**   Operation of the fuel cell based on a proton exchange membrane.

The cathode also consists of a platinum catalyst and its active element is oxygen, usually from air. The reaction at the cathode is

$$\tfrac{1}{2}O_2 + 2e^- + 2H^+ \rightarrow H_2O \tag{15.33}$$

The overall reaction is

$$H_2 + \tfrac{1}{2}O_2 \rightarrow H_2O \tag{15.34}$$

The electrolyte contains a dissolved acid that allows the transport of the protons from the anode to the cathode. Highly concentrated phosphoric acid is used in large stationary fuel cells. The cells operate at 200°C and provide high energy efficiency because the heat released by the fuel cell can be used in space heating. Small, portable, fuel cells use a proton exchange membrane (PEM) electrolyte. Today's fuel cells utilize DuPont's Nafion®, a solid polymer membrane of trifluoromethanesulfonic acid that conducts electricity by mobile protons. This membrane requires dissolved water for its conductivity.

Hydrogen is supplied from a pressurized tank, by metal hydrides similar to those used in the nickel-metal hydride battery or by chemical hydrides such as LiH or $NaBH_4$ that react with water.

A fuel cell that could utilize a widely available liquid fuel would be highly desirable. Methanol is the only such fuel that is presenting some success, but methanol fuel cells are still in the development stage. In this fuel cell, methanol, instead of hydrogen, reacts at the anode and produces CO_2, hydrogen ions and electrons that charge the anode.

$$\text{Anode reaction: } CH_3OH + H_2O \rightarrow CO_2 + 6H^+ + 6e^-$$

$$\text{Cathode reaction: } 3/2O_2 + 6H^+ + 6e^- \rightarrow 3H_2O$$

$$\text{Overall reaction: } CH_3OH + 3/2O_2 \rightarrow CO_2 + 2H_2O \tag{15.35}$$

All other fuel cells require a chemical refining unit that extracts hydrogen from the liquid fuels. This greatly reduces the overall efficiency of the device.

These limitations hamper the wide applications of fuel cells for now.

15.6 ULTRACAPACITORS

Ultracapacitors are the subject of active research and development. They differ from batteries in that energy is not provided by electrochemical reactions but is stored in a capacitor as an electric charge Q with voltage V (see Chapter 11). They differ from conventional capacitors by their ability to store a much larger amount of electrical energy QV.

The amount of charge held by a capacitor is

$$Q = \varepsilon_o \kappa \frac{A}{d} V \tag{15.36}$$

where ε_o is the permittivity of vacuum, κ the dielectric constant, A the area of the capacitor, and d is the distance between the conductive plates.

The development of ultracapacitors is proceeding along two lines: electric double-layer capacitors and high-voltage barium-titanate dielectrics.

Conventional capacitors consist of two metallic plates separated by a solid dielectric of dielectric constant κ (Figure 15.14). In the electric double-layer capacitor the dielectric is replaced by two highly porous, quite conductive, layers separated by a very thin insulator. This construction gives the battery a very small effective distance d and a large area A. The porous material consists presently of activated charcoal, but development efforts aim at replacing it with oriented carbon nanotubes (Chapter 18). Large commercial ultracapacitors have a capacity as high as 5,000 Farads. (By comparison, the capacity of conventional capacitors is measured in microfarads.) Electric double-layer capacitors are capable of sustaining potentials of 2 to 3 V only, so they will have to be connected in series the same way as batteries when higher voltages are required by the application.

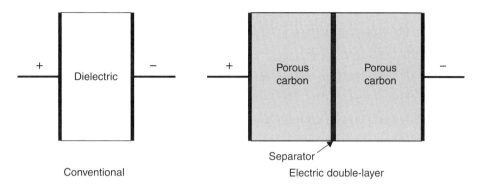

■ FIGURE 15.14 Comparison of the conventional and the electric double-layer capacitor.

Barium titanate is a ferroelectric with a very high dielectric constant $\kappa = 1700$. Conventional barium titanate has a low dielectric strength. Present development aims at providing it with a high dielectric strength with the help of alumina coatings and glass. The high energy density $E = QV$ of these ultracapacitors results from the combination of high κ and high applied voltage V.

In 2008, commercial ultracapacitors store an energy density of 6 W.h/kg while experimental devices offer capacities of 200 W.h/kg, which is fully competitive with the batteries described in this chapter. The lack of electrochemical reactions gives these devices the advantage of high power and very short charging times. Ultracapacitors are used in smoothing out momentary loads and in starters for diesel trucks and railroad locomotives. If successfully commercialized, they will be attractive for electric cars and hybrids because their rapid charging capability makes them suitable for regenerative breaking.

■ SUMMARY

1. Electric batteries are devices that transform the energy released by exothermic chemical reactions directly into electric power.

2. Electrical batteries are electrochemical cells that consist of two electrodes and an electrolyte.

3. The electrolyte is a liquid with a high dielectric constant (water or organic fluid) in which salts, acids, or bases are dissolved as positive and negative ions. The high dielectric constant permits the dissolution of substances in the form of ions. The ions in the electrolyte provide electric conduction between the electrodes.

4. The anode acquires a negative charge because its atoms are dissolved in the electrolyte in the form of positive ions and leave an electron in the anode. This reaction is an oxidation.

5. Positive ions from the electrolyte are neutralized by acquiring electrons from the cathode. This produces a positive charge in the cathode and a negative charge in the electrolyte. This reaction is a reduction.

6. The charges induce an electric voltage between electrolyte and electrode that stops the charge exchange at the electrode at equilibrium. This voltage is the electrode potential.

7. The electrolyte is electrically conductive; it equalizes the charges in the two half cells. There is no electric potential difference in the electrolyte. The open circuit voltage between the electrodes is the difference between the two electrode potentials.

8. Standard electrode potentials are measured against a standard electrode also called the hydrogen electrode. The standard electrode consists of a platinum metal immersed in a 1 molar solution of HCl. The metal under study is immersed in an electrolyte with a 1 molar solution of its own ions. Anodic materials have a negative potential vs. the hydrogen electrode. They are easily oxidized. Cathodic materials obtain a positive potential vs. the hydrogen electrode. They tend to be reduced. Very cathodic metals are the noble metals (Cu, Ag, Au).

9. When an external circuit is established between the electrodes, the battery voltage is lower than its open circuit value, equilibrium is disturbed, an electron current I flows in the open circuit, and an ion current flows in the electrolyte. This discharges the battery which furnishes the power $W = I \cdot V$. The battery is discharged when the material of one or both electrode materials is consumed by the reaction.

10. Some batteries can be recharged: one imposes a voltage larger than the open circuit value. This reverses the electric current in the circuit and electrolyte; it reverses the chemical reactions at the electrodes which return to their initial state. These are the secondary or rechargeable batteries. They are also called accumulators.

11. Some batteries cannot be recharged because the reverse reaction is not practical (if gases are evolved) or dangerous (especially in lithium batteries).

12. Practical batteries have electrodes with a large surface area: they are made of porous material or fine powders in order to deliver a sufficiently large discharge current.

13. Commercial primary batteries have zinc or lithium anodes and porous MnO_2 cathodes. In alkaline batteries, the electrolyte is KOH. In lithium batteries the electrolyte is a polar organic liquid in which lithium salts are dissolved.

14. Commercial secondary batteries include the lead-acid battery that is used in most automobiles; the nickel-metal hydride battery that powers many portable tools and hybrid automobiles; and the lithium ion battery which is in use in electronic and photographic equipment.

15. The lead-acid battery consists of a lead and a lead oxide electrode and a sulfuric acid electrolyte. The discharge converts both electrolytes to lead sulfate and the sulfuric acid to water.

16. In the nickel-metal hydride battery, the anode is a mixture of metals that store hydrogen by forming a hydride, the cathode is NiOOH, and the electrolyte is KOH.

17. The lithium ion battery consists of electrodes in which the lithium can be intercalated. In the discharged state, the anode is graphite and the cathode is $LiCoO_2$. During the charge cycle, Li is

removed from the cathode in the form of Li^+ ions which are transferred to the graphite where they are reduced to Li. The reverse reaction occurs during discharge. Lithium ion batteries are also produced with $LiMn_2O_4$ cathodes.

18. Fuel cells derive their energy from the oxidation of hydrogen to form water. Their active materials are hydrogen gas and oxygen from air that react on metallic catalysts.

19. Ultracapacitors are not based on electrochemical reactions but on their ability to accumulate a very large electric charge. Electric double-layer capacitors differ from conventional capacitors in that the dielectric between the plates is replaced by a highly porous electric double layer separated by a thin insulator. Other ultracapacitors are based on increasing the dielectric strength barium titanate, which possesses a large dielectric constant. Ultracapacitors are capable of high charge and discharge rates that make them attractive for regenerative braking in electric cars and hybrids.

■ KEY TERMS

A
alkali battery, 438
anode, 428

B
barium titanate, 451, 453
batteries, 428
battery discharge, 432
bobbin, 442, 443
buttons, 442

C
carbon nanotubes, 450
cathode, 428
charging, 432
corrosion, 437
cylindrical cell, 442

D
dielectric constant, 450
dielectric strength, 453
diethyl carbonate, 446
dimethyl carbonate, 446
dissociation, 429

E
electric cars, 451
electric double layer capacitor, 450, 453

electrochemistry, 428
electrolyte, 427, 428, 429, 431, 432, 433, 434, 435, 437, 438, 439, 440, 441, 442, 443, 444, 445, 446, 448, 449, 451, 452, 454, 455
electroplating, 437
equilibrium potential, 431, 432, 455
ethyl carbonate, 446

F
Faraday, 433
Faraday's constant, 433, 455
free energy, 431
fuel cells, 448

G
galvanic cell, 428

H
half reaction, 433
hydrogen electrode, 434, 448, 452

I
intercalatation, 446
internal resistance, 432, 438

J
jelly roll configuration, 442

K
KOH, 444, 452

L
lead-acid battery, 428, 443
lead-acid cell, 428
Leclanché battery, 438
$Li/SOCl_2$ battery, 443
$LiBr$, $LiClO_4$, 441
$LiCl_2SO_2$, 441
$LiCoO_2$, 446
lighting and ignition, 428
$LiNi_{(1-x)}Co_xO_2$, 446
$LiPF_6$, 446
lithium batteries, 441
lithium ion, 441, 446, 448, 452, 455
lithium ion battery, 445

M
membrane, 431
metal hydride, 445
metallic grids, 443
methyl ethyl carbonate, 446
Mischmetal, 445

■ REFERENCES FOR FURTHER READING

[1] J. Bard, L.R. Faulkner, *Electrochemical Methods: Fundamentals and Applications*, 2nd ed, Wiley, New York (2000).

[2] J.O'M. Bockris, A.K.N. Reddy, *Modern Electrochemistry*, 2nd ed, Plenum Press, New York (1998).

[3] R.M. Dell, D. Rand, *Understanding Batteries (Paperback)*, Royal Society of Chemistry (2001).

[4] D. Linden, T.B. (eds) Reddy, *Handbook of Batteries*, 3rd ed, Publisher, New York (2002).

[5] R. O'Hayre, S-W. Cha, W. Colella and F.B. Prinz, *Fuel Cell Fundamentals*, Wiley, New York (2005).

[6] S. Srinivasan, *Fuel Cells: From Fundamentals to Applications*, Springer, New York (2006).

■ PROBLEMS AND QUESTIONS

15.1. Alessandro Volta built the first electrical battery by piling 40 cells consisting of a zinc plate, a cardboard disk soaked with brine, and a silver plate. Compute the voltage obtained by this pile. Sketch the arrangement of the plates and indicate the positive and negative ends. Compute the voltage obtained with zinc and copper plates.

15.2. Could you make a battery with two anodic metals such as cadmium and zinc? If yes, what would be the voltage of such a cell, and which electrode would be positive?

15.3. It was stated that water can dissolve a minute amount of lead (or any metal) and that this dissolution stops quickly because an electric potential stops the dissolution. So why can you dissolve a spoonful of salt in the water?

15.4. Describe an electrochemical cell. Define the cathode and the anode.

15.5. What is an electrolyte?

15.6. Why is water an excellent electrolyte?

15.7. What type of material can be a solid electrolyte?

15.8. What properties do organic electrolytes need to be effective?

15.9. What is the role of the ions dissolved in the electrolyte in an electrochemical cell?

15.10. What is the equilibrium potential, and how does it come about?

15.11. Which electrode is dissolved in an electrochemical reaction?

15.12. What defines a standard electrochemical potential?

15.13. What is the standard electrode?

15.14. What is a noble metal?

15.15. Define the open circuit voltage of a battery.

15.16. Explain how the discharge of a battery is analogous to the burning of oil or wood.

15.17. Why are some batteries not rechargeable? Is it a matter of scientific principle or of practical obstacles?

15.18. Why are the electrodes of batteries made of porous material?

15.19. What material is used as the cathode in all primary batteries?

15.20. What is the electrolyte in a lithium battery?

15.21. Why is lithium so desirable as an anode?

15.22. What is the exothermic reaction in the lead-acid battery?

15.23. Why does one use lithium ion batteries instead of lithium secondary batteries?

15.24. Describe the reactions in the lithium ion battery.

15.25. What is the exothermic reaction in the nickel-metal hydride battery?

15.26. Define or describe a fuel cell. What is the reaction in the fuel cell?

15.27. What determines the stored charge in a battery? Write down Faraday's law.

15.28. What determines the stored power in a battery?

15.29. Calculate the stored charge in an alkali battery that has an anode mass of 30 g.

15.30. Compare the theoretical and the practical specific energy of a battery.

15.31. Design an apparatus to electroplate nickel onto iron.

15.32. How does one gold plate a metal?

15.33. What is the problem with overcharging a secondary battery?

Part 4

Environmental Interactions

The second law of thermodynamics states that nature tends to its most probable state, which is that of maximum entropy. The most probable state of iron is not a bulldozer; it is a heap of rust. If we wish our machines and buildings to stay operative for a long period, we must know how nature tends to degrade materials and, with this knowledge, devise methods that prevent or slow down the degradation. This is the subject of Chapter 16. The human body is a particularly aggressive environment. In addition to preventing the corrosion of materials we use in medicine, we must prevent these materials from poisoning us or being rejected by the body. We can do better and use the degradation of a material to advantage; this is the case of resorbable sutures and the scaffolds that steer the growth of bone or other missing tissues. Chapter 17 reviews the materials currently used in medicine.

Corrosion and Wear

Corrosion is the slow degradation of materials due to chemical interaction with the environment. All metals, except gold, corrode; they do so because they tend to a thermodynamically more stable state: that of an oxide or hydroxide. Corrosion in liquid media occurs by electrochemical reactions as in batteries. Standard electrochemical potentials of metals are measured in a standard cell where the reference is the hydrogen electrode. Metals that establish a negative potential with respect to the standard are anodic and tend to be corroded. Metals that establish a positive potential are noble and tend not to corrode. When two metals are in electrical contact in an electrolyte, the more noble metal promotes the corrosion of the more anodic metal. The more conductive the liquid, the more rapid is the corrosion. Corrosion can be prevented by the use of sacrificial anodes that corrode instead of the metal they protect. Changes in oxygen content of water can establish corrosive cells; this is most important in crevice corrosion. The latter is avoided by judicious design. In gaseous corrosion, the corrosion product is usually a solid that stays on the surface of the metal. If this oxide is hard and dense, it forms a protective barrier that prevents the access of the oxygen to the metal. At high temperatures, the oxygen and metal ions diffuse through the oxide layer. At moderate temperatures, corrosion can be prevented by coating the metal (i.e., painting it).

Wear is a mechanical degradation (removal) of material. It occurs by sliding of one surface on another (adhesive wear), by penetration of hard particles (abrasive and erosive wear), by contact fatigue, and by cavitation. When two metals rub against each other, the softer metal wears; its wear resistance is proportional to its hardness. Abrasive wear is thousands of times faster than adhesive wear.

LEARNING OBJECTIVES

After studying this chapter, the student will be able to:

1. Describe the cathode and the anode reactions in an electrochemical cell.

2. Define noble (cathodic) and anodic metals.

3. Use the standard electromotive force series and the galvanic series to control corrosion.

4. Design a cathodic protection system and explain how it functions.

5. Design to avoid crevice corrosion.

6. Describe the mechanisms of gaseous corrosion.

7. Describe protective oxides and select metals that form them.

8. Describe linear, parabolic, and logarithmic oxidation rates.

9. Evaluate the temperature-dependence of oxidation.

10. Describe the main wear mechanisms and when they occur.

11. Select materials for minimum wear.

16.1 SOME QUESTIONS

Observe an ordinary carbon steel knife (not stainless steel): it remains essentially unchanged when left in a dry drawer; its oxidation is slow. Now leave this knife one night in water, or simply wet on a table: in the morning it is rusted. You have observed two different types of corrosion, namely dry oxidation and wet corrosion. We start this chapter with wet corrosion.

Since iron rusts when left in humid environments, how can engineers build skyscrapers with steel beams sunk in humid soil and expect them to remain safe for many years? How, generally, do we select the proper materials and how do we design our structures to prevent or slow down corrosion and what mistakes must we avoid? As usual, the answer lies in knowing how the phenomenon occurs and what drives it. A brief review of the corrosion of the Statue of Liberty in New York harbor and the repairs undertaken to prevent further damage will illustrate the phenomena we are about to study.

16.2 THE ELECTROCHEMICAL NATURE OF CORROSION IN LIQUIDS

The Statue of Liberty, situated on a small island in New York harbor, is constantly exposed to salty humidity (see Figure 16.1). It consists of a copper skin held up by an iron armature. This constitutes a giant battery where an electrochemical potential of 0.25 V is established between the two metals and drives the dissolution of the anodic iron.

The engineers building the statue knew that the copper had to be electrically insulated from the iron to avoid corrosion. Shellac-impregnated asbestos cloth was used for this purpose. After a hundred years, the insulation was soaked with saltwater, which is electrically conductive. With an electric current now flowing between the copper and the iron, the copper drove the corrosion of the iron armature so that the latter had to be replaced. Using their knowledge of galvanic corrosion and the corrosion potentials shown in Table 16.1, the engineers considered aluminum bronze, Cupro-Nickel, and 316L Stainless Steel to replace the iron because these metals do not form a battery with copper. They settled on stainless steel for its superior mechanical properties. This example shows that aqueous corrosion is an electrochemical reaction. We will use the concepts we discussed in Section 15.2 to study it.

■ FIGURE 16.1 The Statue of Liberty in New York Harbor.

16.2.1 **The Mechanisms of Electrochemical Dissolution**

Oxidized iron (Fe^{++}) is in a lower energy state than metallic Fe. When we immerse an iron rod in water (Figure 16.2), some iron atoms are dissolved in the water as Fe^{++} ions. Two electrons remain in the metal for each ion formed; a positive electric charge is established in the water and a negative charge in the iron. When the electric potential difference resulting from this charge separation reaches a certain value E_{Fe}, the Fe^{++} ions are attracted back to the metal, the dissolution stops, and equilibrium is reached. The reaction can be written as

$$Fe \leftrightarrow Fe^{++} + 2e^- \qquad (16.1)$$

The forward reaction, $Fe \rightarrow Fe^{++}$, is an **oxidation** reaction; the reverse, $Fe^{++} \rightarrow Fe$, is a **reduction** reaction.

Table 16.1 Galvanic Series in Seawater.

Saturated Calomel Half-Cell	Reference
Graphite	+0.2
Titanium	0
316 stainless steel (passive)	−0.05
Silver	−0.15
Silver solder	−0.15
Monel (70 Ni, 30 Cu)	−0.2
Nickel	−0.2
430 stainless steel	−0.25
Lead	−0.25
Tin	−0.3
Bronze (Cu-Sn)	−0.3
Copper	−0.35
Aluminum bronze	−0.37
Brasses (Cu-Zn)	−0.4
18Cr-8 Ni (active)	−0.5
Steel or iron	−0.6
Cast iron	−0.65
2024 aluminum	−0.8
Cadmium	−0.7
Zinc	−1.0
Magnesium	−1.6

If the iron and water are in a system in which an electric current can remove the electron charge from the iron and the positive space charge from the water, the corrosion reaction proceeds. In most practical cases, this occurs when the iron is in contact with another metal as shown in Figure 16.3. If the electrode potential E_{Sn} is smaller, a cell potential

$$E_{Cell} = E_{Fe} - E_{Sn}$$

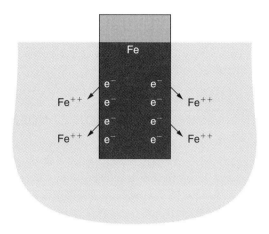

■ **FIGURE 16.2** Dissolution of iron in water: iron atoms are ionized to Fe^{++} and leave two electrons in the iron. The space charge establishes an electrochemical potential E_{Fe} that stops the reaction.

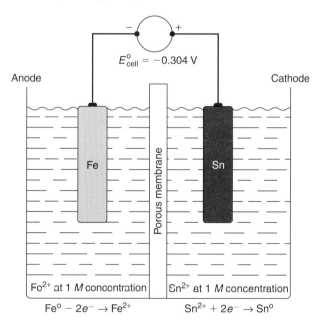

■ **FIGURE 16.3** Cell potential developed by addition of the half-cell reactions for iron and tin electrodes immersed in 1 M aqueous solutions of their respective salts (e.g., $FeCl_2$ and $SnCl_2$).

can be measured between the two metals. When the metals are brought into contact, the electronic charge is drained from the iron and Oxidation Reaction (16.1) proceeds; the iron is dissolved.

The dissolution rate of the iron is proportional to the ionic current that flows in the liquid according to Faraday's law, Section 15.2.4. This current depends on the electromotive force that is established and on the ionic conductivity of the liquid. The electromotive force and the conductivity of the half cell depend on the concentration of ions dissolved in the electrolyte. Table 15.1 shows the standard electrochemical potentials that correspond to 1 molar concentrations (i.e., 1 mole per liter). These are defined as the standard electrochemical potentials.

16.3 ELECTRODE POTENTIALS IN VARIABLE ION CONCENTRATIONS

In practical corrosion the concentration of ions in the liquid is much smaller than 1 molar, and the electrode potentials differ from the standard electromotive forces tabulated in Table 15.1. The relationship between electrode potential and ion concentration is given by the **Nernst equation**:

$$E = E^o + (0.059/n) \cdot \log (M^{n+}) \quad \text{(at 25°C)} \tag{16.2}$$

Here, E^o is the standard potential valid for 1 M concentrations, n is the charge on the ions, and M^{n+} is the actual concentration of the electrolyte in moles. For a 1 M solution ($M^{n+} = 1$), $E = E^o$.

When the concentration is low ($M^{n+} < 1$), the logarithm is negative. Also, the potential is more negative, the electrode is more anodic, and the substance corrodes more than with a 1 molar solution.

At large concentrations ($M^{n+} > 1$), the logarithm is positive, the electrode potential is more positive, and the metal corrodes less. This can be understood intuitively—when the concentration of ions in the liquid electrolyte is large, more ions are driven back to the metal which corrodes less (when the ion concentration is lower, the back current of ions is smaller, more ions go into solution, and corrosion is more rapid).

A similar equation can be written for the full cell potential, when A^{n+} and C^{n+} are the molar concentrations of the anodic and cathodic electrolytes.

$$E_{cell} = E^o_{cell} + (0.059/n) \cdot \log [(A^{n+})/(Cn^+)] \tag{16.3}$$

This equation follows from Equation (16.2) as follows:

$$E_{cell} = E_{Anode} - E_{cathode} = E^o_A + (0.059/n) \cdot \log(A^{n+}) - [E^o_C + (0.059/n)\log(C^{n+}).$$

EXAMPLE 16.1   *What cell potential would arise in a cell containing an Fe electrode in a 0.001 M FeCl₂ solution, and a Cu electrode in a 1.5 M CuCl₂ solution?*
ANSWER   Employing Equation (16.3) and Table 15.1, we know that $E^o_{cell} = -0.777\text{V}$, $(Fe^{2+}) = 10^{-3}$, $(Cu^{2+}) = 1.5$ and $n = 2$. Substituting,

$$E_{cell} = -0.777 + (0.059/2) \log[(10^{-3})/1.5] = -0.871\,V.$$

EXAMPLE 16.2 *If we build an electrolytic cell with Fe electrodes and a 0.001 M FeCl$_2$ solution on one half-cell and a 1 M FeCl$_2$ solution in the other, what potential will be established and which electrode will corrode?*

ANSWER $E_{cell} = 0.059 \cdot \log[10^{-3}] = 0.059 \cdot (-3) = -0.177$ V.

From Equation (16.2), the electrode potential with the lower concentration is the more negative ($E = E^{o} + (0.059/2) \times \log(10^{-3}) = E^{o} - 0.177$ V) and corrodes if the two electrodes are electrically connected.

16.4 **CATHODES IN AQUEOUS CORROSION**

Electrochemical corrosion is a concern every time two different metals are joined and are in contact with sufficiently conductive water. In practical cases no ions are initially present, yet corrosion of the anodic metal proceeds. There are several types of cathodes that establish an electromotive force and drive the corrosion.

The hydrogen cathode has already been described in Section 15.2.5. When the electrolyte is an acid solution, hydrogen ions H$^+$ are present. These are reduced at the cathode and hydrogen gas is formed.

$$2H^+ + 2e^- \rightarrow H_2$$

The oxygen cathode operates in aerated water or in oxidizing acids. Oxygen and water react at the cathode and form hydroxyl ions:

$$O_2 + H_2O + 2e^- \rightarrow 2(OH)^- \tag{16.5}$$

These ions react with the positive metal ions dissolved from the anode. In the case of iron, the reaction is

$$Fe^{2+} + 2(OH)^- \rightarrow Fe(OH)_2. \tag{16.6}$$

The reaction continues in the water to form Fe(OH)$_3$, which is rust.

The standard potential of the oxygen cathode is included in Table 15.1. It is $+0.401$ V with respect to the hydrogen electrode.

The water electrode operates when oxygen is present together with an acid; the H$^+$ ions from the acid react with oxygen to form water. The reaction at the cathode is

$$O_2 + 4H^+ + 4e^- \rightarrow 2H_2O \tag{16.7}$$

Its standard potential against the hydrogen electrode is $+1.229$ V. Thus the oxygen and the water electrode form an electromotive force with iron and are more potent in driving its oxidation than the hydrogen electrode.

16.5 FARADAY'S LAW—CORROSION RATE

The concepts used for the corrosion rate are exactly the same as for the storage capacity of a battery, Section 15.2.4. When an electric current flows, one singly charged ion carries $q = 1.6 \times 10^{-19}$ coulomb (A·sec) of charge. The transfer of one mole corresponds to a charge $F = q \cdot N_A = 96{,}500$ A·s. (N_A is Avogadro's number). $F = 96{,}500$ Clb is Faraday's constant.

The dissolution of a mole of material at the anode of a metal with valence n requires a total charge of nF. An identical amount of reaction (e.g., plating of metal, hydrogen evolution, etc.) occurs at the cathode. The weight loss w (in grams), at the anode, is

$$w = ItM/nF \tag{16.4}$$

where M is the atomic or molecular weight. (This is Equation (15.9) written differently with $Q = It$.) The quantity w/t may be viewed as a **corrosion rate**. If corrosion occurs uniformly over a given area, the weight loss can be easily converted to an equivalent reduction in metal thickness. Common units for corrosion rates are g/m^2-year, μm/year, or inch per year (ipy); these measures, of course relate to a current density (A/m^2) rather than to a total current.

EXAMPLE 16.3   *(a) What is the open circuit potential of the cell in Figure 16.3 if both anode and cathode electrolytes are 1 molar? (b) If 1 mA of current flows, how long will it take for 0.5 mg of the anode to dissolve? (c) If the total anode area is 100 cm^2, what is the metal thickness loss if corrosion is uniform? (d) Under what conditions will the potential reverse?*
ANSWER   (a) Because Fe lies below Sn in Table 15.1, we may assume it is the anode. By a reaction similar to Equation (15.13), Fe$^\circ$ + Sn^{2+} = Fe^{2+} + Sn$^\circ$, and

$$E^\circ{}_{cell} = E^\circ(Fe) - E^\circ(Sn) = -0.440 - (-0.136) = -0.304 \text{ V}$$

(b) The loss of metal is an application of Faraday's law. For Fe,

$M = 55.9$, and $n = 2$. Substitution into Equation (16.4) yields

$t = nFw/IM = 2 \times 96{,}500 \times 0.0005/0.001 \times 55.9 = 1730$ s or 28.8 min

(c) The loss of Fe is 0.0005np g/100 cm^2 = 5×10^{-6} g/cm^2. Dividing by the density of Fe (7.86 g/cm^3), the thickness of Fe lost is $5 \times 10^{-6}/7.86 = 0.636 \times 10^{-6}$ cm = 6.36 nm

(d) The potential reverses when $E_{cell} = 0$. If Sn now becomes anodic or active, the above reaction and cell potential reverses. From Equation (16.3), $E_{cell} = 0 = -0.304 + 0.059/2 \times \log [(Fe^{2+})/(Sn^{2+})]$. Solving, $[(Sn^{2+})/(Fe^{2+})] = 4.95 \times 10^{-11}$.

The cell potential is zero when the Sn^{2+} concentration is 4.95×10^{-11} molar. It becomes positive and tin corrodes only when the Sn^{2+} concentration is lower than that value.

This example illustrates that Sn is normally cathodic to Fe, and this accounts for its use in "tin can" containers.

16.6 **MANIFESTATIONS OF CORROSION**

It has already been noted that wet corrosion requires a metal anode and a cathode that are electrically connected and contacted by an electrolytic medium, so that an electric current flows. These requirements are physically manifested in many ways so that it is often difficult to recognize all of the disguises corrosion displays. It is instructive to enumerate some of the common types of corrosion first as model textbook cells and then in the form they assume in actual practice. We start with galvanic action.

16.6.1 **Galvanic (Two Metal) Corrosion**

This type of corrosion involves two different metals and corresponds to the electrochemical cells. Electric contact between metals with different electrode potentials generates a current that corrodes the more anodic metal. This is the case of the Statue of Liberty.

For the standard electrode potentials of Table 15.1, one immerses each metal in a 1 M solution of its own ions. This is not necessary for corrosion; the two metals can be immersed in the same electrolyte, for instance seawater. In this case, the electrode potentials are not identical to those of Table 15.1. The **galvanic series** in seawater is shown in Table 16.1. Like the electromotive force series, this table orders metals with respect to their anodic and cathodic tendencies. The order of the metals is not necessarily the same as in Table 15.1. Indeed, Zn and Al have switched places relative to Table 15.1, as have Ni and Sn. The galvanic series is usually more appropriate in engineering design than the electromotive series.

Surface Area of the Corroding Metals An important aspect in galvanic corrosion concerns the area of the electrodes. In the electric circuit of a corrosion system it is necessary that the same **current** flow through the anode and the cathode portions. The **current densities** (current/area) and corrosion rates vary inversely with anode size. Small anodes in contact with large cathodes will thus corrode more severely than large anodes in contact with small cathodes. For example, small steel rivets that are used to clamp large plates of copper will corrode through readily in a saltwater environment. On the other hand copper rivets clamping large steel plates would survive much longer, because the iron anode current density is small and the copper cathode does not corrode. These effects are illustrated in Figure 16.4.

Corrosion Protection through Sacrificial Anodes (Cathodic Protection) The preferential corrosion of small anodes in contact with large cathodes is utilized for the protection of steel. Underground steel pipes and ship hull steels are protected against corrosion this way. In the case of ships more strongly anodic zinc or magnesium alloy plates, which are easily replaced, are bonded to the much larger hull, which now assumes a cathodic character. The Zn and Mg become **sacrificial** anodes that corrode and protect the steel. Submerged piping and steel columns are commonly protected by impressing voltages that make them cathodic with respect to a sacrificial anode (Figures. 16.5 and 16.6).

Table 16.1 explains the choice of metals examined and selected for the repair of the Statue of Liberty: they are slightly cathodic with respect to copper.

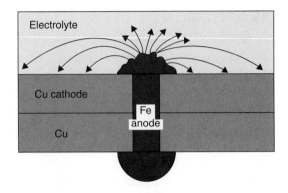

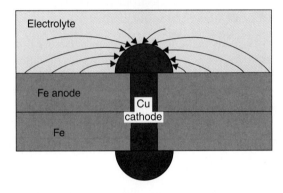

■ **FIGURE 16.4** Effect of electrode area. Top: an anodic iron rivet corrodes rapidly in contact with a large copper cathode. Bottom: The small copper rivet does not corrode, and the corrosion rate of the iron plate is low.

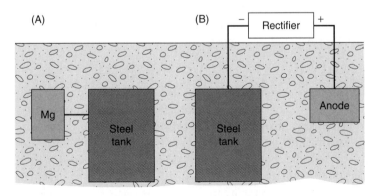

■ **FIGURE 16.5** Cathodic protection with a sacrificial anode. (A) By more anodic material. (B) By impressed potential making the tank cathodic. The electric connections are coated copper wires.

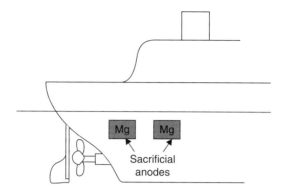

■ **FIGURE 16.6**   Cathodic protection of a ship by sacrificial anodes of Mg or Zn. The latter are easily replaced.

16.6.2 **Single-Metal Corrosion**

Caused by the Metal   When you immerse a piece of iron into acid or even water, it does not need a second metal to form a battery in order to corrode. In practice most cases of corrosion involve a single metal. A single metal corrodes because it contains both anodes and cathodes. These result from nonuniform distributions of alloying elements, precipitates, inclusions, and so on, as well as structural heterogeneities represented by grain boundaries and dislocations. These crystal defects are more energetic than other atoms and therefore more easily dissolved in the electrolyte. They form areas that are anodic with respect to the rest of the metal.

As an example, consider a zinc metal rod dipped in dilute HCl acid (Figure 16.7A). Zn dissolves while H_2 gas bubbles off. Closely separated areas of the same rod are different electrically; the acid exploits these differences and activates anodic and cathodic regions. At anodes, Zn^{2+} ions are released, while H_2 is discharged at cathodes. Another example involves preferential corrosion of cold-worked areas of a nail relative to undeformed regions (Figure 16.7B). Cold working introduces a high concentration of defects that are more energetic than atoms in the unstressed shank, so the former regions become anodic to the latter.

A related form of attack, known as **stress corrosion**, affects some metals that are simultaneously stressed in tension while immersed in a corrosive medium. Metal atoms are made particularly reactive by tensile stresses. The stress concentration at a crack tip makes it anodic and causes local corrosion that leads to crack growth and failure of the metal. Only specific environments and metals are effective in inducing stress corrosion cracking. Stainless steel in chloride solutions and brass in ammonia environments are the classic examples of metals that suffer such attack.

Grain boundaries are important regions for corrosion attack. Atoms are more energetic there than in grain interiors, making grain boundaries anodic relative to the bulk metal. Such preferential corrosion is the mechanism responsible for the controlled etching of metals that delineates grain boundaries for metallographic observation (Figure 16.7C). A far less desirable form of grain boundary corrosion occurs in certain stainless steels that contain more than ~ 0.03 wt% carbon. When these are heated and then slowly cooled through the temperature range 800–500°C, as during welding, chromium carbides (Cr_6C_{23}) nucleate and precipitate at grain boundaries (Figure 16.7D). This locally depletes the surrounding

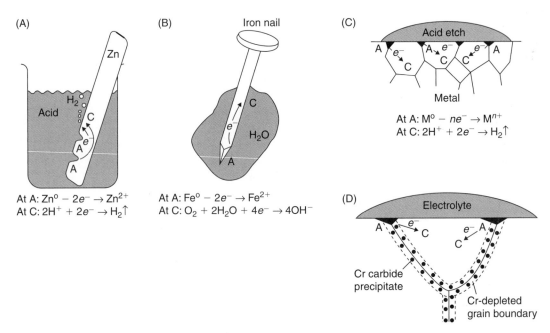

(A)

Zn

Acid

H_2

C

e

A

A

At A: $Zn^0 - 2e^- \rightarrow Zn^{2+}$
At C: $2H^+ + 2e^- \rightarrow H_2\uparrow$

(B)

Iron nail

C

e

H_2O

A

At A: $Fe^0 - 2e^- \rightarrow Fe^{2+}$
At C: $O_2 + 2H_2O + 4e^- \rightarrow 4OH^-$

(C)

Acid etch

A   e^-   A   e^-   e^-   A
     C   C   C

Metal

At A: $M^0 - ne^- \rightarrow M^{n+}$
At C: $2H^+ + 2e^- \rightarrow H_2\uparrow$

(D)

Electrolyte

A   e^-   C   e^-   A
             C

Cr carbide
precipitate

Cr-depleted
grain boundary

■ **FIGURE 16.7**   Corrosion action within a single metal. (A) Local anodes and cathodes develop on Zn dipped in acid. (B) Plastically deformed regions of nail are anodic relative to undeformed areas. (C) Preferential corrosion at anodic grain boundaries relative to cathodic bulk grains. (D) Grain boundaries depleted of Cr corrode relative to the stainless-steel matrix.

region below the 12 wt% Cr level that conferred the "stainless" quality to iron in the first place. The steel is now "sensitized" and susceptible to **intergranular corrosion**. Stainless steels can be spared this problem by lowering the carbon content, or by tying it up in the less harmful form of titanium carbide or niobium carbide. Rapid cooling of the steel from an elevated temperature will also help.

Caused by the Electrolyte   Single metals can be perfectly homogeneous and free of defects and yet still corrode. The reason is due to inhomogeneous distributions of electrolytes that divide the same metal into anodic and cathodic regions. Let us consider the **concentration cell** of Figure 16.8 in which iron is used for both electrodes. There is a lower concentration of Fe^{2+} ions in the left half-cell relative to the right half-cell. Thus we expect the left electrode to become anodic, as described by the Nernst equation.

$$E_{cell} = (0.059/n) \cdot \log[C_L/C_H] \qquad (16.8)$$

Here C_L is the low concentration and C_H the higher concentration at the cathode. (This equation is Equation (16.3) modified by taking $E^\circ_{cell} = 0$ since the same metal is in both half-cells.)

A practical example occurs when discs or propellers rotate in an electrolyte (Figure 16.9). At the outer radius the high linear velocity of the fluid disperses any buildup of ions. This region therefore becomes anodic with respect to the metal closer to the shaft where the liquid is less disturbed and the ionic concentration is higher. As a result outer regions suffer corrosion.

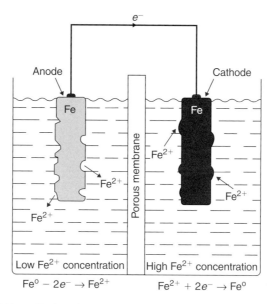

■ FIGURE 16.8 Concentration cell. Identical Fe electrodes are immersed in electrolytes containing different concentrations of Fe^{++}.

■ FIGURE 16.9 Bronze (Cu-10 wt% Sn) pump impeller exposed to 1% sulfamic acid at 30°C, suffering acid corrosion failure in a few hours. From E.D.D. During, *Corrosion Atlas; A Collection of Illustrated Case Histories*, Elsevier, Amsterdam (1988).

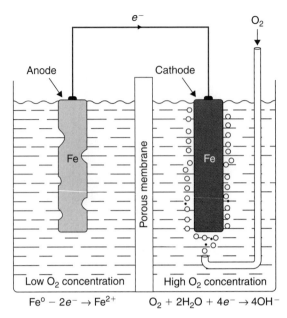

$$Fe^\circ - 2e^- \rightarrow Fe^{2+} \qquad O_2 + 2H_2O + 4e^- \rightarrow 4OH^-$$

■ FIGURE 16.10   Oxygen concentration cell. Corrosion occurs at the electrode exposed to the lower oxygen concentration.

Another important variant of the concentration cell, but this time consisting of different oxygen levels, is exemplified in Figure 16.10. The iron electrode that is least exposed to O_2 becomes the anode according to the Nernst equation (16.2) applied to oxygen.

In this case the anode reaction is still

$$Fe^\circ = Fe^{2+} + 2e^-$$

But, at the oxygen-rich cathode, hydroxyl ions are produced according to the reaction

$$O_2 + 2H_2O + 4e^- = 4OH^- \qquad (16.5)$$

The iron anode loses mass but remains bright. The combination of Fe^{2+} and OH^{-1} yields $Fe(OH)_2$ near the cathode, and with additional O_2, the oxidation to the Fe^{+3} state occurs through the reaction

$$4Fe(OH)_2 + O_2 + 2H_2O = 4Fe(OH)_3 \text{ (rust)} \qquad (16.9)$$

The anode loses mass and rust is deposited at the cathode.

This is the reason for the widespread practice of deaeration or removal of oxygen from boiler and industrial waters. Another implication is corrosion below the waterline in submerged steel piers. The surface of the water forms an oxygen cathode, and the submerged surface loses mass.

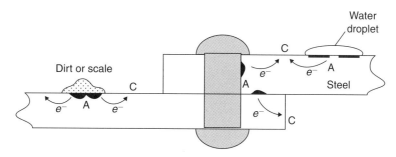

■ **FIGURE 16.11** Crevice corrosion. In the anodic crevices (A) the reaction is $Fe \longrightarrow Fe^{++} + 2e^-$ whereas the reaction $O_2 + 2H_2O + 4e^- \longrightarrow 4OH^-$ occurs on the cathodic steel (C). Similar reactions take place under and near water droplets and dirt.

Crevice Corrosion   A corrosion problem occurs when steel surfaces are covered with dirt, scale, or other pieces of steel, such as a washer, riveted or welded sections, and so on, that trap electrolyte in between (Figure 16.11). Such **crevice corrosion** is caused by **oxygen starvation**, and those regions so deprived become anodic, Equation (16.2), and corrode relative to the cathodic regions exposed to air. An area effect is also at play here: In the oxygen-starved crevices there is a small anode in contact with a very large cathode so that the material in the crevice corrodes rapidly.

16.7 OTHER FORMS OF CORROSION

In practice, corrosion does not always assume the forms just presented. A number of case histories illustrated in Figures 16.12 to 16.15 attribute degradation to the harmful influence of certain chemical species (e.g., chlorine, hydrogen, and ammonia) and stress.

16.7.1 Chlorine

Chlorine in salt form is not only a relatively abundant element in marine and industrial environments, but also a chemically reactive one. Stainless steels and aluminum alloys are particularly susceptible to chloride attack as seen in Figure 16.12. Through reaction of salt and water, enough chloride ions are released on metal surfaces to cause either localized **pitting corrosion** in the case of stainless steel or uniform attack in the case of aluminum. Metals are not the only materials that suffer from the presence of chlorine. Absorption of HCl caused the rubber gasket shown in Figure 16.13 to swell and crack.

16.7.2 Hydrogen

Hydrogen is a harmful element in virtually all major metals. In fact, great efforts are made to degas steel, aluminum, and copper alloy melts prior to mechanical forming operations. Despite this, hydrogen enters metals from ubiquitous corrosive environments and can cause blisters and cavities in extreme cases (Figure 16.14). Hydrogen probably enters the metal in atomic form, and the eventual formation of H_2 in the metal is accompanied by microcrack nucleation and development of internal

■ **FIGURE 16.12** Corrosion by chlorine (saltwater). Top: Pitting corrosion of stainless steel gas cooler pipe. Bottom: Uniform corrosion of aluminum roofing in an atmosphere contaminated by saltwater. From E.D.D. During, *Corrosion Atlas; A Collection of Illustrated Case Histories*, Elsevier, Amsterdam (1988).

stresses. The penetration of hydrogen can also render steels very brittle; this is known as hydrogen embrittlement. The mechanisms of this degradation are not yet well known.

16.7.3 Ammonia

Environments containing ammonia gas and solutions of ammonium ions are common in industry. Brasses are susceptible to ammonia corrosion particularly in the presence of stress. An example of ammonia attack is shown in Figure 16.15. The progressive build-up of corrosion products containing the complex $Cu(NH_3)_4^{2+}$ ion occurred in the valve, and sudden brittle fracture occurred in the bolt. The bolt was under stress and crack propagation occurred across grains with the assistance of corrodent penetration.

■ **FIGURE 16.13** Swelling and cracking of an ethylene-propylene rubber flange seal as a result of exposure to HCl. From E.D.D. During, *Corrosion Atlas; A Collection of Illustrated Case Histories*, Elsevier, Amsterdam (1988).

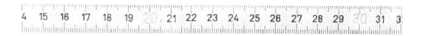

■ **FIGURE 16.14** Formation of cavities during cold hydrogen attack of unalloyed steel. The environment was water with absorbed carbon dioxide. From E.D.D. During, *Corrosion Atlas; A Collection of Illustrated Case Histories*, Elsevier, Amsterdam (1988).

16.8 **PREVENTING CORROSION THROUGH DESIGN**

This chapter has focused attention on the role of materials and their selection when combating corrosion in different environments. But such considerations will go for naught if the structural design is not made correspondingly corrosion safe. A number of simple examples illustrating recommended

■ **FIGURE 16.15**   Ammonia corrosion of a pressure gauge valve. From E.D.D. During, *Corrosion Atlas; A Collection of Illustrated Case Histories*, Elsevier, Amsterdam (1988).

and not recommended design practices are shown in Figure 16.16 without comment. It is left as a challenge to indicate reasons for the poor design.

16.9 GASEOUS OXIDATION

A universal response of metal surfaces exposed to an oxygen-bearing atmosphere is to oxidize. In air, the oxidized metal is not removed from the surface. The oxidation product is usually a solid; it may be an adherent film that protects the underlying metal from further attack, or it may be a porous layer that flakes off and offers no protection. Both types of oxides grow by similar mechanisms that involve mass transport of metal and oxygen ions.

Consider the formation of an oxide layer shown in Figure 16.17. Two simultaneous ion transport processes occur during oxidation. At the metal-oxide interface, neutral metal atoms lose electrons and become ions that migrate through the oxide to the oxygen-oxide interface. The released electrons also travel through the oxide and serve to reduce oxygen molecules to oxygen ions at the surface. These oxidation-reduction reactions are

$$M^{\circ} = M^{n+} + ne^{-} \text{ (at metal-oxide interface)} \tag{16.10a}$$

and

$$1/2O_2 + 2e^{-} = O^{2-} \text{ (at } O_2\text{-oxide interface)} \tag{16.10b}$$

Ionic mass transfer through a solid oxide is very much slower than through a liquid electrolyte. This is why elevated temperatures are required to generate oxide layers of appreciable thickness. If the metal

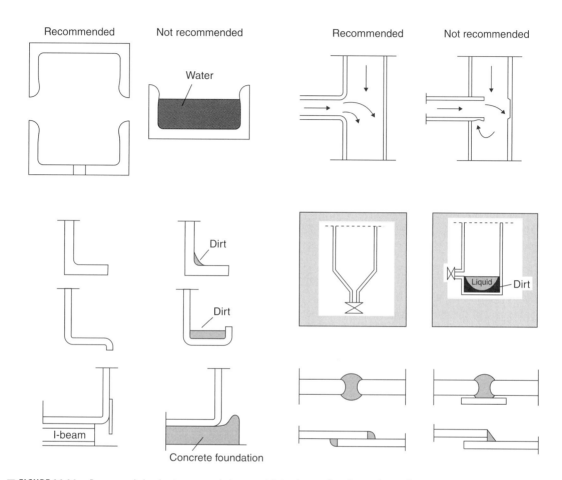

■ **FIGURE 16.16**   Recommended and not recommended structural design features from the standpoint of corrosion.

cations migrate more rapidly than oxygen anions, the oxide grows at the outer surface. Oxides of Fe, Cu, Ni, Cr and Co grow this way. On the other hand, oxide forms at the metal-oxide interface when oxygen migrates more rapidly than metal. This is the case in oxides of Ti, Zr, and Si.

16.9.1 Protective Oxide Layers

The physical integrity of the oxide coating is the key to whether the underlying metal will suffer further degradation or not. If the oxide is dense, it protects the metal from further oxidation. But if it is porous and spalls off, the exposed metal surface will suffer further deterioration. Whether the oxide is dense or porous can frequently be related to the ratio of the oxide volume produced to the metal consumed when it oxidizes. The quotient, known as the **Pilling-Bedworth ratio** (PBR), is given by

$$\text{oxide volume/metal volume} = (M_o/\rho_o)/(aM_m/\rho_m) = M_o\rho_m/aM_m\rho_o \tag{16.11}$$

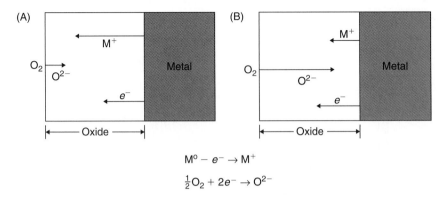

$$M^{\circ} - e^{-} \rightarrow M^{+}$$

$$\tfrac{1}{2}O_2 + 2e^{-} \rightarrow O^{2-}$$

■ **FIGURE 16.17** Oxidation of solids as a result of ionic transport. (A) Metal migrates more rapidly than oxygen in the oxide. (B) Oxygen migrates more rapidly than metal in the oxide.

where M and ρ are the molecular weight and density, respectively, of the oxide (o) and metal (m), and a is the number of metal atoms per oxide molecule M_aO_b. The ideal ratio is PBR $= 1$; the volume of the oxide formed is equal to that of the metal consumed. In this case, the oxide film sits on top of the metal and is stress free. If PBR is between 1 and 2, the oxide tends to be coherent with the metal and is protective; further growth can only occur by ionic diffusion. The most notable protective oxides are those of Al, Si, Cr, Ni, and Ti. But if PBR is too large (i.e., >2) the oxide may be subjected to large compressive stresses, causing it to buckle and flake off. If the ratio is less than 1, that is, if the volume of the oxide is smaller than the volume of the metal that was consumed, the oxide may be porous or may split off and afford little protection to the metal underneath, giving oxygen direct access to the metal surface and cause the oxide to grow linearly in time. Oxides of the alkali metals Li, Na, and K are of this type.

In addition to the Pilling-Bedworth ratio, another important consideration in assessing oxide (or nitride, carbide, sulfide, etc.) coating integrity is the difference in thermal expansion coefficient of the oxide (α_o) and the substrate metal (α_m).

Assume that metal and oxide are unstressed when the latter forms at elevated temperature. If $\alpha_o > \alpha_m$ the oxide layer will contract more than the metal on cooling to ambient temperature. Since the oxide adheres to the metal, the oxide and metal must have a common length; the layer is placed in tension and the substrate in compression at room temperature as shown in Figure 16.18A. If, however, the reverse is true of the expansion coefficients, the oxide is in compression upon cooling. The latter is generally the case for many practical oxides (Figure 16.18B). If the residual stresses exceed the fracture stress, either cracking or buckling of the oxide may occur.

16.9.2 Oxidation Rates

When the oxide layer is porous or spalls off, the slowest step in the oxidation is the reaction between oxygen and metal. In this case, the oxide film thickness increases **linearly** with time:

$$X_O \sim k{\cdot}t \qquad (16.12)$$

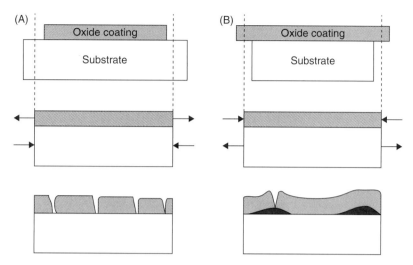

■ **FIGURE 16.18**   Development of stress in oxide coatings upon cooling. At elevated temperatures it is assumed that the coating and metal substrate are unstressed. Top: natural size of oxide and metal upon cooling. Middle: stresses in the oxide and metal because of adhesion of the oxide. Bottom: failure of the oxide when the stresses are too large. (A) If $\alpha_o > \alpha_m$, the oxide contracts more than the substrate and it will be residually stressed in tension. Cracking of the oxide may occur in this case. (B) If $\alpha_o < \alpha_m$, the greater contraction of the metal forces the oxide into compression and possible delamination.

In most cases, the growth of the oxide is limited by the diffusion of the metal and the oxygen through the oxide layer. In this case, as in all diffusion processes, the thickness of the oxide X_O increases with time as

$$X_O^2 \sim Dt \text{ or } X_O \sim \sqrt{Dt} \tag{16.13}$$

This is **parabolic growth**. D is the diffusion coefficient of the migrating ions. Diffusion is a thermally activated process so that D changes with temperature as

$$D(T) = D_o\exp\left(-Q_D/RT\right)$$

This is a very rapid increase with temperature that is described by an Arrhenius curve. Parabolic oxidation is a high-temperature phenomenon. It is of special concern in high-temperature applications such as jet engines.

At low temperatures, diffusion is much too slow for oxide growth. In this case, electrons traverse the very thin oxide by quantum mechanical tunneling and establish an electric field that drives the movement of the ions. The growth rate is **logarithmic**

$$X_O \sim k\cdot\log\left(Ct + 1\right) \tag{16.14}$$

It is easy to see from the logarithmic form that growth is rapid initially and becomes extremely slow after some time. As a result very thin protective native oxides, often less than ~3 nm thick, form on the surfaces of Si, Ti, Al, and Cr, and then virtually stop growing.

In certain materials, for example in copper and bronze, oxidation slowly grows a protective layer called a patina. In the Statue of Liberty, the copper skin acquired its green patina over a period of 25 years. After 100 years, the copper skin had lost less than 0.005 in (about 0.12 mm) in thickness so that the engineers decided to leave it alone. Bronze statues acquire a beautiful dark brown patina that protects them from the weather. Greek and Roman bronze statues remain essentially intact after more than 2000 years.

16.10 **WEAR**

Wear is the gradual loss of material that is due to friction during the relative motion of bodies. Most of the time, friction and wear are undesirable: friction increases the energy necessary to drive machines, and wear deteriorates the geometry of the machine and finally leads to loss of function. Friction is, however, necessary in some applications such as clutches and the maintenance of the grip of nuts and bolts. Wear, especially abrasive wear, is essential in the shaping of mechanical parts in grinding and polishing. Friction and wear are not limited to engineering systems. They arise in the movement of animal joints and lead to arthritis and the degradation of bone-socket interfaces, necessitating replacement prostheses discussed in Chapter 17. We rely on friction when we walk.

Wear is caused by interfacial mechanical forces. Avoiding wear is of crucial significance in machine components such as ball bearings and computer disc heads where high relative velocities between the contacting surfaces (i.e., ball and race, head and disc) occur. Heads, which fly over, but very close to, the surface of computer discs, normally do not contact them except during starting and stopping of the drives. Occasionally, the contact is such that a "crash" occurs with attendant loss of stored information. In this case as well as in bearings, much damage can be produced by very little mass loss.

From an engineering standpoint, three separate categories of behavior can be distinguished based on the relative magnitudes of friction and wear:

1. **Friction and wear are both low.** This is the case in bearings, gears, cams, and slideways.

2. **Friction is high but wear is low.** This combination is desired in devices that use friction to transmit power such as clutches, belt drives, tires, and shoe soles.

3. **Friction is low and wear of one body is high.** This is the situation that prevails in material removal processes such as machining, cutting, drilling, and grinding. In these operations the tools also suffer wear and must be periodically sharpened.

A number of wear mechanisms have been identified, and these will be discussed in turn.

16.10.1 **Adhesive Wear**

Material surfaces that appear smooth on a macroscopic level actually consist of microscopic asperities or peaks. When bodies containing two such surfaces are brought together, contact occurs at relatively few asperities. The asperities deform under loading, and the local stress is approximately equal to the hardness of the softer material. Sliding of one surface on the other produces large shear stresses. If shear occurs at the interface between asperities, there is no wear. Some shear occurs away from the interface because of fatigue or because the interfacial bond strength exceeds the cohesive strength of one (or both) of the contacting bodies. In this case, material is transferred from one surface to the other, usually from the softer to the harder body. With further rubbing the transferred material becomes detached to form loose wear particles. The power consumed in overcoming friction increases the temperature at the contact. When load, velocity, and friction are high, the increase in temperature softens the material and leads to smearing, **galling**, and **seizure** of the surfaces

Three facts of adhesive wear are universally recognized:

1. The volume (V) of wear material is proportional to the distance (L) over which relative sliding occurs.

2. The volume of wear material is proportional to the applied load (F).

3. The volume of wear material is inversely proportional to the hardness (H) or yield stress of the **softer** material.

Neither the total friction force nor the amount of wear depends on the macroscopic area of the rubbing surfaces. The reason is that the real area of contact is established by local deformation of the asperities and is much smaller than the macroscopic or design area.

The Archard equation combines these observations and expresses the volume of material lost as

$$V = kFL/H \tag{16.15}$$

where k is a dimensionless constant known as the wear coefficient, F is the load, that is, the force pressing the two surfaces together, L is the total length of sliding, and H is the hardness of the softer material. Values of k are obtained by measurement. They are listed in Table 16.2. In the presence of

Table 16.2 Approximate Values of the Wear Coefficient in Air.

Unlubricated	k
Mild steel on mild steel	10^{-2} to 10^{-3}
60-40 brass on hardened tool steel	10^{-3}
Hardened tool steel on hardened tool steel	10^{-4}
Polytetrafluoroethylene (PTFE) on tool steel	10^{-5}
Tungsten carbide on mild steel	10^{-6}
Lubricated	**k**
52100 steel on 52100 steel	10^{-7} to 10^{-10}
Aluminum bronze on hardened steel	10^{-8}
Hardened steel on hardened steel	10^{-9}

From S. Kalpakjian, *Manufacturing Processes for Engineering Materials*, 2nd ed., Addison Wesley, Reading, MA (1991).

lubricants, the wear coefficient k decreases by many orders of magnitude relative to the unlubricated case. The reason stems from the development of interfacial incompressible lubricant (fluid) films that keep the bearing surfaces from contacting.

16.10.2 Abrasive Wear

Abrasive wear covers two different situations in practice. Both involve the plowing or gouging-out out of the softer material by a harder material. In the first case a rough, hard surface rubs against a softer material as in the action of a file or abrasive paper on a soft metal. This type of wear can be eliminated by minimizing surface roughness. In the second type of abrasive wear, airborne dust, grit, and the products of adhesive or corrosive wear are trapped between the contacting materials. Sealing and filtration is the only practical precaution against the ingress of airborne abrasive particles. It is estimated that 85% of cylinder wear in automobiles is caused by dust entering through the air intake. Thus replacing the air filter is as important for the life of an engine as replacing the oil.

The volume of material removed by abrasive wear is computed by a formula similar to Equation (16.15), but the wear coefficient k is about twenty thousand times larger than in sliding wear. At equal hardness H and applied load, ceramics abrade much faster than metals because they are brittle. While abrasion of ductile metals involves much plowing, that is, deformation of surface material without removal, the penetration of the hard abrasive fractures the ceramic. Abrasive wear is utilized in grinding and polishing of materials. It is the only feasible machining technique for ceramics. Care must be taken with the grinding of ceramics because of the surface cracks introduced by the abrasive. These cracks are similar to the ones used in the indentation toughness measurement technique described in Section 2.6.2. They are responsible for the low reliability of these materials under tensile stresses. For this reason, ceramics are shaped in several grinding steps with progressively finer abrasives.

16.10.3 Fatigue Wear

Fatigue wear is the loss of surface material due to repeated loading and unloading of the contacting surfaces. Consider the rolling element bearing on a wheel of your car. The weight carried by one wheel is borne at any time by about six rollers in the double bearing. The contact area between rollers and race of the bearing is small, and the contact pressures are above 1 GPa. With a diameter of, say 60 cm (2 ft), the wheel travels 1.8 m per turn. In a total distance driven of 160,000 km (100,000 mi) the wheel has made about 90 million turns. Such bearings are prime candidates for fatigue wear.

Fatigue wear initiates at metallic inclusions, porosity, and microcracks near the surface and eventually results in dislodged particles. The testing of large numbers of roller bearings has revealed that their life is inversely proportional to the cube of the applied load. Methods to prevent fatigue wear are the same as for fatigue fracture: one avoids inclusions of foreign material and one builds the bearings with polished surfaces. Bearing steel is processed by vacuum metallurgy to eliminate oxygen and other gases that would result in oxide inclusions or in porosity. It is a tribute to the quality of bearing steel that failure of wheel bearings is rarely observed.

16.10.4 **Corrosive Wear**

Corrosive wear requires both corrosion and rubbing. In the corrosive atmosphere, either liquid or gaseous, the corrosion products that form at the contacting surfaces are poorly adherent and may act as abrasive particles to accelerate wear.

16.10.5 **Fretting Wear**

Fretting wear occurs under conditions of vibratory motion of small amplitude (in the range of 1–$200\,\mu m$). It is observed in bolted or riveted joints or press-fits that are subject to transverse loading; it is aggravated by vibration. Many sequential damage processes occur during fretting including breakup of protective films, adhesion and transfer of material, oxidation of metal wear particles, and nucleation of small surface cracks.

16.10.6 **Erosion**

Erosive wear is the removal of material from a surface caused by the impingement of free particles propelled by air or liquid. The erosion rate is dependent on the impingement angle as shown in Figure 16.19. The wear rate exhibits opposing trends depending on whether ductile (metal) or brittle (ceramic) surfaces are involved. Low impingement angles stretch ductile materials in tension, causing them to fracture, whereas normal impact peening compresses the surface and strengthens it. Conversely, brittle materials suffer more erosion at normal incidence.

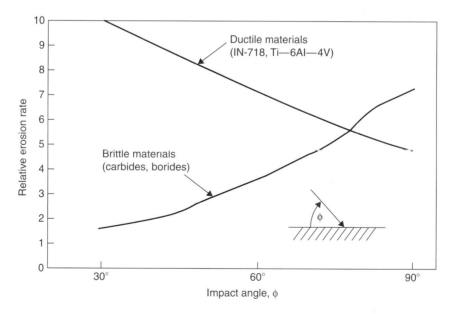

■ **FIGURE 16.19**   Erosion rates of ductile and brittle materials as a function of particle impact angle ϕ.

Erosion degradation is common in earthmoving, mining, and agricultural equipment handling rock, particulates, and slurries. It is a troublesome form of wear in gas turbine engines that ingest airborne particulates. Erosion also occurs during electrical arcing and leads to pitting degradation of electrical contacts. Spark erosion is exploited in spark cutting, a machining technique employed to cut hard metals through controlled incremental erosion.

16.10.7 **Cavitation**

Cavitation is caused by the generation of minute bubbles at the interface between rapidly flowing liquids and solid surfaces. In impellers, turbines, and propellers, water flow causes very low pressure that favors the evaporation of water and formation of bubbles. These bubbles contain water vapor that condenses under pressure; their collapse causes a sharp impact similar to a hammer blow to the surface and erodes it away.

16.10.8 **Wear in Cutting Tools**

Wear of cutting tools has a major impact on the costs sustained by the metal finishing industries, and figures prominently in the price of machined products. An instant in the high-speed cutting action of a tool against a workpiece is frozen in the depiction of Figure 16.20A. High local stresses serve to shear the metal being cut and form the chip. The friction generated between the chip and the tool can result in local temperatures of $1,000\,^\circ$C or more depending on tool and workpiece materials, cutting speeds, and feed rates, extent of lubrication, and so on (Figure 16.20B). Wear craters and material loss, accelerated by abrasive particles, thermal softening, oxidation, and chemical reactions occur on the tool (rake) face, side or flank face, and nose (Figure16.20C). The tool eventually fails and must either be sharpened or replaced because the work dimensions and surface finish no longer meet specifications.

Wear of the tool is inversely proportional to the product of its fracture toughness (K_{Ic}) and hardness. Today TiC, TiN, and Al_2O_3 coatings ($\sim$5–10μm thick), used singly or in combination, are the chief means of combating tool wear.

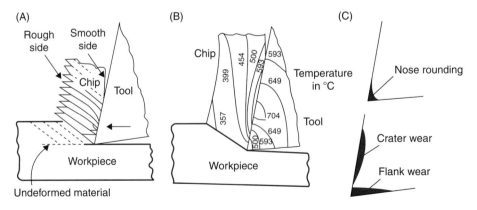

■ **FIGURE 16.20** (A) Cutting action of a tool bit against a metal workpiece in a machining operation. (B) Typical temperature profiles within tool and chip. (C) Tool degradation caused by nose rounding (above) and flank and crater wear (below).

■ SUMMARY

1. Corrosion in water is an electrochemical dissolution where positive ions are dissolved in the liquid and leave the corresponding number of electrons in the metal.

2. The resultant charge separation creates an electric potential difference between liquid and metal. At equilibrium potential, the dissolution stops.

3. The equilibrium potentials of metals are measured in a standard electrolytic cell against a hydrogen electrode (Pt on which $2H^+ + 2e^- \leftrightarrow H_2$). Metals with a negative potential in the cell are anodic and tend to corrode (Zn, Mg). Metals with a positive potential are cathodic or noble and tend not to corrode (Cu, Ag, Pt, Au).

4. When two different metals are in contact, the more noble metal drives the dissolution of the more anodic metal.

5. The galvanic series represents the electrochemical potential of metals in saltwater.

6. A method to combat corrosion is cathodic protection. One makes the metal (iron) cathodic by pairing it with a sacrificial anode: the latter can be a more anodic metal (zinc, magnesium) or iron to which one applies an appropriate electric potential.

7. Single metals can be corroded because anodic areas are created by change in composition or by mechanical defects (cold working).

8. Variations in oxygen content in water form corrosive cells. This is the cause of pitting corrosion, crevice corrosion, and corrosion near the surface of water. The metal exposed to lower oxygen concentration is the one that corrodes.

9. Wet corrosion is minimized by pairing metals with similar galvanic potential, by electrically insulating different metals from each other, by avoiding crevices in design, and avoiding the accumulation of dirt or stagnant fluid.

10. In gaseous corrosion, the corrosion product (e.g., an oxide) is a solid that accumulates at the surface of the metal.

11. When this solid is dense and adherent, it protects the surface from further corrosion (Al, Cr, Ni, Ti). Porous and poorly adherent scales do not protect the metal (rust, Na).

12. At moderate temperatures, gaseous corrosion can be prevented with coatings (paint).

13. At high temperatures, oxidation proceeds by diffusion of ions through the oxide. This is parabolic oxidation where the depth of oxidation increases with the square root of the time. ($X_O \sim \sqrt{Dt}$). The diffusion coefficient D increases rapidly with increasing temperature (Arrhenius law).

14. Wear is the mechanical removal of material.

15. Sliding (adhesive) wear occurs when surfaces slide over each other. It is decreased by lubrication.

16. Abrasive wear results from the penetration of hard particles into the surface. It is thousands of times faster than adhesive wear.

17. The wear volume is proportional to the load and sliding distance and inversely proportional to the hardness. It does not depend on the macroscopic contact area.

18. Fatigue wear results from repeated application of contacts (e.g., in ball bearings and rails). It is increased by the presence of hard particles.

19. Corrosive wear results from sliding in a corrosive medium.

20. Erosive wear results from loose hard particles streaming over the surface.

21. The wear resistance of metals is proportional to their hardness.

22. The wear resistance of ceramics increases with their toughness.

■ KEY TERMS

316L Stainless, 460

A
abrasive wear, *482*
acid, 465, 469
adhesive wear, *480*
Al, 467, 479
Al_2O_3, 484
alloying element, 469
aluminum, 473
ammonia, 474
ammonia corrosion, 474
animal joints, 480
applied load, 481
Archard, *481*
Arrhenius curve, 479
arthritis, 480
asperities, 480
Avogadro's number, 466

B
ball bearings, 480
battery, 460
blisters, 473
brass, 474
bronze, *460, 480*

C
carbon steel, 460
cathodes in aqueous corrosion, 465

cathodic protection, 467
cavitation, *484*
cavities, 473
ceramic, 482
chloride attack, 473
chlorine, 473
clutches, 480
coatings, 484
cold working, 469
computer disc heads, 480
concentration cell, 470, 472
copper, 460, 467, 473, 480
corrosion rate, 466
corrosive wear, *483*
Cr, 479
crevice corrosion, 473
Cupro-Nickel, 460
current, 462, 466, 467
current densities, 467
cutting tools, *484*

D
design, 475
diffusion coefficient, 479
dislocations, 469
dissolution, 461, 463, 466, 485
dry oxidation, 460

E
electrochemical dissolution, 461

electrochemical nature of corrosion, 460
electrochemical potential, 460
electrode potential, 460
electrolyte, 464
electromotive force series, 467
embrittlement, 474
erosion, *483*

F
Faraday, 466
Faraday's constant, 466
Faraday's law, 463
fatigue wear, *482*
filtration, 482
fretting, *483*
friction, *480*

G
galling, 481
galvanic (two metal) corrosion, 467
galvanic series, 467
gaseous oxidation, 476
grain boundaries, 469
grinding, 480

H
hardness, 481, 484
HCl, 469, 473

■ REFERENCES FOR FURTHER READING

[1]   J.O'M. Bockris and A.K.N. Reddy, *Modern Electrochemistry*, 2nd ed., Plenum Press, New York (1998).

[2]   M.G. Fontana and N.D. Greene, *Corrosion Engineering (Materials Science and Engineering)*, 3rd ed., McGraw-Hill, New York (1986).

[3]   E. Rabinowicz, *Friction and Wear of Materials*, 2nd ed., Wiley-Interscience, New York (1995).

[4]   P.R. Roberge, *Handbook of Corrosion Engineering*, McGraw-Hill Professional, New York (1999).

[5]   G. Stachowiack and A.W. Bachelor, *Engineering Tribology*, 3rd ed., Butterworth-Heinemann, Woburn, MA (2005).

■ PROBLEMS AND QUESTIONS

16.1.   A galvanic cell consists of a magnesium electrode immersed in a 0.1 molar $MgCl_2$ electrolyte and a nickel electrode immersed in a 0.005 molar $NiSO_4$ electrolyte. These two half cells are connected through a porous plug. What is the cell emf?

16.2.   What is the potential of a Fe-Cu cell in which the electrolytes consist of 0.01 M $FeCl_2$ and 0.01 M $CuCl_2$? How does it compare with the standard potential?

16.3.   Hot water heaters contain magnesium sacrificial anodes. A 0.4 kg anode lasts 15 years before it is consumed. What (continuous) current must have flowed to produce this amount of corrosion?

16.4.   It is common to attach brass valves onto iron pipes.

    a.   Which metal is likely to be anodic?

    b.   Why do such apparent galvanic couples tend to survive without extensive degradation?

16.5.   A current of $1.1 \, \mu A/cm^2$ is measured to flow when iron corrodes in a dilute salt electrolyte. How much loss of iron, in millimeters, will occur in a year?

16.6.   The silver anodes that are used in electroplating silverware corrode and transfer the silver through the electrolyte to deposit on the cathodes (forks, spoons, etc.). How long would it take a kilogram of silver to dissolve or corrode in this way if a current of 100 A flowed through the plating tank?

16.7.   Explain the following observations concerning corrosion:

    a.   Corrosion is more intense on steel pilings somewhat below the waterline.

    b.   A rubber band, stretched tightly around a stainless steel plate submerged in a salt solution, will slowly but steadily cut through it much like a saw blade.

    c.   Corrosion is often observed near dents in a car fender.

    d.   Stainless steel knives sometimes exhibit pitting in service.

    e.   Corrosion often occurs in the vicinity of very dilute electrolytes.

    f.   Two-phase alloys corrode more readily than single-phase alloys.

16.8.   Localized pitting corrosion is observed on an aluminum plate that is 1.2 mm thick. The pits are 0.2 mm in diameter and appear to extend into the plate without much widening. A hole appeared after 50 days.

    a.   What single pit current density was operative in this case?

    b.   Similar damage developed over an area of $100 \, cm^2$ with a pit density of 25 pits/cm^2. What is the total corrosion current?

16.9.   In seawater which of the following metals corrodes preferentially?

    a.   Titanium and steel.

    b.   Brass and stainless steel.

    c.   Copper and nickel.

    d.   Monel and brass.

    e.   Silver and gold.

16.10.   Use Table 16.1 to justify the choice of the metals to be paired with the copper in the repairs of the Statue of Liberty.

16.11. Which of the two half-cells will be anodic in an electrochemical cell?

    a. $Zn/0.5\,M\ ZnCl_2$ and $Ni/0.005\,M\ NiCl_2$

    b. $Cu/0.15\,M\ CuSO_4$ and $Fe/0.005\,M\ FeSO_4$

    c. $Mg/1\,M\ MgSO_4$ and $Al/0.01\ AlCl_3$

    d. $Ag/0.1\ AgNO_3$ and $Cd/0.05\,M\ CdCl_2$

16.12. In a laboratory experiment, six different alloys (A, B, C, D, E, F) were tested to determine their tendency to corrode. Pairs of metals were immersed in a beaker containing a 3% salt solution. A digital voltmeter displayed the following potentials between the electrodes with the indicated polarities. (Elements at the column tops are connected to the positive meter terminal, and elements at the left side of the rows are connected to the negative terminal.)

	A	B	C	D	E	F
A	X	+0.163 V	−0.130 V	−0.390V	+0.130V	+0.033V
B	X	X	−0.325 V	−0.650 V	−0.013V	−0.098V
C	X	X	X	−0.243 V	+0.163V	+0.039V
D	X	X	X	X	+0.487 V	+0.260V
E	X	X	X	X	X	−0.013V
F	X	X	X	X	X	X

Order the metals from the most anodic to the most cathodic.

16.13. Distinguish between the following terms:

    a. Ohm's law and Faraday's law.

    b. Electromotive force series and galvanic series.

    c. Passive and active electrodes.

    d. Standard and nonstandard electrodes.

16.14. Illustrate through an example the following types of corrosion:

    a. Uniform corrosion attack.

    b. Pitting corrosion.

    c. Grain boundary corrosion.

    d. Stress corrosion cracking.

    e. Crevice corrosion.

    f. Corrosion fatigue.

16.15. Steel screws used as fasteners on aluminum siding underwent severe corrosion even though iron is normally cathodic to aluminum. Provide a possible explanation for this occurrence.

16.16. The famous wrought iron pillar of Delhi, India has not rusted in 1700 years of exposure to the ambient conditions. In addition to iron, the 7.2 m high pillar contains 0.15% carbon and 0.25% phosphorus, and has a magnetic coating that is 50–600 μm thick. Provide possible reasons for the rust-free state of preservation.

16.17. Old homes were equipped with galvanized iron pipes. These are no longer used and have been replaced by copper, and lately, polymer (PVC) pipes. When repairs are needed, plumbers routinely insert copper pipes. What happens in these cases? Even with the iron pipes, valves are usually made of brass or copper. Does this present a major problem?

16.18. Of the half million bridges in the United States, 200,000 are deficient and on average 150–200 suffer partial or total collapse each year. Corrosion of both steel and concrete is responsible for many failures. Suggest some details of actual damage mechanisms.

16.19. The corrosion current density in a dental amalgam (an alloy containing mercury, in this instance Ag-Hg) filling is $1 \mu A/cm^2$.

    a. How many univalent ions are released per year if the filling surface area is $0.12 \, cm^2$?

    b. What is the metal thickness loss if the density and the molecular weight of the filling are $10 \, g/cm^3$ and 160 amu, respectively?

16.20. Prove that the ratio of the volume of oxide produced by oxidation to that of the metal consumed by oxidation = PBR = $Mo\rho_m/aM_m\rho_o$ (Equation (16.11)).

16.21. Two metal wires, one Ti and one Ni, are oxidized completely. One of them will be a hollow tube of oxide, the other a solid rod of oxide. Determine which one will be hollow and which one will be solid. Justify your answer.

16.22. For the following, assume the following densities: $\rho(Al_2O_3) = 3.97 \, g/cm^3$, $\rho(AlN) = 3.26 \, g/cm^3$, $\rho(TiO_2) = 4.26 \, g/cm^3$, $\rho(TiN) = 5.22 \, g/cm^3$.

    a. Predict the relative protective nature of Al_2O_3 and AlN coatings on Al.

    b. Do the same for TiO_2 and TiN coatings on Ti.

16.23. According to the Pilling-Bedworth ratio, which of the following oxide coatings are expected to be protective? The densities are: $\rho(UO_2) = 11.0 \, g/cm^3$; $\rho(U_3O_8) = 8.30 \, g/cm^3$; $\rho(ThO_2) = 9.86 \, g/cm^3$.

    a. UO_2 on U.

    b. U_3O_8 on U.

    c. ThO_2 on Th.

16.24. According to the Pilling-Bedworth ratio, which of the following grown films are expected to protect the underlying semiconductor? $M_{SI} = 28.1$; $M_{SIO_2} = 60.1$; $\rho(Si) = 2.33 \, g/cm^3$; $\rho(SiO_2) = 2.27 \, g/cm^3$ $M_{Ge} = 72.6$; $M_{GeO_2} = 104.4$; $\rho(Ge) = 5.32 \, g/cm^3$; $\rho(GeO_2) = 6.24 \, g/cm^3$ $M_{Si_3N_4} = 140.3$; $\rho(Si_3N_4) = 3.1 \, g/cm^3$.

    a. SiO_2 on Si.

    b. GeO_2 on Ge.

    c. Si_3N_4 on Si.

16.25. A tungsten carbide tool slides over an unlubricated mild steel surface that has a hardness of 1,000 MPa. What distance of travel between the two contacting surfaces is required for a volume loss of $0.0001 \, cm^3$ by adhesive wear if a load of 50 kg is applied?

16.26. Comment on the relative magnitudes of friction and wear that are desired between the following contacting surfaces: (1) A diamond stylus and the grooves of a plastic phonograph record. (2) A ball and race contact in ball bearings.(3) A pick up head flying over (and contacting) the surface of a computer disc. (4) A high speed steel drill and the aluminum plate being drilled. (5) A brake shoe and drum in a car. (6) A copper-copper contact in an electrical switch. (7) A boot sole and rock during mountain climbing. (8) A ski and snow.

16.27. For a precision bearing to lose no more than $10^{-10} \, cm^3$ when its lubricated hardened steel surfaces contact each other for a total length of a mile (1,600 m), what is the maximum load that should be applied to it? Assume a hardness of $1,100 \, kg/mm^2$.

16.28. Why does a carbon-steel knife stay unstained in a drawer but rust when left wet?

16.29. Why does iron dissolve rapidly in hydrochloric acid but transform to rust in water?

16.30. Why is the sequence of potentials in the standard galvanic series in seawater (Table 16.2) different from that in the standard electromotive series (Table 15.1)?

16.31. Describe the problem with the corrosion of the Statue of Liberty, and what were the two possible avenues for its solution?

16.32. What is crevice corrosion and what drives it?

16.33. How are ships protected from corrosion? (There are two ways.)

16.34. How are steel buildings protected from corrosion?

16.35. What is galvanized iron and how is corrosion prevented?

16.36. If galvanic corrosion requires an electrochemical cell, how do single metals corrode?

16.37. When two metals are in contact, which one corrodes? Does it form the anode or the cathode in the cell?

16.38. There is a galvanic cell with two identical metal electrodes, but the electrolyte is more concentrated in one half-cell than the other. In which half-cell does corrosion occur?

16.39. Why do steel pilings corrode more underneath the surface of the water, although the concentration of oxygen is higher at the surface?

16.40. Why does dirt accumulation promote wet corrosion?

16.41. Describe the oxidation kinetics of iron in hot dry air. What phenomenon controls the corrosion rate?

16.42. Why does iron corrode and chromium does not?

16.43. Chromium and gold do not corrode. Is it for the same reason?

16.44. When does a solid oxide layer protect the metal from further oxidation, and when does it not?

16.45. Describe an effective protection against gaseous corrosion.

16.46. Compare adhesive and abrasive wear. Which one is faster? How much faster (approximately)?

16.47. How much does the oxidation rate increase with temperature? What is the name of the graph that describes this dependence?

16.48. Draw an Arrhenius plot of the temperature-dependence of gaseous corrosion. Put the correct units on the horizontal and vertical scales

16.49. Describe erosive wear.

16.50. What material property governs the wear resistance of metals?

16.51. What material property governs the wear resistance of ceramics?

16.52. What is cavitation and what causes it?

16.53. Compare the amount of material removed by wear of two pieces of metal sliding over a stone surface. They weigh the same, but one has double the contact area with the stone.

17

Biomaterials

Biomaterials are used in medical devices, intended to interact with biological systems. They include all types of materials: metals, ceramics, polymers, and composites. Metals are used in artificial hips and other joint prostheses and in tooth implants. Ceramics are found in joint prostheses, in heart valves, and as tooth replacements. Polymers find use in eyeglasses, contact lenses, stents, sutures, skin grafts, and, lately, porous scaffolds that direct the growth of human tissues. While they are not different from other materials, biomaterials are selected and processed to be tolerated by the human body: they must not corrode, they must not be toxic in the body, and they must be biocompatible.

LEARNING OBJECTIVES

After studying this chapter, the student will be able to:

1. Explain the special requirements of biomaterials.

2. Name and describe the metals that are approved for use as biomaterials.

3. Describe the main ceramics used as biomaterials with applications.

4. Select the proper polymer for a medical application.

5. Explain what biomaterials are used as scaffolds for the growth of tissue.

17.1 **BIOMATERIALS**

Biomaterials are defined as materials used in medical devices, intended to interact with biological systems. There is a very large and increasing number of applications of materials in medicine; they are used in prostheses in the skeletal system (artificial hip joints, knees, elbow or shoulder joints, and finger joints; see Figures 17.1 and 17.2) in dentistry (crowns, implants, Figure 17.3, fillings, cements), as heart valves, catheters, artificial hearts, skin repair templates, artificial kidneys, and for vision (eyeglasses, but also contact lenses and lenses implanted in the eye, Figure 17.4).

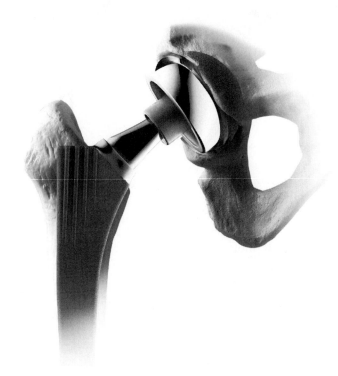

■ **Figure 17.1** Total hip joint replacements. Image Courtesy of Wright Medical Technology, Inc.

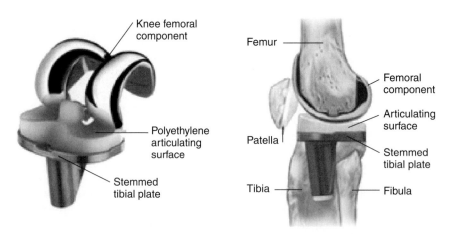

■ **Figure 17.2** Knee prosthesis. Image Courtesy of Wright Medical Technology, Inc.

■ **Figure 17.3**   Titanium tooth implant pins. Image Courtesy of Wright Medical Technology, Inc.

■ **Figure 17.4**   Schematic of an implanted ocular lens. Image Courtesy of Wright Medical Technology, Inc.

All classes of materials, metals, ceramics, polymers, and composites are employed in medical devices. They are similar to those used in other applications, but they are selected and processed to satisfy the special requirements that are imposed by their service in the human body.

17.1.1 Requirements of Biomaterials

Absence of Toxicity   Obviously, biomaterials must not be toxic. This restricts the choice of metals: aluminum and many heavy metals are too toxic for use. Polymers may not leach low-molecular-weight components, for instance the catalysts used in their processing.

Corrosion Resistance   The environment of the body is the same as that of saltwater; it is chemically corrosive. Materials, especially metals, must be selected to maintain their integrity for extended periods.

Biocompatibility   A material used in the human body must not provoke adverse reactions such as blood clotting or adhesion of bacteria. The biocompatibility of a material with the human body is more difficult to define. More often than not, it is based on experience: if the material is tolerated by the body, it is deemed biocompatible. The composition and properties of the surface are important for biocompatibility; therefore several surface treatments have been developed for that purpose.

"No material implanted in living tissue is inert because all materials elicit a response form living tissues.

If the material is toxic, the surrounding tissue dies.

If the material is nontoxic and biologically inactive (nearly inert), a fibrous tissue of variable thickness forms.

If the material is nontoxic and biologically active (bioactive) an interfacial bond forms.

If the material is nontoxic and dissolves, the surrounding tissue replaces it."

(Larry L. Hench and Serena Best, in *Biomaterials Science,* 2nd ed. Academic Press, San Diego, CA 2004, p.154.)

17.2 **METALS**

A small number of metallic alloys have so far been approved for use in medical devices; these are 316L stainless steels, Cr-Co alloys similar to the superalloys used in jet engines, and titanium. Table 17.1 shows some of their applications, and Table 17.2 displays their mechanical properties.

Austenitic **316L stainless steel** is a low-carbon variant of 316. Its low carbon content (<0.03% C) serves to avoid grain boundary corrosion (Section 16.6.2) and provides excellent corrosion resistance in the body's environment. Sufficient strength for this alloy is obtained by cold working. Because 316L is susceptible to corroding when highly stressed, it is mostly utilized for the repair of fractures and other temporary applications.

Cobalt-based alloys are the superalloys Haynes Stellite 21 and 25 that were developed for jet engines. They present excellent corrosion resistance in the human body. These alloys exist as **castable ASTM F75** and **wrought F90**.

F75 alloy is based on Stellite 21. Its composition is 59–69.5% Co, 27–30% Cr, 5–7% Mo, with 1% Mn, 2.5% Ni, 1% Si, 0.75% Fe, and 0.35% C. It is difficult to machine; therefore prostheses are usually prepared in near-net shape by investment (i.e., lost wax) casting or by powder metallurgy, where a fine powder of the alloy is sintered in hot isostatic pressing (HIP) at a pressure of 110 MPa and a temperature of 1,100°C. F75 alloys are used in the manufacture of artificial joints and in dentistry.

F90 is Stellite 25. It is a Co-Cr-W-Ni alloy. Its composition is 19–21% Cr, 14–16% W, and 10% Ni with minor additions of Fe and Mn; the balance is cobalt. It is shaped by hot forging. **F562** is a

Table 17.1 Uses of Metallic Alloys.

316L Stainless steel

Joint replacement
Bone plate for fracture fixation
Heart valve

Cobalt chromium alloys

Bone plate for fracture fixation
Knee prostheses
Stems for hip prostheses

Titanium alloys

Joint replacement
Dental implant for tooth fixation

Table 17.2 Mechanical Properties of Metallic Biomaterials.

Metal	ASTM designation	Condition	Young's modulus (GPa)	Yield strength (MPa)	Tensile strength (MPa)	Elongation %
Stainless Steel	126L	Annealed	190	220	485	40
Co-Cr alloys	F75	Cast/annealed	210	450	660	8
		PM HIP	250	840	1,280	
	F90	Annealed	210	450	950	30
		44% cold worked	210	1,600	1,900	
	F562	Hot forged	232	970	1,200	8
		Cold worked	232	1,500	1,800	
Ti alloys	F67	30% cold worked	110	485	760	25

high-strength alloy, also called MP35 N (trademark of Standard Pressed Steel Technologies); it contains 35% Cr and 35% Mo. MP stands for "multiphase": cold working of this alloy induces transformation from the FCC to the HCP structure. The HCP exists as fine platelets in the FCC grains. This microstructure is an effective impediment for dislocation motion and provides very high strength. Cold working is difficult in relatively large pieces. Because of its strength, this alloy is used in hip implant stems which are fabricated by hot forging.

■ **Figure 17.5**   Surface treatment of lower plate in a knee implant (see Figure 17.2). Image Courtesy of Wright Medical Technology, Inc.

Titanium is used in the form of commercially pure (CP) titanium F67, which contains up to 0.49% oxygen; and an Extra-Low-Interstitial Ti-6Al-4V alloy denominated F136. The addition of Al and V increases the strength and the fatigue limit of the material. Titanium and its alloys are the metals of choice for dental implants (Figure 17.3) and for hip, knee, and shoulder prostheses.

The surface of metallic prostheses is often rendered porous by the partial sintering of metallic beads. This encourages the growth of bone material into the pores and provides good adhesion of the prosthesis to the bone (Figures 17.5 and 17.6).

17.3 **CERAMICS**

Ceramics are hard, chemically inert, but brittle materials. They are utilized in load bearing prostheses where fracture resistance is not critical, such as balls and caps in hip prostheses, and for dental implants where their wear resistance and appearance are desirable. Table 17.3 shows their most important applications.

High-density, high purity (>99.5%) alumina is used in load bearing hip prostheses and dental implants because of its excellent corrosion resistance, good biocompatibility, high wear resistance, and high strength. Alumina is pressed and sintered. In hip implants, the alumina must have very nearly perfect spherical shape. For that purpose, the mating alumina ball and socket are polished together. With time, in service, the friction coefficient of alumina implants approaches that of the natural joint, leading to very low wear. The cups are stabilized by growth of bone into the pores of the alumina. Zirconia is also used as the articulating ball in hip prostheses because of its relatively low elastic modulus and high strength.

■ **Figure 17.6**   Surface treatments of total hip arthroplasty to enhance short and long-term fixation. Image Courtesy of Wright Medical Technology, Inc.

Hydroxyapatite $Ca_{10}(PO_4)_6(OH)_2$, also called hydroxylapatite, is the main constituent of natural bone. In medicine, the ceramic hydroxyapatite can be used as a filler to replace amputated bone or as a coating to promote bone ingrowth into prosthetic implants. Coral skeletons can be transformed into hydroxyapatite by heating to high temperatures; their porous structure allows relatively rapid growth of bone tissue into the ceramic at the expense of initial mechanical strength. The high temperature also burns away any undesirable organic molecules such as proteins. Its mechanical weakness prevents its bulk use in bone replacement prostheses.

Some modern dental implants are coated with hydroxylapatite. The most popular method for depositing coatings of this material is plasma spraying, which produces a porous structure that favors bone ingrowth into the implants.

Bioactive glasses and glass ceramics are used in ear surgery. Common glass, containing a large amount of silica, is bioinert. Bioglasses contain 45–55% SiO_2, 6% P_2O_5, 12–24.5% CaO, and 20–24.5% Na_2O; some contain a small amount of CaF_2. These glasses form a bond with bone.

Pyrolitic carbon is the material of choice for artificial heart valves because of its excellent compatibility with blood (Figure 17.7). Pyrolytic carbon is obtained by pyrolysis (i.e., decomposition by heat) of

Table 17.3 Biomedical Uses of Ceramics.

Alumina

Load bearing orthopedic prostheses
Dental implants
Maxillofacial reconstruction

Hydroxyapatite

Coatings for chemical bonding in dental, orthopedic, and maxillofacial prostheses
Dental implants
Periodontal pocket obliteration
Ear implants
Maxillofacial reconstruction
Bony defect repair

Bioactive glasses and glass ceramics

Coatings for chemical bonding in prostheses
Dental implants, ear implants
Maxillofacial reconstruction

Pyrolytic carbon

Heart valves

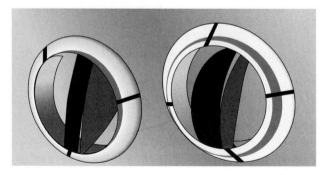

■ **Figure 17.7**   Components of a heart valve, including pyrolytic carbon valves. Courtesy St. Jude Medical.

hydrocarbons such as propane, propylene, acetylene, and methane in the absence of oxygen. It is produced as a coating in a fluidized bed reactor that is heated to temperatures of 1,000–2,000°C. In order to manufacture a heart valve, one introduces a substrate (usually graphite) into the reactor where it is coated with the pyrolytic carbon. The surface of the valve is polished to an Ra of 40–100 nm. Pyrolytic carbon is a hard and very wear-resistant material. It acquires these properties from its turbostratic structure which is similar to that of graphite, but with many more defects. The stacking of the sheets is

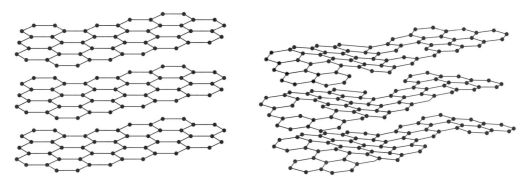

■ Figure 17.8   Structure of graphite (left) and of turbostratic pyrolytic carbon (right).

not ordered as in graphite, and the layers are curved and kinked (Figure 17.8); this prevents them from sliding over each other and provides for the mechanical resistance of the material.

17.4 **POLYMERS**

Polymers are the largest and most versatile class of biomaterials. As a rule, they exhibit excellent biocompatibility, but care must be exercised that they do not leach toxic additives such as polymerization catalysts or plasticisers. Their mechanical properties are well suited for applications as replacements for soft and hard tissues. Functional groups on their backbone or side chains make them adaptable to their environment. Nonpolar polymers, such as polyethylene, PTFE, polypropylene, and so on, are hydrophobic and are virtually indestructible in the human body. Polymers with polar groups are hydrophilic; materials such as poly lactic acid and poly glycolic acid can absorb water to become soft gels. Ester linkages make polymers susceptible to hydrolysis and render them resorbable by the body as nontoxic monomers. In addition, polymers can be spun as fibers and manufactured as textiles. The uses of polymers are summarized in Table 17.4.

17.4.1 **Thermoplastics**

UHMWPE (Ultra-high molecular weight polyethylene) is used in hip and other joint prostheses between the moving parts (see Figures 17.2 and 17.9). The very long molecular chains provide the necessary resistance to creep and wear. Polyethylene wear particles finding their way between the metallic parts and the bone cause inflammation and infections. In order to reduce wear of the polymer, the surfaces of the mating metal or ceramic pieces must be highly polished.

PET (poly ethylene theraphthalate) is the most widely used biopolymer, especially in the form of fibers and textiles.

PTFE (polytetrafluoroethylene) is used in sutures for its low friction. **CoreTex** fabric is composed of a thin, porous fluoropolymer membrane with a urethane coating that is bonded to a fabric, usually nylon or polyester. The pores of the membrane are small enough to make it impenetrable to liquids

Table 17.4 Polymers and Their Uses.

Polymer	Applications
UHMPE: Ultrahigh molecular weight polyethylene	Acetabular component in hip and other joint replacements
PET: Poly(ethylene therapthalate)	Fixation of implants, hernia repair, textiles (Dacron)
PTFE Teflon: Polytethrafluoroethylene	Sutures, catheters, textile: Coretex
PMMA Lucite: Poly(methyl methacrylate)	Lenses, intraocular lenses
Silicone: Poly(dimethyl siloxane)	Finger joints, heart valves, breast implants, ear, chin and nose repairs
PHEMA: Poly(hydroxyethyl methacrylate)	Soft contact lenses

■ **Figure 17.9** Acetabular cup of hip prosthesis with UHWM polyethylene insert. Image Courtesy of Wright Medical Technology, Inc.

while still allowing water vapor to pass through. The result is a material that is breathable and water-proof (Figure 17.10).

PMMA poly(methyl methacrylate) is a transparent polymer with good refractive properties. It is known as Lucite or Plexiglass. It is used for the fabrication of eyeglasses and lenses implanted in the eye (Figure 17.11). The lenses are shaped in a lathe and polished. A modification of PMMA, where the methyl group is replaced with a hydroxyl group, produces PHEMA that can be machined like PMMA but subsequently swollen with water to produce soft contact lenses.

Silicone, poly(dimethyl siloxane) is a rubbery polymer with the structure

$$\text{CH}_3$$
$$\text{-[-Si—O-]-}$$
$$\text{CH}_3$$

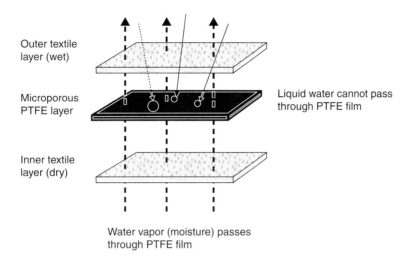

Outer textile
layer (wet)

Microporous
PTFE layer

Liquid water cannot pass
through PTFE film

Inner textile
layer (dry)

Water vapor (moisture) passes
through PTFE film

■ **Figure 17.10**   Schematic of Gore-tex: Microporous PTFE film transmits vapor but prevents passage of liquid.

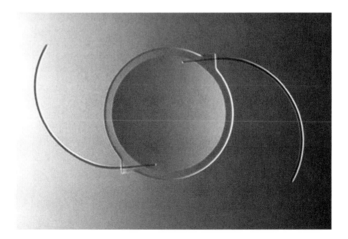

■ **Figure 17.11**   Implantable (intraocular) PMMA lens. The two handles allow the surgeon to fold the lens and introduce it into the eye with a smaller incision. Image Courtesy of Wright Medical Technologies, Inc.

Its stability and excellent flexibility make it ideal for finger joint prostheses, especially in cases of severe arthritis (Figure 17.12 and 17.13). It is also used in breast and testicular implants.

17.4.2 **Medical Fibers and Textiles**

We described the fiber spinning techniques for polymers in Chapter 9. The fibers so obtained have a diameter of 10–500 µm. Often much thinner fibers are required for medical applications.

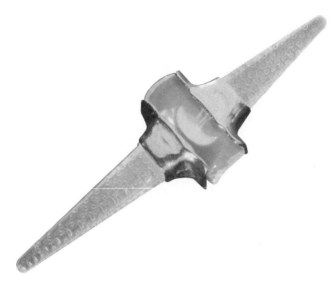

■ **Figure 17.12**   Silicone prosthesis for finger joints. Image courtesy of Wright Medical Technology, Inc.

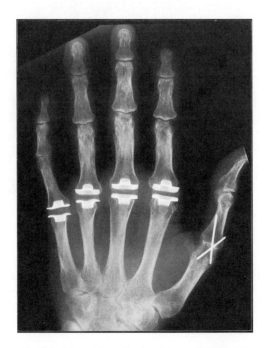

■ **Figure 17.13**   Utilization of silicone rubber finger joint prostheses (top figure) in an arthritic hand. Image courtesy of Wright Medical Technology, Inc.

■ **Figure 7.14** Heart valve graft with gelveave®. Courtesy of St Jude Medical.

These are produced by electrospinning where a voltage of 5–30 kV is applied between the solution and a grounded base. This electric potential overcomes the surface energy of the polymer and accelerates the fine jet toward the grounded base. Filaments as thin as 1 μm to 100 nm are obtained by this method.

These fibers are used to fabricate textiles by weaving, knitting, or braiding. An example of woven polymer is shown in Figure 7.14.

17.4.3 **Hydrogels**

Hydrogels are structures of crosslinked polymers that are swollen with water. They are composed of hydrophilic polymers, such as

poly(vinyl alcohol) PVA.

$$-\!\!\left[CH_2-\underset{\underset{OH}{|}}{CH}\right]_{\!n}\!\!-$$

poly(ethylene glycol) PEG, also called Poly(ethylene oxide) or PEO

$$HO-[CH_2-CH_2-O]_n-H$$

Poly(hydroxyethyl methacrylate) PHEMA.

$$
\begin{array}{c}
\text{H} \quad \text{CH}_3 \\
| \qquad | \\
-\text{C}-\text{C}- \\
| \qquad | \\
\text{H} \quad \text{C}-\text{O}-\text{CH}_2-\text{CH}_2\text{OH} \\
\parallel \\
\text{O}
\end{array}
$$

PVA is crosslinked by glyoxal, glutaraldehyde, or borate. PEO is usually crosslinked by ultraviolet radiation or by the addition of suitable molecules. Radiation crosslinking is preferred because it avoids the introduction of possibly toxic chemicals. PEO is slowly water-soluble and is used in gel form for drug delivery.

PHEMA has been developed specifically for soft contact lenses. It can be ground to shape a lens in its dry form and subsequently swollen with water to make a soft contact lens. In order to limit the swelling by water, PHEMA is often copolymerized with the hydrophobic polymethylmethacrylate PMMA (Lucite). As seen in Chapter 9, PMMA is similar to PHEMA; the CH_2CH_2OH of the latter is replaced by a CH_3 group in PMMA.

17.4.4 **Bioresorbable and Bioerodible Polymers**

Certain polymers can be resorbed in the body. Under the action of water, they dissociate into their mers which are absorbed by the body. This is the case of anhydride polymers, which are derived from another substance by removal of a water molecule. For instance, polylactic acid (PLA),

$$
\begin{array}{c}
\text{CH}_3 \quad \text{O} \\
| \qquad \parallel \\
-\!\!\left[\text{O}-\text{CH}-\text{C}\right]_n\!\!-
\end{array}
$$

is formed from lactic acid

$$
\begin{array}{c}
\text{CH}_3 \quad \text{O} \\
| \qquad \parallel \\
\text{H}-\text{O}-\text{CH}-\text{C}-\text{OH}
\end{array}
$$

by a condensation reaction. It reacts with water to undo the polymerization.

$$
\begin{array}{ccc}
\text{CH}_3 \ \text{O} \quad \text{CH}_3 \ \text{O} & & \text{CH}_3 \ \text{O} \qquad \text{CH}_3 \ \text{O} \qquad\quad \text{O} \\
| \quad \parallel \qquad | \quad \parallel & & | \quad \parallel \qquad\quad | \quad \parallel \qquad\qquad \parallel \\
-\text{O}-\text{CH}-\text{C}-\text{O}-\text{CH}-\text{C}- & \rightarrow & -\!\!\left[\text{O}-\text{CH}-\text{C}-\text{OH} + \text{HO}-\text{CH}-\text{C}-\text{O}\rightarrow -\text{O}-\text{C}- \\
\uparrow & & \\
+\text{H}_2\text{O}
\end{array}
$$

Such a material will dissolve in time as lactic acid, which is resorbed in the body as a nontoxic material. Lactic acid is in fact produced by the body as a consequence of exercise. Aliphatic anhydride polymers, such as PLA, degrade within days; some aromatic polyanhydrides degrade over several years.

Copolymers of aliphatic and aromatic anhydrides can be formulated with intermediate degradation rates. There are a number of bioresorbable polymers, some of which are shown in Figure 17.15. They have in common an ester linkage

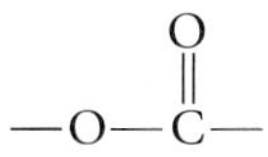

$$-O-\overset{\overset{\displaystyle O}{\|}}{C}-$$

that can be opened by water as seen above with lactic acid.

An important requirement of bioresorbable materials is that they decompose into fragments that are resorbable without damage to the human body. Polyglycolic acid, polylactic acid, polydioxanone, and polycaptolactone have been approved by the Food and Drug Administration (FDA). Some applications of resorbable polymers are shown in Table 17.5.

Polymer scaffolds are placed in the body to direct the growth of missing tissue. They contain pores larger than $100\,\mu m$. The pores constitute a continuous connected network. In time, the desired tissue

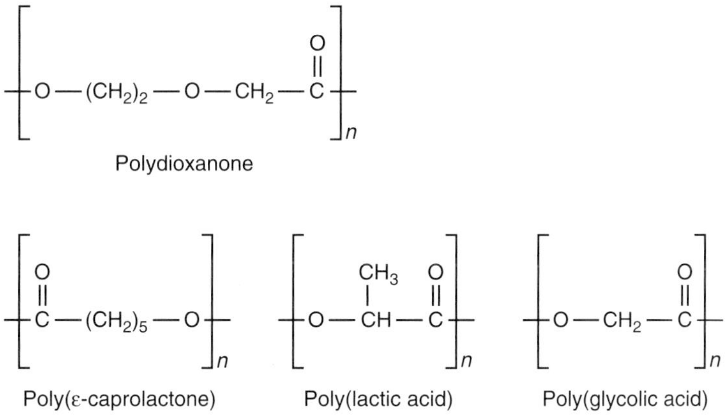

$$\left[O-(CH_2)_2-O-CH_2-\overset{\overset{\displaystyle O}{\|}}{C} \right]_n$$
Polydioxanone

$$\left[\overset{\overset{\displaystyle O}{\|}}{C}-(CH_2)_5-O \right]_n \qquad \left[O-\overset{\overset{\displaystyle CH_3}{|}}{CH}-\overset{\overset{\displaystyle O}{\|}}{C} \right]_n \qquad \left[O-CH_2-\overset{\overset{\displaystyle O}{\|}}{C} \right]_n$$
Poly(ε-caprolactone)   Poly(lactic acid)   Poly(glycolic acid)

■ **Figure 17.15** Chemical structures of synthetic degradable polymers.

Table 17.5 Degradable Polymers and Some Applications.

Degradable polymer	Applications
Poly(glycolic acid), poly(lactic acid) and copolymers	Drug delivery, scaffolds for tissue regeneration, stents, staples, sutures
Polycaprolactone	Long-term drug delivery, orthopedic applications, staples, stents
Polydioxanone	Fracture fixation in non-load-bearing bones, wound clip

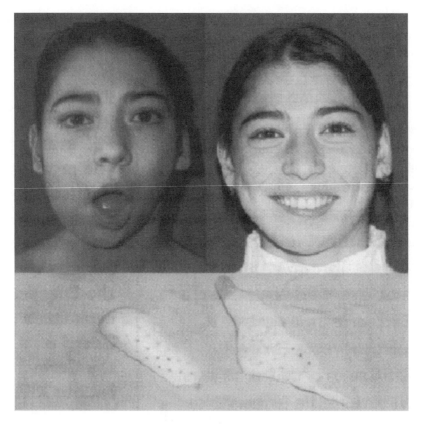

■ **Figure 17.16**   Correction of a jaw in 15-year-old Anarea Djantemiroma using a biodegradable scaffold consisting of poly(lactic acid) into which a growth enzyme was incorporated. Courtesy of Professor Roginsky, Moscow, from *Materials Today*, March 2005, p. 23.

grows in these pores, and the scaffold is slowly resorbed and replaced by the new tissue. Isolated pores, of the sort used in polystyrene or mattresses, cannot be used, as they do not allow the penetration of growing tissue as shown in Figure 17.6.

■ SUMMARY

1. Biomaterials are metals, ceramics, polymers, or composites that are selected and processed for medical use in the human body.

2. Biomaterials must satisfy three requirements: they must not be toxic; they must not corrode in the aggressive environment of the body; and they must be biocompatible, which means they must be tolerated by the body.

3. Biocompatibility is difficult to predict and is established by experience.

4. Only three types of metals have been approved for biomedical use by the Food and Drug Administration (FDA); these are titanium and its alloys, low-carbon 316 stainless steel, and Co-Cr based superalloys.

5. The surface of metallic implants is often made rough or porous to allow bone tissue to grow into the pores and provide a stronger bone. This is achieved by partial sintering of spheres or by plasma spray.

6. Ceramics are chemically inert and easily tolerated by the body. The most important are alumina, titania, hydroxyapatite, glass, and pyrolytic carbon.

7. Hydroxyapatite, a hydrated calcium phosphate, is a natural constituent of bones. As a ceramic, it is used as filler for bone tissue or a coating to promote adhesion to bone. It is too weak to be used for load bearing applications.

8. Bioactive glasses contain calcium, phosphorus, and sodium oxides; they form a bond with bones.

9. Pyrolytic carbon is produced by pyrolysis of hydrocarbons. Its structure is similar to that of graphite but with many defects that prevent the gliding of graphite sheets over each other. It is hard, wear resistant, and very biocompatible. It is used in heart valves.

10. Polymers are versatile and very biocompatible biomaterials. The three main classes are thermoplastics, hydrogels, and bioresorbable polymers.

11. Ultrahigh-molecular weight polyethylene (UHMWPE) is used in joint prostheses. Its long molecules make it resistant to creep and to wear. Wear particles in prostheses move between bone and stem where they cause inflammations.

12. PET is used in textiles and fibers.

13. PTFE is used in sutures because of its low friction and in CoreTex.

14. PMMA (Lucite) is used in eyeglasses and eye lens implants.

15. Silicone rubber is used mainly as finger joint replacement.

16. Hydrogels are crosslinked polymers that absorb water. Examples are polyvinyl alcohol and PHEMA. The latter is used in contact lenses.

17. Bioresorbable polymers are used in sutures and in scaffolds for bone growth. They contain an ester linkage that reacts with water and breaks up the polymer chain (hydrolysis). The monomers are absorbed by the body.

■ KEY TERMS

A
alumina, 498, 509

B
bioactive glasses, 499, 509
biocompatibility, 495, 508

bioresorbable, 506, 509

C
$Ca_{10}(PO_4)_6(OH)_2$, 499
ceramics, 498
Cobalt, 496

coral, 499

corrosion, 495

E
electrospinning, 505

■ REFERENCES FOR FURTHER READING

[1]   M.M. Domach, *Introduction to Biomedical Engineering*, Prentice Hall, Englewood Cliffs, NJ (2003).

[2]   J. Enderle, S.M. Blanchard, and J. Bronzino, *Introduction to Biomedical Engineering*, 2nd ed., Academic Press, San Diego, CA (2005).

[3]   J. Park and R.S. Lakes, *Biomaterials, an Introduction*, 3rd ed., Springer, New York (2007).

[4]   B.D. Ratner, A.S. Hoffman, F.J. Schoen and J.E. Lemons, *Biomaterials Science: An Introduction to Materials in Medicine*, 2nd ed., Academic Press, San Diego, CA (2004).

■ PROBLEMS AND QUESTIONS

17.1.   What requirements does a material need to satisfy in order to be used as a biomaterial?

17.2.   Surgeons first tried Teflon as the sliding surface in hip and knee prostheses because of its low friction. They had to abandon it and use UHMWPE instead. What superior qualities does this material have and why?

17.3.   What are the outstanding properties of titanium that make it such a good biomaterial?

17.4.   316 stainless steel is used in prostheses. Is this a ferritic, martensitic, or austenitic steel?

17.5.   Low carbon content weakens the steel, so why is low-carbon stainless steel used in prostheses?

17.6.   Describe in detail how the stem of a hip replacement is treated to obtain good bonding with the bone. How is the surface processed, and how does the surface look?

17.7.   What is the material of choice for replacement of finger joints? What are the relevant properties of this material?

17.8. How is hydroxyapatite produced?

17.9. What does one do to glass to make it bond to bone? Why is this treatment effective?

17.10. What features make pyrolytic carbon hard and wear resistant?

17.11. How is pyroltic carbon produced?

17.12. Describe the structure and the distinctive properties of Gore-Tex.

17.13. Describe an application that makes use of the low friction of PTFE.

17.14. Compare the composition, properties, and applications of PMMA and PHEMA.

17.15. What type of materials are used for contact lenses? What is the relevant chemical property of these materials?

17.16. Name two applications of bioresorbable polymers.

17.17. Describe the molecular mechanism that operates when a bioresorbable polymer is resorbed in the human body.

17.18. What properties are necessary for a polymer to act as a bone-growth scaffold?

Part 5

Nanomaterials and the Study of Materials

This last section contains two subjects of a more scientific nature. Nanomaterials are presented in Chapter 18. Some aspects and applications of nanomaterials have been included in the previous chapters, notably the high strength resulting from extremely fine grain structures, the ever decreasing size of the elements of integrated circuits, and the use of nanomaterials in exploratory solar cells. Nanomaterials that exist as very small particles possess unique mechanical, electrical, optical, and magnetic properties that are a direct consequence of their small size. They require novel processing and handling techniques; and they present certain novel dangers because they can defeat the body's natural filtering and protection. In Chapter 19, we examine the most important modern techniques with which one measures the chemical composition, the crystal structure, and the morphology of materials. These techniques are very powerful; they now permit the visualization of individual atoms in a material and they can measure the shape, the chemical composition, and the crystal structures of features that are only a few atoms in size.

Chapter **18**

Nanomaterials

Nanomaterials are characterized by a size smaller than 100 nm in one, two, or all three dimensions. Accordingly, they can be small particles, very thin fibers (nanotubes), stacks of very thin planes, or bulk materials with nanometer-size grains. These materials, which are currently the subject of much research, have unique properties that are a direct consequence of their small size. They are expected to provide revolutionary innovations, some of which have already been realized. In this chapter we consider nanocomposites; fullerenes, and carbon nanotubes; metallic nanoparticles; metallic nanorods; and semiconductor quantum dots. We describe their properties, the methods by which they are synthesized, and some of their applications. We finish with the description of the nanometer-thick layers of the GMR read head for magnetic data storage.

LEARNING OBJECTIVES

After studying this chapter, the student will be able to:

1. Define the properties that derive from the small size of nanomaterials.

2. Describe carbon nanotubes, some of their properties, and applications.

3. Describe the synthesis of carbon nanotubes.

4. Define metallic nanoparticles and describe the principle of their synthesis.

5. Describe surface plasmon resonance and some of its applications.

6. Define quantum dots, their outstanding properties, and some applications.

7. Describe the synthesis of quantum dots.

8. Describe the fabrication of a GMR read head for magnetic storage.

9. Name the safety concerns about nanomaterials.

Nanomaterials, with dimensions smaller than 100 nm, are the object of active research. Nanomaterials require processing methods and possess unique mechanical, electrical, optical, magnetic, and chemical properties that are a direct consequence of their small size. These novel properties promise, and have already delivered, important innovations in technology. Their small dimensions also present novel health dangers that need to be investigated.

We shall discuss the synthesis methods, properties, and application of the most important nanomaterials. We will not describe here the reduction to nanometer size of transistor elements in integrated circuits nor of micro electro-mechanical systems (MEMS); they were included in Chapter 13. Some of the applications have been alluded to in the chapters on structural and functional materials.

In view of the high research activity and the large number of companies established to develop methods and applications for nanomaterials, new materials and applications will certainly exist by the time you read this text. It is hoped that the principles presented here will help you understand and evaluate these new developments.

18.1 THE UNIQUE PROPERTIES OF NANOMATERIALS

18.1.1 Mechanical Properties

In Chapter 4 we saw that the reduction of grain size increases the strength of metals. A similar Hall-Petch law is obeyed in WC/Co and other cermet (ceramic-metal) composites. In metals, the benefit of Hall-Petch hardening extends to 50 nm grains; with further decrease in grain size, the material becomes weaker. We also found in Chapter 10 that thin fibers are stronger than bulk material because of the absence of cracks and dislocations. Nanometer-sized filaments, metal rods, and, especially, carbon nanotubes possess extraordinary strength.

18.1.2 Electronic Structure

In the discussion of the electron-band structure of solids, we have seen that every energy band contains as many energy levels as there are atoms in the crystal. In nanoparticles that contain as few as 100 atoms, the energy bands contain a finite number of levels with measurable energy separation instead of the quasi-continuum of macroscopic crystals. In addition, as Figure 18.1 illustrates, the energy band with a small number of atoms is narrower than for a macroscopic crystal. Since the valence band and conduction band of a nanostructure semiconductor are narrower, the gap separating them increases with decreasing size of the particle. This phenomenon is exploited in semiconductor quantum dots in which the optical absorption threshold depends on particle size.

18.1.3 Optical Properties

The free electrons of a metal can be collectively displaced by an electric field. In a small particle this displacement sets up a dipole with the immobile nuclei. If the field is suddenly removed, this displacement vibrates with a frequency that depends on the size of the particle (in a way similar to the oscillations of a pendulum). This phenomenon is illustrated in Figure 18.2. When this vibration frequency

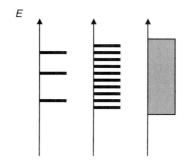

■ **Figure 18.1**   Energy levels for valence electrons for a system of three atoms, a particle with 10 atoms, and a macroscopic solid. Each band contains exactly as many levels as atoms in the particle.

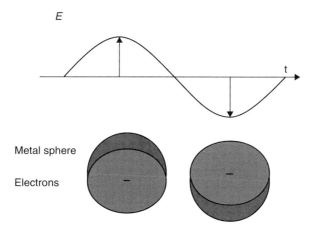

■ **Figure 18.2**   Surface plasmon resonance in a metallic nanoparticle. The electric field of the light displaces the electrons in the particle. When the frequency of the light is equal to the vibration frequency of the electric dipole, resonance absorption of the light occurs.

is equal to the frequency of incoming light, there is a resonance in the absorption and the scattering of the light. This phenomenon is called surface plasmon resonance; it is responsible for the vivid colors of glass in which metal particles are dissolved. (One example is the famous Murano glass.) This phenomenon is used in colloidal metals.

18.1.4 **Magnetic Properties**

Nanometer-sized ferromagnetic particles are too small to contain several magnetic domains; they consist of single magnetic domains. This fact has a profound influence on the magnetic properties of the material. The single domain provides magnetic materials with a higher coercive field, especially if they are elongated, as discussed in Chapter 14; this property is used in magnetic storage media. Giant

Magnetoresistance discussed in Section 14.5.1 also depends on the fabrication of layers a few nanometer in thickness.

18.2 **NANOSTRUCTURED METALS AND COMPOSITES**

In Chapter 6, we have learned that very fine structures are obtained by rapid nucleation and slow growth of crystals. In bulk metallic alloys this was done by rapid cooling from the melt. Rapid cooling of bulk materials is not sufficient for the production of nanometer-sized particles. In one of the first techniques developed, the material is evaporated in vacuum, and the vapor is collected on a surface cooled with liquid nitrogen. The low temperature prevents the diffusion of atoms and forces the formation of nanometer-sized particles. More radical methods for confining the growth of the particles are employed when the size or shape of the particles needs to be controlled.

Industrial **catalysts** consist of metallic nanoparticles that are supported on a highly porous ceramic support. Their preparation starts with a water-soluble salt of the desired metal. A dilute solution of this salt is absorbed in a highly porous ceramic support (often SiO_2/Al_2O_3); evaporation of the solvent or precipitation by change in pH leaves extremely small solid particles. Calcining (heating in air) of the salt produces an oxide of the metal that is later reduced in hydrogen or a reducing atmosphere. The resulting catalyst particles are so small that a large fraction of their atoms reside on their surface and participate in the catalytic reaction.

WC/Co Nanocomposites are prepared by spray drying: One prepares a solution of tungsten and cobalt salts and sprays a mist through a nozzle. Evaporation of the very fine droplets produces small particles containing a mixture of cobalt and tungsten salts. These are then calcined to obtain very small WCo oxides. Carbonization of the particles in acetylene produces nanostructured WC-Co particles. WC/Co composites follow the Hall-Petch law: their hardness increases with decreasing size of the Co phase between WC ceramic grains. WC/Co nanocomposites are used, for instance, in drill bits for integrated circuits where their high wear resistance leads to increased productivity.

18.3 **CARBON NANOMATERIALS**

The first carbon nanomaterials to be discovered, in 1985, were Fullerenes or Bucky balls. Carbon nanotubes were later discovered in Japan. Two-dimensional crystals of graphene consist of single layers of graphite.

18.3.1 **Fullerenes**

Fullerenes consist of 60 carbon atoms bound in a spherical shape by sp^2 hybrid bonds. The name Fullerenes, also Bucky balls or Buckminsterfullerenes, reflects the similarity of their structure to that of geodesic domes made famous by the architect Buckminster Fuller. Fullerenes are produced in an electric arc between graphite rods in an inert atmosphere, preferably helium at 100–200 Torr. In optimized conditions, this process produces 5–15% Fullerenes among other small graphite particles. The majority of Fullerenes consist of 60 carbon atoms (C_{60}), illustrated in Figure 18.3. Smaller quantities of C_{70},

C_{80}, C_{84}, C_{88}, and C_{96} Fullerenes are also obtained. (The subscript indicates the number of atoms.) The Fullerenes are separated by liquid chromatography. Recently, Fullerenes have been produced (Figure 18.3) in a sooty flame produced by burning benzene in low oxygen ambient. No industrial applications of these particles are known to date (ca. 2007). Early attempts at using them as lubricants (molecular ball bearings) have not been successful.

18.3.2 **Graphene**

Graphene consists of single layers of carbon with the structure of a graphite sheet. Graphene sheets have been produced very recently by exfoliation of graphite or by separation of single graphite layers from a crystal with adhesive tape. This material does not have any practical applications at this time.

18.3.3 **Carbon Nanotubes**

These can described as rolled up graphene layers with half Fullerene spheres at their extremities (Figure 18.4). Their diameter is as small as 14 nm, and their length can reach several centimeters. These particles have quite extraordinary properties and are already finding many applications.

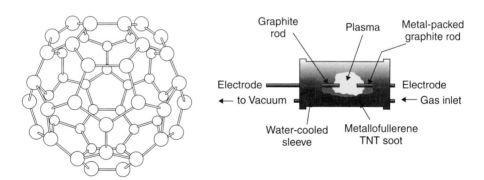

■ **Figure 18.3**   Left: Structure of Fullerene C_{60}. Right: Krätschmer-Huffman arc furnace for the production of Fullerenes. From DiVentra, Evoy, and Heflin, *Introduction to Nanoscale Science and Technology*, Springer, New York (2004).

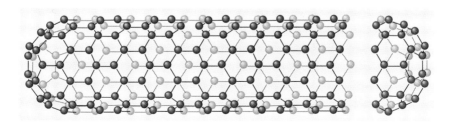

■ **Figure 18.4**   Structure of a carbon nanotube.

Synthesis   The synthesis of carbon nanotubes requires a metallic catalyst on which they grow essentially from the vapor phase (Figure 18.5). Most carbon nanotubes are grown by the vapor-liquid-solid (VLS) method in a CVD reactor. The catalyst is deposited by spin-coating a dissolved metal salt $Fe(NO_3)_3$ onto a silicon or silica substrate. The salt is transformed into iron oxide clusters by heating in air (calcining); reduction of the oxides by hydrogen finally produces the iron catalyst nanoparticles. A hydrocarbon (acetylene, methane, hexane, or benzene) is then admitted into the hot reactor and is pyrolyzed on the metallic catalyst from which the nanotubes grow. The nanotubes can consist of a single wall of graphene (SWNT) or may contain multiple concentric walls (MWNT).

The diameter of the nanotubes is essentially that of the catalyst particles. Multiple wall nanotubes (Figure 18.6) are easier to grow; they require a temperature of 300–800°C. Single wall nanotubes grow at a temperature of 600–1,150°C in an ambient of hydrocarbon gas, hydrogen, and inert gas such as argon. The growth of carbon nanotubes is facilitated by plasma-enhanced CVD (PECVD). Carbon nanotubes are available commercially. The price was above \$180 per gram for SWNT and \$36/g for MWNT (in 2007) but will certainly be lowered as more progress is made.

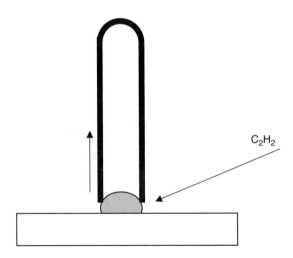

C_2H_2

■ **Figure 18.5**   Schematic of the vapor-liquid-solid (VLS) growth of a carbon nanotube. White: silica substrate; gray: iron catalyst particle; black: carbon nanotube.

■ **Figure 18.6**   Schematic of a multiwall nanotube (MWNT). From DiVentra, Evoy, and Heflin, *Introduction to Nanoscale Science and Technology*, Springer, New York (2004).

Properties   The properties of carbon nanotubes are quite extraordinary. They are the stiffest material measured to date. Their Young's modulus is around 1,000 GPa. (Compare this to 210 GPa for iron and 320 GPa for alumina.) Their tensile strength depends on the method of synthesis. Strengths as high as 150 GPa have been measured. By comparison, carbon fibers have $E = 200 - 400$ GPA and $\sigma_T = 2 - 7$ GPa (Table 10.1). Because all valences are satisfied, carbon nanotubes exhibit excellent thermal stability and chemical inertness. These materials have the highest thermal conductivity of any materials, similar to that of diamond. Electrically, these materials can be metallic or semiconductors, depending on their structure.

Applications

Mechanical: Carbon nanotubes are used to reinforce the polymer matrix in certain high-performance graphite-fiber composites.

Functional: The addition of carbon nanotubes to the graphite anode increases the capacity of the lithium ion battery by about 20% and increases its performance at high currents. Transistors utilizing single-wall nanotubes as the active element are in the development stage. Nanotubes are also used in the development of novel solar cells, in biosensors and in field emission displays.

18.4 **METALLIC NANOMATERIALS**

18.4.1 **Metallic Nanoparticles**

Metallic nanoparticles are synthesized by colloidal growth. The very fine structure is obtained by confining the reagents to a very small space. Figure 18.7 illustrates the process. A water/hydrocarbon mixture

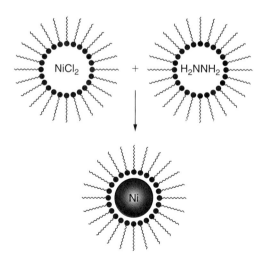

■ **Figure 18.7**   Synthesis of Ni nanoparticles in water/oil emulsions. From A. Boal, *Nanoparticles, Building Blocks for Nanotechnology,* V. Rotello, Ed., Springer, New York (2004) p. 4.

containing a surfactant is stirred to create an emulsion that is so fine that it is transparent. A salt of the metal (for instance $NiCl_2$) is dissolved in the water phase. The surfactant confines the reaction to the interior of the water bubble and guarantees the production of nanometer-size particles. The size of the particles can be controlled by the relative amounts of water and of surfactant. A second emulsion is prepared, in which the water dissolves a reduction agent, such as N_2H_4. One mixes the two emulsions and obtains an emulsion of metallic nanoparticles (Ni in this instance) with diameter less than 10 nm. Other reactions involve the thermal decomposition of metal carbonils. Iron nanoparticles are obtained by the decomposition of $Fe(CO)_5$ in trioctylphosphine oxide (TOPO) containing oleic acid. The latter attaches itself to the iron particles and protects them from oxidation.

It is possible to prepare nanoparticles consisting of two concentric metals. Figure 18.8 shows the synthesis of a particle with an iron core and a gold shell. One first prepares the iron particles by reacting fine $FeSO_4$ and $NaBH_4$ emulsions. Subsequent reaction of $HAuCl_4$ and $NaBH_4$ in larger droplets creates the $Fe_{core}Au_{shell}$ particles shown in the figure.

Metallic particles with a diameter much smaller than the wavelength of light exhibit the phenomenon of **surface plasmon resonance** described above. At the resonance frequency, the particles absorb light

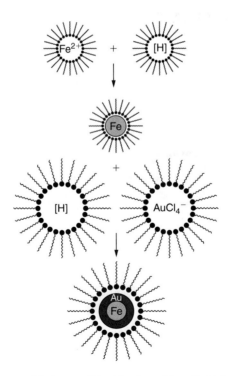

■ **Figure 18.8** Synthesis of $Fe_{core}Au_{shell}$ nanoparticles by sequential reactions in emulstions. From A. Boal, *Nanoparticles, Building Blocks for Nanotechnology*, V. Rotello, Ed., Springer, New York (2004) p. 7.

and transform it into heat with great efficiency. This effect is being considered as a therapeutic technique for cancer because the human body is nearly transparent to the near-infrared light that is used.

By adsorption of specific organic molecules, the nanopatricles can be "functionalized," which means that they can attach themselves to specific biological substances. Because of their small size, the particles are easily introduced into the body, where they are utilized to increase the sensitivity of Magnetic Resonance Imaging (MRI) scans.

18.4.2 **Metallic Nanorods**

Small rods of nanometer diameter can be grown by vapor deposition or by pulsed electrochemical deposition into nanoporous membranes (Figure 18.9). By successive deposition of different metals, it is possible to produce segmented metal nanorods. Such particles are commercially available as nano-barcodes identifying individual pharmaceutical pills.

Silicon nanowires can be grown by the VLS method in the manner described for carbon nanotubes (Figure 18.5): small gold particles serve as the growth catalyst and determine the thickness of the wires. Interaction of silane (SiH_4) vapor with the gold saturates the particle with silicon. The latter then grows as a solid wire with the same diameter as the particle. The application of silicon and other compound semiconductor nanowires is still in the research stade. These wires are contemplated for the production of novel, nanometer-sized, transistors for computers.

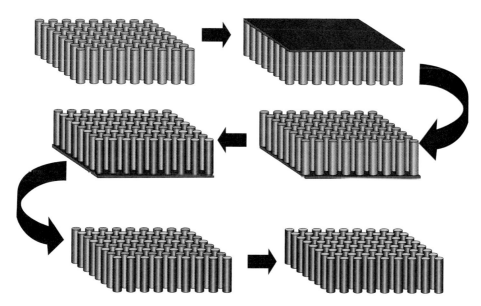

■ **Figure 18.9**   Template synthesis: A conductive seed layer is sputtered on the backside of the template. The nanowires are electrodeposited in the template. The seed layer is removed by physical or chemical means. The template is dissolved to yield a colloid of nanowires in solution. Courtesy of Professor N.V. Myung.

18.5 **SEMICONDUCTOR NANOPARTICLES—QUANTUM DOTS**

Semiconductor quantum dots exploit the electronic structure of nanoparticles. As we have seen in Section 18.1 the width of the energy bands and, therefore, the band gap of the semiconductor, vary with the size of the particle. The band gap defines, among other things, the frequency of the light that is absorbed by excitation of an electron from the valence band to the conduction band or emitted by electron-hole recombination. The smaller the particle, the larger the band gap and the higher the photon energy of light absorbed and emitted. By adjusting the size of the particle, one obtains different colors: red for the larger particles and orange, yellow, green, and blue for progressively smaller particles.

Solid-state physicists offer an alternate description of this effect. It goes as follows: Consider a large semiconductor and its energy bands shown in the left of Figure 18.10. In the conduction band of a solid, the energy levels of the elctrons correspond to their kinetic energy

$$E = p^2/2m \tag{18.1}$$

The bottom of the band corresponds to zero kinetic energy. The momentum p = mv of an electron is related to its wavelength by the de Broglie relation

$$p = h/\lambda \tag{18.2}$$

and its kinetic energy therefore is

$$E = p^2/2m = h^2/2m\lambda^2. \tag{18.3}$$

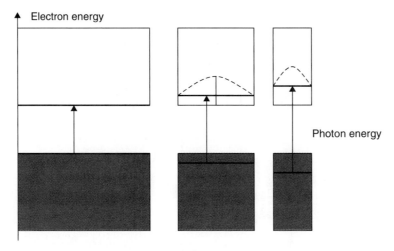

■ Figure 18.10 Schematic of the effect of particle size on the band gap and optical properties of quantum dots.

Since the quantum waves of electrons vanish at the surface of a particle of length d, the shortest wave that can exist has wavelength $\lambda/2 = d$. (Musicians adjust the pitch of their sound by changing the length of the vibrating string in a guitar or of the vibrating air column in a trombone.) In a small solid, the state with zero momentum does not exist and the lowest kinetic energy in the band is

$$E = h^2/8md^2 \qquad (18.4)$$

Thus, the lowest energy level in a small particle is not really at the bottom of the conduction band, but at an energy above that minimum given by Equation (18.4). The smaller the size d, the higher is that energy.

The same argument holds for the holes in the valence band maximum albeit in a less intuitive manner. Thus, the smaller the particle, the larger the energy band gap and the shorter the wavelength of absorbed and emitted light, as shown in the figure. Thus it is possible to tune the wavelength of the absorbed or emitted light by controlling the size of the quantum dot. This quantization of the energy levels is responsible for the name quantum dot.

18.5.1 Synthesis of Quantum Dots

The synthesis of CdS and CdSe nanoparticles (Figure 18.11) utilizes the same method of confinement by micelles as we have described for metals. The particles are produced by the decomposition of organometallic molecules. One first dissolves $Cd(CH_3)_2$ in the surfactant tri-n-octylphosphine (TOP) or tri-n-octylphosphine oxide (TOPO) at $300\,^\circ C$ and then rapidly injects Se powder or the compound $(TMS)_2Se$. The reaction between the two substances produces the desired compound particles.

18.5.2 Applications

Quantum dots are used as biological tags. By adsorbing selected organic molecules, the particles can be made to interact with specific biological materials. Their size is selected to make them absorb and emit in the near infrared, to which the body is reasonably transparent.

They are also used in electro-optic devices, especially in organic LEDs described in Chapter 13, where they act as lumophores. In this capacity they serve as recombination centers for electrons from the cathode and holes from the anode. The color of the emitted light is controlled by the size of the particles. This is illustrated in Figure 18.12.

Efforts are underway to utilize quantum dots in photocells for the conversion of solar energy.

18.6 TWO-DIMENSIONAL SYSTEMS

A two-dimensional system consists of a stack of planes of nanoscale thickness. Such systems can be grown by controlled successive deposition of different materials, either by evaporation, sputtering, chemical vapor deposition, or electroplating. Macroscopic areas can be deposited in thicknesses

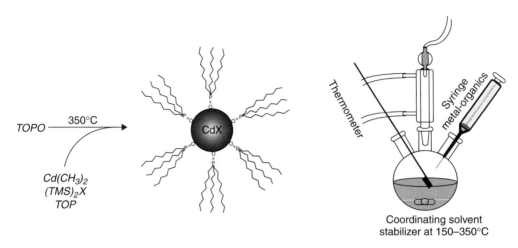

■ **Figure 18.11**   Synthesis of CdSe quantum dots. From Scaff and Emrick, *Nanoparticles, Building Blocks for Nanotechnology,* V. Rotello, Ed., Springer, New York (2004).

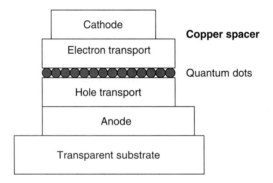

■ **Figure 18.12**   Schematic of a light-emitting diode with quantum dots as lumophores. From S. Coe-Sullivan, *Material Matters*, Sigma Aldrich, Vol. 2(1), p. 14.

smaller than 100 nm. One deposits the large area and subsequently defines the small device area by photolithography and etches away the unwanted areas.

Magnetic memory read heads are fabricated by such methods. They operate on the principle of Giant Magnetoresistance (GMR) in which the magnetic field from the memory disc generates a large change in the electric resistance of the detector. The scientific principle was described in Chapter 14. Giant Magnetoresistance is obtained in a system consisting of alternate layers of magnetic and nonmagnetic materials. The magnetization direction of one layer is fixed, and that of the other layer can be rotated by the magnetic field emanating from the memory disc. This change in magnetization entrains a large change in the electric resistance of the conducting layer. It is necessary to prevent the magnetic field

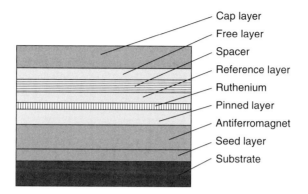

■ **Figure 18.13** Structure of an SAF spin-valve GMR read head. From DiVentra, Evoy, and Heflin, *Introduction to Nanoscale Science and Technology*, Springer, New York (2004).

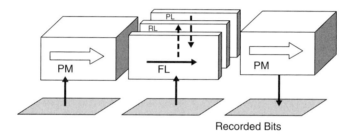

Recorded Bits

■ **Figure 18.14** The magnetic orientation of the free layer is biased perpendicular to those of the pinned and reference layers. It is rotated parallel or antiparallel to that of the reference layer by the magnetic field of the storage medium. From DiVentra, Evoy, and Heflin, *Introduction to Nanoscale Science and Technology*, Springer, New York, (2004).

from the fixed layer from immobilizing the free layer (competing with the field from the memory disc). This is achieved in the Synthetic Antiferromagnet (SAF) Spin Valve device shown in Figure 18.13. The Giant Magnetoresistance occurs in the system consisting of the free layer whose magnetization is rotated by the magnetic field of the disk, the copper spacer, and the reference layer whose magnetization is fixed. The lower structure, namely the ruthenium spacer, the pinned layer, and the antiferromagnet, serve to fix the magnetization of the reference layer. The antiferromagnet fixes the magnetic orientation of the pinned layer.

The ruthenium layer is 8 Å (0.8 nm) thick and produces a strong antiferromagnetic coupling to the reference layer (see Figure 18.14). The pinned and reference layers form a synthetic antiferromagnetic system with a closed-loop magnetic field. As a consequence the magnetization direction of the free layer is only slightly influenced by that of the reference layer and is sensitive to the field of the magnetic storage medium.

The device is fabricated on top of a substrate. A seed layer is first deposited; its function is to stimulate large grain growth in the subsequent layers in order to minimize electron scattering by grain boundaries.

The seed layer consists of tantalum or a NiFeCr alloy. The antiferromagnetic layer is either NiO or PtMn. The ferromagnetic pinned, reference, and free layers consist usually of a CoFe alloy or a CoFe/NiFe bilayer. After successive deposition of the pinned Ru reference layers, the Cu spacer, and the free magnetic layer, the system is capped by a Ta layer for protection.

18.7 **SAFETY CONCERNS**

The very small size of nanoparticles makes them impossible to filter so that they could penetrate the human body through breathing, eating, drinking, and perhaps even by diffusion through the skin. No means for stopping nanoparticles is known at present. It is not yet known whether all nanoparticles, even if not of toxic material, in the lungs, the bloodstream, or in the organs present a health hazard. The source of nanoparticles could be spillage in plants that produce them or destruction of objects that contain them. They could travel easily through air or in water. This problem has not been solved to date.

■ SUMMARY

1. Nanomaterials are smaller than 100 nm in one, two, or all three dimensions. They consist of isolated particles or fine grains in an extended solid.

2. The Hall-Petch law states that the hardness of metals and composites increases with decreasing grain size. This law is followed when the grain size is larger than 50 nm. Smaller grains weaken the material.

3. The band gap of semiconductors increases with decreasing size of nanoparticles. Light absorption and emission can be tuned through the whole visible spectrum by changing particle size.

4. In metallic nanoparticles, collective displacement of the electrons vibrates with a specific frequency that is determined by the particle size. Light with the same frequency is strongly absorbed and transformed into heat. By selection of the particle size, the absorption frequency can be tuned through the visible spectrum and the near infrared.

5. Carbon nanotubes are grown in Chemical Vapor Deposition (CVD) on iron nanoparticle catalysts.

6. Carbon nanotubes are the stiffest materials known ($E \geq 1,000\,\text{GPa}$) and very strong. They have the highest thermal conductivity of any materials. Carbon nanotubes can be metallic or semiconductors.

7. Nanoparticles are grown by reduction of a salt solution in a very fine water/oil emulsion. Their growth is limited by the size of the droplets.

8. Metallic nanorods are grown by vapor or galvanic deposition of the metal into a porous substrate that is subsequently dissolved.

9. A large number of applications of nanomaterials are presently in the research stage.

10. Nanoparticles are too small to be stopped by filters. There is a concern about their unimpeded penetration into the human body and possible health problems. The health effects of nanoparticles have not been established yet.

■ KEY TERMS

A
antiferromagnet, 527

B
biological tags, 525
Bucky balls, 518

C
calcining, 518
catalysts, 518
CdS, 525
CdSe, 525
CoFe, 528
CoFe/NiFe, 528
colloidal growth, 521

E
electronic structure, 516

F
fullerenes, 518, 530
functional applicatioin of
    nanotubes, 521
functionalizing, 523

G
giant magneto-resistance, 526

graphene, 518, 519, 520

H
Hall-Petch, 516, 518, 528

L
lumophores, 525

M
magnetic properties, 517
mechanical applications of
    nanotubes, 521
mechanical properties, 516
metallic nanoparticles, 521

N
nanorods, 523
nanowires, 523
NiFeCr, 528
NiO, 528

O
optical properties, 516

P
PECVD, 520
plasma-enhanced CVD, 520
PtMn, 528

Q
quantum dots, 515, 516, 524,
    525, 530

R
reference layer, 527
ruthenium, 527

S
safety concerns, 528
surface plasmon resonance,
    515, 517, 522
synthesis of carbon
    nanotubes, 520
synthesis of quantum dots, 525

T
tensile strength, 521
TOP, 525
TOPO, 522, 525
tri-n-octylphosphine, 525

W
WC/Co, 516, 518, 530

Y
Young's modulus, 521

■ REFERENCES FOR FURTHER READING

[1] M. Di Ventra, S. Evoy and J.R. Heflin, Jr., (eds) *Introduction to Nanoscale Science and Technology*, Publisher, New York (2004).

[2] A.S. Edelstein and R.C. Cammaratra, (eds) *Nanomaterials: Synthesis, Properties and Applications, (Paperback)*, 2nd ed., Taylor & Francis, London (1998).

[3] G.A. Ozin and A.C. Arsenault, *Nanochemistry, A Chemical Approach to Nanomaterials*, RSC Publishing (The Royal Society of Chemistry), London (2005).

[4] V. Rotello, *Nanoparticles, Building Blocks for Nanotechnology*, Springer, New York (2004).

■ PROBLEMS AND QUESTIONS

18.1. How does nanometer structure influence the mechanical properties of fibers and why?

18.2. How does nanometer structure influence the mechanical properties of bulk materials and why?

18.3. How does the band structure of nanomaterials differ from that of macroscopic solids?

18.4. Describe how the size of particles influences the color of metal particles.

18.5. Describe how the size of particles influences the color of semiconductors.

18.6. Describe the processing of nanometer-size catalyst particles.

18.7. Describe the processing of nanostructured WC/Co composites.

18.8. What is the *sp* hybridization in Fullerenes?

18.9. What is the size of carbon nanotubes, and how are they processed?

18.10. Describe an application of carbon nanotubes, and what specific properties are used?

18.11. What is the principle used in the production of metallic and semiconductor nanoparticles?

18.12. Describe an application of metallic nanoparticles and the specific properties of these particles that are exploited.

18.13. By what confinement method are metallic nanorods produced?

18.14. Find an application of metallic nanorods in the literature or on the web. Describe the physical principles of their function in this application.

18.15. What is a quantum dot, and why does it have that name?

18.16. Describe an important application of quantum dots and explain the physical principles used.

18.17. Describe the fabrication of a magnetic recording read head. Why must the read head be composed of nanomaterials?

18.18. Comment on the reasons why there is opposition to nanomaterials in the public.

Chapter 19

The Characterization of Materials

In this chapter we examine the techniques that are used to study the composition, crystal structure, and microstructure of materials. Examples of the information they provide have been presented in the preceding chapters. There are simple and rapid measurement techniques; the optical microscope quickly reveals the microstructure of a material when its features are not smaller than one micrometer; a handheld X-ray fluorescence device provides an instantaneous measure of the composition of a material. On the other hand, there are powerful instruments such as the Scanning Transmission Electron Microscope, which is capable of providing an image with better than nanometer resolution and at the same time measure the chemical composition and crystal structure of features a few nanometers in size. These probes utilize the interaction of X-rays or electrons with matter. To measure the composition, one exploits the binding energies of core electrons in the atom. The crystal structure is revealed by diffraction of electrons or of X-rays. Scanning probe microscopes investigate the surface by means of a very sharp tip that scans the specimen less than a nanometer above its surface. These instruments not only provide a picture of the surface with atomic resolution, but they also can measure the attractive and repulsive forces between atoms.

LEARNING OBJECTIVES

After studying this chapter, the student will be able to:

1. Select the proper characterization technique for the required information.

2. Distinguish between elastic and inelastic scattering of X-rays and electrons and state what these are used for.

3. Describe the principle of core electron spectroscopy used for the identification of chemical elements.

4. Explain the thinking that explains Bragg's equation of diffraction and derive it from these principles.

5. Perform an X-ray diffraction experiment and identify the material with the help of diffraction data.

6. Prescribe the appropriate measurements in a scanning electron microscope and utilize their results.

7. Compare the information obtained from an SEM, a TEM, and an STEM.

8. Describe how the size, composition, and crystal structure of a phase in a material can be observed and interpret the data.

9. Describe the construction and operation of a scanning tunneling microscope and what information it provides.

10. Describe the construction and operation of an atomic force microscope and what information it provides.

The measurements of the performance of materials, such as their strength and other mechanical properties, their electrical conductivity, and their optical and magnetic properties have been discussed in the relevant chapters. It may be necessary to characterize the materials more fully as quality control when receiving a purchased material, to assess the results of in-house processing, or to find the cause of a failure.

We may need to observe the shape and size of various features: grains in a material, different phases, transistor elements or interconnects in an integrated circuit or the structure of a MEMS; we do this with the help of various kinds of microscopes. Seeing the various phases in a multiphase material, we may need to know the chemical composition of each. For this purpose we examine the binding energies of the core electrons in its atoms; these energies form a precise spectrum that is unique to each element. This technique provides an elemental chemical analysis. For the identification of chemical compounds or the particular structure of a polymorph, we need to measure the crystal structure of the phase; the latter is determined by the diffraction of X-rays or electrons.

The measurement techniques utilize the interaction of X rays or electrons with the material. When these impinge on a solid, they do three things. A number of them traverse the material without interacting with it. Some are scattered by the atoms without loss of energy or wavelength, changing only their direction; this is **elastic scattering**, which is used in microscopy and in the study of crystal structures. Others transmit all or part of their energy to the electrons in the material; this is **inelastic scattering** which we use for chemical analysis.

Light, X-rays, and electrons are rapidly absorbed in metals and in most other materials; therefore they let us analyze the surface or a very thin zone below the surface only. When one needs to study the bulk of a material, one makes a section of the piece, polishes it, sometimes one etches it to reveal grain boundaries or other features, and observes the surface so obtained. The previous chapters contain many examples of such sections.

Table 19.1 Binding Energies of Electrons in Selected Elements (eV).

Electrons	Carbon	Aluminum	Silicon	Iron	Gold	
1s	283.8	1,559.6	1,838.9	7,112.0	80,724.9	K shell
2s	6.4	117.7	148.7	846.1	14,352.8	L shell
2p	6.4	73.3	99.5	721.1	13,733.6	"
		72.9	98.9	708.1	11,918.7	"
3s		2.2	3.0	92.9	3,424.9	M shell
3p		2.2	3.0	54.0	3,147.8	"
					2,743.0	"
3d				3.6	2,291.1	"
					2,205.7	"
4s				3.6	758.8	N shell
4p					643.7	"
					545.4	"
4d					352.0	"
					333.9	"
4f					86.6	"
					82.8	"
5s					107.8	O shell
5p					71.7	"
					53.7	"
5d					2.5	
6s					2.5	P shell

From American Institute of Physics Handbook, pp. 7–158 to 7–165.

19.1 MEASURING CHEMICAL COMPOSITION: CORE ELECTRON SPECTROSCOPY

The binding energies of the core electrons of atoms are an excellent tool for measuring the composition of a material. We repeat Table 1.2 as Table 19.1 here for convenience; it shows the binding energies of the core electrons of some elements.

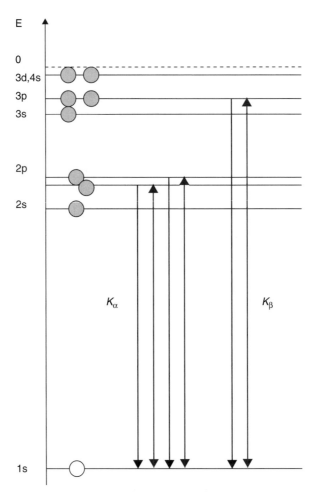

■ **FIGURE 19.1**   X-ray emission: a high-energy electron impinging on the atom ejects the 1s electron. Electrons from the 2p or 3p levels fall into the empty 1s level and get rid of the excess energy by emitting a photon of energy $h\nu = E_{2p} - E_{1s}$ or $h\nu = E_{3p} - E_{1s}$. The black arrows show the electron transition, and the color arrows the energy of the emitted X-rays.

Electrons can be excited from an occupied energy level into an empty one (usually into the continuum of the conduction band) by the impact of a high-energy electron or the absorption of an X-ray. This is illustrated in Figures 19.1 and 19.3.

When an electron in an atom has been excited, it leaves an empty level behind; electrons with higher energy can now fall into this level and get rid of the excess energy by emitting an X-ray. When, for example, the empty level is 1s and a 2p electron falls into it, the photon energy of the X-ray is

$$hv = E_{2p} - E_{1s} \tag{19.1}$$

This is illustrated in Figure 19.1. Let us look at the case of iron; the electron or the X-ray photon exciting the $1s$ electron must have energy larger than $7{,}112.0\,\text{eV}$. When $2p$ electrons fall into the empty $1s$ levels, they emit X-rays with energy

$$h\upsilon = E_{2p} - E_{1s} = (-721.1) - (-7{,}112.0) = 6{,}390.9 \text{ eV}$$

and

$$(-708.1) - (-7{,}112.0) = 6{,}403.9 \text{ eV}$$

The energy $h\upsilon$ of an X-ray is related to its wavelength λ by

$$E = h\upsilon = hc/\lambda \tag{19.2}$$

When we express the photon energy in keV and the wavelength λ in nanometers, Equation (19.2) provides the numerical relationship

$$E(\text{keV}) = 1.24/\lambda\,(\text{nm}) \tag{19.3}$$

The two X-rays emitted have wavelengths

$$\lambda = 1.24/6.3909 = 0.1940 \text{ nm}$$

and

$$\lambda = 1.24/6.4039 = 0.1936 \text{ nm}$$

These transitions are called the K_α transitions. (For conservation of symmetry, emission of an X-ray is possible only when the second quantum number changes by $\Delta l = \pm 1$, namely from p to s orbitals or from s and d orbitals into p orbitals.)

X-rays emitted by electrons falling from the $3p$ levels into the $1s$ level can be computed in a similar way.

$$h\upsilon_{3p-1s} = 7{,}019.1 \text{ and } 7{,}058 \text{ eV}$$

These are the K_β transitions.

When a $2s$ level is vacated by electron impact or X-ray absorption, the transitions from $3p$ to $2s$ have the energy $h\upsilon_{3p-2s} = 792.7\,\text{eV}$.

If the $2p$ level is vacated, the transitions from the $3s$ levels have the energy

$$h\upsilon_{3s-2p} = 615.2 \text{ and } 628.2 \text{ eV}$$

Thus, excitation of a material by high-energy electrons or X-rays produces the emission of a characteristic spectrum of X-rays. These energies can be translated into wavelengths with Equation (19.3).

19.1.1 **The X-Ray Source**

X-rays used in medicine and in the analysis of materials are produced just as described above. A cathode (a hot wire) emits electrons that are accelerated to high energy and impinge on an anode. The latter emits X-rays. One is usually interested in the strongest emission lines, namely the K_α from $2p$ to $1s$. Practical X-ray sources use copper anodes which emit at

$$Cu_{K_\alpha} = 8.02783 \text{ keV at } \lambda = 0.155439 \text{ nm}$$

and

$$8.04778 \text{ keV at } \lambda = 0.1540562 \text{ nm}$$

Other sources use a molybdenum anode which emits X-rays at

$$Mo_{K_\alpha} = 17.3743 \text{ keV at } \lambda = 0.07135 \text{ nm}$$

and

$$17.4934 \text{ keV at } \lambda = 0.070930 \text{ nm}$$

19.1.2 **Energy-Dispersed X-Ray Spectroscopy (EDX) in Electron Microscopes**

Electron microscopes use high-energy electrons to produce high-resolution images of the samples. These electrons impinge on the atoms and generate X-rays as discussed above. These are evaluated in an energy-dispersive analyzer which measures their energy rather than their wavelengths. The analyzer is a reverse biased p-n junction in which each X-ray photon produces an electron-hole of high energy. The electron and the hole get rid of their excess energy by exciting other electron-hole pairs until a number of pairs proportional to the energy $h\nu$ of the X-ray photon is produced. The result is a current pulse with amplitude proportional to $h\nu$. The instrument plots the number and amplitude of the resultant current spikes. The spike amplitude is a measure of the photon energy and identifies the element; the number of spikes is a measure of the number of X-ray photons emitted and represents the concentration of the element. Figure 19.2 is an example obtained in an EDX instrument. An alternate name of this technique is Energy Dispersive Spectroscopy (EDS).

The relationship between X-ray emission intensity and atomic concentration is somewhat complex; it has been measured for all elements and is stored in the computer of the instrument which translates the X-ray data into the composition of the material automatically.

19.1.3 **X-Ray Fluorescence**

X-ray fluorescence is illustrated in Figure 19.3. A high-energy X-ray is absorbed and creates empty core levels by exciting electrons. Other electrons fall into the empty orbital and emit characteristic X-rays according to Equation (19.1). The energy $h\nu$ of the secondary X-rays is measured by EDS, and the composition of the sample is computed as in the previous section. Figure 19.4 shows a commercial handheld X-ray fluorescence instrument. The handheld instrument contains the source of primary X-rays and the EDS analyzer.

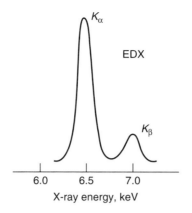

■ FIGURE 19.2 Example of an EDX Spectrum of iron obtained in a scanning electron microscope.

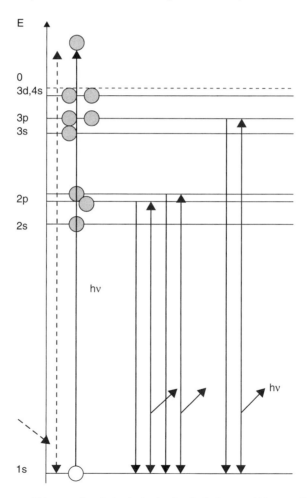

■ FIGURE 19.3 X-ray fluorescence: a high-energy X-ray is absorbed and excites the 1s electron to high energy. Electrons from the 2p and 3p levels fall into the 1s level and get rid of the energy by emitting a photon of energy $h\nu = E_{2p} - E_{1s}$. The energy of the exciting X-ray is shown in dotted color lines and that of emitted X-ray in solid color. Black lines show electron transitions.

■ **FIGURE 19.4** The Thermo Scientific NITON XL3t handheld (X-ray fluorescence) XRF analyzer.

X-ray fluorescence is fast and convenient. It does not require any preparation of the sample to analyze. It measures the **overall** composition of the material within the size of the X-ray beam. Local compositions can be measured in electron microscopes as described in Sections 19.3.2 and 19.3.4.

19.1.4 Electron Energy-Loss Spectroscopy (EELS)

This technique is utilized in the Transmission Electron Microscopes. Some of the high-energy electrons used for microscopy excite electrons in the atoms of the sample as in Section 19.1.1; they traverse the sample and enter into a high-resolution energy analyzer that measures how much energy they lost. This loss is equal to the energy they transferred to electrons that was excited into the empty band; it is characteristic of the element in which the transition occurred. The technique again measures the elemental composition of the sample.

Figure 19.5 shows such an energy loss spectrum for silicon. The sharp rise near 100 eV is due to the excitation of electrons from the $2p$ level of silicon into the empty conduction band (Table 19.1). Figure 19.6 displays the EELS of nitrogen in silicon nitride and oxygen in silicon oxide. We shall visit these in a later example.

There are other measuring methods that utilize transitions of electrons between energy levels, such as Auger Electron Emission and Photoelectron Emission. The interested reader can find them easily in relevant textbooks or on the web.

19.2 DETERMINATION OF THE CRYSTAL STRUCTURE BY DIFFRACTION

The measurement of the crystal structure of a material has several uses. It can determine what phase of a polymorphic material is present; it can identify chemical compounds; it determines whether the material is a single crystal or polycrystalline, and it can measure the internal strains associated with residual stresses. Crystal structure determination is done by means of X-ray or electron diffraction. It uses the wave nature of the X-rays and the electrons.

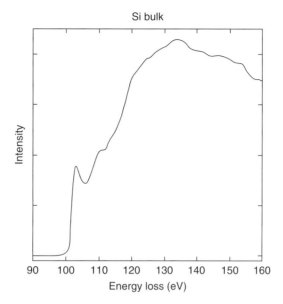

■ FIGURE 19.5 Electron energy loss spectrum of silicon. Courtesy of Peter Ercius, Aycan Yurtsever, and David Muller, Cornell University.

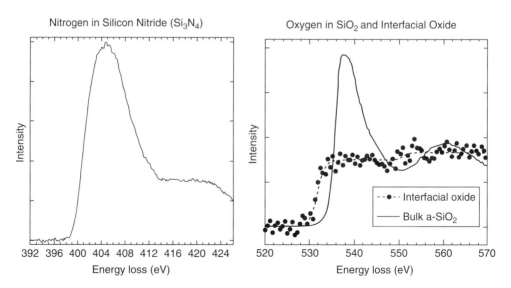

■ FIGURE 19.6 EELS measurements of nitrogen in silicon nitride and oxygen in silicon oxide. Courtesy of Peter Ercius, Aycan Yurtsever, and David Muller, Cornell University.

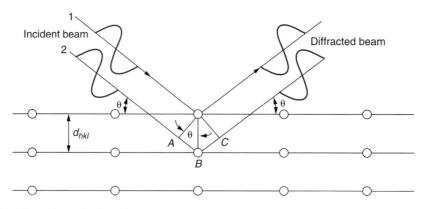

■ FIGURE 19.7   Diffraction by parallel planes of distance d_{hkl}.

19.2.1 **Diffraction**

A crystal is a periodic assembly of atoms. The latter are situated on planes that are stacked at equal distance d from each other. Diffraction of waves by such an array is illustrated in Figure 19.7.

When a plane wave traverses the crystal, part of it is reflected by every plane it encounters as by a mirror. The waves reflected by the planes experience interference: the outgoing wave is the algebraic sum of all the reflected waves. When all the reflected waves are in phase with each other, as in Figure 19.7, the interference is positive, all the reflected signals add to each other, and the outgoing wave is very strong. This occurs when the difference in distance traveled by the waves reflected from two consecutive planes is exactly one wavelength or an integer number of wavelengths, namely $AB + BC = n\lambda$. It is easy to see from the figure that $AB = d_{hkl} \sin\theta$ when d_{hkl} is the distance between the planes with indices (hkl) and θ is the angle of incidence of the wave. The condition for positive interference is

$$2d_{hkl} \sin \theta = n\lambda \tag{19.4}$$

When Equation (19.4) is not satisfied, there is a phase shift in the reflected waves and there will be a plane from which the reflected beam is 180° out of phase with that from the first plane and both reflections will cancel each other. The next plane cancels the reflection from plane #2, and so on: the reflected intensity is zero. Equation (19.4) represents **Bragg's law of diffraction**. It allows us to compute the distance between atomic planes when the wavelength of the wave λ is known and the angle of diffraction is measured. Note, by observing Figure 19.7, that the angle between incoming and diffracted beam is 2θ.

Diffraction can also be used to measure the wavelength of X-rays; one inserts a material with known structure and measures the diffraction angles.

Amorphous materials do not have planes of atoms with regular spacing. In these materials the only interference occurs between waves reflected by neighboring atoms. Wave scattering by amorphous materials presents a single broad peak at an angle given by Equation (19.4) where the distance

(A)

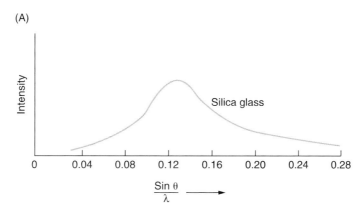

(B)

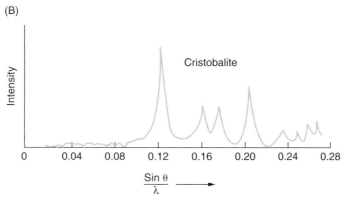

■ **FIGURE 19.8**   Diffraction by (A) amorphous silica glass and (B) cristobalite, a crystalline form of glass.

between planes is replaced by the mean distance between molecules d_{mol}. Figure 19.8 shows the diffraction patterns obtained from amorphous silica and cristobalite, which is one of the crystal structures of silica (Figure 8.14).

Any radiation capable of penetrating the solid can be used for diffraction measurements provided, as Equation (19.4) shows, that the wavelength of the radiation always be smaller than the distance $2d$. We concern ourselves with the two most important diffraction methods, namely X-ray and electron diffraction.

19.2.2 **X-Ray Diffraction**

The experimental geometry and equipment employed for measuring diffraction from powders and polycrystalline materials are shown in Figure 19.9. X-rays are generated by a source described in Section 19.1.1. The more intense K_α X-rays are used; after collimation and filtering to eliminate other wavelengths, they impinge on the specimen. In the measurement, the sample rotates by the angle θ and the collector by the angle 2θ with respect to the source, and the intensity of the reflected beam is collected

(A)

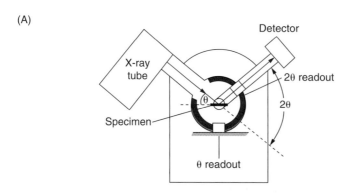

(B)

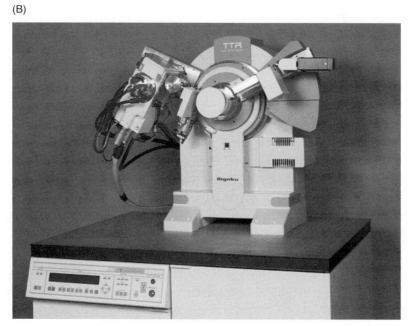

■ **FIGURE 19.9** (A) Geometry of X-ray diffraction. Note that both the X-ray source and the detector rotate with respect to the sample and that the instrument measures the angle 2θ. (B) An actual diffractometer, courtesy of Rigaku.

as the angle increases. A typical diffractometer trace is shown in Figure 19.10; it reveals a number of peaks with varied intensities as a function of angle 2θ. These peaks represent the diffracted X-rays that obey Bragg's law, Equation (19.4).

Note that the geometry used in X-ray diffractometers is similar to that of Figure 19.7. The instrument measures only the distances between planes parallel to the surface of the sample. Most materials are in the shape of a fine powder or a polycrystalline solid. In either case the specimen consists of a large number of small grains randomly oriented, and there are always enough of these grains presenting any crystal plane (*hkl*) parallel to the surface.

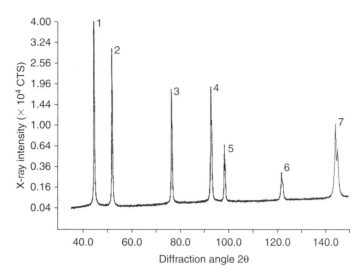

■ FIGURE 19.10   Diffractometer trace of a cubic metal plotted as the intensity of the diffracted signal versus angle 2θ. The peak numbers refer to Example 19.1 where the identity of this metal is revealed. Courtesy of B. Greenberg.

Analysis of X-ray diffraction patterns can be complex. It is relatively simple for cubic crystal structures. In a cubic crystal with lattice parameter (unit cell size) a, the distance d_{hkl} between (hkl) planes is

$$d_{hkl} = \frac{a}{\sqrt{h^2 + k^2 + l^2}} \tag{19.5}$$

Combining this with Bragg's law, Equation (19.4), we obtain

$$\sin^2 \theta = (h^2 + k^2 + l^2) \cdot \lambda^2 / 4a^2 \tag{19.6}$$

Diffraction allows us to distinguish between BCC and FCC crystals. Let us consider diffraction from the (100) plane of the BCC structure shown in Figure 19.11. It satisfies Bragg's relation (19.4) with $d_{100} = a$. The BCC structure has planes going through the sides of the unit cells, but it also has a plane going through the atom in the center of the cube. The rays from neighboring (100) planes (i.e., 1 and 3) a distance a apart, are in phase but are 180° out of phase with the ray scattered from the center plane 2. Since there are equal numbers of type 1 and 2 planes, the net diffracted intensity vanishes for the (100) plane in BCC materials.

Plane 2 has coordinates (200) since it intersects the x axis at $a/2$. The three planes can be considered as three (200) planes with distance $a/2$. The 200 diffraction peak does not disappear because it satisfies Bragg's relation with $d_{200} = a/2$. Thus, in the diffraction pattern, the 100 diffraction beam vanishes but the 200 beam is strong.

The (100) beam is not the only one suppressed because of the center atom. Table 19.2 provides a list of the planes which do and do not diffract in single cubic, BCC and FCC lattices. In BCC the sum $h + k + l$ must equal 2, 4, 6, 8, 10, and so on. In FCC the h, k, and l indices must all be even or odd.

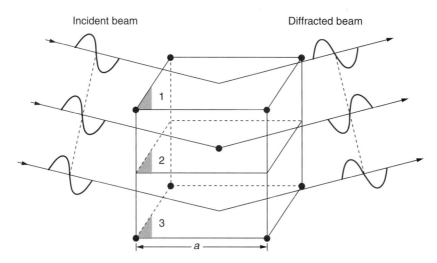

■ **FIGURE 19.11**   Reflected X-rays from the {100} planes of a BCC unit cell. As a result of adding the rays reflected from neighboring planes, the (100) reflection vanishes.

Table 19.2 Diffracting Planes in Cubic Crystals.

1. Rules governing the presence of diffraction peaks.

Lattice	Reflections present	Reflections absent
Simple cubic	All (hkl) planes	None
BCC	$(h + k + l)$ = even	$(h + k + l)$ = odd
FCC	(h,k,l) all odd or all even	(h,k,l) mixed even and odd

2. Miller indices of diffracting planes in BCC and FCC crystals.

Cubic planes (hkl)	$h^2 + k^2 + l^2$	Diffracting planes	
		BCC	FCC
{100}	1		
{110}	2	110	
{111}	3		111
{200}	4	200	200
{210}	5		
{211}	6	211	
{220}	9	220	220
{221}	9		
{310}	10	310	
{311}	11		311
{222}	12	222	222

The interpretation of diffraction patterns is much more difficult for compounds that have complex crystal structures. Practically all known crystals have been analyzed, and their diffraction patterns, namely the angles 2θ and the relative intensities of their peaks, are tabulated. These are stored in the computers of modern diffraction instruments; the latter then identify the material involved automatically. They are also capable of computing the relative amounts of different phases in a multiphase material.

EXAMPLE 19.1   *Identify the cubic metal element that gave rise to the diffractometer trace shown in Figure 19.10. Copper K_a radiation was used with $\lambda = 0.15405\,nm$.*

ANSWER   In order to identify the metal, values for 2θ must be read from the trace and halved so that both sinθ and $\sin^2\theta$ can be evaluated. Then from Equations (19.4) and (19.5), d_{hkl} can be calculated, and $h^2 + k^2 + l^2$ ratios obtained. Finally, a is extracted from Equation (19.6). The results are best tabulated as follows:

LINE	2θ	θ	sinθ	d_{hkl}	$\sin^2\theta$	$h^2 + k^2 + l^2$	(hkl)	a (nm)
1	44.5	22.26	0.3788	0.2033	0.1435	3	(111)	0.3522
2	51.93	25.97	0.4379	0.1760	0.1918	4	(200)	0.3518
3	76.37	38.19	0.6182	0.1246	0.3822	8	(220)	0.3524
4	93.24	46.62	0.7268	0.1060	0.5283	11	(311)	0.3515
5	98.43	49.22	0.7572	0.1018	0.5734	12	(222)	0.3524
6	121.9	60.96	0.8742	0.0881	0.7644	16	(400)	0.3524
7	144.6	72.28	0.9526	0.0809	0.9074	19	(331)	0.3525

The key to the analysis is identifying the crystal structure. Because the ratios of $\sin^2\theta$ for the first two diffraction lines is $0.1435/0.1918 = 0.7482$ or ~3/4, the structure is FCC. The average value of a is determined to be 0.3522 nm and corresponds well with the accepted lattice parameter of nickel, that is, $a = 0.3524$ nm.

19.2.3 Electron Diffraction

The electron beam in the electron microscope is perfectly suited for diffraction measurements. Electrons, like photons, are waves at the same time as they are particles. The wavelength of an electron depends on its kinetic energy eU according to the de Broglie relation

$$\lambda = \frac{h}{p} = \frac{h}{m_o v} = \frac{h}{\sqrt{2m_o eU}} \qquad (19.7)$$

Here, h is Planck's constant, m_o and e are the mass and electric charge of the electrons, and U is the electric potential through which the electron has been accelerated. Inserting the values of the

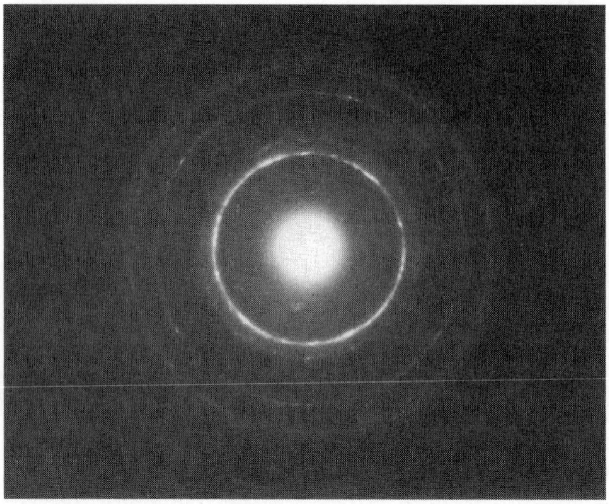

■ **FIGURE 19.12** Typical electron diffraction pattern from polycrystalline iron obtained in an electron microscope.

constants, one obtains the wavelength λ of the electron, measured in nanometers, as a function of the acceleration voltage U measured in volts.

$$\lambda(\text{nm}) = \frac{1.226}{\sqrt{U(V)}} \approx \sqrt{\frac{1.5}{U}} \tag{19.8}$$

In the electron microscope, the accelerating voltage of the electrons is typically 100 kV to ensure adequate transmission through the sample. At that voltage, the wavelength is $\lambda = 0.00388$ nm; therefore the diffraction angles in the electron microscope are small. For instance, with $d = 0.2033$ nm, the first diffraction peak of nickel occurs at an angle $2\theta = 1.214°$ compared to $44.5°$ in X-ray diffraction. Figure 19.12 shows a typical electron diffraction pattern obtained in an electron microscope.

Electron diffraction in the transmission electron microscope demands elaborate sample preparation, as we shall discuss below. It has the advantage of being able to analyze the structure of very small areas in the sample, which is important in multiphase materials.

19.3 MICROSCOPY

Microscopes find many uses in the characterization of materials. The optical microscope is affordable, simple to use, and provides almost instantaneous information on any feature larger than a micrometer. Electron microscopes are powerful machines that provide information not only on the shape of features, but also on chemical composition and crystal structure. They are expensive and, in some cases, require elaborate sample preparation. The new scanning probe microscopes (the atomic force microscope and the scanning tunneling microscope) can have atomic resolution in three dimensions, require modest sample preparation, and operate in air. They are used extensively in nanotechnology.

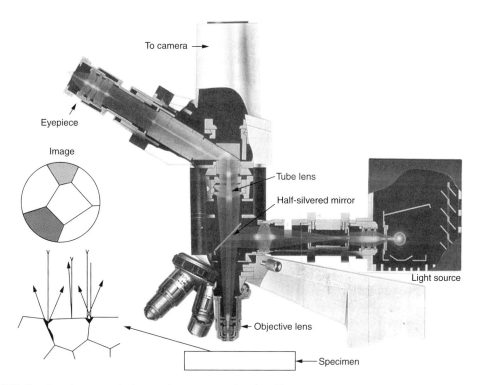

■ **FIGURE 19.13**   Optical microscope. Light enters the microscope and is reflected by a mirror down toward the specimen where it is reflected and passes through the objective lens and the eyepiece. Optical microscopes are also equipped with a camera, as shown. Courtesy of Olympus Optical Company.

19.3.1 **The Optical Microscope**

The optical microscope is the simplest and most convenient instrument. Since materials are usually not transparent, one utilizes a reflection microscope shown in Figure 19.13. If one wishes to observe the microstructure of a sample, one needs to polish its surface and etch it. Etching attacks grain boundaries preferentially and makes them visible; it also creates surface structures that provide differently oriented grains with different reflectivity so that a brass sample, for instance, appears as in Figure 19.14.

The maximum practical magnification of optical microscopes is ×1000 with a resolution of about 0.5 μm. At the higher magnification (×500 and ×1,000), the depth of field of the optical microscope is very small so that only flat specimens are easily observed. This small depth of field can be used to measure relief: one focuses on the high and the low features and measures the difference in vertical position of the microscope on a precision knob inscribed in micrometer height.

19.3.2 **The Scanning Electron Microscope (SEM)**

Figure 19.15 is a schematic of a scanning electron microscope. Electrons are emitted by a cathode (usually a hot filament, or a field-emission source); they are accelerated and shaped into a thin beam that

■ FIGURE 19.14 The microstructure of brass containing 70 wt% Zn and 30 wt% Cu observed with an optical microscope (×100). Courtesy of G. V. Vander Voort, Buehler Ltd.

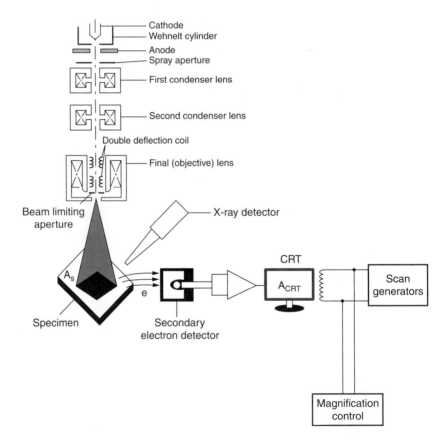

■ FIGURE 19.15 Schematic of the scanning electron microscope.

is focused on a point on the specimen. When these electrons impinge on the specimen with energy of 10–30 kV, they cause the emission of secondary electrons into vacuum; they produce X-rays as we discussed above; or they are scattered back without losing energy. The primary beam scans the surface of the specimen; at each position the secondary electron or the backscattered electron current is measured and displayed as intensity at the equivalent position on a monitor similar to a television screen. The magnification of the microscope is the ratio of the size of the monitor image to that of the area scanned on the sample. The resolution is that of the diameter of the scanning electron beam. Contrast in the picture is provided by variations in the amount of secondary or backscattered current.

The narrow electron beam provides the scanning electron microscope with a large depth of field that permits the observation of irregular surfaces. The benefits of such depth of field are apparent in Figures 2.24, 2.26, 2.25, 8.4, 8.23, 12.21, and 17.4. It is also possible to place the sample at an angle with respect to the probing electron beam. Figure 19.16, for example, shows a lamella that has been lifted from the surface of silicon nitride by friction. By tilting the sample, the electron beam provides a "perspective" image (Figure 19.16A); it can look at the surface of the lamella (Figure 19.16B) and at

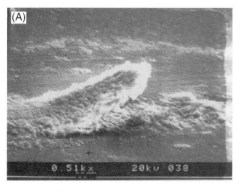

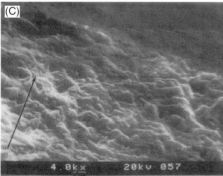

■ **FIGURE 19.16** Observation of a wear sample in oblique beam incidence in the STM. (A) Overview: a lamella is lifted from the surface. (B) The outer surface has been worn by microfracture. (C) The lower surface of the lamella shows it was lifted by intergranular fracture.

the specimen surface underneath the lamella (Figure 19.16C). Figure 19.16B shows that wear causes microfracture on top of the lamella, but that the latter separates from the piece by intergranular fracture (Figure 19.16C).

The X-rays produced by electron impact can be analyzed in an EDX system and provide an elemental chemical analysis as described in Section 19.1.3. There are two ways one can measure composition. In one variant, one keeps the electron beam stationary on the selected spot and collects the whole EDS spectrum. This provides the chemical analysis of the spot. In the other variant, one selects the X-ray energy corresponding to a given element and scans the sample: this provides a map of the distribution of this element on the sample. The SEM analyzes the surface of the specimen. X-ray spectroscopy provides chemical analysis of a depth of about 5 μm below the surface.

Sample preparation is more elaborate than in optical microscopy, but is still relatively simple. To begin with, the sample surface must be clean. An insulating specimen must be coated with a very thin metallic film to prevent charging. (The white areas in Figure 19.16A are caused by charging.) The film is deposited by sputtering (Section 7.5.3B). The coating must be just thick enough to provide conduction; a thick film hides the finer details of the sample.

19.3.3 **The Transmission Electron Microscope (TEM)**

The transmission electron microscope is capable of very high resolution. The short wavelength of the electrons (λ = 0.00388 nm for 100 kV electrons) confers a high resolution to the instrument, enabling features as small as 0.2 nm, such as crystal planes and single atoms, to be resolved. This has been used, for instance, in the image of a dislocation in Figure 3.19. In addition, the TEM permits the determination of crystal structure by diffraction, a chemical analysis by EDS and by the electron energy loss spectroscopy (ELS).

As its name implies, the transmission electron microscope forms an image with electrons that are **transmitted through the sample**. It analyzes the entire thickness of the specimen. But electrons can typically penetrate only through 100 nm to about 1 μm of a solid with the range of operating voltages available (100–300 kV normally), and the specimen must be thinned to these thicknesses prior to observation. The samples are first cut to thin slices and polished to a few micrometers thickness; the final thickness of 100 nm is achieved by ion milling, a technique by which high-energy ions bombard the sample and eject its atoms.

The capabilities of the instrument are illustrated by the analysis of wear particles from silicon nitride after rubbing in humid air (Figure 19.17). Figure 19.17A shows a TEM picture of the particle. Figure 19.17B is an electron diffraction pattern from the sample; it contains diffraction spots characteristic of silicon nitride and a stronger diffuse halo that indicates the presence of amorphous material. By forming the TEM image, using only the electrons in the diffraction pattern (i.e., by inserting a shield that blocks the electrons in the diffuse halo) one obtains the dark-field image, Figure 19.17C, in which the bright spots represent the crystallites. One increases the magnification further and places the beam on the amorphous area; measurement of the electron energy losses in this area yields the EELS spectrum given in Figure 19.17D. The latter shows the presence of silicon and oxygen, but no nitrogen: the amorphous material is silicon oxide. Placing the beam on a crystallite, one obtains the energy-loss

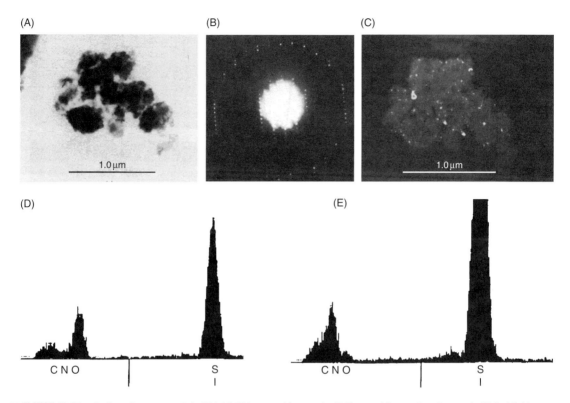

■ FIGURE 19.17   Analysis of a wear particle by TEM. (A) TEM image of the particle. (B) Electron diffraction from the particle. (C) Dark field image using the diffracted beams only. (D) Energy-loss spectrum from the amorphous region. (E) Energy-loss spectrum from the crystallites.

spectrum, Figure 19.17E, that has silicon and nitrogen peaks, showing that the crystallites are silicon nitride. (The small oxygen peak is most probably due to amorphous material in the proximity of the crystallite.) The measurements are interpreted the following way: rubbing silicon nitride in humid air produces very small silicon nitride particles as well as amorphous silicon oxide by a chemical reaction.

19.3.4 The Scanning Transmission Electron Microscope (STEM)

This instrument combines properties of the two previously described instruments. It is shown schematically in Figure 19.18. The probing electron beam is narrowly focused and scans the sample as in the SEM, but the sample is very thin and the scanning beam traverses the specimen as in the TEM. X-ray emission spectroscopy and energy loss spectroscopy provide chemical analysis. Like the TEM, this instrument is capable of measuring the shape, crystal structure and chemical composition of a piece of material a few nanometers in lateral size and about 100 nm thick. The resolution of this instrument is not as good as that of the TEM, but the scanning beam provides the STEM with a great range of capabilities. The instrument is used for the study of organic materials where the TEM would not ordinarily provide enough contrast. Sample preparation is the same as for the TEM and represents a major part of the skill and effort of microscopy.

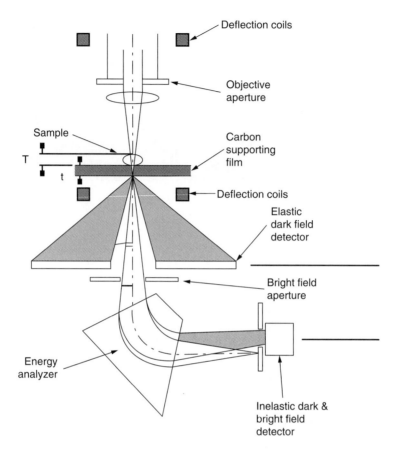

■ **FIGURE 19.18** Schematic diagram of a Scanning Transmission Electron Microscope (STEM) showing the lens, aperture, and deflection systems. Elastically scattered electrons are collected by an annular detector and provide the elastic dark-field signal. A spectrometer deflects those electrons that have lost energy at a larger angle than the unscattered electrons, thus facilitating the acquisition of the inelastic dark-field signal. The coherent bright-field signal arising from unscattered and low-angle elastically scattered electrons is collected through a small aperture placed on the optical axis. The various signals can be collected in parallel and processed online as the probe is scanned over the sample.

19.3.5 The Scanning Probe Microscopes (SPM)

The scanning probe microscopes consist of a very sharp point that scans the specimen at a very small distance from its surface. The relative position and motion of the scanning tip and the specimen can be controlled with a precision of about a tenth of the size of an atom (~0.02 nm) by means of piezoelectric actuators. When the tip approaches the surface to within 1 nm, two important phenomena take place. (1) A current can flow between tip and surface by the quantum-mechanical tunnel effect; this phenomenon is utilized in the Scanning Tunneling Microscope (STM). (2) The tip is attracted to the surface by the inter-atomic forces described in Chapter 1; this phenomenon lies at the base of the atomic force microscope. The atomic force microscope can also be utilized to manipulate materials: it can deposit extremely small amounts of material, or it can cut extremely small grooves with very high precision.

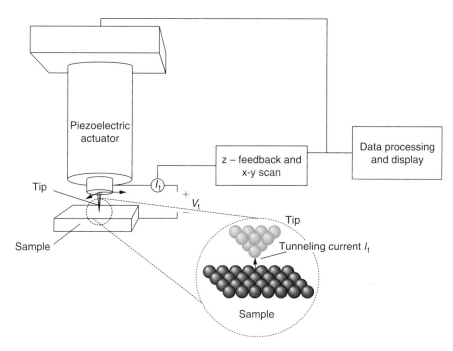

■ **FIGURE 19.19** Schematic of a scanning tunneling microscope. The tip is at the end of a piezoelectric cylinder that produces its vertical and lateral movement. A small voltage between tip and specimen causes a tunneling current between the atoms of the probe and the specimen that are nearest to each other.

19.3.6 **The Scanning Tunneling Microscope (STM)**

The scanning tunneling microscope is the first one invented, in 1981 by Gerd Binnig and Heinrich Rohrer of IBM's Zurich Laboratory. The invention garnered the two a Nobel Prize in Physics (1986).

It utilizes a sharp tip of diameter around 100 nm that scans the surface at a distance of about 1 nm. An electric potential applied between the conducting tip and specimen causes an electric current to flow by tunnel effect. This current is very sensitive to the specimen-tip distance so that the variation in height caused by the presence of single atoms on the specimen can be detected. The active "tip" is a single atom, namely the atom on a small asperity on the tip closest to the surface.

Figure 19.19 is a schematic of the STM. The position and movement of the tip is achieved by a piezoelectric tube. Piezoelectricity was described in Chapter 11: a piezoelectric material increases or decreases in length when an electric potential difference is applied to it. Figure 19.20 shows the basic shape of the cylinder, the potential differences applied, and the resultant deformations. These deformations, controlled by application of suitable voltages, can be extremely small. The piezoelectric material is a ceramic; it is hard and sturdy so that the positioning of the specimen is free of vibrations. In this manner, the surface of the specimen can be controllably approached to within atomic distance of the tip extremity.

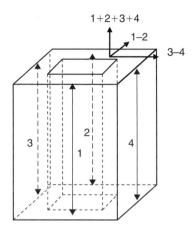

■ **FIGURE 19.20**   Operation of the piezoelectric transducer. A voltage applied to any side changes its length. Apply positive voltage (color arrow) to side 1 and negative voltage to side 2; side 1 elongates and side 2 shortens, and the resultant deformation moves the top face backward (arrow 1−2). Similarly, +V applied to side 3 and −V applied to side 4 moves the bottom to the right. When the same voltage +V is applied to all four sides, the cylinder elongates and the top moves up.

The small tunneling junction and precise positioning of the tip make it possible to image individual atoms. Figure 19.21, for instance, shows the atoms on a gold surface.

19.3.7 **The Atomic Force Microscope (AFM)**

A schematic of the AFM is shown in Figure 19.22. It consists of a sharp tip scanning the surface with the help of a piezoelectric cylinder, similar to the STM. When the tip approaches the surface to within 1 nm, it is attracted to the specimen by inter-atomic bond forces (see Figure 1.19). The attractive force bends the cantilever supporting the tip by a very small amount. A laser beam is reflected from the top of the cantilever toward four photodiodes. Figure 19.23 shows the principle: When the beam is centered, all four diodes produce the same current. Even a very small deformation of the cantilever moves the reflected beam on the photodiodes and changes the relative signals on the photodiodes.

The signals are processed by a computer. In the usual mode, the computer controls the voltages on the piezoelectric so that the distance between tip and specimen remains constant. The specimen is now displaced laterally by the piezoelectric so that the probe point scans its surface; during the scan, the specimen is moved vertically in order to keep the distance from surface to point constant. The vertical position vs. the scan position represents a high-resolution profile of the surface. The AFM is sensitive enough to detect the presence of single atoms. Figure 19.24 shows the surface structure of graphite observed by AFM. The scanned area can be as large as 1 mm square; in this case, the instrument measures the surface topography, such as steps on the surface or the presence of very small objects.

Figure 19.25 displays the flexibility the data display in the AFM. It shows the surface profile of a DVD disc. The tip scans the surface and measures the variation of height represented by the optical storage bits, and the computer displays the data as desired. In Figure 19.25A, the height variation is shown in a

■ **FIGURE 19.21**   Atomically resolved STM image of clean Au(100). This image is made with an Omikron Low Temperature STM by Erwin Rossen, Technical University, Eindhoven, 2006.

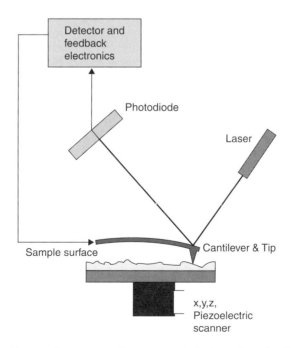

■ **FIGURE 19.22**   Operation of the atomic force microscope. The specimen is displaced vertically and laterally by a piezoelectric cylinder. The movement of the probe tip is measured by the displacement on a photoelectric quadrant of a laser beam reflected from its surface.

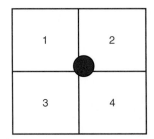

■ **FIGURE 19.23**   Detection of tip movement by a reflected beam on a quadrant of photodiodes. The signal strength on the diodes is $2 > 1 = 2 > 3$.

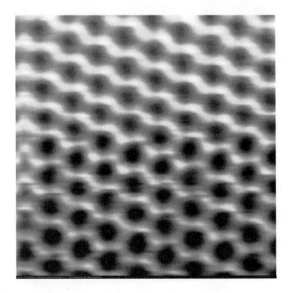

■ **FIGURE 19.24**   AFM image showing the hexagonal crystal structure of graphite.

three-dimensional perspective display. In Figure 19.25B, the surface is shown in two dimensions, and the height is indicated by color. Figure 19.25C shows the variation in height along a single scan, with the output of the scan shown on the right.

The instrument is also capable of measuring the friction between tip and surface by detecting the torsion of the cantilever carrying the tip; the torsion is measured by the displacement of the laser beam on the photoelectric quadrant as well. The AFM can be used to deposit material on the surface in a controlled fashion or to machine the surface. It can also measure the atomic force as a function of interatomic distance, as in Figure 1.19, or it can explore chemical forces by adsorbing the desired molecules on the surface of the tip.

The web site www.afmuniversity.org gives a comprehensive description of the operation and capabilities of the atomic force microscope.

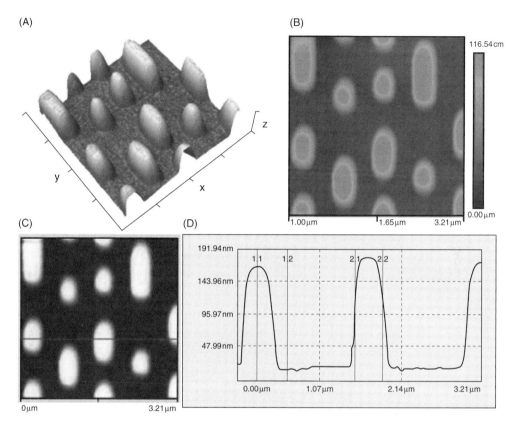

■ **FIGURE 19.25** Utilization of the atomic force microscope. A DVD disc surface. (A) Three-dimensional image. (B) Two-dimensional image. (C) Profile measurement of the surface. Courtesy of Pacific Nanotechnology Inc.

■ SUMMARY

1. The characterization of materials uses elastic and inelastic scattering of X-rays and electrons. Elastic scattering occurs without change in energy or wavelength; it is used in microscopy and in diffraction. Inelastic scattering involves a loss of energy that is transmitted to the electrons in the atoms; it is utilized in the measurement of chemical composition of a material.

2. The identity, amount, and in some cases position of elements are measured by core electron spectroscopy. Tightly bound electrons can be excited into the conduction band by collision with an electron or by absorbing an X-ray of sufficient energy. Higher lying electrons can fall into the resultant empty level; they get rid of their excess energy by emitting an X-ray.

3. X-ray fluorescence is a simple technique that does not require sample preparation. It excites the material with high-energy X-rays and measures the energy of the lower-energy X-rays emitted by the electrons in the atoms.

4. Energy-dispersive spectroscopy of X-rays is performed in the electron microscopes. A high-energy electron ejects electrons from the atom, and one measures the energy of the resultant X-rays. In the scanning electron microscopes (SEM and TEM), the technique also determines the spatial distribution of the elements.

5. Electron Energy Loss Spectroscopy (EELS) is performed in the Transmission Electron Microscopes (TEM and STEM). One measures the loss of energy of the electrons that have traversed the sample: this energy loss is a measure of the binding energy of the atomic electrons and identifies the elements.

6. Crystal structure is measured by X-ray or electron diffraction. Diffraction is based on elastic scattering: a diffraction beam exists when the waves reflected by parallel planes are all in phase. This occurs when the difference in distance traveled by the reflected waves is an integer number of wavelengths. It is expressed by Bragg's formula.

7. Interference of reflections from planes inside the unit cell leads to the disappearance of certain diffraction beams. This systematic extinction is used to determine the structure of the unit cell (BCC or FCC in cubic materials).

8. The optical microscope is the simplest and most convenient of the microscopes. It has maximum magnification of $\times 1,000$. Its depth of field is small; this feature can be used to measure the height or depth of surface features.

9. The scanning electron microscope (SEM) operates in vacuum. Sample preparation is simple, but electric insulators must be coated with a thin metallic film to avoid charging. The resolution of the SEM is about 1 nm. The instrument has a large depth of field that allows it to observe irregular specimens or to observe them at an oblique angle with the surface. The SEM provides the identity, concentration, and special distribution of elements.

10. The transmission electron microscopes operate in vacuum and demand the preparation of very thin samples. They detect the identity and concentration of elements by EDX and by Electron Energy Loss Spectroscopy (EELS). The TEM is capable of atomic resolution. The scanning transmission electron provides spatial distribution of elements. It permits imaging with electrons having lost selected energies and is used for organic and polymer samples.

11. The scanning probe microscopes consist of a very fine tip that scans the sample at a very small distance from the surface. The movement of the tip is performed by a piezoelectric tube. They are also capable of manipulating the surface at an atomic scale.

12. The scanning tunneling microscope measures the tunneling current between tip and specimen when a small electric potential is established between them. It is capable of seeing individual atoms.

13. The atomic force microscope measures the atomic attraction forces between tip and specimen. Usually, an electric feedback maintains the tip-to-sample constant. The microscope then acts as a very-high-resolution profilometer.

KEY TERMS

■ REFERENCES FOR FURTHER READING

[1] C. Barrett, and R.B. Massalski, *Structure of Metals: Crystallographic Methods, Principles, and Data (International Series on Materials Science and Technology) (Paperback)*, 3rd ed, Pergamon Press, Elmsford, NY (1980).

[2] D. Bonnell, *Scanning Probe Microscopy and Spectroscopy: Theory, Techniques and Application*, 2nd ed, Wiley-VCH, New York (2000).

[3] W. Clegg, *Crystal Structure Determination (Oxford Chemistry Primers, No. 60) (Paperback)*, Oxford University Press, Oxford, UK (1998).

[4] J. Goldstein, D.E. Newbury, D.C. Joy, C.E. Lyman, P. Echlin, E. Lifshin, L.C. Sawyer and J.R. Michael, *Scanning Electron Microscopy and X-ray Microanalysis*, 4th ed, Springer, New York (2003).

[5] A. Guinier, *X-Ray Diffraction: In Crystals, Imperfect Crystals, and Amorphous Bodies (Paperback)*, Dover Publications, Mineola, NY (1994).

[6] G.R. Lachance, F. Claisse, *Quantitative X-Ray Fluorescence Analysis: Theory and Application*, Wiley, New York (1995).

[7] E. Meyer, H. Hug and R. Bennewitz, *Scanning Probe Microscopy: The Lab on a Tip*, Springer, New York (2003).

[8] P. West, web site available at www.afmuniversity.org.

[9] D.B. Williams, and C.B. Carter, *Transmission Electron Microscopy: A Textbook for Materials Science (4 Vol. Set. Paperback)*, Springer, New York (2004).

■ PROBLEMS AND QUESTIONS

The next several problems are based on the electron energy levels for the elements listed in the following table and illustrated in Figures 19.1 and 19.3.

Element	K	L₁	L₂	L₃	M₁	M₂,₃	M₄,₅
Cr	5,989	695	584	575	74.1	42.5	2.3
Cu	8,979	1,096	951	931	120	73.6	1.6
Mo	20,000	2,865	2,625	2,520	505	400	229
W	69,525	12,100	11,544	10,207	2,820		

19.1. a. Create an energy level diagram for Cu.

b. What is the energy of the photon emitted in the $M_3 \rightarrow K$ electron transition in Cu?

    c. In what range of the electromagnetic spectrum (visible, infrared, X-ray, etc.) does this photon lie?

    d. Repeat parts b and c for the $M_{4,5} \rightarrow M_{2,3}$ transition.

19.2. a. Will electrons that travel at velocities of 4.7×10^7 m/s have enough kinetic energy to eject the K electron from Cr? Will they eject Mo K electrons?

    b. Will a photon with a wavelength of 0.161 nm have enough energy to eject a K electron from Cu? Will it eject the Cr K electron?

19.3. The metals listed in the previous problem are all used commercially as targets in X-ray generating tubes.

    a. Calculate the photon wavelength corresponding to the $L_3 \rightarrow K$ electron transition in each metal. This transition gives rise to the so-called $K\alpha_1$ X-ray.

    b. X-ray tubes with Cu, Cr, Mo, and W targets got mixed up in a laboratory. In order to identify them, they were operated sequentially and the $K\alpha_1$ wavelengths were measured in an EDX system. The first tube tested yielded a wavelength of 0.0709 nm. What is the target metal?

19.4. Moseley's law of atomic physics suggests that the energy of $K\alpha_1$ X-rays varies as $(Z-1)^2$ where Z is the atomic number of the element. Plot the $K\alpha_1$ X-ray energies for Ti, Cr, Cu, Mo, and W vs. $(Z-1)^2$ so that a straight line results. Based on your plot, what are the energy and wavelength of $K\alpha_1$ X-rays in Sn?

19.5. Compute and draw an X-ray diffraction pattern similar to that of Figure 19.10 for BCC and FCC iron at 914 °C. Since you will not be able to compute the amplitude of the diffraction peaks, draw them as vertical lines of equal height. The results of Problem 3.14 are useful here.

19.6. Describe how X-ray diffraction can be used to measure residual stresses at the surface of a solid.

19.7. Fluorescent X-ray analysis from an automobile car fender revealed a spectrum with lines at 5.41 keV (intense) and 5.95 keV (less intense), and a weaker line at 8.05 keV. Interpret these findings.

19.8. During the Renaissance the white pigment used in oil paints was lead oxide. In the nineteenth century zinc oxide was used as well. In more recent times titanium oxide has been the preferred choice. A painting suspected of being a forgery is examined by EDX methods and yielded 75 keV $K\alpha_1$ X-rays from a region painted white. Pending further investigation what, if anything, can you infer about the painting's age? (Hint: see Problem 19.4.)

19.9. The chemical composition of a series of binary Cu-Ti alloys is calibrated by measuring the relative intensity of fluorescent X-rays emitted. For pure Cu the rate of X-ray emission is measured to be 1562 (photon) counts per second (cps), while 2,534 cps were detected from pure Ti. The alloy yields a rate of 656 cps for Cu plus 1,470 cps for Ti X-rays. What is the overall alloy composition? (In all cases the same sample and measurement geometry was employed, and a linear composition calibration is assumed.) Is the composition measured in weight or atomic percent?

19.10. EDX provides an elemental analysis of the sample but does not provide information on the phase (solid, liquid, gas) or state of chemical bonding (metallic, ionic, covalent). Why? Such information, however, can be obtained in the STEM. Why?

19.11. Describe the X-ray source.

    a. How is it built (what are its essential components)?

    b. How is it operated?

    c. What physical phenomena are involved?

19.12. Describe X-ray fluorescence.

    a. What are its essential components?

    b. How is it operated?

    c. What information does it provide?

    d. What physical phenomena are involved?

19.13. Describe the formation of an image in the scanning electron microscope (SEM). If the microscope produces a picture of size 10×15 cm (4×6 in), how large is the scanned area on the sample at linear magnification of $20,000 \times$? How fine must be the focusing of the electron beam for this magnification to be effective (i.e., resolution of 0.1 mm on the picture)?

19.14. Describe EDX in the SEM.

    a. What are its essential components?

    b. How is it operated?

    c. What information does it provide?

    d. What physical phenomena are involved?

19.15. Describe electron energy loss spectroscopy.

    a. What are its essential components?

    b. How is it operated?

    c. What information does it provide?

    d. What physical phenomena are involved?

19.16. Describe an X-ray diffractometer.

    a. What are its essential components?

    b. How is it operated?

    c. What information does it provide?

    d. What physical phenomena are involved?

19.17. Compare the two main methods for the measurement of crystal structures.

    a. What is the probing radiation in each?

    b. Which one determines the structure of the different phases in a material?

    c. Which one is more convenient to use?

19.18. a. What is the fundamental thought that establishes Bragg's law of diffraction?

    b. On the basis of this thought, derive Bragg's equation.

19.19. Show that a plane of atoms reflects all radiation like a mirror at all angles of incidence. (Use an argument similar to the one that establishes Bragg's law).

19.20. Describe or sketch an atomic force microscope. What phenomenon is used to control the position of the probe with 0.02 nm resolution? What physical phenomenon is used in the measurement?

19.21. Describe how the AFM can be used to measure the interatomic forces and establish a graph similar to Figure 1.19 or 1.20.

19.22. Describe elastic and inelastic scattering of radiation. In what techniques is elastic scattering used? What techniques make use of inelastic scattering?

19.23. What properties of atoms allow the measurement of chemical composition by radiation?

19.24. Describe an experiment that would verify the de Broglie equation, $\lambda = h/mv$. Section 19.2 contains all the information you need.

Answers to Selected Problems

CHAPTER 1

1.8. a. $4.996 \cdot 10^{22}$
    b. $1.23 \cdot 10^{24}$
    c. $2 \cdot 10^{-7} = 0.2 \, \text{ppm}$
1.10. 868 g Ni, 132 g Al
1.18. 3′2″ (3.17 feet = 95 cm)

CHAPTER 2

2.1. a. 2.0011 m
    b. 3.58 m
2.3. 2.36 m
2.5. b. 1.33 kJ/kg
    c. 432 kJ/kg
2.7. a. $35.3 \cdot 10^6 \, \text{psi}$
    b. 73.5 ksi
    c. 125 ksi
    d. 50
    e. 33
2.10. 1.8 mm
2.14. $E_c = 79.8 \, \text{kJ/mol}$ $n = 3.91$
2.16. b. 400 MPa
    c. 251,000 cycles
    d. 158,000 cycles

CHAPTER 3

3.1. 0.263 nm
3.2. Ni: 0.125 nm

3.3. Ni: $9.14 \cdot 10^{22}$ atoms/cm^3

3.4. Ni: 8.90 g/cm^3

3.6. (110) plane, [111] direction

3.13. for the ($\bar{1}\,\bar{2}\,\bar{2}$) plane, the directions are [$01\bar{1}$], [$\bar{2}10$] and [$\bar{2}01$]; note that all three are perpendicular to the [$\bar{1}\,\bar{2}\,\bar{2}$] direction.

3.14. BCC at 914°C: a = 0.2898 nm, ρ = 7.619 g/cm^3
FCC at 914°C: a = 0.3550 nm, ρ = 8.29 g/cm^3

CHAPTER 4

4.5. Steel: σ_Y = 3,456 MPa

CHAPTER 5

5.1. 731 kg

5.9. At 1200°C: L (20 at% As); 100%L
At 1000°C: L (14 at% As), GaAs (50 at% As); 83%L, 17% GaAs
At 200°C: L (0.5 at% As), GaAs (50 at% As); 61% L, 39% GaAs
At 20°C: Ga (0 at% As), GaAs (50 at% As); 60% Ga, 40% GaAs

5.11. At 2800°C: L (50 wt% Re); 100% L
At 2551°C: L (43 wt% Re), β (54 wt% Re); 36% L, 64% β
At 2448°C: α (45 wt% Re), β (54 wt% Re); 44% α, 56% β
At 1000°C: α (38 wt% Re), β (59 wt% re); 43% α , 57% β

5.16. a. Mullite or γ.41.8 wt%, b: Lower.

5.17. a. 9.49 g,
b. 8.36 g,
c. 1.64 g,
d. 1.13 g,
e. 0.51 g

CHAPTER 6

6.10. R_C 53

6.11. c is reduced from 3 mm to 0.33 mm.

CHAPTER 7

7.8. 5.81%W, 4.57%Cr, 1.16%V, 7.4%C , 81%Fe.

7.48. a. 0°C: D = $6 \cdot 10^{-18}$ cm^2/s; 200°C: D = $3 \cdot 10^{-11}$ cm^2/s; 700°C: D = $8 \cdot 10^{-7}$ cm^2/s
b. 15, x = 8.9 mm; 165, x = 35.8 mm; 605, x = 69.3 mm; 1hr, x = 0.537 mm.

7.50. T = 1,136 K = 863°C.

CHAPTER 9

9.3. 1240 mers

9.4. a. butadiene 0.557 mol, isoprene 0.443 mol

    b. 5.33 g

9.6. 24,150 amu

9.9. a. $-12°C$

    b. $0.86 \, g/cm^3$

CHAPTER 10

10.1. a. 32 GPa

    b. 4.40 GPa

10.4. 6.87%

10.5. a. 4.2 mm

    b. 80%

10.7. a. $33.9 \, GPA.cm^3/g$

    b. 18%

10.8. 5

CHAPTER 11

11.1. A double winding of a wire. Length of the double winding: about 1.5 cm.

11.9. For copper and constantan, the thermal contributions to resistivity are similar, for copper, $\rho_{th} = 0.067.T \, n\Omega m$, for constantan, $\rho_{th} = 0.05.T \, n\Omega m$.

11.11. 0.94 mm/s

11.14. a. 8.06×10^{18} cells

    b. $8 \, \Omega cm$

11.15. a. $0.04 \, \Omega cm$

    b. 636°C

11.19. DC: d = 2.9 m

    AC: d = 1 mm

11.21. 10^{11} Hz

CHAPTER 12

12.1. <380 nm

12.5. c. For A, $\rho \approx 0.015 \, \mu\Omega cm$, for B, $\rho \approx 0.04 \, \mu\Omega cm$, for C, $\rho \approx 0.04 \, \mu\Omega cm$

12.7. a. Misfit $= -1.43 \cdot 10^{-3}$, AlAs in compression, GaAs in tension

    b. Misfit $= -6.34 \cdot 10^{-3}$, InP in compression, CdS in tension

    c. Misfit $= -2.82 \cdot 10^{-3}$, ZnSe in compression, GaAs in tension

CHAPTER 13

13.2. a. $2.02 \cdot 10^8$ m/s

    b. $42.5°$

    c. No

13.4. 43.7μm

13.6. Cu: at 0.5μm, R = 0.624, at 0.95μm, R = 0.987

    Au: at 0.5μm, R = 0.504, at 0.95μm, R = 0.980

13.9. 0.36

13.11. At 0.45μm, R = 0.0193

    At 0.60μm, R = 0.00553

    At 0.77μm, R = 0.0168

13.13. a. 4.83 ms

    b. 176

13.16. a. 12:1

    b. 33.3:1

13.17. For Si, $\lambda = 1.11 \mu$m

    For GaP, $\lambda = 0.555 \mu$m

    For ZnSe, $\lambda = 0.481 \mu$m

13.20. 0.904

13.22. Maximum cell power = $\frac{1}{4} I_{sc} V_{sc}$

13.24. a. 6.25 microwatt

    b. 5 milliwatt

CHAPTER 14

14.1. a. 3000

    b. 16.5 V

14.5. a. 800 kJ/m^3

    b. 200 kJ/m^3

14.6. a. 800 kJ/m^3

    b. 50 kJ/m^3

14.9. b. $5 \cdot 10^4$ kJ/m^3

    c. H = 10^5 A/m

14.11. a. $2.64 \cdot 10^{-7}$ °C

    b. $4.17 \cdot 10^5$ hours 9 (= 47 years, 7 months, 7 days)

14.13. a. $2.52 \cdot 10^5$ A/m

    b. $1.26 \cdot 10^5$ A/m

CHAPTER 15

15.1. 62.5 V, Ag positive

15.2. 0.36 Cd positive

CHAPTER 16

16.1. $-2.075\,\text{V}$

16.5. $0.504\,\text{mil}\ (= 0.0128\,\text{mm} = 12.8\,\mu\text{m})$

16.6. 2.48 hours

16.8. a. $8.03 \cdot 10^{-4}\,\text{A/cm}^2$

    b. $6.31 \cdot 10^{-4}\,\text{A}$

16.11. a. Zn

    b. Fe

    c. Mg

    d. Cd

16.19. a. $2.36 \cdot 10^{19}$ ions

    b. $5.23 \cdot 10^{-3}\,\text{cm} = 52.3\,\mu\text{m}$

16.22. a. PBR of Al_2O_3/Al = 1.28, PBR of AlN/Al = 1.26

    b. PBR of TiO_2/Ti = 1.76, BR of TiN/Ti = 1.12

16.25. 612 m

16.27. 2.05 kg

CHAPTER 19

19.2. a. Yes for Cu, no for Mo

    b. No for Cu, yes for Cr

19.3. Mo

19.9. 42 at% Cu, 589 at% Ti

Index

PROPERTIES OF SELECTED ELEMENTS (AT 20°C)

Element	Symbol	Atomic number	Atomic weight (amu)	Crystal structure[a]	Lattice constant(s) [a(nm), c(nm)]	Atomic radius (nm)	Ionic radius (nm)	Valence	Density (g/cm³)
Aluminum	Al	13	26.98	FCC	0.4050	0.143	0.057	3 +	2.70
Argon	Ar	18	39.95						
Arsenic	As	33	74.92	Rhomb		0.125	0.04	5 +	5.72
Barium	Ba	56	137.33	BCC	0.5019	0.217	0.136	2 +	3.5
Beryllium	Be	4	9.012	HCP	0.229, 0.358	0.114	0.035	2 +	1.85
Boron	B	5	10.81	Rhomb		0.097	0.023	3 +	2.34
Bromine	Br	35	79.90			0.119	0.196	1 -	
Cadmium	Cd	48	112.41	HCP	0.298, 0.562	0.148	0.103	2 +	8.65
Calcium	Ca	20	40.08	FCC	0.5582	0.197	0.106	2 +	1.55
Carbon (gr)	C	6	12.011	Hex	0.246, 0.671	0.077	~0.016	4 +	2.25
Cesium	Cs	55	132.91	BCC	0.6080	0.263	0.165	1 +	1.87
Chlorine	Cl	17	35.45			0.107	0.181	1 -	
Chromium	Cr	24	52.00	BCC	0.2885	0.125	0.064	3 +	7.19
Cobalt	Co	27	58.93	HCP	0.251, 0.407	0.125	0.082	2 +	8.85
Copper	Cu	29	63.55	FCC	0.3615	0.128	0.096	1 +	8.94
Fluorine	F	9	19.00				0.133	1 -	
Gallium	Ga	31	69.72	Ortho		0.135	0.062	3 +	5.91
Germanium	Ge	32	72.59	Dia Cub	0.5658	0.122	0.044	4 +	5.31
Gold	Au	79	196.97	FCC	0.4079	0.144	0.137	1 +	19.3
Helium	He	2	4.003						
Hydrogen	H	1	1.008				0.154	1 +	
Iodine	I	53	126.91	Ortho		0.136	0.220	1 -	
Iron	Fe	26	55.85	BCC	0.2866	0.124	0.087	2 +	7.87
Lead	Pb	82	207.2	FCC	0.4950	0.175	0.132	2 +	11.3
Lithium	Li	3	6.94	BCC	0.3509	0.152	0.078	1 +	0.53
Magnesium	Mg	12	24.31	HCP	0.321, 0.521	0.160	0.078	2 +	1.74
Manganese	Mn	25	54.94	Cubic	0.8914	0.112	0.091	2 +	7.43
Mercury	Hg	80	200.59			0.150	0.112	2 +	14.2
Molybdenum	Mo	42	95.94	BCC	0.3147	0.136	0.068	4 +	10.2
Neon	Ne	10	20.18						
Nickel	Ni	28	58.69	FCC	0.3524	0.125	0.078	2 +	8.9
Niobium	Nb	41	92.91	BCC	0.3307	0.143	0.069	4 +	8.6
Nitrogen	N	7	14.007			0.071	0.015	5 +	
Oxygen	O	8	16.00			0.060	0.132	2 -	
Phosphorus	P	15	30.97	Ortho		0.109	0.035	5 +	1.83
Platinum	Pr	78	195.08	FCC	0.3924	0.139	0.052	2 +	21.4
Potassium	K	19	39.10	BCC	0.5344	0.231	0.133	1 +	0.86
Silicon	Si	14	28.09	Dia Cub	0.5431	0.117	0.039	4 +	2.33
Silver	Ag	47	107.87	FCC	0.4086	0.144	0.113	1 +	10.5
Sodium	Na	11	22.99	BCC	0.4291	0.186	0.098	1 +	0.97
Sulfur	S	16	32.06	Ortho		0.104	0.174	2 -	2.07
Tin	Sn	50	118.69	Tetra		0.158	0.074	4 +	7.30
Titanium	Ti	22	47.88	HCP	0.295, 0.468	0.147	0.064	4 +	4.51
Tungsten	W	74	183.85	BCC	0.3165	0.137	0.068	4 +	19.3
Uranium	U	92	238.03	Ortho		0.138	0.105	4 +	19.0
Vanadium	V	23	50.94	BCC	0.3028	0.132	0.061	4 +	6.1
Zinc	Zn	30	65.39	HCP	0.266, 0.495	0.133	0.083	2 +	7.13
Zirconium	Zr	40	91.22	HCP	0.323, 0.515	0.158	0.087	4 +	6.49

[a] FCC=Face-centered cubic; Rhomb=rhombohedral; BCC=body-centered cubic; HCP=hexagonal close-packed; Hex=hexagonal; Ortho=orthorhombic; Dia cub=diamond cubic; Tetra=tetragonal